Krause · Jäger (Eds.)
High Performance Computing in Science and Engineering '02

Springer-Verlag Berlin Heidelberg GmbH

Egon Krause · Willi Jäger

Editors

High Performance Computing in Science and Engineering '02

Transactions of the High Performance Computing Center
Stuttgart (HLRS) 2002

With 258 Figures, 155 in Color

 Springer

Editors

Egon Krause
Aerodynamisches Institut der RWTH Aachen
Wuellnerstraße zw. 5 u. 7
52062 Aachen, Germany
e-mail: ek@aia.rwth-aachen.de

Willi Jäger
Institut für Wissenschaftliches Rechnen
Universität Heidelberg
Im Neuenheimer Feld 368
69120 Heidelberg, Germany
e-mail: jaeger@iwr.uni-heidelberg.de

Cataloging-in-Publication Data applied for

Bibliographic information published by Die Deutsche Bibliothek

Die Deutsche Bibliothek lists this publication in the Deutsche Nationalbibliografie;
detailed bibliographic data is available in the Internet at <http://dnb.ddb.de>.

Front cover figure: A spaghetti representation of the magnetic fields in a Herbig-Haro jet simulated on NEC SX-5 (Theorie-Gruppe Landessternwarte Königstuhl).

Mathematics Subject Classification (2000): 65Cxx, 65C99, 68U20

http://www.springer.de
ISBN 978-3-642-63947-0   ISBN 978-3-642-59354-3 (eBook)
DOI 10.1007/ 978-3-642-59354-3

© Springer-Verlag Berlin Heidelberg 2003
Originally published by Springer-Verlag Berlin Heidelberg New York in 2003
Softcover reprint of the hardcover 1st edition 2003

Typeset by the authors. Edited by Kurt Mattes, Heidelberg.
Cover design: *design & production* GmbH, Heidelberg

SPIN: 10865614   46/3142LK - 5 4 3 2 1 0 – Printed on acid-free paper

Preface

Prof. Dr. Egon Krause

Aerodynamisches Institut
RWTH Aachen
Wüllnerstr. zw. 5 u. 7, D-52062 Aachen

Prof. Dr. Willi Jäger

Interdisziplinäres Zentrum für Wissenschaftliches Rechnen
Universität Heidelberg
Im Neuenheimer Feld 368, D-69120 Heidelberg

The Fifth Results and Review Workshop on High Performance Computing in Science and Engineering was held at the High Performance Computing Center Stuttgart (HLRS) September 30th – October 1st, 2002. 40 projects processed at the HLRS and at the Scientific Supercomputing Center Karlsruhe (SSC) were selected for presentation at the workshop and, after an internal review, prepared for publication in this fifth volume of the transactions of the HLRS. The results reported were obtained during the time after the last workshop in October 2001. The projects were initiated at the universities in Aachen, Bayreuth, Belfast, Berlin, Bielefeld, Braunschweig, Cottbus, Darmstadt, Erlangen-Nürnberg, Essen, Freiburg, Göttingen, Greifswald, Halle-Wittenberg, Hamburg-Harburg, Heidelberg, Hohenheim, Jena, Karlsruhe, Kiel, Konstanz, Mainz, Marburg, Montpellier, München, Münster, Rome, the Saarland, Salzburg, Stuttgart, Tübingen, Ulm, Worcester, Würzburg, and Zürich. Several projects are carried out in cooperation with institutes of the Max Planck Society in Stuttgart, the German Research Center of Aero- and Astronautics in Braunschweig and Stuttgart, the Geo-Research Center in Potsdam, the Los Alamos National Laboratory, the Alfred-Wegener Institute of Polar and Maritime Research, and the Research Center Karlsruhe.

The main thrust of all projects processed on the systems of the HLRS and the SCC is directed towards solving problems in fluid dynamics, and of these most of them are initiated and carried out at Stuttgart University: They comprise aero-elastic problems of helicopter rotor blades, numerical simulation of coal-fired furnaces, and direct simulation with massively parallelized codes of flame kernels in spark-ignited turbulent mixtures. Direct three-dimensional simulation is also used to determine the heat transfer from droplets moving with transient velocity. Other studies are concerned with the simulation of three-dimensional flow and heat transport in high-temperature nuclear reactors. Two-phase flows, investigated with parallelized solutions, and problems of pneumatic transport are also of major concern. The well-known problem of the three-dimensional boundary-layer transition induced by superposed steady and traveling cross-flow vortices. Last but nor least, supersonic combustion is studied in a joint project with the DLR Stuttgart.

The Technical University of Braunschweig together with the DLR Braunschweig study the problem of fluid-structure interaction on high-lift devices at low Mach numbers. At Karlsruhe University the flow in a low-pressure turbine, influenced by incoming wakes is simulated, and at the RWTH Aachen the flow around the Phoenix configuration is numerically investigated. At the University Erlangen-Nürnberg the Lattice-Boltzmann method is used to analyze the flow in fixed-bed reactors. Turbulent wall-bounded flows with and without adverse pressure gradient and separating turbulent boundary layers are investigated with large-eddy and direct simulations at the TU München.

A wide range of problems is also covered in physics. At the University of Münster, a three-component reaction diffusion model is developed to simulate the replication of dissipative solitons by many-particles interaction. The localization of the surface exciton at the Si(111)-(2x1) surface due to self-trapping is studied in another project. At Konstanz University simulations of nano-structures in reduced geometry are under way. A self consistent auxiliary particle theory for strongly correlated Fermion systems is developed at the University of Karlsruhe; studies of the thermodynamics and dynamics of correlated electron systems are jointly carried out at Stuttgart University and the Max Planck Institute for Solid State Research in Stuttgart; also at Stuttgart University, the strain distribution in quantum dot nano-structures is computed by means of atomic simulations. Excitonic and local-field effects in optical spectra are obtained in real-space time-domain calculations at Jena University. A joint project of exact diagonalization studies of spin, orbital, and lattice correlations in CMR manganites is under way at the Universities Bayreuth, Erlangen, and Greifswald. At the University of Bielefeld, the temporal quark and gluon propagators were calculated on a large isotropic lattice, with the spectral function extracted from them. The importance of the intermediate range order in silicates is investigated with molecular dynamics simulations in a joint investigation of the Universities Mainz and Montpellier, France. Convection driven spherical dynamos are studied at Bayreuth University. At Würzburg University the influence of the half-breathing $O(?, 0)$ mode in the three-band Hubbard model is investigated with Quantum Monte Carlo simulations. In astro-physics, the magnetized Herbig-Haro flows of low-mass stars are numerically simulated at the Landessternwarte Heidelberg.

In the fields of climatic, meteorological, and oceanic research, regional climate changes for central Europe are studied at the University at Cottbus. The projects in the geo-sciences are concerned with the development of a thermal evolution model for the earth crust, a joint project of the University in Jena and the Los Alamos National Laboratory, USA; inertial instabilities in precession driven flow are investigated at Göttingen University.

In chemistry, the iron(III) catalyzed Michael reaction is studied at the University of Technology Berlin, where also accurate ab initio calculations for vanadium oxide clusters are carried out. The individually selecting configuration interaction method is implemented and applied at the Research Center Karlsruhe. Structural trends and transitions in water clusters are investigated at Kiel University.

In structural mechanics, geo-technical boundary value problems are modeled with a finite element method at the TU Hamburg-Harburg.

In computer sciences, at Karlsruhe University, collective operations with SKaMPI are benchmarked, whereas at Freiburg University, an efficient divergence cleaning for three-dimensional MHD simulations is developed. At Salzburg University, different granularities for parallel wavelet packet video coding using block-based motion compensation and the performance of the corresponding MPI implementations on the HLRS Cray T3-E are investigated. In bio-informatics at Heidelberg University the internal protein dynamics are studied with computer simulations.

The above brief description of the projects shows the large number of universities and research centers engaged in high-performance computing at the facilities in Stuttgart and Karlsruhe. High-performance computing is almost accepted everywhere. Analysis of the performance reports of the HLRS and the SCC show, that the demand for additional computing power remains increasingly high. The seemingly never-ending demand is confirmed by the developments in the USA and Japan. Both countries have reacted to the demand with hard- and software developments, that will bring dramatic changes in the supercomputing landscape. While German centers are announcing the installation of systems with a performance of several TFLOPS, the USA and Japan are increasing the pace in the supercomputer race in a way never seen before. Last year the US Advanced Simulation and Computing Initiative (ASCI) has successfully installed a 12.3 TFLOPS system, the ASCI White, putting the USA ahead of her Japanese competitor in 2001. With a 30 TFLOP system (ASCI Q) planned to be ready in 2002 US researchers will be provided with state of the art technology to take the lead in computational sciences

With this advancement the US initiatives have triggered a counter offensive response from Japan. Although announced already two years ago the installation of the Japanese Earth Simulator system hit the US scientific community like a shock wave, comparable to the effect of the launching of the Soviet Union's Sputnik satellite in 1957. With its 40 TFLOPs the Earth Simulator is about a factor of 3 faster than ASCI White. What is even more impressive is the fact that in real applications the Earth Simulator outperforms all other systems by at least a factor of 7 in sustained performance. This puts the Earth Simulator on top of every top list for years to come.

The impact of these developments on high-performance computing in Germany and especially on the HLRS has not been fully analyzed yet. The answer, that Germany does not have to simulate weapons as the USA have, or track typhoons as in Japan, is too simple. The aerospace sciences, which helped to conquer a large portion of the world market in aviation and space transportation, are faced with enormous multidisciplinary problems of aero-structural dynamics, aero-acoustics, and combustion. Internationally competitive advancement of these fields alone would justify the installment of a system comparable to the Earth Simulator in Germany. It is also not a satisfying solution that the HLRS together with its partner centers Jülich (NIC) and Munich (LRZ) have initiated a coordinated program to increase the overall in Germany available performance by linking the existing systems. New plans have to be drawn up, if the widening gap in high-performance computing between the USA, Japan, and Europe is to be controlled and eventually closed in the future.

The articles published in this volume to a certain extent demonstrate the state of the art in high-performance computing in Germany. The authors were encouraged to emphasize computational techniques used in solving the problems investigated. The importance of the newly computed results for the specific disciplines, as interesting they may be from a scientific point of view, were not in the focus here.

The pace making and continued support of the Land Baden-Württemberg in promoting and supporting high-performance computing is gratefully acknowledged. The Deutsche Forschungsgemeinschaft (DFG) supported many of the projects processed on the systems of the HLRS and the SSC. Also, the increasing activities of the WiR, strengthening scientific investigations on a large scale in the State of Baden-Württemberg deserve mentioning. Finally, we thank the Springer Verlag for publishing this volume and thus helping to position the local activities into an international frame. We hope that this series of publications is contributing to the promotion of high performance scientific computing.

Stuttgart, July 2002

W. Jäger
E. Krause

Contents

Computational Fluid Dynamics (CFD)

Physics

PD Dr. Hans-Peter Nollert and Prof. Dr. Hanns Ruder

Theoretische Astrophysik, Universität Tübingen, Auf der Morgenstelle 10,
D-72076 Tübingen

The articles contained in this section represent work from highly advanced areas of physics, with a high standard of mathematical formalism and complexity. It seems appropriate that these complex problems are attacked with the most advanced computational algorithms and the most complex computer hardware available. Over the last decades, it has become obvious that this combination of competence in fundamental formalism on the one hand and in computer technology on the other hand offers a unique tool in solving problems that could not be handled successfully before. Accordingly, the work presented here consists of long-term projects that have successfully developed and progressed over several years.

Among the projects collected in this section, the group of Gähler et al. has made the most extensive use of the resources available for simulations using their molecular dynamics code IMD: Their account has been charged for more than half the total sum in this section. They succeeded in performing simulations of strain distributions in quantum dot structures for systems of realistic size, for a larger variety of sizes than had ever been done before. The results of these simulations even agree with experiments!

Most of the remaining resources have been used for geophysics: Simitev and Busse continued the study of convection driven dynamos in rapidly rotating spherical shells with an exhaustive parameter study for a large range of Prandtl numbers. Studying inertial instabilities in precession driven flows, Lorenzani and Tilgner have found a cycle involving a "resonant collapse": A laminar flow becomes unstable, developing small scale instabilities which quickly dissipate the energy and re-laminarize the flow, allowing the cycle to eventually start all over again. The flow in the earth's liquid core, driven by the precession of the earth's rotation axis, has a bearing on the rotation of the earth and on the earth's magnetic field.

Using some of the crumbs that remain, Liehr et al. present results of realistic simulations of the replication of dissipative solitions by many particle interactions, using an MHD code on unstructured tetrahedral threedimensional meshes. They demonstrate that controlling the divergence of the magnetic field is crucial in obtaining their results. Camenzind and Thiele present simulations of magnetized Herbig-Haro flows of low-mass stars, based on the conservative form of the MHD equations. High spatial resolution achieved on the NEC SX-5 allows insight into the phenomenon of current filamentation and the complex internal structure. A highlight of this article are the many impressive pictures visualizing the results.

At the other end of the length scale, in the sub-nuclear world, Karsch et al. simulate temporal propagators and quasiparticles in hot quantum chromo dynamics. Using lattice Monte-Carlo simulations they determined the dispersion relation for quarks and gluons and extracted the temperature dependent quasiparticle masses.

Even with the impressive scientific results presented here, scientists encounter more and more the limitations of the computers currently available at the HLRS: Increased CPU power, memory size, speed of communication networks, and availability of resources to researchers are mentioned as crucial for achieving the next complexity of modelling and simulations in computational physics, with better resolution and more detailed calculations to cover physical effects missed on the current scale.

Computation of Strain Distributions in Quantum Dot Nanostructures by Means of Atomistic Simulations

Franz Gähler, Christopher Kohler, Johannes Roth, and Hans-Rainer Trebin

Institut für Theoretische und Angewandte Physik, Universität Stuttgart, D-70550 Stuttgart, Germany

Abstract. Strain distributions around Ge quantum dots embedded in a Si matrix are computed by means of classical molecular dynamics simulations using the molecular dynamics code IMD. The Tersoff potential is employed in order to model covalent bonds. Two crystal lattice structures are considered, the cubic and hexagonal diamond structure. The distributions of the planar strain are studied for a large number of system sizes and lattice misorientations. In a second part, the scaling of IMD is analyzed for different parallelization schemes and machine architectures.

1 Introduction

Quantum dots (QDs) in semiconductors are of great technological importance because of their peculiar optical and electronic properties. In the growth of several layers of planar arrays of QDs, a vertical correlation is observed which leads to regular three dimensional superstructures of QDs. Several forms of such vertical self-orderings have been observed: Perfect vertical alignment [1], vertical anticorrelation with different forms of stackings [2, 3], and vertical correlation with lateral shift for the case of misoriented samples [4].

The reason for these vertical correlations of QDs is usually attributed to the strain arising from the lattice mismatch between the material of the QDs and the embedding material. The tensile strain in the vicinity of the QDs favours the nucleation of QDs in succesive layers at positions where the strain is enlarged. In order to control the growth of QD superlattices, the knowledge of the strain distributions around the QDs is thus of importance. Several theoretical methods have been applied so far to study the correlation of QDs driven by the strain: Continuum elasticity theory [5–7], the finite element method [8, 9], and classical molecular dynamics (MD) simulations [5, 10–15]. MD simulations have the advantage of providing the strain distributions on the atomic level and are the most realistic methods. Quantum mechanical simulations are not feasible in this case due to the large dimensions of the systems amounting to at least several nanometers.

In this article, we report on large-scale MD simulations of strain distributions in QD structures. Similar simulations have been performed previously focusing on different aspects of the strain and stress due to the QDs [5, 10–15].

The present article deals particularly with the influence of the lattice orientation as well as of the elastic anisotropy on the strain distributions. In order to gain insight into the dependence on the anisotropy, we have performed simulations of Ge/Si QDs with two different crystal structures: cubic diamond structure and hexagonal diamond structure. These structures can be considered as model systems for many semiconductors.

Most of the results presented in this article have been obtained on the CRAY-T3E. There is, however, a trend to new hardware architectures, like large clusters with cheap commodity hardware, or clusters with large multi-processor nodes. Moreover, on machines with shared memory multiprocessor nodes new parallelization options become available. For this reason, we have undertaken a comparison of the performance and scaling of our MD code IMD for several different machine architectures and parallelization schemes. These results are expected to be relevant also for other MD codes using short range interactions.

This article is organized as follows. In section 2, we describe the methods of the simulations. The results of the simulations are presented in section 3. In section 4, we present an analysis of the parallel performance of our MD program IMD for different parallelization schemes and machine architectures. Finally, in section 5, we summarize our work.

2   Simulation Methods

Typical semiconductors used in the fabrication of QDs are group IV elements (e.g. Si, Ge) or compounds of group III-V elements (e.g. GaAs, InGaN, InP). These are predominantly covalent materials. For their modeling in classical MD simulations, many-body potentials have to be used which favour specific bond angles. In the present simulations, the Tersoff potential has been employed. This potential depends on a number of parameters which have to be determined by fitting to experimental data. Parameterizations for several semiconductors have been given in the literature. In the present work, we have performed simulations of Ge QDs in a Si matrix. The corresponding parameters are taken from Reference [16].

We have used two different crystal lattice structures which are representative for many semiconductors: The cubic diamond structure and the hexagonal diamond structure. The former has a cubic elastic anisotropy similar to that of the zincblende structure while the latter is similar to the wurtzite structure with a hexagonal elastic anisotropy.

For the simulations, we have employed a starting configuration that is analogous to the experimental setup in the epitaxial growth of QDs. The simulation box consists of a substrate of Si atoms above which a thin wetting layer of Ge atoms is placed. The QD of Ge atoms has the form of a truncated square pyramid and is situated above the wetting layer. A spacer layer (SL) of Si atoms covers the wetting layer and the QD. A part of this configuration can

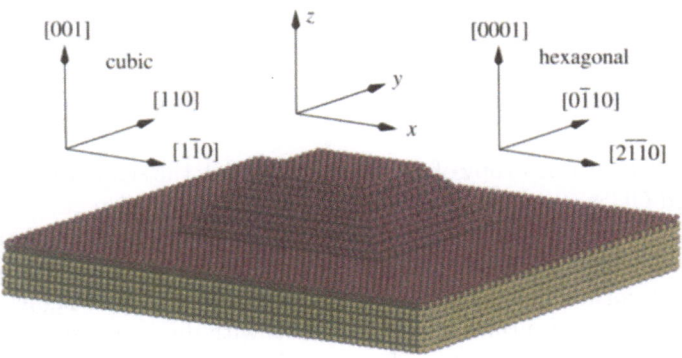

Fig. 1. Starting configuration of a QD simulation. The Si atoms (yellow) belong to the substrate. The wetting layer and the QD consist of Ge atoms (red). The capping spacer layer of Si atoms is not shown in this figure.

be seen in Figure 1 where the upper part of the substrate and the uncovered QD is shown. At the sides of the simulation box, periodic boundary conditions are imposed which means that the system effectively consists of an infinite two dimensional horizontal superlattice of QDs. At the bottom of the simulation box, some layers of the substrate are held fixed during the simulation while at the top of the SL, free boundary condition are used. We have performed simulations with different sizes of the simulation box varying the width from 18 nm to 50 nm and the height from 14 nm to 40 nm. The simulation box for the largest system contained about four milion atoms. The QD always has a fixed size of 12 nm width and 3 nm height. We have also used simulation boxes with different misorientations of the lattices. The crystallographic directions indicated in Figure 1 for the cubic and hexagonal diamond lattice correspond to the systems without misorientation. In the case of misoriented systems, the lattice is tilted about the [110] and the [$2\bar{1}\bar{1}0$] directions for the cubic and hexagonal structures, respectively.

The relaxed states of the QD systems are obtained by MD simulations where during the dynamic evolution of the systems in the NVE ensemble, the velocities of all particles are set to zero whenever the velocity in the configuration space of the system is in the direction opposite to the force vector. The MD program package IMD [17] has been used for these simulations. Depending on the system size, an MD run took 4–6 hours on the CRAY-T3E where usually 128 processors have been used.

From the displacements of the atoms of the relaxed configurations relative to the starting configuration, the strain tensor has been computed using a utility program that is also part of IMD. In the following we discuss only the planar strain $\epsilon_{xx} + \epsilon_{yy}$ since this is the most important part of the strain for the explanation of the vertical correlations in QD systems.

3   Results

3.1   Quantum Dots in the Cubic Diamond Lattice

Due to the elastic anisotropy of the matrix material, the strain distributions of the relaxed QD systems display a complex behaviour with regions of positive and negative strain. In order to gain a qualitative insight into the distributions of planar strain, it is convenient to visualize the surfaces of vanishing planar strain which form the boundaries of the regions of different signs of strain. Figure 2 shows these surfaces for QD systems with diamond lattice for angles θ of misorientation in the range $0° \leq \theta \leq 90°$. The system widths are about 23 nm and the SL thicknesses in all cases are 28 nm. Simulations of systems with smaller SL thicknesses have shown that the strain distributions can be obtained by truncating the system of 28 nm SL thickness (apart from small boundary effects at the SL surface). Thus, from the shape of the surfaces in Figure 2, the sign of the planar strain on the SL surfaces of systems of different height can be read off.

Generally, the planar strain has a strong decay and a characteristic change of sign as a function of increasing distance from the QD. In the case of the misorientation angle $0°$, Figure 2 (a) shows that the strain is oscillating with the distance frome the QD. Directly above the QD (in between the curved surfaces) the strain is tensile. Above the upper surface and below the lower surface, the strain is compressive. For misorientation angles greater than zero, the two surfaces merge to a tube that is tilted in the direction opposite to the misorientation (Figures 2 (b)-(d)). For misorientation angles $\approx 50°$, the positive planar strain is confined within a nearly vertical tube. In these cases, there is no oscillation of the strain (Figures 2 (e),(f)). Increasing the misorientation angle further, the tube is bent in the direction of the misorientation where at different heights of the tube, necks between the tubes of different QDs in the QD superlattice form (Figures 2 (g)-(i)).

Using simulation boxes of different heights, that is, different SL thicknesses, the correlation behaviour of QDs in several vertical layers can be predicted by determining the positions of maximal planar strain on the top of the SL. Depending on the misorientation angle and the SL thickness, we thus obtain a variety of possible correlations which partly have been predicted by continuum elasticity calculations and also found experimentally.

The oscillatory behaviour of the strain above the QDs is caused by the finite horizontal extent of the simulation box. For the case of a QD system without misorientation, the profile of the planar strain along a vertical line through the middle of the QD is shown in Figure 3. In order to characterize this profile quantitatively, we have fitted the following function to the profile:

$$\epsilon(z) = \epsilon_0 \frac{\cos\left[2\pi\left(\frac{z-z_0}{p}\right)\right]}{1 + \left(\frac{z-z_0}{h}\right)^\alpha}. \tag{1}$$

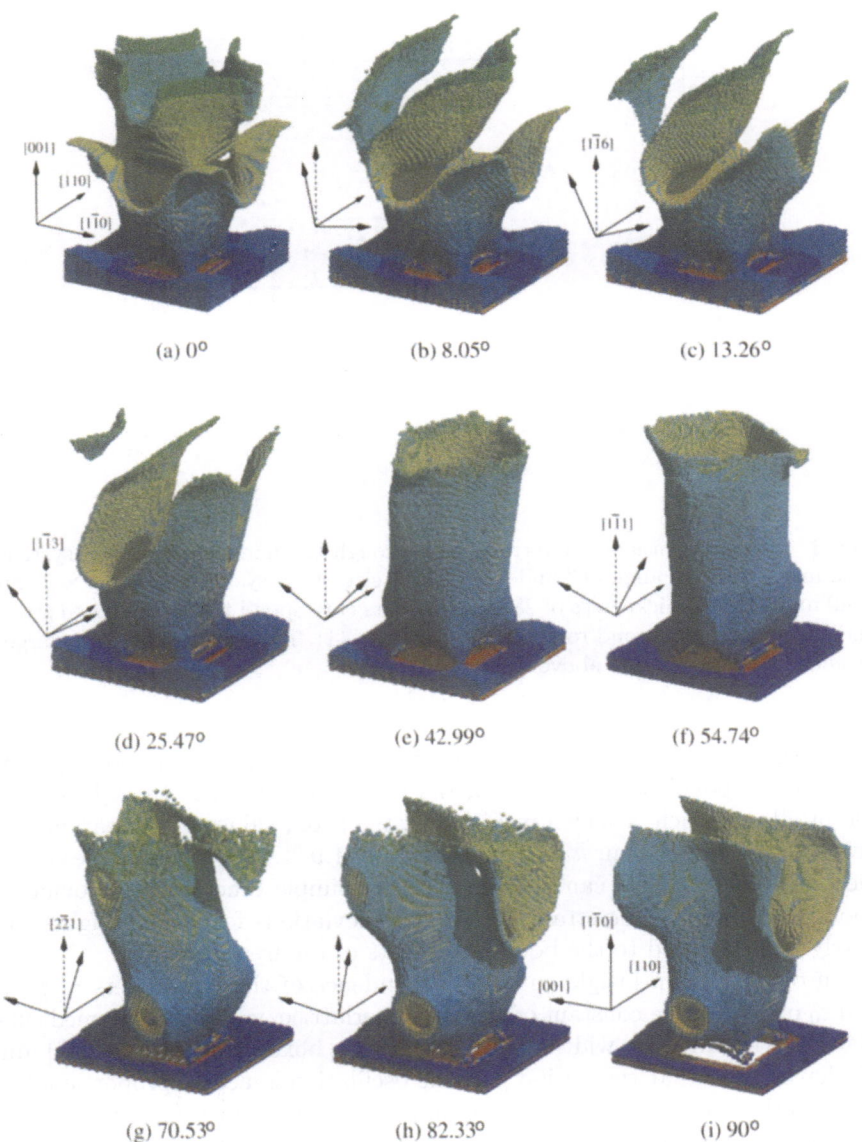

Fig. 2. Surfaces of vanishing planar strain in a Ge/Si quantum dot system with cubic diamond lattice for different misorientation angles. The size of the simulation box is 23 nm × 23 nm × 35 nm. The tilt axis is the [110] direction and the growth direction is indicated by a dashed arrow.

This function describes an oscillation of the strain and a decay by a power law. The exponent α characterizes the strength of the decay of the strain and p is the period of the oscillation. The paramter h corresponds to the height

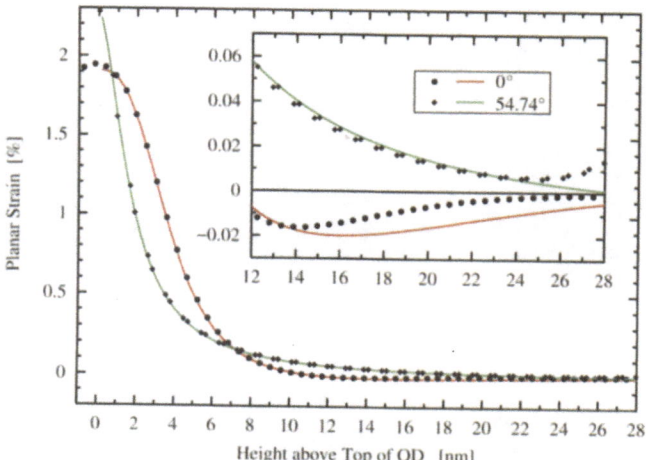

Fig. 3. Planar strain along a vertical line through the middle of a QD in systems with misorientation angles 0° and 54.74°. The size of the systems is 23 nm × 23 nm × 40 nm with SL thicknesses of 28 nm. The dots correspond to the simulated strain and the curves correspond to the fitting function (1). The inset shows the enlarged profiles for larger heights above the dot.

where the strain has fallen off to half its maximal value ϵ_0 at the height z_0 which corresponds to the height of the top of the QD. For the QD system the profile of which is shown in Figure 3, the fitted values of the parameters are $\alpha = 2.9$, $p = 45$ nm, $h = 4$ nm, and $\epsilon_0 = 1.9$ %. The fitted curve is also plotted in Figure 3. It can be seen that the simple function (1) provides a good description of the strain profile. The deviations for larger heights can partly be attributed to the boundary effects of the free surface.

In order to gain insight into the dependence of the parameters α, p, h, and ϵ_0 on the lattice constant of the QD superlattice, we have performed simulations with different widths of the simulation boxes from 18 nm to 50 nm. We have found that the period p of the oscillation is nearly proportional to the superlattice constant while α, h, and ϵ_0 are unchanged for a wide range of superlattice constants.

3.2   Quantum Dots in the Hexagonal Diamond Lattice

The surfaces of vanishing planar strain of QD systems with hexagonal diamond structure for several misorientation angles θ between 0° and 90° are shown in Figure 4. It can be seen that the strain distributions are different from the ones of the QD systems with cubic diamond lattice given in Figure 2. In the case of misorientation angle 0° (Figure 4 (a)), the tensile planar strain above the QD is confined to a narrow tube. There is no oscillation of

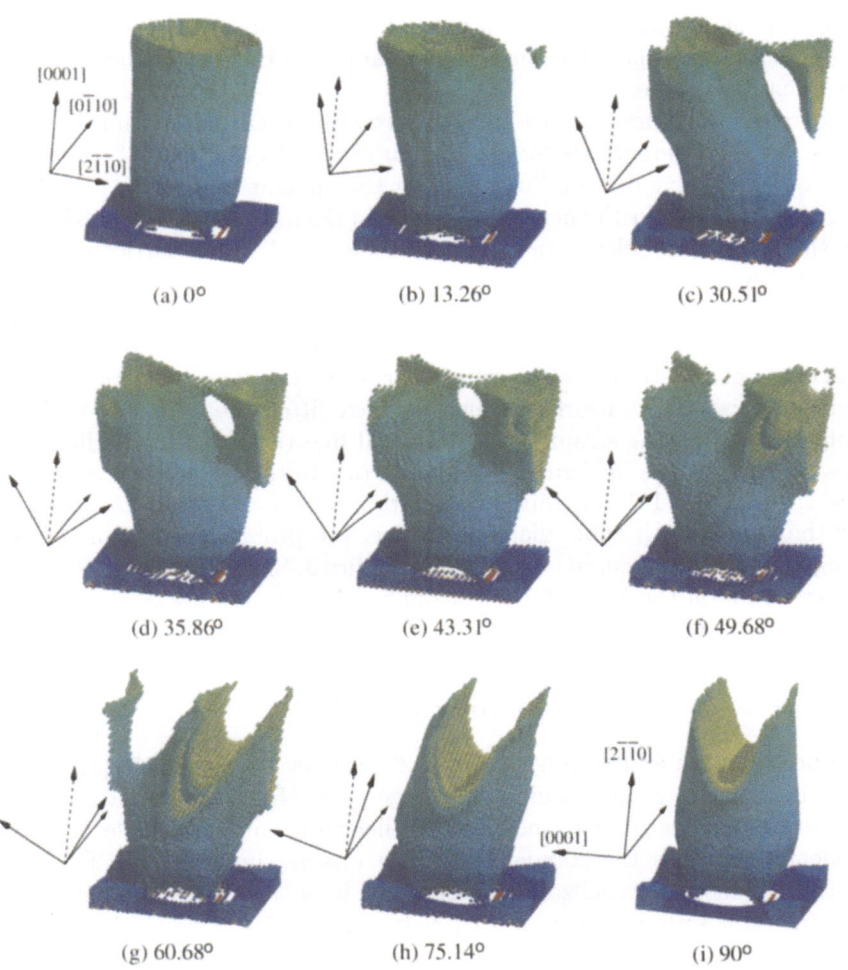

Fig. 4. Surfaces of vanishing planar strain in a Ge/Si quantum dot system with hexagonal diamond lattice for different misorientation angles. The size of the simulation box is 25 nm × 25 nm × 40 nm. The tilt axis is the [0$\bar{1}$10] direction and the growth direction is indicated by a dashed arrow.

the strain along the vertical direction. For finite misorientation angles, the tube is bent in the direction of the misorientation (Figures 4 (b), (c)). For larger misorientations, the tubes of neighbouring QDs of the QD superlattice merge. For intermediate misorientation angles, the tubes of QDs in the direction perpendicular to the tilt axis are connected (Figures 4 (d)-(f)) whereas for large angles the tubes in the direction of the tilt axis are connected (Figures 4 (g)-(i)). The double sheets of surfaces in Figures 4 (e)-(g) are due to internal relaxations of the hexagonal diamond lattice which cause an internal

displacement of sublattices. As in the case of the cubic diamond lattice, we
expect a variety of possible vertical correlations of QDs as a function of the
misorientation angle.

Comparing Figures 2 and 4, it can be seen that there is a certain similarity
of the shapes of the surfaces for the angular range $54.74° < \theta < 90°$ (Figures 2
(f)-(i)) in the cubic diamond case and for the angular range $0° < \theta < 43.31°$
(Figures 4 (a)-(e)) in the hexagonal diamond case. This can be explained by
the similarity of the elastic properties of the cubic diamond lattice along the
<111> directions and the hexagonal diamond lattice along the [0001] direc-
tion. The reason for this is that the orientation of the coordination tetrahedra
in both structures are similar along the respective lattice directions. The vertical decay of the strain is also very similar in the two types of lattices for the
considered ranges of misorientations. We have fitted the function (1) to the
profiles of the planar strain along a vertical line through the middle of the
QDs for the systems in Figures 2 (f) and 4 (a). In the former case we obtain
$\alpha = 1.7$ and $h = 1.7$ nm while in the latter case $\alpha = 1.6$ and $h = 1.8$ nm.
For the system with cubic diamond lattice, the profile of the planar strain
along with the fitted curve is included in Figure 3. Since in both systems, the
planar strain on the SL surface is strongly concentrated above the position
of the QD, a perfect vertical correlation can be expected.

4   Performance and Scaling of IMD

The performance and scaling of IMD had last been analyzed in [17]. Since
then, IMD has evolved considerably. Large parts of the communication routines have been rewritten, and new parallelization methods have been implemented. Besides the standard Message Passing Interface (MPI), which
is still the main parallelization workhorse, it is now also possible to use
OpenMP [18], either instead of MPI or in combination with MPI. OpenMP
is a standard set of compiler directives, which instruct the compiler to set
up several threads to share the work. OpenMP is suitable for multiprocessor
machines with a *shared* memory. A combination of OpenMP and MPI can be
used on clusters of several such multiprocessor nodes; inside the nodes the
parallelization is via OpenMP, whereas the inter-node communication is via
MPI. Of course, in all these cases it is also possible to use just plain MPI, also
for intra-node communication. It is a priori not clear which parallelization
scheme is more efficient. OpenMP might have an advantage in that the work
sharing is dynamic inside a node, whereas in a pure MPI scheme it is static.
This might alleviate load balancing problems, which are hard to resolve in
a pure MPI scheme (IMD uses a geometric domain decomposition). Also,
with OpenMP it is not necessary to make copies of certain data; the same
data can be read by all threads in a node. On the other hand, if OpenMP
is combined with MPI, all the communication of a node is done by a single
thread, whereas with MPI every process does its own communication. This
might well turn out to be a bottleneck.

There is currently a trend towards building massively parallel clusters with cheap commodity hardware, typically running under Linux. A crucial point for such machines is the communication network between the cluster nodes. It would be interesting to know how an MD code like IMD performs with different types of networks, such as 100 Mbit Ethernet, or the more expensive, but faster Myrinet, and how this compares to real supercomputers. A high performance network makes up for a considerable fraction of the total system cost, and the question is whether this is a good investment.

The performance tests were run on the T3E at HLRS, the Hitachi SR8000 at HLRS, with both pure MPI communication and with a combination of OpenMP and MPI, the Kepler cluster [19] of the Sonderforschungbereich 382 in Tübingen, and the Volvox cluster at HLRS. Both the Kepler cluster and the Volvox cluster are similarly configured, with dual processor Pentium III nodes of 650 MHz and 1000 MHz, respectively. While the Kepler cluster has more nodes than the Volvox cluster, on the latter we could compare the communication performance for different networks (Ethernet and Myrinet 2000). On Kepler, only Myrinet was available. On the other hand, on Kepler only the gcc compiler was installed, whereas on Volvox we could use the Intel compiler, which produces significantly fast code (almost by a factor of 2).

For the performance tests we used a simple fcc crystal with tabulated pair interactions. We should emphasize, however, that the performance characteristics and scaling behaviour is not significantly different for different types of *short range* interactions. Therefore, the general trends are relevant also for Tersoff interactions, or EAM interactions. We determined two kinds of performance scaling, with samples of a fixed size, and with samples of a size proportional to the number of processors. In the first case, a given problem of fixed size is solved with the help of more and more processors, whereas in the latter case more and more processors are used to solve larger and larger problems. In both cases, two sample sizes were used. One should keep in mind that the communication overhead is proportional to the surface of the block of material a processor has to deal with. Also, the processes have to be synchronized after each time step. The more work per time step a processor has to do, the lower are the synchronization costs. Therefore, for a given sample size there is a maximum number of processors which is feasible. If one goes beyond this number, the parallelization overhead becomes overwhelming.

The results of the performance tests are shown in Figures 5 (samples of constant size) and 6 (samples of constant size per processor). In an ideal world (perfect scaling), all curves would be horizontal. The curves for the T3E come close to this ideal, which shows that the T3E still has an excellent network. Except for the big sample of constant size, the scaling of the Hitachi SR8000 is less perfect. Also, the pure MPI mode on the SR8000 is 10-20% more efficient than the OpenMP+MPI mode. In addition to the reasons discussed above, this may also be due to the still fairly new OpenMP implementation (both in the compiler and in IMD). So far, we could not make comparisons for other machines where both MPI and OpenMP are available.

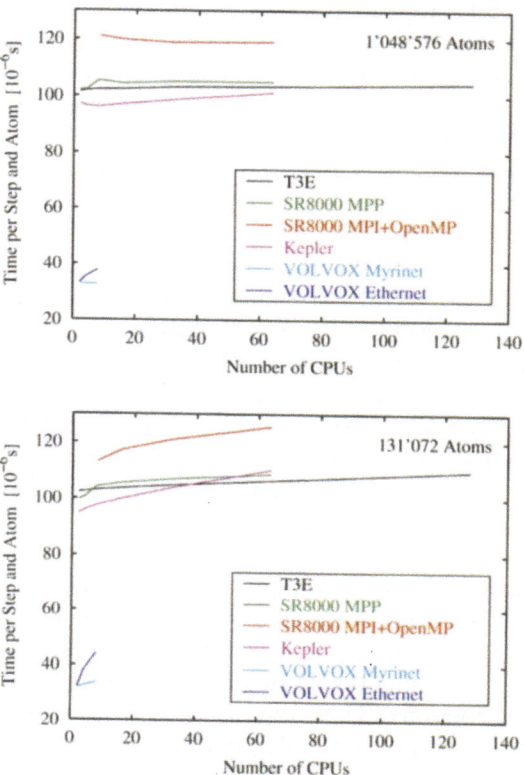

Fig. 5. Performance and scaling of IMD on different machines and architectures for a fixed total number of atoms. These atoms are distributed on a varying number of CPUs. The results for two sample sizes are shown.

So, for the moment a pure MPI scheme seems preferable. This may be different, however, if the atom distribution is very inhomogeneous in the sample. One might think of the simulation of an indentor, which is pushed into the surface of a block of atoms, and which consists itself of atoms whose movements have to be simulated. In such a situation, it is very difficult to divide the system into many blocks of the same geometric size and the same number of atoms. One can easily make such a division into four blocks, but not more. If the different blocks contain largely different numbers of atoms, there will be severe load balancing problems in a pure MPI scheme. If one divides the sample into four blocks only, and uses for each a multiprocessor node with 16 or even 32 processors, one can use a large number of processors without load balancing problems, because with OpenMP the work sharing between the different threads in a node is dynamic. The results for the clusters show that with the faster Myrinet a scaling performance similar to that of the SR8000 can be achieved. With Ethernet, however, the performance is much

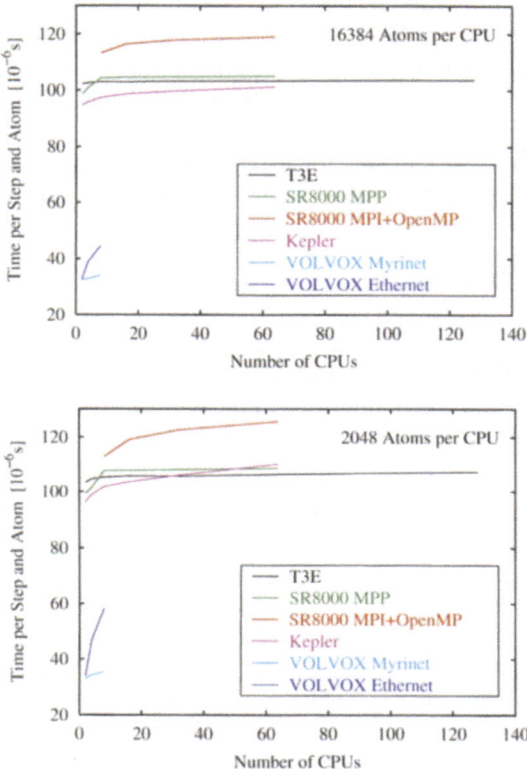

Fig. 6. Performance and scaling of IMD on different machines and architectures for a fixed number of atoms per CPU. The results for two sample sizes are shown.

worse already for moderate CPU numbers. Therefore, communication over Ethernet is clearly not suitable for massively parallel MD simulations, where the different MPI processes have to be synchronized after every time step.

5   Conclusions

In this article, we have reported on large-scale MD simulations of QD structures performed on the CRAY-T3E using the MD program package IMD. We have performed simulations of strain distributions in QD structures for systems of realistic size where the dependence of the strain on the form of the simulation box has been investigated for a larger variety of sizes than have been done so far. The results of these simulations are in agreement with experiments and theoretical strain computations using continuum elasticity theory. Furthermore, we have predicted new vertical correlations in QD superlattices for semiconductor systems of cubic (diamond, zincblende) and hexagonal (wurtzite) lattices.

In a second part, the performance and scaling of IMD has been analyzed for different parallelization schemes and machine architectures. A pure MPI parallelization scheme is usually preferable over a combination of OpenMP+MPI, even though the latter can be more efficient in particular cases with load balancing problems. A massively parallel cluster built from commodity hardware is suitable for IMD (and presumably other MD codes), *provided* it is equipped with a fast network such as Myrinet.

Acknowledgements

This work was supported in part by the Deutsche Forschungsgemeinschaft (DFG) through the Graduiertenkolleg "Grenzflächen in kristallinen Materialien" (GRK 285) and the Sonderforschungsbereich 382 "Verfahren und Algorithmen zur Simulation physikalischer Prozesse auf Höchstleistungsrechnern".

References

1. Q. Xie, A. Madhukar, P. Chen, and N. P. Kobayashi, Phys. Rev. Lett. 75, 2542 (1995)
2. M. Strassburg *et al.*, Appl. Phys. Lett. **72**, 942 (1998)
3. G. Springholz, V. Holý, M. Pinczolits, and G. Bauer, Science 282, 734 (1998)
4. A. Fantini, F. Phillipp, C. Kohler, J. Porsche, and F. Scholz, to appear in J. Crystal Growth
5. C. Pryor, J. Kim, L. W. Wang, A. J. Williamson, and A. Zunger, J. Appl. Phys. **83**, 2548 (1998)
6. V. A. Shchukin, D. Bimberg, V. G. Malyshkin, and N. N. Ledentsov, Phys. Rev. B **57**, 12262 (1998)
7. V. Holý, G. Springholz, M. Pinczolits, and G. Bauer, Phys. Rev. Lett. **83**, 356 (1999)
8. T. Benabbas, P. Francois, Y. Androussi, and A. Levebvre, J. Appl. Phys. **80**, 2763 (1996)
9. T. Benabbas, Y. Androussi, and A. Levebvre, J. Appl. Phys. **86**, 1945 (1999)
10. Y. Kikuchi, H. Sugii, and K. Shintani, J. Appl. Phys. **89**, 1191 (2001)
11. M. A. Makeev and A. Madhukar, Phys. Rev. Lett. **86**, 5542 (2001)
12. I. Daruka, A.-L. Barabási, S. J. Zhou, T. C. Germann, P. S. Lomdahl, and A. R. Bishop, Phys. Rev. B **60**, R2150 (1999)
13. W. Yu and A. Madhukar, Phys. Rev. Lett. **79**, 905 (1997); **79** 4939(E) (1997)
14. K. Scheerschmidt, D. Conrad, H. Kirmse, R. Schneider, and W. Neumann, Ultramicroscopy 81, 289 (2000)
15. C. Kohler (preprint 2002)
16. J. Tersoff, Phys. Rev. B **39**, 5566 (1989)
17. J. Stadler, R. Mikulla, and H.-R. Trebin, Int. J. Mod. Phys. **C8**, 1131 (1997)
18. http://www.openmp.org
19. http://kepler.sfb382-zdv.uni-tuebingen.de/kepler/index.shtml

Parameter Dependences of Convection Driven Spherical Dynamos

R. Simitev and F. H. Busse

Institute of Physics, University of Bayreuth, D-95440 Bayreuth

Abstract. Recent results are presented for convection driven dynamos in rotating spherical fluids shells which are obtained from computations carried out at the Stuttgart Supercomputing Center. Studies of the dependence on the Prandtl number P indicate that dynamo action disappears with increasing P unless the magnetic Prandtl number P_m is also increased. Relaxation oscillations of convection coupled to magnetic torsional oscillations are found at low Prandtl numbers and various types of reversals of dipolar dynamos have been identified.

1 Introduction

Considerable progress has been made in the past years in simulating numerically the process of the generation of magnetic fields by convection in rotating spherical fluid shells. While in earlier work often particular values of the parameters of the problem have been chosen which seemed to provide the optimal compromise between applicability to the Earth's core and computational efficiency, it now appears that extrapolations to conditions of planetary cores are best obtained on the basis of known dynamo properties over an extended domain in the parameter space. In this connection it should be mentioned that the often employed assumption that all diffusivities are replaced by the same eddy diffusivity caused by turbulent motions at small numerically unresolved scales is too simple. From experimental measurements (Ahlers and Xu, 2001) as well as from theoretical considerations (Eschrich and Rüdiger, 1983) it is evident that even in highly turbulent systems diffusivity ratios such as the Prandtl number are not necessarily equal to unity.

In this paper we thus intend to analyze the dependence of convectively driven dynamos on the Prandtl number P and on the magnetic Prandtl number P_m as well as on the Coriolis number and on the Rayleigh number which are the more commonly considered parameters. One of the major goals of studying parameter dependences is the possible discovery of approximate scaling relationships which would permit the elimination of one or more parameters from the problem. In the magnetostrophic approximation the Prandtl number dependence is dropped which is justified in the case of large P (Glatzmaier and Roberts, 1995). For large P_m the magnetostrophic approximation can also be expected to hold as we shall discuss in section 6. More interesting from a geophysical point of view are situations with small values of P and P_m which, unfortunately, are also difficult to explore from a numerical point of view. Some new phenomena found in this parameters regime will be reported in this paper.

We start in section 2 with a brief review of the mathematical formulation of the problem and the numerical methods used for its solution. Computational aspects are discussed in section 3. Results on convection and convection driven dynamos with Prandtl numbers of the order unity or larger are presented in section 4, whereas the situation at low Prandtl numbers is described in section 5. Possibilities for the magnetostrophic approximation are discussed in section 6 and special topics such as reversals are addressed in section 7. An outlook on future research is given in section 8.

2   Mathematical Description of the Problem and Numerical Methods Employed for Its Solution

The analysis of finite amplitude convection in rotating spherical shells and its dynamo action is based on the standard formulation used in earlier work by the authors (Busse et al., 1998; Grote et al., 1999, 2000b, 2001). Instead of the general temperature distribution used in the last of these papers we shall focus on the homogeneously heated sphere as the basic static state of the problem. The gravity field is given by $\boldsymbol{g} = -\gamma d\boldsymbol{r}$ where $\boldsymbol{r}$ is the position vector with respect to the center of the sphere and r is its length measured in units of thickness d of the shell. In addition to d, the time d^2/ν, the temperature $\nu^2/\gamma\alpha d^4$ and the magnetic flux density $\nu(\mu\varrho)^{1/2}/d$ are used as scales for the dimensionless description of the problem where ν denotes the kinematic viscosity of the fluid, κ its thermal diffusivity, ϱ its density and μ is its magnetic permeability. The density is assumed to be constant except in the gravity term where its temperature dependence given by $\alpha \equiv (d\varrho/dT)/\varrho =$ const. is taken into account. The general representation in terms of poloidal and toroidal components can be used for the velocity field $\boldsymbol{u}$ and the magnetic flux density $\boldsymbol{B}$,

$$\boldsymbol{u} = \nabla \times (\nabla v \times \boldsymbol{r}) + \nabla w \times \boldsymbol{r} \ , \tag{1a}$$

$$\boldsymbol{B} = \nabla \times (\nabla h \times \boldsymbol{r}) + \nabla g \times \boldsymbol{r} \ . \tag{1b}$$

As in the earlier work 5 equations for v, w, h, g and the deviation Θ of the temperature from the static distribution can be obtained,

$$[(\nabla^2 - \partial_t)L_2 + \tau\partial_\varphi]\nabla^2 v + \tau Q w - L_2\Theta = -\boldsymbol{r}\cdot\nabla\times[\nabla\times(\boldsymbol{u}\cdot\nabla\boldsymbol{u} - \boldsymbol{B}\cdot\nabla\boldsymbol{B})] \tag{2a}$$

$$[(\nabla^2 - \partial_t)L_2 + \tau\partial_\varphi]w - \tau Q v = \boldsymbol{r}\cdot\nabla\times(\boldsymbol{u}\cdot\nabla\boldsymbol{u} - \boldsymbol{B}\cdot\nabla\boldsymbol{B}) \tag{2b}$$

$$\nabla^2\Theta + RL_2 v = P(\partial_t + \boldsymbol{u}\cdot\nabla)\Theta \tag{2c}$$

$$\nabla^2 L_2 h = P_m[\partial_t L_2 h - \boldsymbol{r}\cdot\nabla\times(\boldsymbol{u}\times\boldsymbol{B})] \tag{2d}$$

$$\nabla^2 L_2 g = P_m[\partial_t L_2 g - \boldsymbol{r}\cdot\nabla\times(\nabla\times(\boldsymbol{u}\times\boldsymbol{B}))] \tag{2e}$$

where ∂_t and ∂_φ denote the partial derivatives with respect to time t and with respect to the angle φ of a spherical system of coordinates r, θ, φ and where the operators L_2 and Q are defined by

$$L_2 \equiv -r^2 \nabla^2 + \partial_r(r^2 \partial_r)$$

$$Q \equiv r \cos \theta \nabla^2 - (L_2 + r\partial_r)(\cos \theta \partial_r - r^{-1} \sin \theta \partial_\theta)$$

The Rayleigh number R, the Coriolis parameter τ, the Prandtl number P and the magnetic Prandtl number P_m are defined by

$$R = \frac{\alpha \gamma \beta d^6}{\nu \kappa}, \quad \tau = \frac{2\Omega d^2}{\nu}, \quad P = \frac{\nu}{\kappa}, \quad P_m = \frac{\nu}{\lambda} \tag{3}$$

where λ is the magnetic diffusivity. We assume stress-free boundaries with fixed temperatures,

$$v = \partial_{rr}^2 v = \partial_r(w/r) = \Theta = 0 \quad \text{at} \quad r = r_i \equiv \eta/(1-\eta)$$

$$\text{and at} \quad r = r_o = (1-\eta)^{-1} \tag{4a}$$

where η is the radius ratio of the spherical shell. Throughout this paper the case $\eta = 0.4$ will be assumed. For the magnetic field electrically insulating boundaries are usually used such that the poloidal function h must be matched to the function $h^{(e)}$ which describes the potential fields outside the fluid shell

$$g = h - h^{(e)} = \partial_r(h - h^{(e)}) = 0 \qquad \text{at } r = r_i \text{ and } r = r_o. \tag{4b}$$

But computations with a conducting inner core, $0 < r < r_i$, have also been done. The numerical integration of equations (2) together with boundary conditions (4) proceeds with the pseudo-spectral method as described by Tilgner and Busse (1997) which is based on an expansion of all dependent variables in spherical harmonics for the θ, φ-dependences,

$$v = \sum_{l,m} V_l^m(r, t) P_l^m(\cos \theta) \exp\{im\varphi\} \tag{5}$$

with analogous expressions for w, Θ, h and g. P_l^m denotes the associated Legendre functions. For the r-dependence expansions in Chebychev polynomials are used. For further details see also Busse et al. (1998).

The standard numerical resolution is given by 33 collocation points in the radial direction and spherical harmonics up to the order 64. But cases up to 65 collocation points and spherical harmonics up to the order 128 are often used.

3   Implementation of the Numerical Algorithm on the Stuttgart Supercomputer

The functions $V_l^m(r,t), W_l^m(r,t), T_l^m(r,t), H_l^m(r,t)$ and $G_l^m(r,t)$ corresponding to the fields v, w, Θ, h and g, respectively, are being stored in r, l, m-space.
In order to allow fast transforms from normal space to Chebychev expansion
and vice versa, the N collocation points in the radial direction are chosen
to lie at $x_n = \cos\left(\pi \frac{n-1}{N-1}\right)$ where x is defined by $x \equiv 2r - (r_o + r_i)$. The
dynamic equations are converted into a system of ODEs in time through the
enforcement of the full equations at every collocation point. The decomposition in terms of Chebychev polynomials is thus merely used to compute
radial derivatives.

The main reason for this "pseudo-spectral" method is the fact that in a
pure spectral method the computation of the nonlinear terms would be very
expensive both in terms of CPU time and in memory consumption. On the
other hand, spectral methods provide a better convergence behavior than
finite difference or finite element methods. Thus, the goal is to benefit from
both techniques as much as possible. In our preference for the pseudo spectral
method we feel confirmed by the comparison done by Fornberg and Merrill
(1997) and by the predominance of the pseudo-spectral approach in a recent
dynamo benchmark study (Christensen et al., 2001).

Time stepping is performed by a combination of an Adams-Bashforth second order scheme treating all the right hand sides of equations (2) explicitly,
whereas the terms at the left hand sides are included in an implicit Crank-
Nicolson step. At the beginning of a time integration an Euler step is used
to start up the scheme.

In order to take advantage of the multiprocessor architecture the code
is parallelized in azimuthal direction, i.e., all coefficients sharing a common
index m are stored at the same processor. This means that for a typical
calculation usually 64 processors are being used. All communications between
them are done with the help of the *Message Passing Interface* (MPI).

At the beginning of each time step all fields and their first and second
derivatives are given in r, l, m-space. At this stage the spectral coefficients of
v, w, Θ, g, h are stored such that coefficients with identical m are stored at
the same processor.

The calculation of the fields v, w, Θ, g, h requires adding associated Legendre functions and performing a Fourier transform. The summation over l in
(5) is implemented as matrix vector multiplications and obviously parallelizes
over m. The summation over m is the Fourier transform which requires inter-
processor communication. Before the actual execution of the FFT, data are
redistributed such that individual processors contain all data with a given index l. The fast Fourier transform algorithm can then be executed locally and
in parallel for separate l and r. The nonlinear terms are now easily obtained
since they only involve multiplications of local data. The transformation back

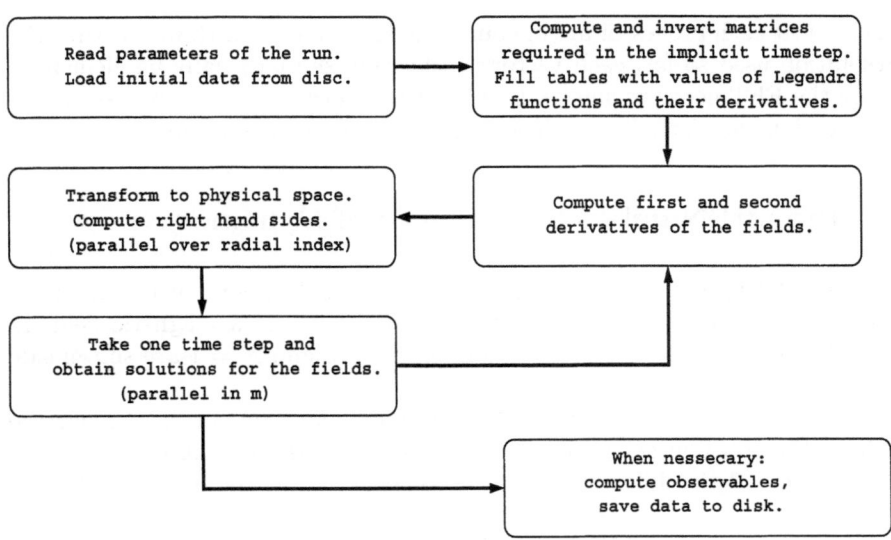

Fig. 1. Diagram of steps in the computational process for solving equations (2) in time.

into the r, l, m-space is performed with a FFT followed by a Gauss quadrature. These are technically the same operations as for the first transformation. At the end of the FFT the original data distribution is restored, i.e. all variables at a given m are collected in the storage of individual processors. The Gauss quadrature is again expressed in terms of matrix vector multiplications which run independently on all processors for different m.

Once the nonlinear terms have been obtained, they can be combined as required by the Adams-Bashforth scheme and added to the terms of the implicit time step involving the variables at the present moment in time only. To complete the time step, a set of N linear equations must be solved for every l, m. Boundary conditions are also included in this set of equations. The coefficients in these equations are independent of m and are collected in separate matrices which are inverted during initialization and multiplied with vectors containing the spectral coefficients of v, w, Θ, g, h during the actual time step. These multiplications separate again in m and involve only local data for each processor. The discretized equations are formulated such that the updated fields are obtained in the n, l, m-space where the radial derivatives can be conveniently computed. A fast cosine transform brings the variables back into the r, l, m-space ready for use in the next time step. For further details we refer to the analogous numerical treatment of the problem of non–magnetic convection by Tilgner and Busse (1997). The diagram of Fig. 1 provides an overview of these computational steps. In summary, the computational burden lies mostly in matrix-vector multiplications, followed by fast cosine and Fourier transforms. The matrix vector multiplications carried out at each processor are of course readily vectorized. However, with the resolution used so far, each vector is relatively short (usually 64 elements). Only the FFT needs to shuffle data between processors. Interprocessor communication thus contributes little to the CPU time expenditure.

4   Prandtl Number Dependence of Dynamos

Before entering the discussion of dynamos we shall briefly outline the influence of the Prandtl number P on convection without a magnetic field. In Fig. 2 typical examples of convection for $P = 15$ and $P = 1$ are shown side by side.

While the columnar form of the convection eddies do not differ much the differential rotation generated by convection is most strikingly different. In the case $P = 1$ the differential rotation is generated by the Reynolds stresses

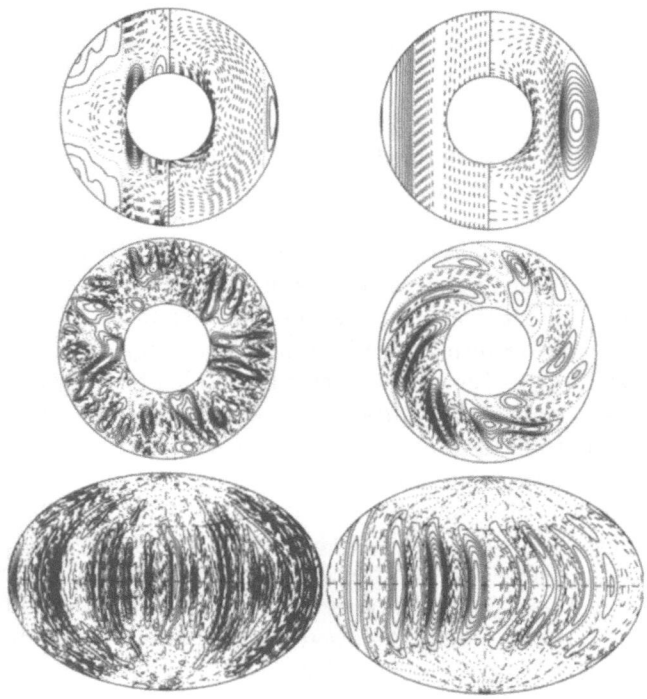

Fig. 2. Convection in rotating spherical fluid shells in the cases $\tau = 5 \cdot 10^3, R = 8 \cdot 10^5, P = 15$ (left column) and $\tau = 10^4, R = 4 \cdot 10^5, P = 1$ (right column). Lines of constant mean azimuthal velocity $\bar{u}_\varphi$ are shown in the left halves of the upper circles and isotherms of $\bar{\Theta}$ are shown in the right halves. The plots of the middle row show streamlines, $r\partial v/\partial \varphi = $ const., in the equatorial plane. The lowermost plots indicate lines of constant u_r in the middle spherical surface, $r = r_i + 0.5$.

of convection and obeys the geostrophic balance perfectly, i.e. it varies only with distance from the axis. On the other hand, in the case $P = 15$ the differential rotation is much weaker and no longer satisfies the geostrophic balance. Instead it obeys primarily a thermal wind relationship as can be seen by comparison with the plot of the axisymmetric component of Θ. The strong decay of the energy density $\bar{E}_t$ of the differential rotation is caused primarily by the general decrease of the amplitude of the convection velocity with increasing P. The energy densities of different components of the velocity field are defined by

$$\bar{E}_p = \frac{1}{2}\langle|\,\nabla \times (\nabla v \times r)\,|^2\rangle, \quad \bar{E}_t = \frac{1}{2}\langle|\,\nabla\bar{w} \times r\,|^2\rangle \tag{6a}$$

$$\check{E}_p = \frac{1}{2}\langle|\,\nabla \times (\nabla\check{v} \times r)\,|^2\rangle, \quad \check{E}_t = \frac{1}{2}\langle|\,\nabla\check{w} \times r\,|^2\rangle \tag{6b}$$

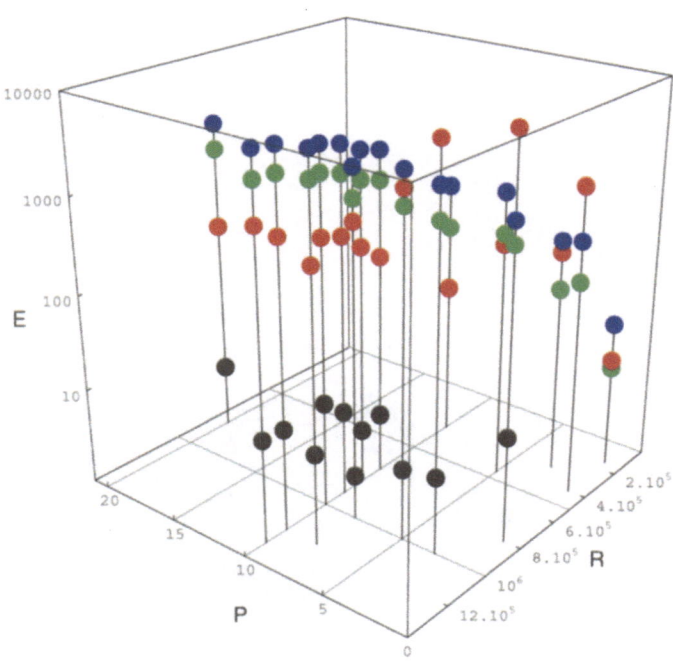

Fig. 3. Dependences of energy densities of the axisymmetric toroidal (red), axisymmetric poloidal (black), non-axisymmetric toroidal (blue) and non-axisymmetric poloidal (green) components of motion on R and P in the case of $\tau = 5.10^3$. The energy densities have been multiplied by P^2 and thus are measured in terms of the thermal scaling

where the angular brackets indicate the average over the fluid shell and where the bar indicates the average over the azimuthal coordinate φ. Thus $v = \bar{v} + \check{v}$ holds where $\check{v}$ denotes the non-axisymmetric component of v.

If thermal scaling κ/d is used instead of the viscous scaling ν/d the fluctuating component of the velocity field show much less variation with P as shown in Fig. 3. But the differential rotation still decreases with increasing P because the tilt of the streamlines of the convection columns as shown in the equatorial plane plots of Fig. 2 is much weaker for larger Prandtl numbers.

An overview of the Prandtl number dependence of convection driven dynamos is given in Fig. 4. The results displayed here have been obtained for the particular value $5 \cdot 10^3$ of the Coriolis parameter τ. But from earlier work (see, for example, Busse et al., 1998) and several computations carried out for $\tau = 10^4$ we expect that these results are representative for a fairly large regime of the parameter τ. A most important result is the property that for $P \geq 5$ the value of the magnetic Prandtl number P_m must always exceed a value of the order P in order that dynamo action can be achieved. The

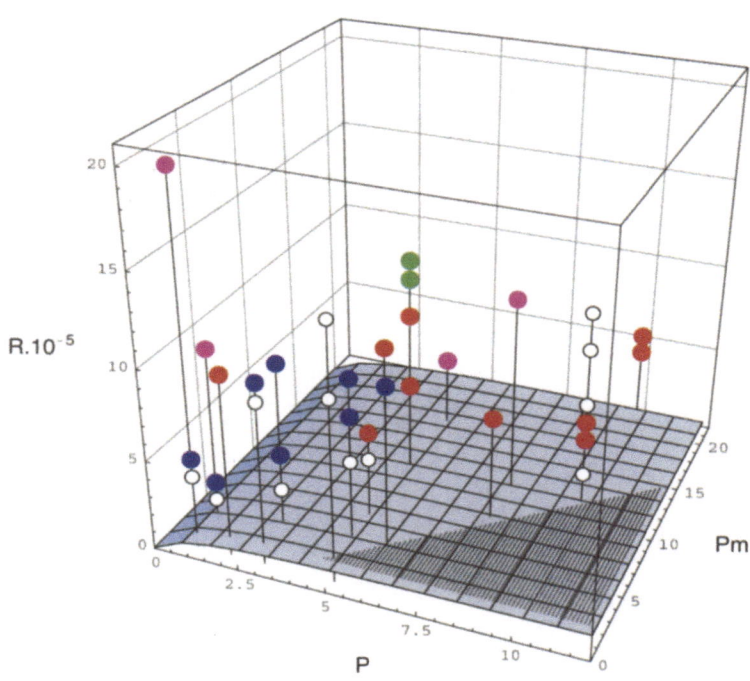

Fig. 4. Dynamo solutions indicated by red (dipolar), blue (quadrupolar), green (hemispherical) and purple (mixed symmetry) balls in the $R - P - P_m$ parameter space. No dynamo solution could be obtained for values of P, P_m in the shaded region.

numerous attempts to obtain dynamos in the darkly shaded region of Fig. 4 have not been displayed. But the obvious difficulty to reach high Rayleigh numbers with sufficient numerical resolution is not the cause of disappearing dynamo action as is evident from the results obtained in borderline cases such as $P = P_m = 10$. It is found in this case that dynamos are obtained for intermediate values of the Rayleigh number R of the order $5 \cdot 10^5$. At such a value of R the magnetic Reynolds number is high enough to permit dynamo action, but is not so high that flux expulsion from the convection eddies becomes a dominant effect. The cause for the increase of the critical value of P_m with increasing P lies in the decline of the differential rotation with increasing P. Indeed, it has been pointed out frequently in previous studies (Grote et al., 2001; Grote and Busse, 2001a) that the interaction between the differential rotation generated by the Reynolds stresses of convection and the magnetic field is the major feature of convection driven dynamos with Prandtl numbers of the order unity or less. Speaking in terms of mean-field-magnetohydrodynamics (Krause and Rädler, 1980), we typically find in this regime that the dynamos are $\alpha\omega$-dynamos for which the differential rotation plays an essential role. For Prandtl numbers larger than unity the nomenclature of mean-field-magnetohydrodynamics is no longer useful since the

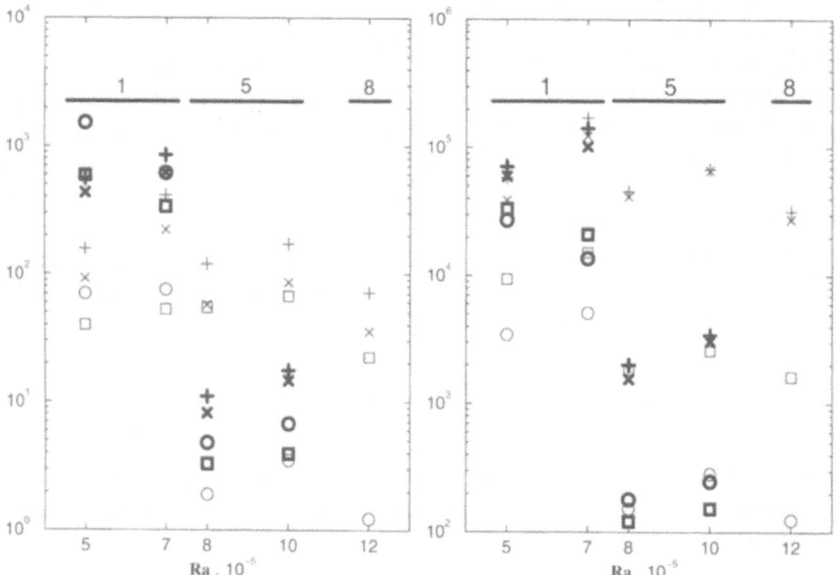

Fig. 5. Magnetic (heavy symbols) and kinetic (light symbols) energies on the left side and corresponding dissipations (right side) have been plotted for $P = 1$, 5 and 8. Energy densities and dissipation densities of axisymmetric field components are indicated by circles (poloidal) and squares (toroidal) while plus-signs and lying crosses denote the corresponding quantities for fluctuating poloidal and toroidal components, respectively.

axisymmetric large scale components of the magnetic field become small in comparison with the small scale non-axisymmetric components as can be seen in Fig. 5 where energies of different components have been plotted. For the magnetic energies $\bar{M}_p, \bar{M}_t, \check{M}_p, \check{M}_t$ definitions analogous to relationships (6) are used with h and g replacing v and w. Another feature that can be noticed in Fig. 4 is the preference for dipolar dynamos with increasing Prandtl number. The property that mixtures of dipolar and quadrupolar components become predominant for high Rayleigh number dynamo is just a consequence of the onset of convection in the polar regions of the fluid shell. As a result the convection velocity field looses its approximate symmetry with respect to the equatorial plane and thus dynamos can no longer be clearly separated into those of dipolar and those of quadrupolar symmetry.

5   Low Prandtl Number Regime

At low Prandtl numbers higher values of the Coriolis parameter τ can be reached since the stabilizing effect of rotation increases with $P\tau$ But the increasing fluctuations in time cause numerical difficulties because of the small time steps that are required. Therefore the systematic study of convection with and without dynamo action has not proceeded as far as it has done in the regime of Prandtl numbers of the order unity and larger. A typical example of convection without magnetic field is shown in Figs. 6a and 6b.   The time dependence of $\bar{E}_t$ exhibits the relaxation oscillations well known from earlier studies at values of P of the order one (Grote et al., 2000; Grote and Busse, 2001a; Grote et al., 2002). The energies $\check{E}_t$ and $\check{E}_p$ of the fluctuating components of the velocity field do not decay to zero, however, in the interval where $\bar{E}_t$ is large because the Rayleigh number for onset of convection in the polar regions is exceeded. Only the convection columns outside the tangent cylinder touching the inner core at its equator participate in the relaxation oscillations. Since the polar convection dominates the heat transport, the Nusselt numbers which are defined by

$$Nu_{i,o} = 1 - \frac{P}{r_{i,o}} \left.\frac{\partial \bar{\bar{\Theta}}}{\partial r}\right|_{r=r_i,r_o} \tag{7}$$

exhibit a weaker influence of the relaxation oscillations than at $P \approx 1$. In particular the Nusselt number Nu_o measured at the outer boundary is nearly constant in time.

The highly time dependent nature of dynamic processes in convecting low Prandtl number fluids is also reflected in the convection driven dynamos. A typical example is shown in Fig. 7a where the intermittent character of dynamo action is evident. From a strongly convective state in which the predominantly dipolar magnetic field suppresses the differential rotation the system changes into a state of weak convection with strong differential rotation. The magnetic field is still predominantly dipolar as can be seen from

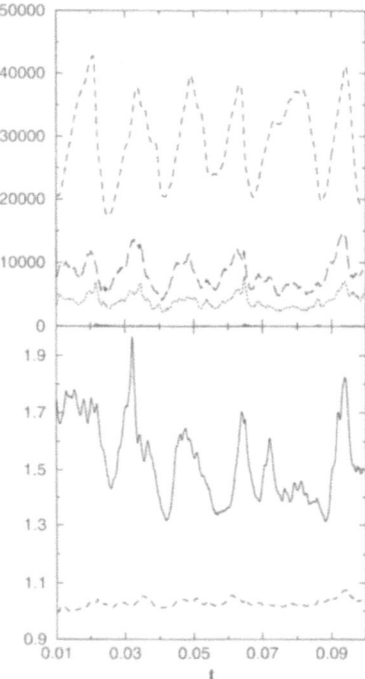

Fig. 6. a) Relaxation oscillation of convection in the case $\tau = 2 \cdot 10^4$, $R = 6 \cdot 10^5$, $P = 0.1$. A section of a much longer time series is shown with the energies $\bar{E}_t$ (short dashed line), $\check{E}_t$ (long dashed line) and $\check{E}_p$ (dotted line) displayed in the upper graph and the Nusselt number Nu_i (solid line) and Nu_o (dashed line) in the lower graph.

the time sequence of plots shown in Fig. 7b. Oscillations in the magnetic field strength are clearly apparent which show some correlation with the spatial structure of the differential rotation, but not with its energy $\bar{E}_t$. It is of interest to see that the columnar convection weakens when the poloidal magnetic field is weak. When the poloidal field grows it inhibits the differential rotation and thereby facilitates the columnar convection. The polar convection appears to be 180° out of phase with equatorial convection. The heat transport carried through the polar sections thus alternates in time with the heat flux transferred through the equatorial region.

The dynamo oscillations become much more pronounced as the Coriolis parameter τ and the Rayleigh number R are increased. In Fig. 8a the strong nearly periodic oscillations are clearly seen. They represent a striking coherent structure of a highly turbulent system. The oscillations are similar to the relaxation oscillations of convection in the absence of a magnetic field in that the convection columns can grow in amplitude only if the differential rotation is sufficiently weak. While viscous diffusion leads to the decay of

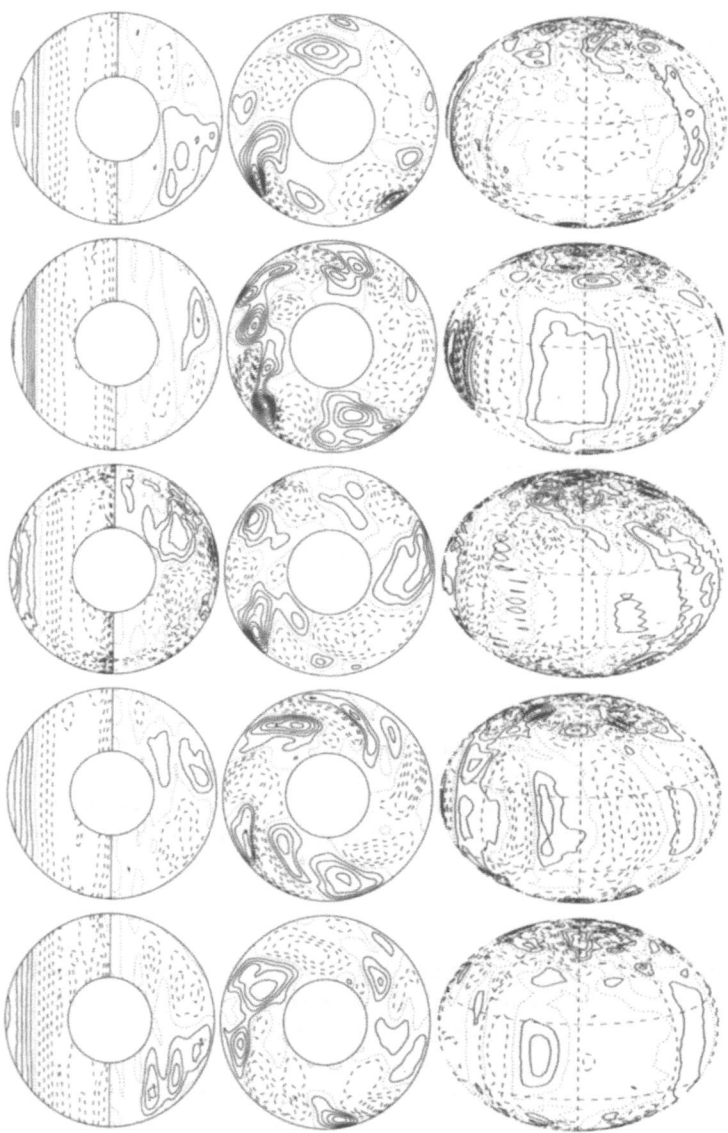

Fig. 6. b) Time sequence of equidistant plots (top to bottom) covering the time span from $t = 0.057$ to $t = 0.073$ of the time series of Fig. 6a. The left half of the left circle in each row indicates lines of constant $\bar{u}_\varphi$, the right half displays meridional streamlines, $r \sin \theta \partial v / \partial \theta = $ const.. The middle circle shows streamlines, $r \partial v / \partial \varphi = $ const., in the equatorial plane and the plot on the right side shows lines of constant u_r on the spherical surface $r = r_i + 0.5$.

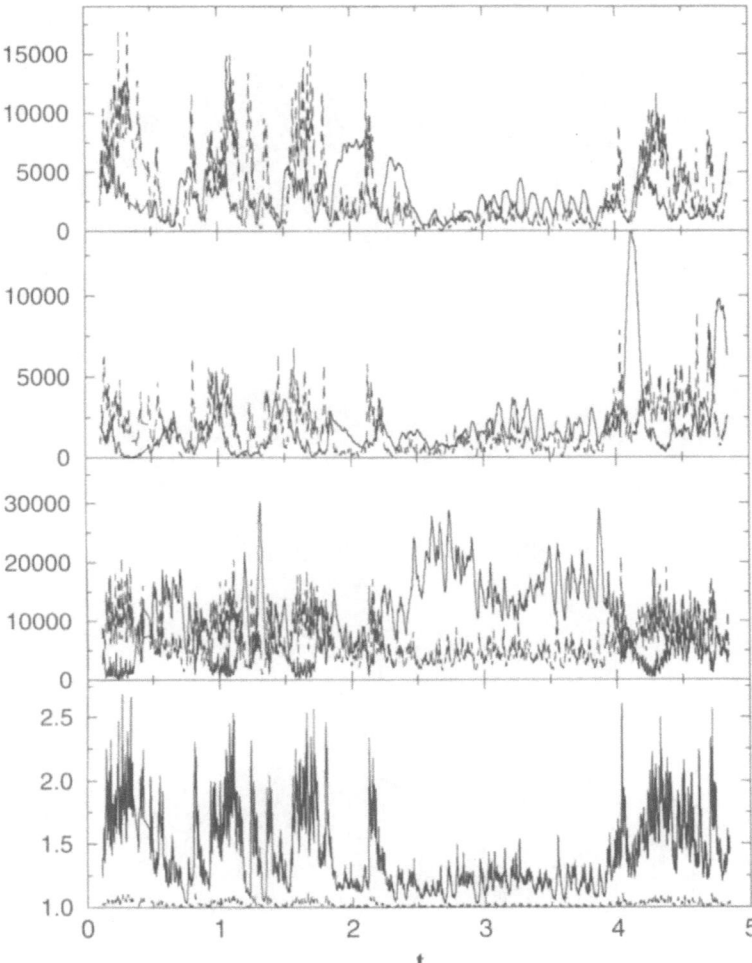

Fig. 7. a) Time series of a convection driven dynamo with $\tau = 3 \cdot 10^4$, $R = 8.5 \cdot 10^5$, $P = 0.1$, $P_m = 1$. The first, second and third plot from the top show energy densities of the dipolar and quadrupolar components of the magnetic field and of the velocity field, respectively. The mean toroidal and fluctuating toroidal energy densities are indicated by solid and dashed lines, respectively. The lowermost plot shows the Nusselt number Nu_i (solid line) and Nu_o (dashed line).

$\bar{E}_t$ in the non-magnetic case, the axisymmetric poloidal component of the magnetic field brakes the differential rotation in the case of Fig. 8a. Thus a period of about 0.02 is seen instead of the period 0.1 resulting from the viscous decay. From Fig. 8b it is apparent that the radial dependence of the differential rotation and not only its amplitude changes throughout the oscillation period. Similarly the distribution of the azimuthal component $\bar{B}_\varphi$

Fig. 7. b) Time sequence of equidistant plots (top to bottom) covering the time span from $t = 2.24$ to $t = 2.69$ of the time series of Fig. 7a. The left half of the first circle in each row indicates lines of constant $\bar{u}_\varphi$ while the right half displays meridional streamlines, $r \sin \theta \partial v / \partial \theta = $ const.. The second circle in each row shows streamlines, $r \partial v / \partial \varphi = $ const., in the equatorial plane. The oval plot exhibits lines of constant u_r on the surface $r = r_i + 0.5$. The last circle in each row indicates lines of constant $\bar{B}_\varphi$ in its left half and meridional field lines, $r \sin \theta \partial h / \partial \theta = $ const., in its right half.

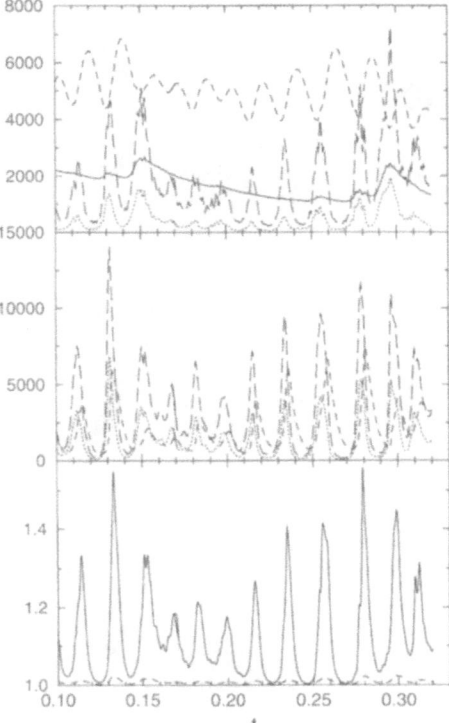

Fig. 8. a) Time series of magnetic (upper graph) and kinetic (middle graph) energy densities and of Nusselt numbers (lower graph) in the case $\tau = 10^5, R = 2 \cdot 10^6, P = 0.1, P_m = 1$. Axisymmetric poloidal (solid line) and toroidal (short dashed lines) components and non-axisymmetric poloidal (dotted lines) and toroidal (long dashed line) components of the energy densities are shown. The component $\bar{E}_p$ is too small to be noticeable in the middle graph. Nu_i (solid line) and Nu_o (dashed line) are displayed in the lower graph.

of the magnetic flux density changes owing to the stretching of the meridional field lines by the differential rotation. The period of 0.02 corresponds roughly to that expected for a standing torsional Alfvén wave. There thus appears to be a resonance between the convection oscillation and a torsional Alfvén wave. A more detailed investigation of the torsional oscillation will be of geophysical interest since it has long been believed (Braginsky, 1980) that they play an important role in the geomagnetic secular variation.

6   On the Magnetostrophic Approximation

Among the various advection terms in the basic equation (2) for convection driven dynamos the advection of momentum appears to be the least important. The advection of the temperature provides the essential nonlinearity

Fig. 8. b) Time sequence of equidistant plots (top to bottom) covering the time span from $t = 0.234$ to $t = 0.258$ of figure 8a. The left half of the first circle in each row indicates lines of constant $\bar{u}_\varphi$, while the right half shows meridional streamlines, $r \sin\theta \partial \bar{v}/\partial\theta = \text{const.}$. The middle circle exhibits streamlines, $r\partial v/\partial\varphi = \text{const.}$, in the equatorial plane. The right circle shows lines of constant $\bar{B}_\varphi$ in its left half and meriodional field lines, $r \sin\theta \partial\bar{h}/\partial\theta = const$, in its right half.

for the dependence of the convection amplitude on the Rayleigh number and the advection of the magnetic flux density represents an intrinsic part of the equation of magnetic induction. But in the equation of motion the Coriolis force term usually exceeds by far the other inertial terms and it is not surprising that this property is used as an argument for the neglection of the term $\partial \boldsymbol{u}/\partial t + \boldsymbol{u} \cdot \nabla \boldsymbol{u}$. This so-called magnetostrophic approximation is formally justified in limit of large Prandtl numbers. When the thermal time scale d^2/κ is used instead of the viscous one and when the magnetic flux density is scaled with $\sqrt{\varrho\mu_0}\kappa\sqrt{P}/d$, the basic equations assume the form

$$P^{-1}\frac{Du}{Dt} + \tau k \times u = -\nabla\pi + \Theta r + \nabla^2 u + (\nabla \times B) \times B \qquad (8a)$$

$$\frac{D\Theta}{Dt} = Ru \cdot r + \nabla^2\Theta \qquad (8b)$$

$$\frac{\kappa}{\lambda}\left(\frac{DB}{Dt} - B \cdot \nabla u\right) = \nabla^2 B \qquad (8c)$$

where D/Dt is the material derivative, $D/Dt = \partial/\partial t + u \cdot \nabla$. The temperature has been scaled by β_0/R in this case. In the magnetostrophic approximation the term $P^{-1}Du/Dt$ is neglected and thus one of the dimensionless parameters of the problem has disappeared.

There is another way in which a magnetostrophic approximation can be obtained. Through the use of the magnetic diffusion time scale, d^2/λ, and the scale $\sqrt{\varrho\mu_0}\lambda\sqrt{P_m}/d$ for the magnetic flux density the basic equation can be transformed into

$$P_m^{-1}\frac{Du}{Dt} + \tau k \times u = -\nabla\pi + \Theta r + \nabla^2 u + (\nabla \times B) \times B \qquad (9a)$$

$$\frac{\lambda}{\kappa}\frac{D\Theta}{Dt} = Ru \cdot r + \nabla^2\Theta \qquad (9b)$$

$$\frac{DB}{Dt} - B \cdot \nabla u = \nabla^2 B \qquad (9c)$$

For large P_m it appears to be justified to drop $P_m^{-1}Du/Dt$ from equation (9a) whereby a magnetostrophic approximation is obtained with one parameter less than in the original equations. While convection without magnetic field seems to become nearly independent of the Prandtl number P for $P \geq 10$, this does not seem to be the case for convection driven dynamo which depend rather sensitively on P even for $P > 10$. Although the differential rotation decreases in amplitude rapidly with increasing P it continues to excert important influence on the dynamo process. In the case of equations (9) the magnetostrophic approximation appears to have a similarly restricted range of validity.

7   Oscillatory Dipolar Dynamos and Reversals

Dipolar dynamos with low magnetic field strength usually do not possess the oscillatory character that is characteristic for quadrupolar and hemispheric dynamos (Grote et al., 2000). But there are some situations where a cyclical behavior can also be realized in the case of dipolar dynamos. At high Rayleigh numbers after convection has also appeared in the polar regions of the spherical shell quadrupolar dynamos are replaced by predominantly

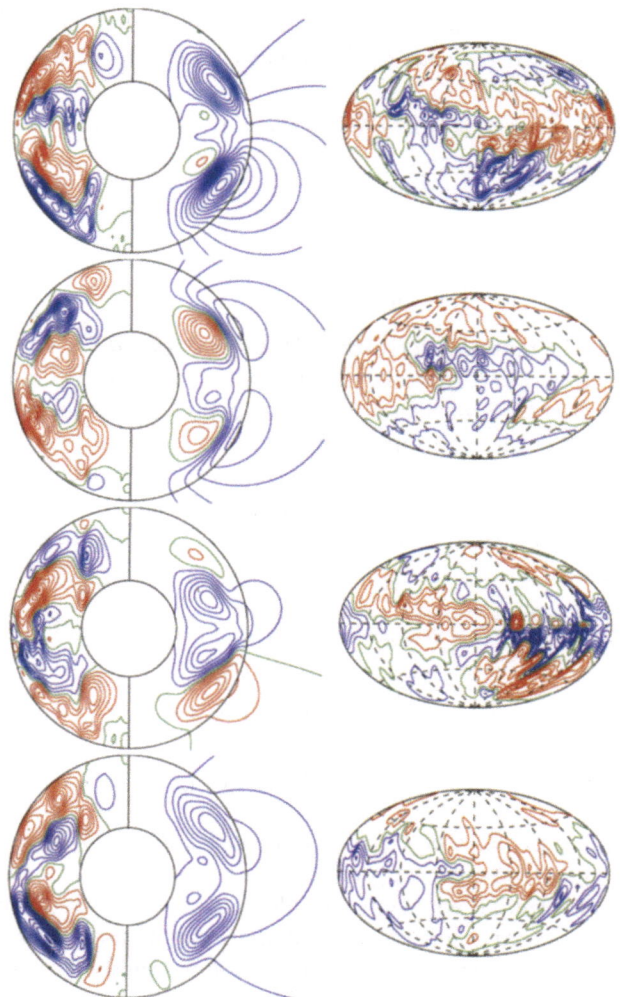

Fig. 9. A chaotically oscillatory dynamo of predominantly dipolar character for $R = 5.4 \cdot 10^5, \tau = 10^4, P = 1, P_m = 4$. A sequence of plots equidistant in time (from top to bottom with $\Delta t = 0.08$) is shown with lines of constant $\bar{B}_\varphi$ in the left halves and meridional field lines, $r \sin\theta \partial h/\partial\theta = $ const., in right halves of the circles on the left side. On the right lines of constant B_r are shown on the spherical surface $r = 1.3 r_0$.

dipolar dynamos which, however, tend to continue to exhibit the oscillatory behavior of the quadrupolar dynamo as shown in figure 17 of Busse (2002). Since R is of the order 20 times its critical value the dynamo process is highly chaotic. But except for the different symmetry with respect to the equatorial plane the oscillations resemble those of the quadrupolar case. The dynamo

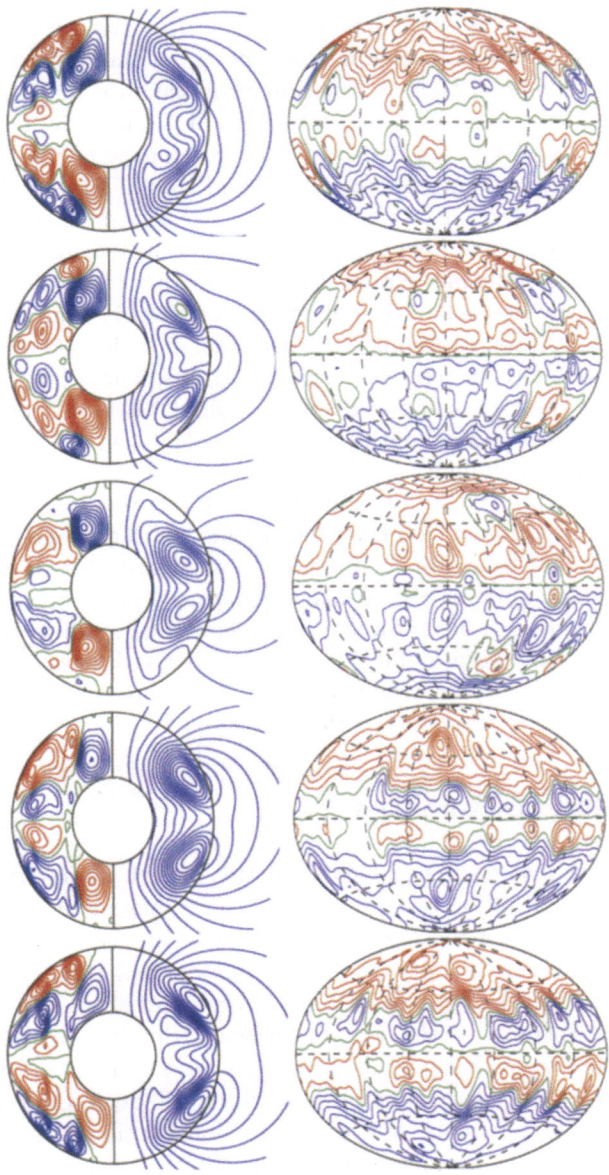

Fig. 10. A time sequence (top to bottom with $\Delta t = 0.06$) of plots for a partially oscillating dipolar dynamo with $\tau = 5 \cdot 10^3, R = 6 \cdot 10^5, P = P_m = 5$. The left halves of the circles in the left column show lines of constant $\bar{B}_\varphi$, the right half indicates field lines, $r \sin \theta \partial \bar{h} / \partial \theta = $ const.. The right column displays lines of constant B_r on the spherical surface $r = r_o + 0.3$.

oscillation of Fig. 9 is quite different. While the flux tubes of alternating polarity of the mean azimuthal component $\bar{B}_\varphi$ of the magnetic field propagate towards higher latitude as in the quadrupolar case, an oscillation cannot be detected in the axisymmetric component of the poloidal field outside the sphere. This property is caused by the predominance of the $m = 1$ component of the field as shown on the right hand side of Fig. 9. The oscillatory process is nearly 180° out of phase on opposite sides of the sphere. At larger distances from the sphere an equatorial dipole slowly drifting relative to the rotating frame of reference will be the main feature of the field.

Another type of oscillatory dynamo is exhibited in Fig. 10. As is typical for dynamos at Prandtl numbers of the order 5 or larger, strong magnetic flex tubes surround the axis in the polar regions. These flux tubes together with the axisymmetric component of the poloidal field do not change much in time while the mean azimuthal magnetic field $\bar{B}_\varphi$ in the equatorial region outside the tangent cylinder executes the usual nearly periodic oscillations as shown in the figure. Little information about this oscillation can be gained by watching the poloidal field from the outside which exhibits a steady dipole with only a minor oscillatory modulation at low latitudes

Besides these more or less regular oscillatory dipolar dynamos numerous chaotic dynamos have been found which show much more irregular oscillations such that only occasionally a reversal is observed. Further computations will be needed to obtain a clearer picture of the dependence of the statistics of reversals on the parameters of the problem.

8   Concluding Remarks

The computational results presented in this report represent highlights of a systematic exploration of convection driven dynamos in rapidly rotating spherical shells. The attention has been focused on the influence of Prandtl numbers different from unity. It has become evident that even a slight increase of P to values of the order 5 has a profound effect on the differential rotation and thereby on the dynamo process. Similarly, a lowering of the Prandtl number to values of the order 0.1 gives rise to new forms of coherent phenomena of the turbulent systems which still need to be explored more systematically. Of particular interest are torsional oscillations for which some evidence exists in the geomagnetic secular variation.

Further studies are also needed for the determination of the parametric dependence of aperiodic reversals, their typical properties and the frequency with which they occur. It appears that from these studies and those of other groups a much better understanding of the geodynamo will soon become available.

References

Ahlers, G., and Xu, X., Prandtl-Number Dependence of Heat Transport in Turbulent Rayleigh-Bénard Convection, *Phys. Rev. Lett.* **86**, 3320–3323 (2001)

Braginsky, S.I., Magnetic Waves in the Core of the Earth II, *Geophys. Astrophys. Fluid Dyn.* **14**, 189–208 (1980)

Busse, F.H., Grote, E., and Tilgner, A., On convection driven dynamos in rotating spherical shells, *Studia geoph. et geod.* **42**, 211–223 (1998)

Busse, F.H., On convection driven dynamos in rotating spherical shells, *Phys. Fluids* **14**, 1301–1314 (2002)

Christensen, U., Aubert, J., Cardin, P., Dormy, E., Gibbons, S., Glatzmaier, G.A., Grote, E., Hankura, Y., Jones, C., Kono, M., Matsushima, M., Sakuraba, A., Takahashi, F., Tilgner, A., Wicht, J., and Zhang, K., A numerical dynamo benchmark, *Phys. Earth. Planet. Int.* **128**, 25–34 (2001)

Eschrich, K.-O., and Rüdiger, G., A Second-order Correlation Approximation for Thermal Conductivity and Prandtl Number of Free Turbulence, *Astron. Nachr.* **304**, 171–180 (1983)

Fornberg, B., and Merrill, D., Comparison of finite difference- and pseudospectral methods for convective flow over a sphere, *Geophys. Res. Lett.* **24**, 3245–3248 (1997)

Glatzmaier, G.A., and Roberts, P.H., A three-dimensional convective dynamo solution with rotating and finitely conducting inner core and mantle, *Phys. Earth Plan. Int.* **91**, 63–75 (1995)

Grote, E., and Busse, F.H., Computation of Convection Driven Spherical Dynamos, pp. 13–25 in "High Performance Computing in Science and Engineering '98", E. Krause and W. Jäger, eds., Springer-Verlag, Berlin, Heidelberg, 1999

Grote, E., and Busse, F.H., Dynamics of Convection and Dynamos in Rotating Spherical Fluid Shells, *Fluid Dyn. Res.* **28**, 349–368 (2001a)

Grote, E., and Busse, F.H., Dynamics of Convection and Dynamos in Rotating Spheres, pp. 13–36 in "High Performance Computing in Science and Engineering 2000", E. Krause and W. Jäger, eds., Springer-Verlag, Berlin, Heidelberg, 2001b

Grote, E., Busse, F.H., and Simitev, R., Buoyancy Driven Convection in Rotating Spherical Shells and its Dynamo Action, pp. 12–34 in "High Performance Computing in Science and Engineering 2001", E. Krause and W. Jäger, eds., Springer-Verlag, Berlin, Heidelberg, 2002

Grote, E., Busse, F.H., and Tilgner, A., Convection driven quadrupolar dynamos in rotating spherical shells, *Phys. Rev. E* **60**, R5025–R5028 (1999)

Grote, E., Busse, F.H., and Tilgner, A., Regular and Chaotic Spherical Dynamos, *Phys. Earth. Planet. Int.* **117**, 259–272 (2000)

Krause, F., and Rädler, K.-H., Mean-field magnetohydrodynamics and dynamo theory, Akademie-Verlag, Berlin and Pergamon Press, Oxford, 1980

Kuang, W., and Bloxham, J., An Earth-like numerical dynamo model, *NATURE*, **389**, 371–374 (1997)

Tilgner, A., and Busse, F.H., Finite amplitude convection in rotating spherical fluid shells, *J. Fluid Mech.* **332**, 359–376 (1997)

Inertial Instabilities in Precession Driven Flow

S. Lorenzani and A. Tilgner

Institute of Geophysics, University of Göttingen, 37075 Göttingen

Abstract. The flow of incompressible fluid inside an ellipsoidal shell with imposed rotation and precession is investigated by direct numerical simulation. The flow becomes unstable and eventually turbulent at large enough precession rates. The mechanisms behind these transitions are relevant for geophysical problems.

1 Introduction

Precession driven flow is one of the basic problems in the field of rotating fluids. Without the precessing motion of the container, the fluid would simply be entrained by viscous forces at the boundaries until it rotates uniformly like a solid body. Precession disturbs the solid body rotation. Perturbations of uniformly rotating fluids, in particular due to precession, have already been studied in the past for a number of reasons: Fluids in rotation are the rule rather than the exception in engineering flows (e.g. in turbomachinery). One case in which the role of precession is particularly obvious is the problem of attitude control of satellites. Spacecrafts are frequently stabilized by spinning them about an axis. But they also contain liquid fuel. When the orientation of the spin axis needs to be changed during some maneuver, the reaction of the liquid fuel must be taken into account. The engineering community was led by this problem to perform experiments with precessing containers which are very much the same as those motivated by geophysical problems ([20], [21]).

Vortices in shear layers and turbulent flows may also be regarded as being locally in a state of uniform rotation. The instability of such a vortex, if stretched or deformed into an ellipsoidal rather than a circular cross section, is observed in a variety of transitional flows ([12], [7], [10]). The same type of distorted vortex also occurs in a precessing ellipsoidal vessel.

Geophysics has been the main motivation for the study of precession driven flow in its simplest form. The earth's liquid core and the flow driven in it by the precession of the earth's rotation axis has a direct bearing on two geophysical problems: the rotation of the earth and the generation of the earth's magnetic field. For both problems, it is important to know whether the flow is laminar or in an unstable or turbulent state. This report presents some findings on the nature of the instability of precession driven flow. It is shown how viscous corrections to Poincaré's inviscid solutions lead to instability. These instabilities may lead to small scales which are numerically unresolvable and which may obscure inertial instabilities. Inertial instabilities do not require viscosity for their existence and are predicted by theory to

occur in precession driven flow. It is therefore necessary to change to free slip
boundaries in order to find this type of instability. The numerical methods
required to do so are presented here as well as first results obtained with
these boundary conditions.

2   Solutions of the inviscid equation of motion

Without precession, the fluid settles to a motion of uniform rotation in unison
with the rotation of the container. Such a flow has constant vorticity. When
the container starts to precess, the liquid tends to maintain its initial motion
due to inertia. Viscous and pressure torques exerted by the boundaries on
the fluid act to align the vorticity of the flow with the rotation axis of the
mantle. As the mantle's axis is continuously moving in inertial space due
to precession, this alignment is never quite reached. Poincaré [13] assumed
that the flow maintains a spatially uniform but time dependent vorticity
throughout its evolution. Uniform vorticity flows are indeed solutions of the
inviscid equation of motion and are commonly called Poincaré flows. These
solutions are derived using the Lagrangian formalism in ref. [13] (parts of
which are translated into English in the final pages of the book by Lamb [6]).

Figure 1 summarizes the main geometrical properties of the Poincaré solution: Streamlines lie on ellipsoidal surfaces. In addition, streamlines are
confined to planes. Streamlines are therefore ellipses with identical elliptic-
ities lying in these planes. The normal to these planes is different from the
line running through the centers of the ellipses.

Application to earth's parameters ($\Omega_z = \cos 23.5°/(26000 \cdot 365)$, $\Omega_x =$
$\sin 23.5°/(26000 \cdot 365)$, $c/a = 399/400$) predicts that the angle between the
rotation axes of the mantle and the core is 1.7×10^{-5} rad.

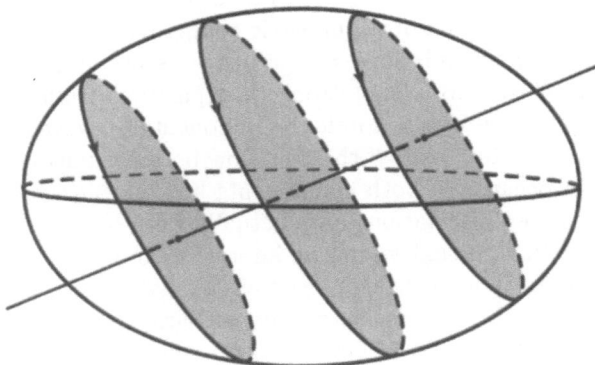

Fig. 1. Sketch of the Poincaré flow. Three streamlines are shown together with the
normal to the planes to which streamlines are confined.

3   Viscous effects

We have seen in the previous section that the inviscid equation of motion does not uniquely determine a solution. If viscosity is taken into account, one needs to find solutions which satisfy the no slip boundary conditions. The first analytical attempts to include viscous effects represented the full solution as a Poincaré solution modified near the boundaries by a viscous boundary layer ([16], [15]). In a linear theory which assumes zero Rossby number and small Ekman number, a particular orientation of the Poincaré flow is selected. Busse [1] extended the previous theory by including non-linear effects and determined the flow from an expansion in Ekman and Rossby numbers. The non-linear effects introduce modifications of the Poincaré solution in the interior of the fluid in the form of a differential rotation. This correction is a second order effect in the sense that its amplitude is proportional to the square of the precession rate. Crucially however, the correction contains a singularity in the limit of zero Ekman number: In a spherical container of radius 1, it diverges at a distance cos 30° from the rotation axis of the fluid. One therefore does not recover the Poincaré solution from the full Navier-Stokes equation for an Ekman number tending to zero. This conclusion is only obtained when non-linearity is taken into account.

The divergence found in the theory has its counterpart in the real system in the form of a cylindrical shear layer coaxial with the rotation axis of the fluid. Experiments have revealed a shear layer at the location predicted by theory and also additional weaker layers. These shear layers are also reproduced in numerical simulations [9].

The numerical results concerning the modifications of the basic flow due to viscous effect are detailed in the references ([17], [18], [19], [9], [8]). Here, we are only interested in the effect of these modifications on the stability of the flow.

Instability is easily detected because it breaks a symmetry of the flow. The laminar flow is centrosymmetric with respect to the origin, whereas the unstable modes contribute antisymmetric components.

According to the simulations, at least two essentially independent instability mechanisms coexist. The strongest departures from a solution with constant vorticity are in the toroidal components of wavenumber 0 and 1 with respect to the fluid axis. If the bulk flow becomes unstable it could presumably do so because of both components but only instabilities triggered by the wavenumber 1 deviations have actually been observed. Experimental reports suggest that the axisymmetric internal cylindrical shear layers cause an instability even though the employed visualization methods cannot ascertain whether these shear layers merely act as tracers or whether they actually trigger an instability. In all the simulations discussed so far, the deviations in the basic state from a flow with uniform vorticity which have a wavenumber equal to one outweight the axisymmetric deviations. However, as the Ekman number is decreased, the viscous corrections contributing to the wavenumber one deviations diminish, whereas the axisymmetric shear layer connecting the critical latitudes becomes more and more singular. An instability of that shear layer is thus plausible at low E. In addition, the boundary layer becomes unstable independently if its Reynolds number is large enough (certainly if it is larger than 100). An impression of the boundary layer flow is given in Fig. 2.

The unstable modes vary seemingly continuously in going from ellipsoidal to spherical containers so that the container shape does not matter. In the sphere, the Poincaré solution is a solid body rotation which is a stable flow, so that viscous corrections to the Poincaré flow must be responsible for the instabilities.

The range of parameters accessible numerically is limited mainly by the onset of the boundary layer instability accompanied by unresolvably small scales. Figure 2 presents a well resolved case, but at slightly more extreme parameters, the simultaion is underresolved and runs into numerical instability.

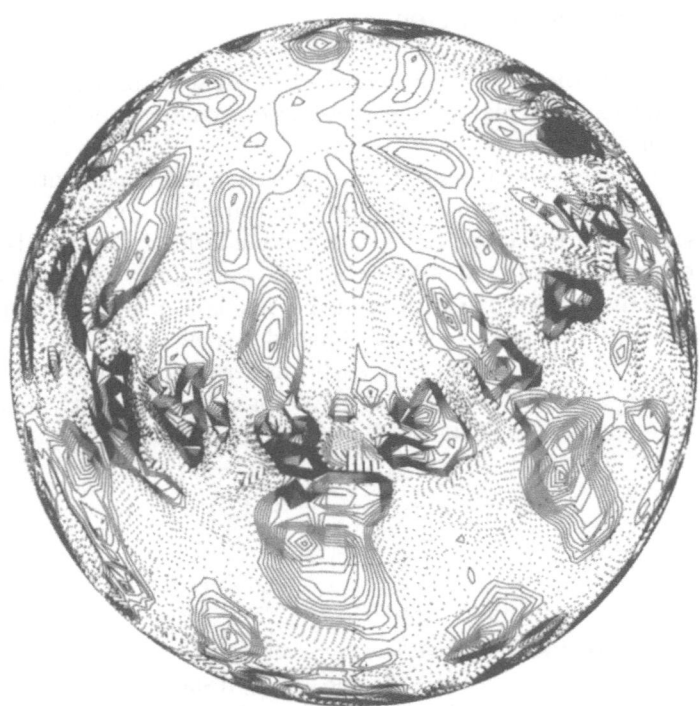

Fig. 2. The radial component of velocity of the antisymmetric contribution at a distance of 0.02 from the outer boundary for $\alpha = 30°$, $e = 0.04$, $E = 2 \times 10^{-5}$ and $\Omega = -1.8 \times 10^{-2}$. Continuous and dashed contour lines mark positive and negative values, respectively.

The energy contained in the unstable modes has always stayed small compared with the total energy and has not been large enough for dynamo action. In the attempts at simulating a precession driven dynamo made so far, the initial magnetic field got converted by interaction with the flow into a strong toroidal field which eventually decayed.

In order to make further progress numerically, one needs to get rid of the boundary layers. The most elegant way to do so is to use stress free boundary conditions. In that case, the Ekman layers disappear in favor of much weaker boundary layers which exert virtually no influence on the interior flow. The laminar flow is then practically identical to Poincaré's solution. No slip boundaries also reproduce this solution in the limit of vanishing Ekman number and sufficiently far away from the cylinder joining the critical latitudes.

Stress free boundary conditions are of course not realized in experiments, but they are appropriate for describing the upper boundary of an atmosphere. A numerical code implementing these boundary conditions could thus directly be used to investigate tidally driven flow as well.

4 Stress free boundaries

The numerical method has been described in great detail elsewhere [9], so that the features particular to the stress free boundary conditions will be emphasized here.

The most convenient reference frame for the numerical computation is the frame attached to the container. Within this frame, two coordinate systems will be used: The original one in which the boundaries are ellipsoids of revolution and the computational one in which the boundaries are spherical. The first system will be described with primed symbols. Consider incompressible fluid of kinematic viscosity ν in an ellipsoidal shell rotating with angular frequency ω_D about the z-axis. The shell furthermore executes precessional motion characterized by the precession vector $\omega_D \Omega_p \hat{\boldsymbol{\Omega}}_p$ (hats denote unit vectors). The boundaries of the shell are given by:

$$\frac{x'^2}{a^2} + \frac{y'^2}{a^2} + \frac{z'^2}{c^2} = 1 \tag{1a}$$

$$\frac{x'^2}{(\eta a)^2} + \frac{y'^2}{(\eta a)^2} + \frac{z'^2}{(\eta c)^2} = 1 \tag{1b}$$

$\eta < 1$ and both boundaries have the same ellipticity $e = 1 - c/a$. Units of length and time are chosen as $(1-\eta)a$ and $1/\omega_D$, respectively. Using the same primed symbols as above to denote the dimensionless lengths, the equation of motion for the velocity $\boldsymbol{u}'(\boldsymbol{r}', t)$ reads in a frame of reference attached to the shell:

$$\frac{\partial}{\partial t} \nabla' \times \boldsymbol{u}' + \nabla' \times \{(2(\hat{\boldsymbol{z}}' + \boldsymbol{\Omega}_p) + \nabla' \times \boldsymbol{u}') \times \boldsymbol{u}'\} = E\nabla'^2 \nabla' \times \boldsymbol{u}' + 2\hat{\boldsymbol{z}}' \times \boldsymbol{\Omega}_p \tag{2}$$

$$\nabla' \cdot \boldsymbol{u}' = 0 \tag{3}$$

The Ekman number E is defined by $E = \nu(\omega_D(1-\eta)^2 a^2)^{-1}$. The computational coordinate system is now introduced by the transformations:

$$x = x' \qquad y = y' \qquad z = \frac{z'}{1-e} \tag{4}$$

If the velocities are transformed likewise,

$$u_x = u_x' \qquad u_y = u_y' \qquad u_z = \frac{u_z'}{1-e} \tag{5}$$

one obtains again a solenoidal vector field, $\nabla \cdot \boldsymbol{u} = 0$. The boundaries are now given by:

$$x^2 + y^2 + z^2 = r_o^2 \tag{6a}$$

$$x^2 + y^2 + z^2 = r_i^2 \tag{6b}$$

with $r_i/r_o = \eta$ and $r_o - r_i = 1$. In this new formulation, the problem lends itself to a spectral discretization in spherical harmonics. However, the equation of motion in the unprimed system is more complicated:

$$\frac{\partial}{\partial t}\nabla \times \boldsymbol{u} - E\nabla^2\nabla \times \boldsymbol{u} + \boldsymbol{L}$$
$$= -\nabla \times \boldsymbol{N} + e\nabla \times (N_z\hat{\boldsymbol{z}}) + 2(1-e)\hat{\boldsymbol{z}} \times \boldsymbol{\Omega}_p \tag{7}$$

The nonlinear terms have been grouped together as well as the linear terms which vanish for $e = 0$:

$$\boldsymbol{N} = (2(\hat{\boldsymbol{z}}' + \boldsymbol{\Omega}_p) + \nabla' \times \boldsymbol{u}') \times \boldsymbol{u}' \tag{8}$$

$$\boldsymbol{L} = e(e-2)\frac{\partial}{\partial t}\nabla \times (u_z\hat{\boldsymbol{z}}) - E\frac{e(2-e)}{(1-e)^2}\frac{\partial^2}{\partial z^2}\nabla \times \boldsymbol{u}$$
$$- E\left[e(e-2)\nabla^2 - \left(\frac{e(e-2)}{1-e}\right)^2\frac{\partial^2}{\partial z^2}\right]\nabla \times (u_z\hat{\boldsymbol{z}}) \tag{9}$$

The precession axis $\hat{\boldsymbol{\Omega}}_p$ forms the angle α ($0 < \alpha < \pi/2$) with the z-axis and is time dependent in the chosen system of reference:

$$\hat{\boldsymbol{\Omega}}_p = \sin\alpha\cos t \ \hat{\boldsymbol{x}} - \sin\alpha\sin t \ \hat{\boldsymbol{y}} + \cos\alpha \ \hat{\boldsymbol{z}} \tag{10}$$

The solenoidal vector field $\boldsymbol{u}$ can be written in terms of poloidal and toroidal scalars Φ and Ψ:

$$\boldsymbol{u} = \nabla \times \nabla \times (\Phi\hat{\boldsymbol{r}}) + \nabla \times (\Psi\hat{\boldsymbol{r}}) \tag{11}$$

which are then decomposed in spherical harmonics:

$$\Phi = r \sum_{l=1}^{L} \sum_{m=-l}^{l} V_l^m(r,t) P_l^m(\cos\theta) e^{im\varphi}$$

$$\Psi = r^2 \sum_{l=1}^{L} \sum_{m=-l}^{l} W_l^m(r,t) P_l^m(\cos\theta) e^{im\varphi} \qquad (12)$$

and into Chebycev polynomials T_n as

$$V_l^m(r,t) = \sum_{n=0}^{N_r-1} v_{l,n}^m(t) T_n(x) \quad , \quad W_l^m(r,t) = \sum_{n=0}^{N_r-1} w_{l,n}^m(t) T_n(x) \qquad (13)$$

with $x = 2(r - r_i) - 1$. The collocation points are placed in direct space at

$$r_j = r_i + \frac{1}{2}(1 + \cos\pi\frac{j-1}{N_r - 1}), \quad j = 1..N_r$$

so that a fast cosine transform can be used to switch between physical and spectral space. Equations (14a, 14b) are enforced at every collocation point and the spectral representation in radius is merely used to compute derivatives. Operating with $\hat{\boldsymbol{r}}\cdot$ and $\hat{\boldsymbol{r}} \cdot \nabla\times$ on (7) one obtains two equations for $V_l^m(r,t)$ and $W_l^m(r,t)$:

$$\frac{\partial}{\partial t}\mathcal{D}_l V_l^m - E \cdot \mathcal{D}_l^2 V_l^m - \frac{r}{l(l+1)}[\hat{\boldsymbol{r}} \cdot \nabla \times \boldsymbol{L}]_l^m$$

$$= \frac{r}{l(l+1)}[\hat{\boldsymbol{r}} \cdot \nabla \times \{\nabla \times \boldsymbol{N} - e\nabla \times (N_z\hat{\boldsymbol{z}})\}]_l^m \qquad (14a)$$

$$\frac{\partial}{\partial t}W_l^m - E\left(\frac{\partial^2}{\partial r^2} + \frac{4}{r}\frac{\partial}{\partial r} + \frac{2 - l(l+1)}{r^2}\right)W_l^m + \frac{1}{l(l+1)}[\hat{\boldsymbol{r}} \cdot \boldsymbol{L}]_l^m$$

$$= -\frac{1}{l(l+1)}[\hat{\boldsymbol{r}} \cdot \{\nabla \times \boldsymbol{N} - e\nabla \times (N_z\hat{\boldsymbol{z}})\}]_l^m + (1-e)[f]_l^m \quad (14b)$$

with:

$$\mathcal{D}_l = \frac{\partial^2}{\partial r^2} + \frac{2}{r}\frac{\partial}{\partial r} - \frac{l(l+1)}{r^2}$$

$$f = \frac{1}{2}\Omega_p \sin\alpha \ [iP_1^1 e^{i(\varphi+t)} + 2iP_1^{-1} e^{-i(\varphi+t)}]$$

$[\ ]_l^m$ denotes the l, m-component of the quantity in the square bracket.

We now turn to the boundary conditions. A normal to the fluid boundary is given by:

$$\boldsymbol{n'} = (x', y', z'/(1-e)^2)$$

$$\boldsymbol{n} = (x, y, z/(1-e))$$

The condition that no fluid traverses this boundary becomes:

$$\boldsymbol{u}' \cdot \boldsymbol{n}' = u_x x + u_y y + u_z z = \boldsymbol{u} \cdot \boldsymbol{r}$$

In terms of the poloidal and toroidal scalars, this means that V_l^m must be zero at the boundaries, which is a condition that is also implemented for no slip boundaries. The stress at the boundaries is given by $\boldsymbol{Tn}'$ (to within a factor of 2) with

$$T_{ij} = \frac{\partial}{\partial x_i'} u_j' + \frac{\partial}{\partial x_j'} u_i'.$$

One finds that

$$\boldsymbol{Tn}' = \boldsymbol{t}_s + e \frac{2-e}{(1-e)^2} z \frac{\partial}{\partial z} \boldsymbol{u} + \hat{\boldsymbol{z}} [\frac{e}{1-e} r \frac{\partial}{\partial z} \boldsymbol{u} - e(\boldsymbol{r} \cdot \nabla) u_z + e^2 \frac{e-2}{(1-e)^2} z \frac{\partial}{\partial z} u_z]$$

$\boldsymbol{t}_s$ is the stress exerted on a spherical surface, this term survives for $e = 0$:

$$t_{s,i} = (\frac{\partial}{\partial x_i} u_j + \frac{\partial}{\partial x_j} u_i) x_j \tag{15}$$

At the boundaries (where $u_r = 0$) this reads in polar coordinates:

$$\hat{\boldsymbol{\varphi}} \boldsymbol{t}_s = r^2 \frac{\partial}{\partial r} (\frac{u_\varphi}{r})$$

$$\hat{\boldsymbol{\theta}} \boldsymbol{t}_s = r^2 \frac{\partial}{\partial r} (\frac{u_\theta}{r})$$

$$\hat{\boldsymbol{r}} \boldsymbol{t}_s = 2r \frac{\partial}{\partial r} u_r$$

The tangential directions to the fluid surface are given by $\hat{\boldsymbol{\varphi}}$ and $\hat{\boldsymbol{\tau}}$ with

$$\hat{\boldsymbol{\tau}} = \frac{e}{1-e} \cos\theta \sin\theta \hat{\boldsymbol{r}} + [1 + \frac{e}{1-e} \cos^2\theta] \hat{\boldsymbol{\theta}}.$$

The first boundary condition, $\hat{\boldsymbol{\varphi}} \boldsymbol{Tn}' = 0$ therefore becomes:

$$r \frac{\partial}{\partial r} (\frac{u_\varphi}{r}) + e \frac{2-e}{(1-e)^2} (\cos^2\theta \frac{\partial}{\partial r} - \frac{\cos\theta \sin\theta}{r} \frac{\partial}{\partial \theta}) u_\varphi = 0 \tag{16}$$

and a similar but lengthy expression is deduced from $\hat{\boldsymbol{\tau}} \boldsymbol{Tn}' = 0$.

The time step procedure is similar to the one for no slip boundaries, except that the boundary conditions now couple the poloidal and toroidal scalars and different l. Let $y(t)$ be a vector containing the coefficients $v_{l,n}^m(t)$ and $w_{l,n}^m(t)$. If equations (14a) and (14b) are discretized in time using implicit Euler steps and second order Adams-Bashforth steps for the linear and nonlinear terms, respectively, (14a) and (14b) can be written for a time step of size h in the form

$$\boldsymbol{M} y(t + h) = \boldsymbol{M}' y(t) + \frac{h}{2} [3 \boldsymbol{nl}(y(t)) - \boldsymbol{nl}(y(t - h))] \tag{17}$$

where M and M' represent matrices and nl all nonlinear terms and the Coriolis force. For the implicit part of the time step, linear systems of the form

$$M_1 y + M_2 y = A \tag{18}$$

need to be solved, where M_1 represents the discretization of the first two terms in (7), M_2 the remaining terms of the left hand side of (7), and A the result of the Adams-Bashforth scheme plus the value of $M'y$ at the previous time step (the right hand side of (17)). The boundary conditions are included in M_1. The dynamic equations for the poloidal field at the two inner and outermost collocation points as well as the equation for the toroidal scalar on the boundaries are replaced with equations expressing the boundary conditions. M_1 thus is a sparse matrix since apart from those boundary conditions, it consists of small blocks of size $N_r \times N_r$. It is therefore advantageous to invert M_1 with the help of the Woodbury formula [14]. The matrices necessary for the inversion of M_1 are computed during initialization and stored for later use. In order to solve the full system, a Jacobi type iteration is employed:

$$y_{n+1} = M_1^{-1}(-M_2 y_n + A) \tag{19}$$

where y_n is the n-th iterate of the solution. Each inversion of M_1 invokes the Woodbury formula.

The operation load per time step is of course increased compared with the no slip version because of the couplings introduced by the boundary conditions. However, at equal resolution, one can reach much more extreme parameters, especially regarding the Ekman number. At a typical resolution ($N_r = 129$, $L = 128$, $M = 64$) the CPU time per time step has increased by a factor 3.7. Because of the more complicated structure of M_1, memory requirements have increased by a factor 13.8.

5   Inertial instability

It has been known for some time from theory ([12], [3]) that flows with elliptically distorted streamlines undergo inertial instability. The Poincaré solution possesses elliptical streamlines so that inertial instability must be expected. The elliptical streamlines of the Poincaré flow lie in planes. A normal to these planes passing through the center of one of the ellipses does not pass through the center of any other ellipse (see Fig. 1) which shows that there must be shearing motion between adjacent planes. This shear also leads to inertial instability [5].

Two paths have been followed to predict the critical parameters for the onset of instability ([3], [5]). The first one is a pertubation approach in which distortions of the ground state from a solid body rotation are assumed small. In the second approach, it is assumed that the unstable modes can be described by a combination of low order polynomials. This yields sufficient conditions for instability. Both calculations start from the Euler equation and the damping effect of viscosity is introduced in a somewhat empirical fashion at the end.

From theory, inertial instability is expected for $\alpha = 90°$, $\Omega = 0.1$, $E = 10^{-4}$ and $e = 0.15$. These parameters have been chosen for a test run and instability indeed occurred with growth rates and wave numbers compatible with theory. Because the theory is a linear stability analysis, it can not predict the saturation behavior. For the parameters chosen above, the flow reaches an oscillatory state in which the energy contained in the instability varies by an order of magnitude (see Fig. 3).

This behavior can be classified as a "resonant collapse": The laminar flow becomes unstable. Once they have attained a large enough amplitude, the unstable modes themselves develop further instabilities, so that the flow suddenly contains small scales which quickly dissipate the energy contained in the unstable modes. After the flow has relaminarized, the cycle can begin anew. Similar behavior has been observed in experiments ([11], [10], [2]), in which inertial modes after a growth phase desintegrate into turbulence and decay.

According to our understanding of inertial instabilities in precession driven flow, the state of precession driven flow in the Earth's core is uncertain because commonly accepted values of the viscosity of the core put the flow close to its stability limit. Fluid viscosity is however one of the least well constrained material properties of the core. Let us assume that the core is unstable. Precession then possibly drives the geodynamo or at least contributes to the secular variation of a convectively driven dynamo. In the latter case,

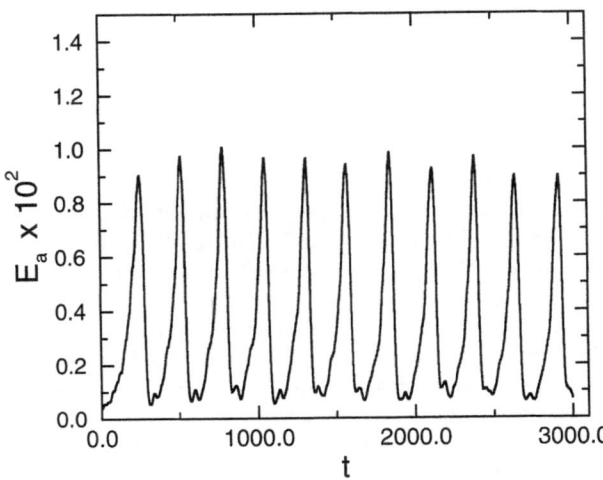

Fig. 3. Time evolution of the energy E_a contained in the velocity components antisymmetric with respect to reflection at the origin for $\alpha = 90°$, $e = 0.15$, $E = 10^{-4}$ and $\Omega = -0.1$. E_a is zero for a laminar flow.

the collapse phenomenon is of special interest. When collapse occurred in our simulations, the growth rate after the collapse was within a factor of 2-3 from the growth rate the same modes have during a linear growth phase starting from a Poincaré solution as described by perturbation theory. Using an upper bound for the growth rate of an inertial instability given in [5] applied to Earth's numbers gives a growth rate of $(20000yr)^{-1}$. Viscosity also acts to slow this growth. If collapses play a role in the Earth's core, they could manifest themselves in variations of the magnetic field with a time constant of 20000 years or longer.

References

1. F.H. Busse. Steady fluid flow in a precessing spheroidal shell. *J. Fluid Mech.*, 33:739–751, 1968.
2. C. Eloy, P. Le Gal, and S. Le Dizès. Experimental study of the multipolar vortex instability. *Phys. Rev. Lett.*, 85:3400–3403, 2000.
3. E.B. Gledzer and V.M. Ponomarev. Instability of bounded flows with elliptical streamlines. *J. Fluid Mech.*, 240:1–30, 1992.
4. H. Goldstein. *Classical Mechanics*. Addison-Wesley, Reading, Mass., 1980.
5. R.R. Kerswell. The instability of precessing flow. *Geophys. Astrophys. Fluid Dyn.*, 72:107–144, 1993.
6. H. Lamb. *Hydrodynamics*. Cambridge University Press, Cambridge, 1932.
7. M.J. Landmann and P.G. Saffman. The three-dimensional instability of strained vortices in a viscous fluid. *Phys. Fluids*, 30:2339–2342, 1987.
8. S. Lorenzani and A. Tilgner. Fluid instabilities in precessing spheroidal cavities. *J. Fluid Mech.*, 447:111–128, 2001.
9. S. Lorenzani and A. Tilgner. Precession driven flow in ellipsoidal cavities. In E. Krause and W. Jäger, editors, *High performance computing in Science and Engineering 2000*. Springer, 2001.
10. V.W.R. Malkus. An experimental study of global instabilities due to the tidal (elliptical) distortion of a rotating elastic cylinder. *Geophys. Astrophys. Fluid Dyn.*, 48:123–134, 1989.
11. R. Manasseh. Breakdown regimes of inertia waves in a precessing cylinder. *J. Fluid Mech.*, 243:261–296, 1992.
12. R.T. Pierrehumbert. Universal short-wave instability of two-dimensional eddies in an inviscid fluid. *Phys. Rev. Lett.*, 57:2157–2159, 1986.
13. H. Poincaré. Sur la précession des corps déformables. *Bull. astronom.*, 27:321–356, 1910.
14. W.H. Press, S.A. Teukolsky, W.T. Vetterling, and B.P. Flannery. *Numerical Recipes*. Cambridge University Press, Cambridge, 1986.
15. P.H. Roberts and K. Stewartson. On the motion of a liquid in a spheroidal cavity of a precessing rigid body. II. *Proc. Camb. Phil. Soc.*, 61:279–288, 1965.
16. K. Stewartson and P.H. Roberts. On the motion of a liquid in a spheroidal cavity of a precessing rigid body. *J. Fluid Mech.*, 17:1–20, 1963.
17. A. Tilgner. Magnetohydrodynamic flow in precessing spherical shells. *J. Fluid Mech.*, 379:303–318, 1999.
18. A. Tilgner. Non-axisymmetric shear layers in precessing fluid ellipsoidal shells. *Geophys. J. Int.*, 136:629–636, 1999.

19. A. Tilgner and F.H. Busse. Fluid flows in precessing spherical shells. *J. Fluid Mech.*, 426:387–396, 2000.
20. J.P. Vanyo and P.W. Likins. Measurement of energy disipation in a liquid-filled, precessing, spherical cavity. *Trans. ASME J. Appl. Mech.*, 38:674–682, 1971.
21. J.P. Vanyo and P.W. Likins. Rigid-body approximations to turbulent motion in a liquid-filled, precessing, spherical cavity. *Trans. ASME J. Appl. Mech.*, 39:18–24, 1972.

Replication of Dissipative Solitons by Many-Particle Interaction

Andreas W. Liehr, Andrei S. Moskalenko, Michael C. Röttger,
Jürgen Berkemeier, and Hans-Georg Purwins

Institute for Applied Physics, Corrensstr. 2/4, 48149 Münster, Germany
http://www.uni-muenster.de/Physik/AP/Purwins/struktur/

Abstract. We are investigating a three-component reaction-diffusion model, which has been established as phenomenological model for pattern formation processes in direct current semiconductor-gas-discharge systems. Concerning two-dimensional systems we are able to reproduce the experimentally observed phenomena of replication of dissipative solitons by many-particle interaction. In three-dimensional systems these phenomena lead to the formation of complex molecules consisting of single dissipative solitons.

1 Dissipative Solitons in Reaction-Diffusion-Systems

One of the most popular and effective concepts in physics is the concept of *particles*. Typically particles are part of a conservative system and their behavior is directed by conservation laws such as energy and momentum conservation. Recently it has been shown that well localized solitary patterns, which are commonly observed in dissipative systems, can also be interpreted as particles [1], although they cannot generally be described on the basis of conservation laws. Such structures appear in the nature as different phenomena: as nerve pulses [2], as concentration spots of reagents in chemical reactions [3,4], as intensity bulbs of light in optical systems [5,6], and as current filaments in semiconductor [7–9] or gas-discharge devices [10]. We refer to the particle-like structures in dissipative systems as dissipative solitons (DSs). They are different from conventional solitons, which originate from the observation of a solitary wave made by Scott Russell in 1834 [11] and which can be found as solutions of one-dimensional Korteweg-de Vries, nonlinear Schrödinger and sin-Gordon equations [12], so far as to them the mathematical theory of inverse scattering transformation cannot be applied.

Modelling dissipative systems often leads to nonlinear reaction-diffusion equations, in case of optical systems with cross-diffusion terms. One of the basic experimental systems, which are used to investigate properties of DSs, are planar gas-discharge devices with a high ohmic semiconductor layer [13,14]. In this system processes with soliton number conservation, such as scattering and formation of moving or rotating molecule-like bound states of DSs, as well as processes without soliton number conservation, such as annihilation of DSs and generation of new DSs, were observed [14,15]. For this experimental system a three-component reaction-diffusion model has been established

[1,16] originating from a phenomenological interpretation of pattern formation processes on the basis of an electrical equivalent circuit [17]:

$$\dot{u} = D_u \Delta u + f(u) - \kappa_3 v - \kappa_4 w + \kappa_1 - \frac{\kappa_2}{||\Omega||} \int_\Omega u \, d\Omega, \tag{1}$$

$$\tau \dot{v} = D_v \Delta v + u - v, \tag{2}$$

$$\theta \dot{w} = D_w \Delta w + u - w, \tag{3}$$

$$u = u(\boldsymbol{r}, t), \quad v = v(\boldsymbol{r}, t), \quad w = w(\boldsymbol{r}, t), \quad \boldsymbol{r} \in \mathbb{R}^1, \mathbb{R}^2, \mathbb{R}^3$$

$$D_u, D_v, D_w, \tau, \theta, \kappa_2, \kappa_3, \kappa_4 \geq 0.$$

Concerning one- and two-dimensional systems the activating component u refers to the charge carrier multiplication in the gas, whereas the inhibiting component v corresponds to the voltage drop at the high ohmic semiconductor layer due to the current flow. The third equation describing the dynamics of the inhibiting component w has been introduced phenomenologically in order to model more than one stable localized moving DSs [16] and could be related to a temperature field or another high ohmic layer, which, due to the phenomenological approach, has not been identified in the experimental system by now [1]. The global feedback term of Eqn. (1) results from the voltage drop at the internal resistance of voltage source or the voltage drop of a shunt resistor in the circuit of the experimental system and limits the maximum current density, which, under certain circumstances, imposes a limitation on the maximum number of simultaneously existing DSs [18].

Concerning three-dimensional domains Eqns. (1)–(3) may describe a chemical system defined in $\mathbb{R}^3$.

Here we investigate the generation of DSs without the restricting feedback term and therefore set $\kappa_2 = 0$. The well localized solitary structures arising as solutions of Eqns. (1)–(3) due to excitations of the homogeneous background state are a result of an interplay between an activator-inhibitor mechanism and diffusion [19] and have a well defined self-organized shape, which depends only on the system parameters. A cross-section of a moving three-dimensional DS is shown in Fig. 1. In the depicted case the inhibitor v has a large time-constant τ and therefore reacts slowly on changes of the activator u. This is the reason for the asymmetry of the moving DS as well as the reason for its movement [20]. The inhibtor w has a relatively small time-constant θ and a big diffusion constant D_w stabilizing the localized structure [1,16]. In the depicted case the tails of the DS decay in an exponential manner to the homogeneous background state. It is found that the interaction of DSs of this kind is purely repulsive [1]. Another important class of DSs, which are typically found for parameters close to the Turing bifurcation of a homogeneous distribution [21], exhibit tails decaying oscillatorily to the background state. Analytical and numerical investigations show that these oscillating tails correspond to regions of attractive and repulsive interaction [1,15,22].

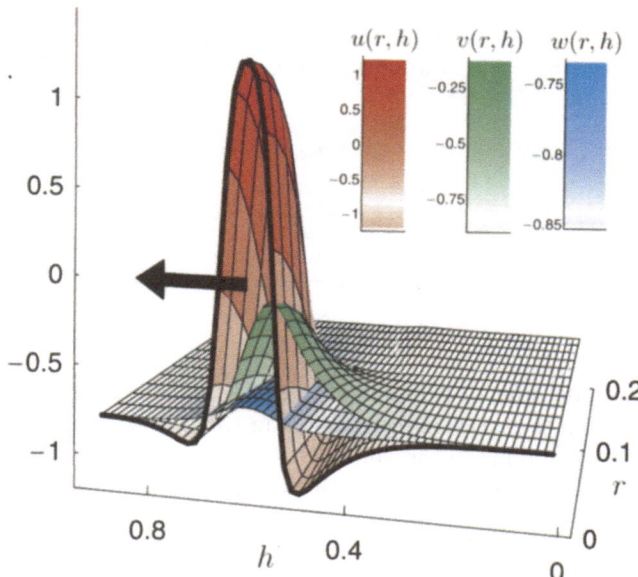

Fig. 1. Three-dimensional moving dissipative soliton simulated in cylindrical coordinates (r, h, ϑ). Due to the rotational symmetry concerning the direction of motion the angular component ϑ can be neglected. This is very useful for checking the existence and stability of moving dissipative solitons for a given set of parameters. The components u, v and w are shown as red, green and blue surfaces, respectively. The activator u is pushed by the slow inhibitor v and is stabilized by the fast inhibitor w. Parameters: $\tau = 48.0$, $\theta = 0.5$, $D_u = 1.5 \cdot 10^{-4}$, $D_v = 1.86 \cdot 10^{-4}$, $D_w = 9.6 \cdot 10^{-3}$, $\lambda = 2.0$, $\kappa_1 = -6.92$, $\kappa_2 = 0$, $\kappa_3 = 8.5$, $\kappa_4 = 1.0$, $\Omega = [0, 0.466] \times [0, 0.932]$, $\Delta x = 0.0155$, $\Delta t = 0.01$.

DSs can interact with each other as well as with inhomogeneities [1] and boundaries. Depending on parameters and initial conditions scattering, formation of bound states, merging, and generation are observed [1,15,23]. Here we report on the replication of dissipative solitons, which has been observed experimentally in the two-dimensional case by Astrov and Purwins in 2001 [14]. A similar phenomenon in three-dimensional systems lead to the formation of complex three-dimensional molecules consisting of individual dissipative solitons.

2   Numerical Methods

2.1   Three-Component Reaction-Diffusion System

The time-dependent partial differential equations of the three-component reaction-diffusion system (1)–(3) are solved on two- and three-dimensional domains with cyclic or no-flux boundary conditions. They are approximated by finite differences in space and a Crank-Nicholson time stepping scheme. Both discretisation lengths in space and time are constant. The resulting set of equations is solved with a successive over-relaxation algorithm whereby the discretisation points are taken account of in red-black order. After each time step the activator and inhibitor distributions are disturbed by noise of amplitude 10^{-9} and $0.5 \cdot 10^{-9}$, respectively. The noise has a rectangular probability density distribution, is uncorrelated in space and time, and has the function of driving the system away from unstable solutions. These could be either small DSs [24], or stationary DSs, which are, for certain parameters, unstable against moving DSs [20].

While two-dimensional problems with 200×200 discretisation points are solved on local workstations, typical three-dimensional problems with grids of $70 \times 70 \times 70$ up to $128 \times 128 \times 128$ discretisation points are solved using 64 nodes of the Cray T3E. In the latter case the solution of the problem is parallelized by dividing the domain in sub-domains with minimal internal boundaries. Each sub-domain is assigned to one node, whereby communication between the nodes is realized via the Message Parsing Interface (MPI) [25]. Speed-up and scale-up of the parallel program have been measured and show good results (Fig. 2). Typical parameter sets of the three-component reaction-diffusion system leading to dissipative soliton solutions exhibit a separation of space and time scales (Fig. 1). This is due to the fact that on the one hand the activating component of a DS is locally stabilized by a wide spreading and fast reacting inhibitor (e.g. w), and on the other hand its motion is initialized by another inhibitor which is slow compared to the time scale of the activator. These scale separations in space and time require small discretisation lengths in time and space and therefore lead to large computation costs.

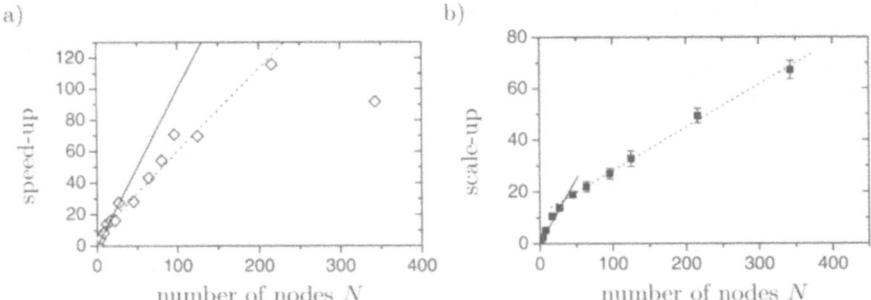

Fig. 2. Performance of the parallel solver for the three-dimensional three-component reaction-diffusion system (1)-(3) on the Cray T3E. a) Optimal speed-up is denoted by a solid line, while the dotted line is a least square fit with slope 0.533 to the measured speed-up for less than 216 nodes. b) Scale-up of the parallel solver with least square fits to the measured scale-up with slope of 0.5 and 0.167 for $N < 40$ and $40 < N < 400$, respectively.

Typical problems like the one presented in [1] are computed on three-dimensional grids with $\Delta x = 0.028$ and $\Delta t = 0.001$ on 128 nodes in 14.8 hours. This corresponds approximately to a computing time of 0.167 seconds on one node per time unit and discretisation point.

2.2 Two-Component Reaction-Diffusion System with Non-Local Feedback

The parameter limit $D_v \to 0$ and $\theta \to 0$ of the three-component reaction-diffusion system (1)-(3) without global feedback ($\kappa_2 = 0$) has been proposed by Or-Guil et al. in 1998 [20] and has the advantage of allowing an analytical description of the dynamics of DSs near the onset of propagation at $\tau = 1/\kappa_3$. In this parameter limit the three-component reaction-diffusion system can be transformed to a two-component reaction-diffusion system with non-local feedback, where the fast inhibitor $w(u)$ is computed as a function of the activator u. This diminishes the separation of time scale and enables an increase of the time discretisation length (Tab. 1).

For the task of computing the fast inhibitor w as a function of the activator u we have combined a parallel three-dimensional multi-grid solver with the Gauß-Seidel iteration of the time dependent equations for $u(r, t)$ and $v(r, t)$, such that $w(u)$ is computed at each step of the Gauß-Seidel iteration. Simulations like the ones presented in Fig. 5 are computed on three-dimensional grids with $\Delta x = 7.81 \cdot 10^{-3}$ and $\Delta t = 1.0$ on 64 nodes in 5.2 hours. This refers approximately to a computing time of $9.3 \cdot 10^{-4}$ s seconds per discretisation point and time unit. In conclusion we state that the enhanced program enables the simulation of larger problems with less computational costs compared to simulations of the three-component model (Tab. 1).

Table 1. Characteristics of simulations concerning DS interaction. In order to get a basis for a rough comparison of the computing costs for the these very different simulations, we have approximated the computing time on one node per time unit and discretisation point.

	Three-component reaction-diffusion system	Two-component reaction-diffusion system with non-local feedback
Computing time	14.8 h	5.2 h
Nodes	128	64
Simulated time	80	680
Δt	0.001	1.0
Grid	80^3	124^3
Reference	[1,23]	Fig. 5
Estimated computing time on one node per time unit and discretisation point	0.167 s	$9.3 \cdot 10^{-4}$ s

3   Replication of Dissipative Solitons in Two Dimensions

In the following we present a two-dimensional example of complex DS inter-action, which includes the formation of molecule-like bound states, the gen-eration of a DS via self-replication and symmetry breaking. For the presented simulation the reaction-diffusion equations have been solved with parameters being primarily introduced in the context of molecule-formation phenomena [1,15]:

$$D_u = 1.1 \cdot 10^{-4}, \ D_v = 0, \ D_w = 9.64 \cdot 10^{-3}, \ \lambda = 1.01,$$
$$\kappa_1 = -0.1, \ \kappa_2 = 0, \ \kappa_3 = 0.3, \ \kappa_4 = 1, \ \tau = 3.47, \ \theta = 0, \tag{4}$$
$$\Omega = [0,1] \times [0,1], \ \Delta x = 0.005, \ \Delta t = 0.1, \ \text{no-flux boundary condition.}$$

For these parameters DS solutions can be found that exhibit tails which decay slowly and in an oscillatory manner to the homogeneous background state u_0. In order to discuss the exemplary simulation we present in Fig. 3 four subsequent snapshots (middle column of Fig. 3) visualizing the activator dis-tribution $u(r, t)$ as gray-scale images. Additionally vectors being proportional to the shift $\alpha(t)$ between the centers of activator u and slow inhibitor v are shown. These shifts determine the direction and the velocity of unperturbed motion of the DSs [1]. The snapshots are supplemented by intersections of activator u and fast inhibitor w distributions along the symmetry axis at $x = 0.5$ (right column of Fig. 3). This symmetry axis is plotted in the gray-scale images of the middle column as a broken line. The intersections of the right column are combined with gray-scales which correspond to the local amplitudes of the activator concentration depicted in the snapshots of the middle column. The left column of Fig. 3 shows the the overall time develop-ment of a quantity $\hat{u}(t)$, which indicates the conservation, annihilation and generation of DSs, such that an increase (decrease) of $\hat{u}(t)$ by $\hat{u}(t = 0)/N_0$ indicates the generation (annihilation) of a DS, if N_0 is the number of initially existing DSs. Therefore $\hat{u}(t) \approx 1$ refers to the conservation of the number N_0 of initially existing DSs. The indicator $\hat{u}(t)$ is computed by:

$$\hat{u}(t) = \frac{\int_\Omega \left(u(r,t) - u_0 \right)^2 \mathrm{d}r}{\int_\Omega \left(u(r,0) - u_0 \right)^2 \mathrm{d}r}, \tag{5}$$

where u_0 corresponds to the homogeneous background state. Within these indicator diagrams a diamont denotes the moment t of the snapshot $u(r, t)$, and the intersections $u(0.5, y, t)$, $w(0.5, y, t)$ in each row.

The simulation starts with four DSs being placed symmetrically to the $x = 0.5$ and $y = 0.5$ axes in the domain. The DSs denoted with 1 and 3 move towards the point $P_1 = (0.4, 0.5)$, and the DSs denoted with 2 and 4 move towards $P_2 = (0.6, 0.5)$, such that close to P_1 and P_2 2-soliton-interactions occur. In this scenario the DSs are fast enough to overcome the

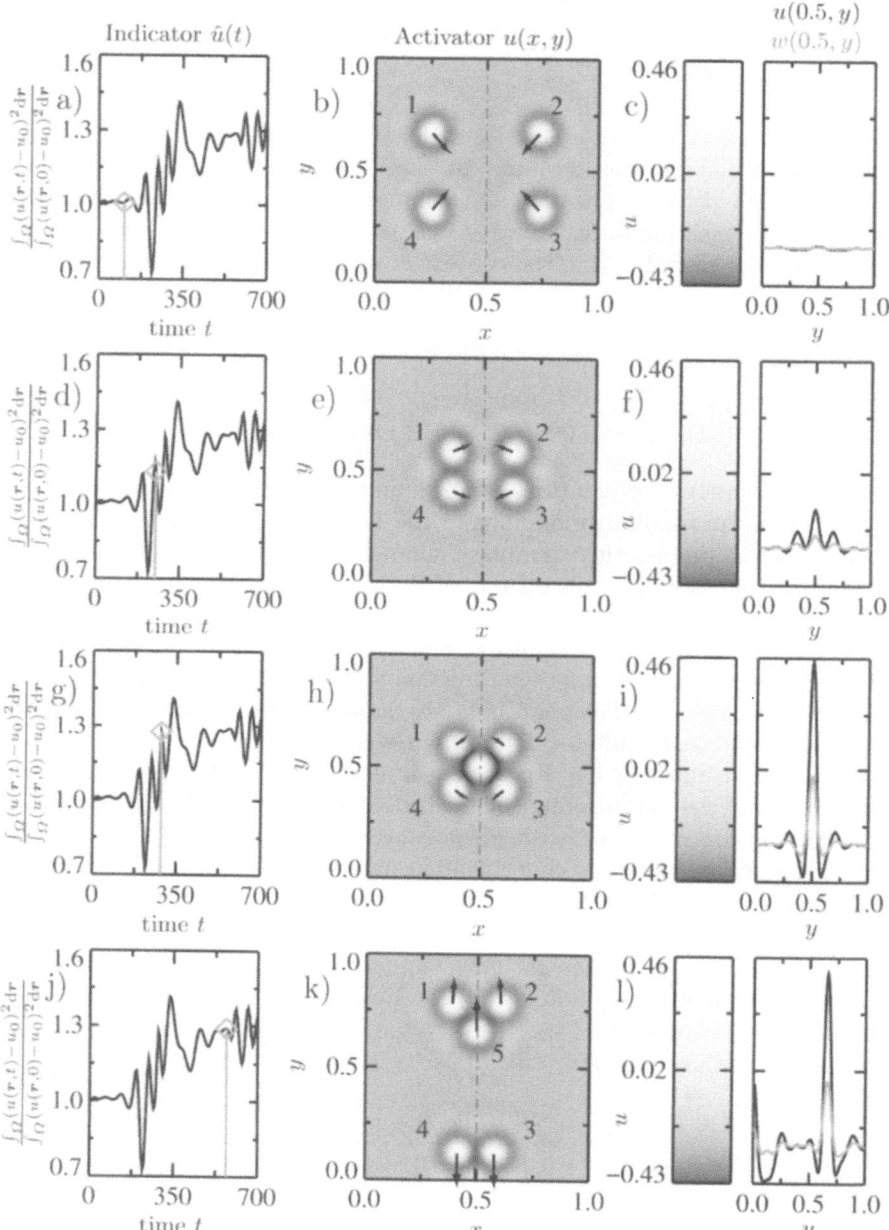

Fig. 3. The replication of a DS by 4-soliton-interaction. The resulting 5-soliton-cluster decays under symmetry breaking into a 2-soliton-molecule and a 3-soliton-molecule. The figure shows four snapshots of the simulation (rows), whereby a diamond within the indicator diagram (left column) denotes the point in time of the snapshot. The second column shows the activator distribution $u(x,y)$ with vectors visualizing the approximated direction of motion. The third column shows the associated gray scale and a cross-section of u and w at $x = 0.5$. Parameters as in (4).

repulsive interaction at the secondary minima of their oscillating tails and reach the first bound state. This leads to the formation of two molecule-like bound states of DSs each consisting of two individual DSs (Fig. 3e). Both molecules move towards each other whereby their tales overlap in the center of the domain. The intersection in Fig. 3f) shows the superposing oscillatory tails in detail. In the center of the domain a local activator maximum is formed due to the influence of four DSs. Additionally, two smaller maxima are created due to the interaction of the tails of two DSs.

This local increase of activator concentration causes a deflection of the DSs such that DS 1 and 2 (3 and 4) move towards the collision point located above (below) the center of the domain (Fig. 3e). The next snapshot Fig. 3h) shows that a fifth DS has ignited in the middle of the four original DSs and a transient bound state consisting of five DSs has formed. The generation is accompanied by an increase in the indicator $\hat{u}(t)$ (Fig. 3g). The generated DS also has oscillating tails (Fig. 3i) which interact with the neighbouring DSs, but, however, do not ignite additional DSs. We also like to note, that in this snapshot (Fig. 3h) the initial bound states of DS 1 and 3 (2 and 4) are broken such that their motion is guided by the oscillating tales of the new generated DS. Until now all discussed snapshots reflect the inital symmetry concerning the $x = 0.5$ and the $y = 0.5$ axis. But at least the symmetry with regard to the $y = 0.5$ axis is broken, which is shown in Fig. 3k). Here the 5-DS bound state has decayed into a bound state consisting of two DSs moving towards the lower domain boundary and a bound state consisting of three DSs moving to the upper domain boundary. At the depicted snapshot the two-soliton bound state has already reached the lower domain boundary, where due to the no-flux boundary condition, a generation scenario occurs, which is comparable to the one described before. While the two-DS bound state has moved close to the lower boundary, the bound state consisting of DSs 1,2 and 5 has only reached the middle of the upper domain half.

In order to investigate the symmetry breaking and the velocity difference of the two-soliton and the three-soliton bound state, we will take a closer look at the dynamics of the interacting DSs. Therefore in Fig. 4 we have plotted two characteristic variables of the dynamics of each DS as functions of time. The first is the absolut value of the shift $\alpha(t)$ between the center of the activator distribution u and slow inhibitor distribution v corresponding to the individual DS, where $\alpha(t)$ is the so-called amplitude of propagator mode [1]. The second characteristic variable is the angle $\phi(t)$ of the direction of motion, which is estimated from the amplitude of propagator mode $\alpha(t) = (\alpha_x(t), \alpha_y(t))$ such that $\alpha_x(t)/|\alpha(t)| = \cos \phi(t)$ and $\alpha_y(t)/|\alpha(t)| = \sin \phi(t)$ (Fig. 4b). Additionally Fig. 4a) shows a gray-scale image of the activator distribution $u(x, y)$ at $t = 437.5$, where the interacting DSs are labelled by numbers, and activator $u(x, y) = -0.2$ contour lines at $t = 17.5$. Vectors visualize the direction of motion.

In the beginning DS 1 moves with a motion angle of $-\frac{\pi}{3}$ from the upper left quarter of the domain towards its collision point close to the middle of the

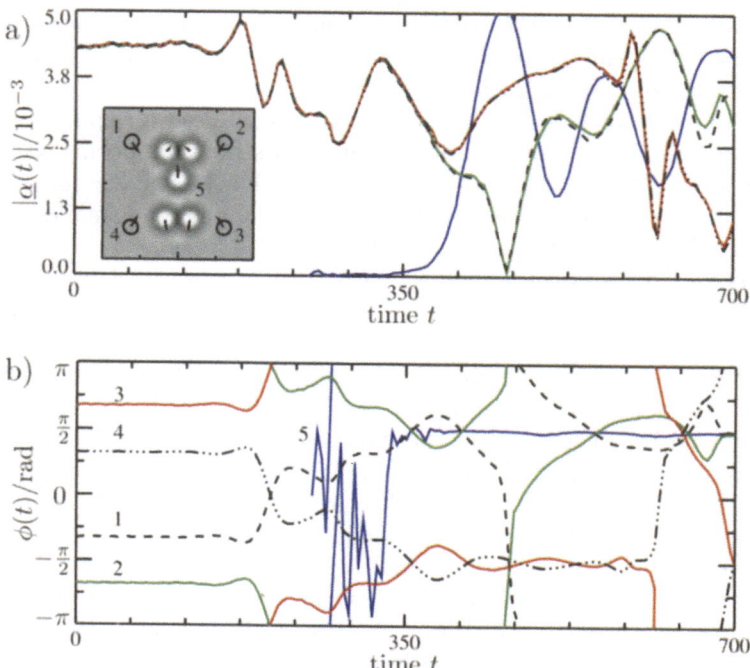

Fig. 4. Dynamics of the interacting DSs presented in Fig. 3. a) Time evolution of the absolute value of the shift $\alpha(t)$ between the centers of activator distribution u and inhibitor distribution v for the individual DSs. The gray-scale image within Fig. a) shows the activator distribution u at t=437.5 and contour lines of the activator $u(x, y) = -0.2$ at t=17.5. Vectors denote directions of motion and numbers identify the individual DSs. Fig. b) Time evolution of the angle $\phi(t)$ of the direction of motion, which is approximated from $\alpha(t) = (\alpha_x(t), \alpha_y(t))$ such that $\alpha_x(t)/|\alpha(t)| = \cos \phi(t)$ and $\alpha_y(t)/|\alpha(t)| = \sin \phi(t)$. The numbers within the gray-scale image of Fig. a) and the numbered curves of Fig. b) refer to the same DSs. That applies also to the line styles of Fig. a) and b). Curves vanishing at the top (bottom) border of diagram b) and reemerging at the bottom (top) border reflect a discontinuity of the inverse function used for the calculation of $\phi(t)$.

domain. DSs 2, 3, and 4 exhibit angles of $-\frac{2\pi}{3}, \frac{2\pi}{3}$, and $\frac{\pi}{3}$, respectively (Fig. 4b, $t \in [0, 100]$). The symmetry of the initial condition is reflected in Fig. 4a) by the identity of $|\alpha(t)|$ for all DSs, which can hardly be discriminated with respect to drawing accuracy. This symmetric behaviour persists even while DSs 1-4 are deflected by the self-replicated fifth DS and their direction angles change signs. (Fig. 4a and b, $t \in [100, 340]$).

After its generation DS 5 is stationary and therefore is an unstable solution of the three-component reaction-diffusion-system (1)-(3) for the chosen system parameters (4) [1,20]. The propagator amplitude $\alpha(t)$ of DS 5 is just slightly excited and therefore the direction of motion changes rapidly (Fig.

4b, $t \in [250, 350]$), until it relaxes at $t \approx 340$ to an angle of $\frac{\pi}{2}$. This is also the time when the symmetry with regard to the $y = 0.5$ axis is broken, such that the curves of $|\alpha(t)|$ for the 1,2-DS molecule and the 3,4-DS molecule branch off. The symmetry breaking and the direction of motion of DS 5 is caused by the noise being applied to the system.

While the excitation of the propagator mode of DS 5 (blue curve in Fig. 4a) and the excitations of the propagator modes of DSs 3 and 4 (solid red and overlaying broken black curves) increase, DSs 1 and 2 are slowed down and the amplitudes of their propagator modes decrease (solid green and overlaying broken black curves). Obviously the new generated DS is within a repulsive region regarding DSs 3 and 4, and within an attractive region concerning DSs 1 and 2. While the repulsive interaction leads to a mutual acceleration of DSs 3-5, and the attractive interaction between DSs 1,2,5 accelerates DS 5, too, DSs 1 and 2 are slowed down, because they have to pull DS 5. The velocity vectors of gray-scale image in Fig. 4a) show an extreme case of this configuration at $t = 437.5$, where the propagator mode amplitudes of the 1,2-DS molecule are close to zero, while the propagator mode of DS 5 is strongly excited and, on the other hand, pushes the 1,2-DS molecule. In the following an alternating acceleration and slowing down of DS 1, 2 and DS 5 is observed, which leads to a smaller average velocity of the 3-soliton-molecule and explains, why the 2-soliton-molecule reaches the domain boundary some time before the 3-soliton-molecule.

The described scenario is very similar to the experimental observations reported by Astrov and Purwins in 2001 [14], where three DSs come close together and replicate a fourth DS. This generation mechanism can be understood in the context of the presented simulation, which explains the replication as a result of the nonlinear superposition of oscillating tails. Experimental evidence for oscillating tails of dissipative solitons has recently been given in the d.c. gas-discharge system [26]. We like to note, that the reported generation mechanism of replication is different from the commonly known mechanism of self-replication [27,28], where individual DSs split due to an internal instability into two DSs, which themselves start to split again.

4   Replication of Dissipative Solitons in Three Dimensions

In order to simulate the replication of DSs in three dimensions we have chosen the following parameters, which, like the two-dimensional case, lead to DS solutions with oscillating tails:

$$
\begin{aligned}
& D_u = 1.1 \cdot 1.3 \cdot 10^{-4}, \ D_v = 0, \ D_w = 9.64 \cdot 10^{-3}, \ \lambda = 0.95, \\
& \kappa_1 = -0.08, \ \kappa_2 = 0, \ \kappa_3 = 0.25, \ \kappa_4 = 1, \ \tau = 4.5, \ \theta = 0, \\
& \Omega = [0, 1.4] \times [0, 1.4] \times [0, 1.4], \ \Delta x = 0.011, \ \Delta t = 1.0, \\
& \text{cyclic boundary condition.}
\end{aligned}
\tag{6}
$$

The initial conditions have been set up from radial symmetric stationary DS solutions of parameters (6) such that three DSs are starting from the vertices of an equilateral triangle with sides of length 0.8 (Fig. 5a). The slow inhibitors v of each dissipative soliton have locally been shifted with respect to the activator by vectors of length $8.2 \cdot 10^{-4}$ such that the motion of each DS is directed towards the point of collision $\boldsymbol{P_c} = (0.75, 0.58, 0.51)$. The fast inhibitors w of the individual DSs are centered around the activator distributions of the particles.

In order to visualize the time evolution of the field equations (1)–(3) Fig. 5 shows four subsequent snapshots of activator iso-surfaces $u(x, y, z) = 0.0$, which are coloured with respect to the slow inhibitor concentration $v(x, y, z)$. Here red (blue) colour indicates a large (small) inhibitor concentration. Vectors denote the direction of motion of the DSs. To give an overview of the dynamics of the presented simulation Fig. 5e) shows the time evolution of the

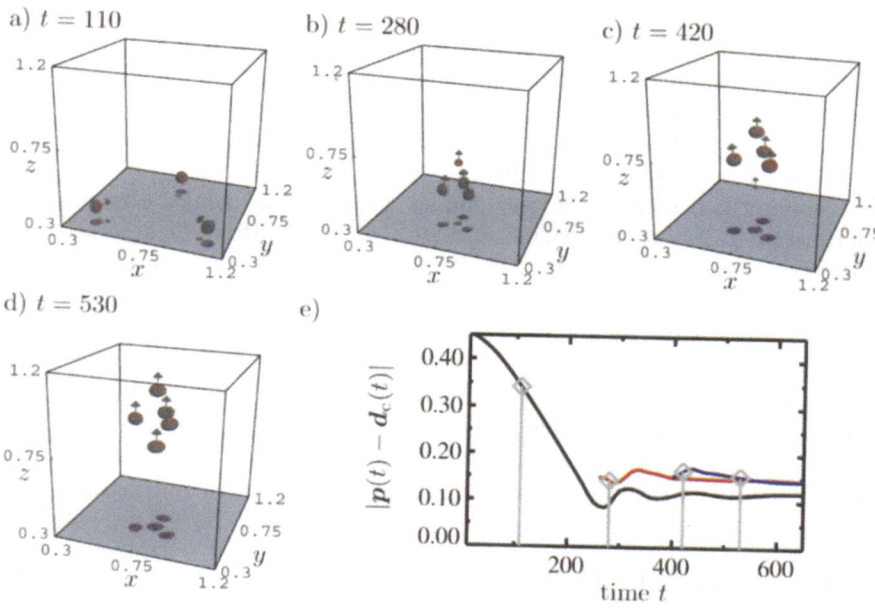

Fig. 5. Replication of three-dimensional dissipative solitons. Fig. a)–d) show iso-surfaces of activator $u(x, y, z) = 0.0$, which are coloured with the respective concentration of the slow inhibitor v. A large (small) inhibitor concentration is indicated by red (blue). Vectors denote the direction of motion of the dissipative solitons, which is approximated by the shift of the slow inhibitor v with respect to the activator u of the individual dissipative soliton. Fig. e) shows the distance $d(t) = |\boldsymbol{p}(t) - \boldsymbol{d_c}(t)|$ of the dissipative solitons at position $\boldsymbol{p}(t)$ to the center $\boldsymbol{d_c}(t)$ of the initial dissipative solitons. The black curve refers to the initial dissipative solitons, the red curve to the fourth dissipative soliton and the blue curve to the fifth dissipative soliton. Parameters as in (6). Noise is not applied to the system.

distance $d(t) = |\boldsymbol{p}(t) - \boldsymbol{d}_{\mathrm{c}}(t)|$ of the DSs at position $\boldsymbol{p}(t)$ to the center $\boldsymbol{d}_{\mathrm{c}}(t)$ of the initial DSs. The dynamics of the initial DSs is reflected by black curves, which coincide due to the choice of the center $\boldsymbol{d}_{\mathrm{c}}(t)$, whereas the dynamics of the new generated ahead (behind) moving DS is plotted as red (blue) curve, respectively.

While approaching their equilibrium velocity at $t = 280$ the DSs reach the closest distance to the center $\boldsymbol{d}_{\mathrm{c}}(t)$ of $d = 0.088$, and a fourth DS ignites at a distance of $d = 0.146$ to $\boldsymbol{d}_{\mathrm{c}}(t)$ (Fig. 5b). In the following the direction of motion of the DSs turn parallel to the z-axis and the distance between the particles relaxes in an oscillatory manner to a steady state corresponding to minimal interaction (Fig. 5e). During this process the initial DSs come close together for another time ($d = 0.11$) and a fifth DS ignites at a distance of $d = 0.153$ to the center $\boldsymbol{d}_{\mathrm{c}}(t)$ of the initial DSs (Fig. 5e). The double pyramid cluster of five DSs continues travelling parallel to the z-axis and the particles relax towards an equilibrium distance of $d = 0.118$, while the distance of the ahead and behind running DSs to the center reaches an equilibrium distance of $d = 0.145$.

The simulation shows that several individual DSs can generate a limited number of new DSs. The mechanism seems to be related to the distance between the DSs, because the replication mechanism only occurs if the initial DSs come close to each other and overcome a critical distance. If they form a bound state and relax to a steady configuration, which is observed for $t > 400$ in the presented simulation, no replication phenomena occur.

5   Conclusion and Outlock

We have demonstrated that DSs as solutions of three-component reaction-diffusion systems can be generated by a mechanism of replication in the process of many-particle interactions. In two-dimensional systems the effect is comparable to that observed by Astrov and Purwins in a d.c. gas-discharge system [14]. Concerning three-dimensional systems the effect of replication of DSs is reported for the first time. The simulations imply that the mechanism of replication is initialized, if the dissipative solitons undershoot a critical distance. The reported phenomena of replication is different to the commonly observed mechanism of self-replication, where an individual DS splits up into two DSs [27,28]. The replication of DSs supplements the hitherto known generation mechanism of DSs [18] and is subject of future research.

Acknowledgment

We would like to thank the High-Performance Computing-Center Stuttgart (HLRS) for granting us access to their Cray T3E system and the Deutsche Forschungsgemeinschaft (DFG) for their support. We also like to thank Daniela Kempa for fruitful discussions on the topic.

References

1. BODE, M.; LIEHR, A. W.; SCHENK, C. P.; PURWINS, H.-G.: Interaction of dissipative solitons: particle-like behaviour of localized structures in a three-component reaction-diffusion system. In: *Physica D* 161 (2002), Nr. 1-2, S. 45–66

2. HODGKIN, A. L.; HUXLEY, A. F.: A quantitative description of membrane current and its application to conduction and excitation in nerve. In: *Journal of Physiology* 117 (1952), S. 500–544

3. OUYANG, Q.; CASTETS, V.; BOISSONADE, J.; ROUX, J. C.; KEPPER, P. D.; SWINNEY, H. L.: Sustained patterns in chlorite-iodide reactions in a one-dimensional reactor. In: *Journal of Chemical Physics* 95 (1991), Nr. 1, S. 351–360

4. ROTERMUND, H. H.; JAKUBITH, S.; VON OERTZEN, A.; ERTL, G.: Solitons in a surface reaction. In: *Physical Review Letters* 66 (1991), Nr. 23, S. 3083–3086

5. STEGEMAN, G. I.; SEGEV, M.: Optical Spatial Solitons and Their Interaction: Universality and Diversity. In: *Science* 286 (1999), Nr. 5444, S. 1518–1523

6. SCHÄPERS, B.; FELDMANN, M.; ACKEMANN, T.; LANGE, W.: Interaction of Localized Structures in an Optical Pattern-Forming System. In: *Physical Review Letters* 85 (2000), S. 748–751

7. BEL'KOV, V. V.; HIRSCHINGER, J.; NOVÁK, V.; NIEDERNOSTHEIDE, F.J.; GANICHEV, PRETTL, S. D.; W.: Pattern formation in semiconducters. In: *Nature* 397 (1999), Nr. 4, S. 398

8. AOKI, K.: *Nonlinear Dynamics and Chaos in Semiconductors*. Bristol and Philadelphia : Institute of Physics Publishing, 2001

9. SCHÖLL, Eckehard: *Cambridge Nonlinear Science Series*. Bd. 10: *Nonlinear Spatio-Temporal Dynamics and Chaos in Semiconductors*. Cambridge : Cambridge University Press, 2001

10. PURWINS, H.-G.; ASTROV, Yu.; BRAUER, I.: Self-Organized Quasi Particles and Other Patterns in Planar Gas-Discharge Systems. In: DING, M. (Hrsg.); DITTO, W. L. (Hrsg.); PECORA, L. M. (Hrsg.); SPANO, M. L. (Hrsg.): *The 5th Experimental Chaos Conference*. Singapore: World Scientific, 2001, S. 3–13

11. RUSSELL, John S.: Report on Waves. In: *Report of the fourteenth meeting of the British Association for the Advancement of Science*. York 1844, 1845, S. 311–390, Fig. XLVII–LVII

12. REMOISSENET, Michel: *Waves Called Solitons: Concepts and Experiments*. 3. Berlin : Springer, 1999

13. AMMELT, E.; ASTROV, Yu.; PURWINS, H.-G.: Stripe Turing Structures in a Two-Dimensional Gas Discharge System. In: *Physical Review E* 55 (1997), Nr. 6, S. 6731–6740

14. ASTROV, Yuri A.; PURWINS, Hans-Georg: Plasma Spots in a Gas Discharge System: Birth, Scattering and Formation of Molecules. In: *Physics Letters A* 283 (2001), S. 349–354

15. LIEHR, A. W.; MOSKALENKO, A. S.; ASTROV, Yu. A.; BODE, M.; PURWINS, H.-G.: *Rotating Bound States of Dissipative Solitons*. 2002. – submitted to Physcial Review Letters

16. SCHENK, C. P.; OR-GUIL, M.; BODE, M.; PURWINS, H.-G.: Interacting pulses in three-component reaction-diffusion-systems on two-dimensional domains. In: *Physical Review Letters* 78 (1997), S. 3781–3783

17. PURWINS, H.-G.; KLEMPT, G.; BERKEMEIER, J.: Temporal and spatial structures of nonlinear dynamical systems. In: *Festkörperprobleme* 27 (1987), S. 27–61

18. LIEHR, A. W.; BODE, M.; PURWINS, H.-G.: The Generation of Dissipative Quasi-Particles near Turing's Bifurcation in Three-Dimensional Reaction-Diffusion-Systems. In: KRAUSE, E. (Hrsg.); JÄGER, W. (Hrsg.): *High Performance Computing in Science and Engineering 2000*, Springer, 2001, S. 425–439

19. OR-GUIL, M.; AMMELT, E.; NIEDERNOSTHEIDE, F.-J.; PURWINS, H.-G.: Pattern formation in activator-inhibitor systems. In: DOELMAN, A. (Hrsg.); VAN HARTEN, A. (Hrsg.): *Pitman Research Notes in Mathematics Series* Bd. 335. Longman, 1995, S. 223–237

20. OR-GUIL, M.; BODE, M.; SCHENK, C. P.; PURWINS, H.-G.: Spot bifurcations in three-component reaction-diffusion systems: The onset of propagation. In: *Physical Review E* 57 (1998), Nr. 6, S. 6432–6437

21. TURING, A. M.: The chemical basis of morphogenesis. In: *Phil. Trans. Roy. Soc. B* 237 (1952), S. 37–72

22. SCHENK, C. P.; SCHÜTZ, P.; BODE, M.; PURWINS, H.-G.: Interaction of self-organized quasiparticles in a two-dimensional reaction-diffusion-system: The formation of molecules. In: *Physical Review E* 57 (1998), Nr. 6, S. 6480–6486

23. SCHENK, C. P.; LIEHR, A. W.; BODE, M.; PURWINS, H.-G.: Quasi-Particles in a Three-Dimensional Three-Component Reaction-Diffusion System. In: KRAUSE, E. (Hrsg.); JÄGER, W. (Hrsg.): *High Performance Computing in Science and Engineering '99*, Springer, 2000, S. 354–364

24. OHTA, T.; MIMURA, M.; KOBAYASHI, R.: Higher-Dimensional Localized Patterns in Excitable Media. In: *Physica D* 34 (1989), S. 115–144

25. Message Passing Interface Forum: *MPI: A Message-Passing Interface Standard*. 1995. – URL: http://www.hlrs.de/organization/par/services/models/mpi/mpi-11.ps.gz

26. ASTROV, Yu. A.; PORTSEL, L. M.; MARCHENKO, V. M.; LIEHR, A. W.; PURWINS, H.-G.: *Dissipative solitons and their interaction in the d.c. gas-discharge system*. 2002. – in preparation

27. WILLEBRAND, H.; NIEDERNOSTHEIDE, F.-J.; AMMELT, E.; DOHMEN, R.; PURWINS, H.-G.: Spatio-Temporal Oscillations During Filament Splitting in Gas Discharge Systems. In: *Physics Letters A* 153 (1991), Nr. 8, S. 437

28. LEE, Kyoung-Jin; McCORMICK, William D.; PEARSON, John E.; SWINNEY, Harry L.: Experimental observation of self-replicating spots in a reaction-diffusion system. In: *Nature* 369 (1994), S. 215–218

The Structure of Magnetic Herbig–Haro Jets

Max Camenzind[1] and Markus Thiele[1]

Landessternwarte Königstuhl, D–69117 Heidelberg, Germany

Abstract. We present simulations of magnetized Herbig–Haro flows of low–mass stars. These slightly overdense jets propagate with about 300 km/s through the inner part of a molecular cloud. The magnetic field is injected from the central star and evolves a complicated time–dependent structure driven by pinch and kink modes. The high spatial resolution achieved on the NEC SX–5 provides for the first time insight into the phenomenon of current filamentation and the complex internal structure of pinch and kink modes.

1 Introduction

Almost 50 years ago, George Herbig and Guillermo Haro independently discovered a number of compact nebulae with peculiar spectra near dark clouds. Schwartz (1975) and Raymond (1979) demonstrated that these objects were shock–excited nebulae. Later workers showed that the large range of excitation conditions requires bow shocks and other complex morphologies. By the early 1980s, several Herbig-Haro (HH) objects were shown to be highly collimated jets of partially ionized plasma moving away from young stars at speeds of 100 to over 1000 km/s.

Today, well over 300 individual HH objects or groups are known. Many individual HH objects consist of separate knots or bow shocks, others consist of highly linear chains or jets (Fig. 1). Most show evidence of being a part of or excited by a highly collimated flow from a young star. Low–mass stars, such as our Sun at the age of one million years, drive their angular momentum loss by means of heavy jets which are known as Herbig–Haro flows. Though this phenomenon has already been detected in the 50es, the main physical processes are still largely unknown. The only mechanism which can explain the acceleration and collimation of the outflowing stellar plasma is a magnetic gating process, in which the rapidly rotating magnetosphere of the central star or its accretion disk generates a coupling between the rapid rotation and the outflowing plasma (Camenzind 1997).

It is apparent that outflows from young stellar objects are an integral part of the star formation process. Most stars undergo a phase that lasts for over 10^5 years during which energetic mass loss occurs in the form of numerous eruptions. These jets may become less collimated with increasing age. Jets have ejection velocities of order of several hundred kilometers per second for low mass stars, and in excess of 1000 km/s for high luminosity sources that will evolve into O, B, and A stars. Jet densities range from $n = 100$ to over 10^5 cm^{-3}, and the ionization fractions vary from way below 1% to 10%.

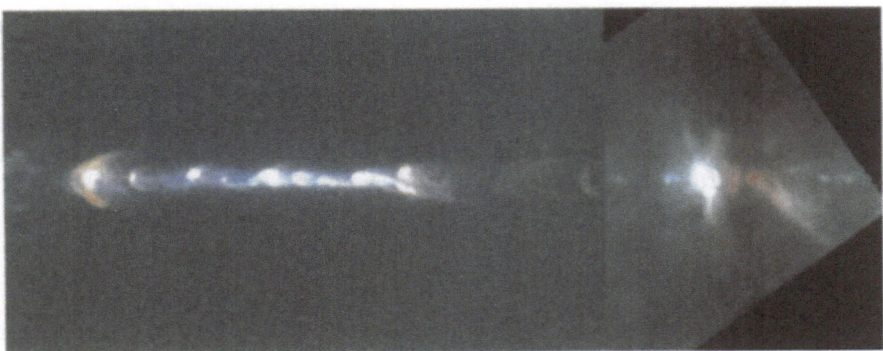

Fig. 1. The Herbig–Haro jet HH 111 (HST, Reipurth) is driving a jet of seven parsecs long. The star at the right end generates a collimated gas flow with a bow shock at the left hand side. The beam gas and the bow shock plasma cool by the emission of atomic lines. A striking fact is that the beam stays highly collimated over a long distance.

The shock cooling times are short (few to thousands of years), which lead to a very rich variety of structures resulting from a combination of cooling and hydrodynamic instabilities and time–dependent variations in the outflow parameters. The multiple bow shocks and S-shaped point symmetry seen in some sources almost certainly requires variations in the mass ejection velocity to produce internal working surfaces and precession or irregular wobbling of the jet.

Many outflows can be traced for parsecs from their exciting sources. HST has been used to image the bright inner portions of these outflows, where the flow takes the form of a jet. The HST observations with their 0.05 to 0.1 arcsec angular resolution, resolve for the first time the cooling length in some shocks in Herbig-Haro flows. In some flows there is evidence for Balmer–line shocks traced by pure $H\alpha$ emission that are well separated from the downstream cooling regions (Bacciotti et al. 2000). With HST, we can measure the proper motions of individual knots on exposures taken less than a year apart. Since the time required to measure the proper motions is likely to be less than the cooling time, it should be possible for the first time to uniquely disentangle true proper motion from photometric variability resulting from intensity variations due to the cooling of distinct fluid elements.

2 The mathematical model

Since magnetic fields are an essential ingredient in modelling Herbig–Haro flows, the mathematical transscription of the problem is given by Newtonian magnetohydrodynamics (shortly called MHD). We investigate jets collimated by magnetic fields which propagate into the cold surrounding molecular clouds. The one–component formulation of MHD is a simplification in this region, since in general we should rely on both neutral and ionized particles. In the beam and cocoon of the jet, the one–component description is a fairly good approximation. What is more important in this part of the jet is cooling by means of Bremsstrahlung and line emission. In the shocked regions, temperatures upto a few hundred thousand Kelvin are reached. For this reason, cooling has to be included in the MHD code (Thiele 2000).

2.1   Basic equations

In the one–component description, the basic variables of MHD are the mass density ρ, the 3–velocity v, pressure P, internal energy e and the magnetic fields B. They evolve in time according to the following set of equations (in Gaussian units)

$$\frac{\partial \rho}{\partial t} + \nabla \cdot (\rho v) = 0 \tag{1}$$

$$\frac{\partial \rho v}{\partial t} + \nabla \cdot (\rho v v) = -\nabla P - \frac{1}{8\pi} \nabla B^2 + \frac{1}{4\pi} (B \cdot \nabla) B \tag{2}$$

$$\frac{\partial e}{\partial t} + \nabla \cdot (ev) = -P \nabla \cdot v - \mathcal{K} \tag{3}$$

$$\frac{\partial B}{\partial t} = \nabla \times (v \times B). \tag{4}$$

The equation of state is given by an adiabatic index Γ, $P = (\Gamma - 1)e$. The energy equation contains explicitly a cooling function $\mathcal{K}$ that follows from Bremsstrahlung emission and line cooling. One could essentially also include heating terms, e.g. due to magnetic reconnection processes. In order to treat cooling self–consistently, one has to add a chemical network due to generation and depletion of individual atomic species (see Thiele 2000).

2.2   Discretisation

The MHD equations are of the conservative form. From a modern point of view one would then apply techniques including Riemann solvers. NIRVANA is a finite volume code modeled according to algorithms developped for the ZEUS3D–code (Ziegler 1995). It is second order accurate and explicit in time–stepping. Artificial viscosity is included in the momentum and energy equation. This dissipates high frequency noise and damps overshooting in shock regions, at the cost of smearing the shocks over a few grid cells. The extension NIRVANA_C includes the chemical network in a time–implicit way, since cooling times and cooling lengths can become quite short (Thiele & Camenzind 2001).

NIRVANA_C has been ported to NEC SX–5 by OPEN_MP–like methods (tests and performance have been discussed in Thiele & Camenzind 2001). There is no parallelised version available for distributed memory machines. Parallelisation for NEC SX–5 is in progress.

2.3   Initial conditions and parameters

We assume that Herbig–Haro jets are formed by plasma injected into the rapidly rotating magnetosphere of the central star (Camenzind 1997). As a consequence, the beam of the jet carries a toroidal and poloidal magnetic field given by the analytic solution of the transverse force–equilibrium (Camenzind 1997)

$$B_\phi = -\frac{R}{R_c}\frac{B_0}{1+(R/R_c)^2} \quad , \quad z = 0, \quad 0 \le R \le R_j \tag{5}$$

$$B_z = \frac{B_0}{1+(R/R_c)^2} \qquad , \quad z = 0, \quad 0 \le R \le R_j. \tag{6}$$

This expression contains the core–radius R_c as a characteristic length–scale and a magnetic field strength B_0. The poloidal magnetic field is constant near the axis and decays very fast beyond the core–radius, while the toroidal component, which is responsible for the confinement, increases linearly with radius, but decays beyond the core–radius. The beam radius R_j for a realistic jet is much bigger than the core–radius. This is one of the crucial points for the resolution of Herbig–Haro jets. The radial force equilibrium requires then a suitable radial profile for the thermal pressure (Thiele & Camenzind 2002). The inclusion of the poloidal field in the initial condition poses a certain technical problem – $\nabla \cdot \boldsymbol{B} = 0$ has to be satisfied everywhere. A solution to this problem is described in Thiele & Camenzind (2002). With this formulation we want to avoid a pre–magnetisation of the external medium, as is assumed usually in MHD simulations for jets. The initial beam velocity v_z also has a certain profile, corresponding to internal shear. This guarantees an approximately constant magnetosonic Mach number over the beam cross section.

Hydro simulations of jets are determined by three dimensionless numbers: (i) the density contrast $\eta = \rho_b/\rho_M$ between beam and external medium, (ii) the internal sonic Mach–number M_b, and (iii) the pressure ratio between internal and external pressure at the injection region. Another important quantity is the density profile of the external medium, which is in general not constant, since jets propagate in complicated molecular structures. Besides the profile of the magnetic field in the beam, the plasma beta, $\beta = P_{\mathrm{Gas}}/P_B$, and the magnetosonic Mach–number M_{ms} of the beam are now additional crucial parameters in MHD simulations. Herbig–Haro jets are slightly over-dense, $\eta = 5$ in the simulation, are magnetically dominated, $\beta = 0.35$, and the magnetosonic Mach–number follows from the collimation, $M_{ms} \simeq 5$. Since molecular clouds are usually turbulent and inhomogeneous, both external density and pressure vary in time and space. We therefore included a stochastic variation in the external pressure.

2.4   Computer resources

In Thiele & Camenzind (2001) we have described pure hydro simulations including cooling (Fig. 2). When magnetic fields are essential, the state vector is now 8–dimensional instead of being 5–dimensional in the pure hydro case. In addition, 3 dimensions are crucial for the development of various instabilities in the magnetic case. For this reason, MHD simulations with high spatial resolution and including a chemical network are still beyond the capacity of the NEC SX–5. With 10 GB of memory we could however simulate a high–resolution MHD jet without cooling. Cooling would be important in the knots formed along the beam. We also could only transfer the data of a few time–steps to the local environment, since the computer capacity at the Landessternwarte is not yet able to handle these enormous data flows. What is lacking is a typical Terabyte server for the local storage of 3D data.

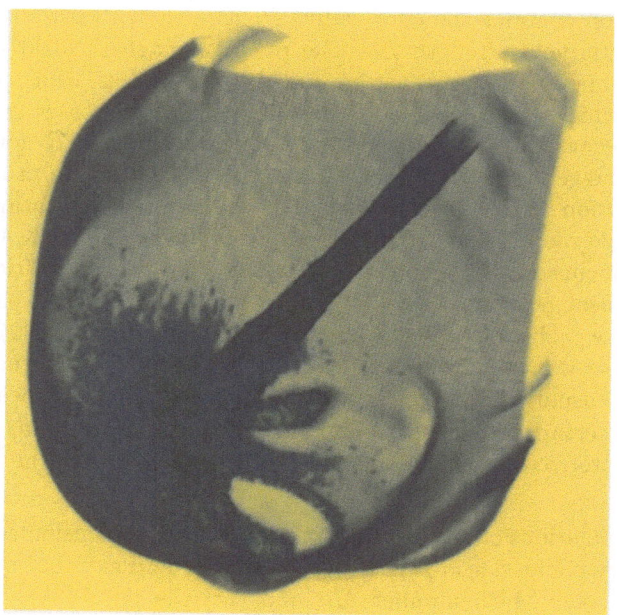

Fig. 2. A 3D simulation of a pure hydro jet with its beam and bow–shock structure.

3   Results

The results of a 2.5D calculation are presented in Thiele & Camenzind (2002). There we have shown that the nose–cone feature typically found in 2D calculations critically depends on the magnetic field topology of the precollimation

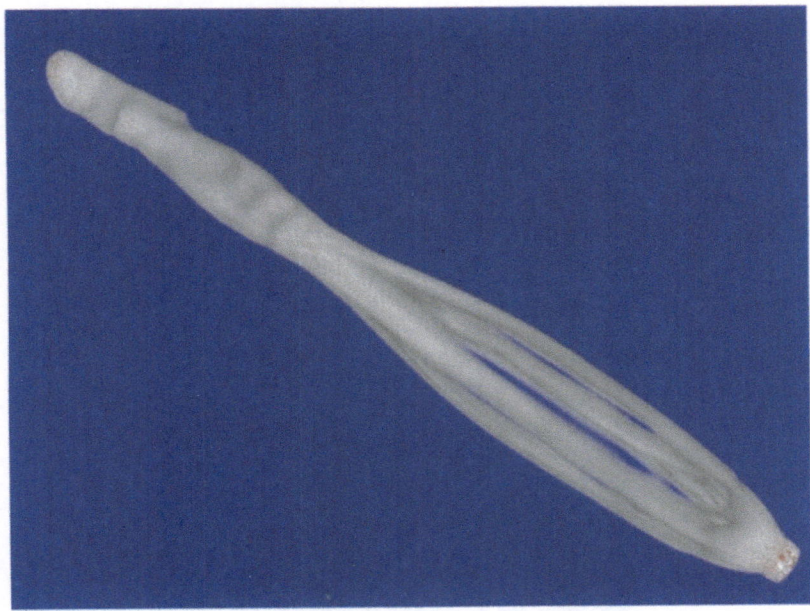

Fig. 3. Current filamentation in the beam in a 3D MHD simulation of Herbig–Haro flows. Due to overpressure, the beam expands initially and recollimates after a certain distance. From here on, internal instabilities, such as the pinch and kink modes develop.

mechanism. The most prominent features we have however found, are knot–like structures along the jet axis which develop in the head–region of the jet. We also have shown that shocks associated with these internal knot–like structures mimic bow–shock like features as seen in many Herbig–Haro flows. In Figs. 3–4 we present a few results from the most recent 3D MHD simulation without cooling. In Fig. 3 the tracer particles of the beam are shown. Due to overpressure injection, the beam rapidly expands and current filamentation occurs. After a certain distance the beam recollimates and a first pinch knot is formed. Beyond this distance, various internal knots are excited due to pinch and kink instabilities.

In Fig. 4 we show the visualisation of the magnetic field structure of this 3D MHD jet. It is nicely seen that magnetic fields are closed, i.e. $\nabla \cdot \boldsymbol{B} = 0$ indeed. The fields are carried along the beam by the plasma flow, pinch and kink modes are excited along the beam. Contrary to the expectation of many people, these instabilities do not disrupt the beam. Obviously, MHD is able to solve topological problems, as shown by the complicated behaviour of the magnetic field lines in the head of the jet.

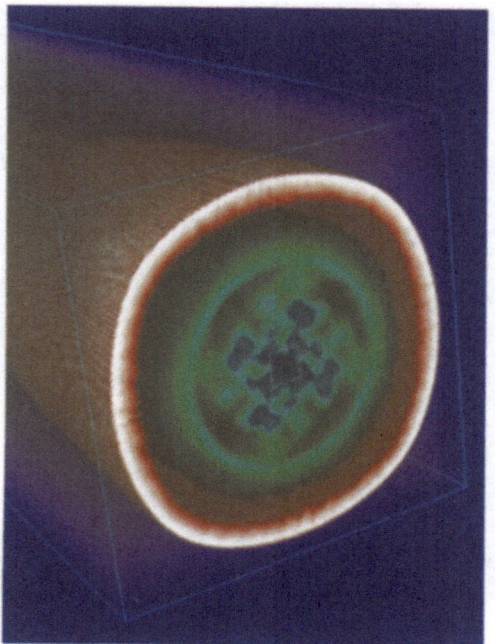

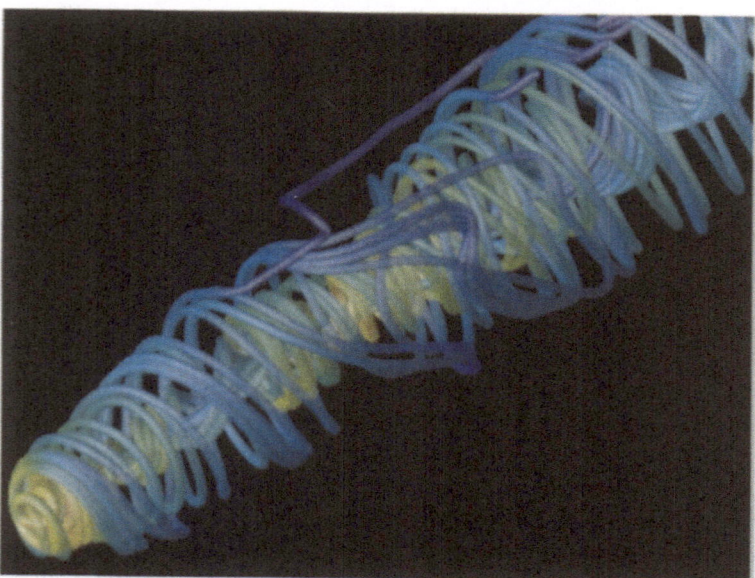

Fig. 4. A 3D simulation of the propagation of a magnetized Herbig–Haro jet in the *spaghetti* representation (bottom) and density structure (top). The jet propagates from top–right to bottom–left. Magnetic fields are only injected by the beam–plasma, the external medium is not magnetized. Turbulence in the molecular cloud excites internal pinch and kink modes, which appear in the observations as propagating knots. These knots have a speed of about half the beam speed.

4   Outlook

NIRVANA_C(P) on the NEC SX–5 is an efficient solver for 3D MHD problems in Astrophysics. Time evolution of protostellar jets could be followed from 10 AU to 10000 AU, 100000 AU would be required for the most extended jets. This corresponds to a time–evolution from about one year to 1000 yeras, 10000 years would be required for large jets. High spatial resolution is required for the investigation of internal knots. Plasma in these knots is rapidly cooling, simulations including cooling are presently beyond the capacity of NEC SX–5. This problem must be attacked in the future with higher performance.

3D magnetic models for Herbig–Haro flows reveal the main characteristics of observed jets. Though the fields are so weak that they are not measurable they play a dominant role in the dynamics. Information is now exchanged over fast magnetosonic and Alfvèn waves, and no longer over the slow sonic waves.

References

[2000] Bacciotti, F., Mundt, R., Ray, T.P., Eisloeffel, J., Solf, J., Camenzind, M.: Hubble Space Telescope STIS Spectroscopy of the Optical Outflow from DG Tauri: Structure and Kinematics on Subarcsecond Scales. ApJ Lett **537** (2000) 49 –

[1997] Camenzind, M.: Energetics, Collimation and Propagation of Galactic Protostellar Outflows, in IAU Symp. **182** *Herbig–Haro Flows and the Birth of Low Mass Stars*, ed. B. Reipurth and C. Bertout, Kluwer (Dordrecht), p. 241–258

[1999] Camenzind, M.: Imprints of Magnetic Fields, in *Formation and Propagation of Young Stellar Jets*, eds. E. Guenther et al., ASP Conf. Series **188**, 129 –

[2000] Thiele, M.: Numerische Simulationen protostellar Jets. Dissertation, University of Heidelberg 2000

[2001] Thiele, M., Camenzind, M.: Propagation of Herbig–Haro jets through inhomogeneous molecular clouds, in *High Performance Computing in Science and Engineering 2001*, Springer–Verlag (Heidelberg), 80 – 91

[2002] Thiele, M., Camenzind, M.: Knot production in magnetized Herbig–Haro jets. A&A **381** (2002) L53–56

[1995] Ziegler, U.: Dissertation, University Würzburg 1995

Temporal Propagators and Quasiparticles in Hot QCD

F. Karsch[1], E. Laermann[1], P. Petreczky[1], S. Stickan[1], and I. Wetzorke[2]

[1] Fakultät für Physik, Universität Bielefeld, Universitätsstraße 25,
   33615 Bielefeld, Germany
[2] NIC/DESY Zeuthen, Platanenalle 6, 15738 Zeuthen, Germany

1 Introduction

Strongly interacting matter undergoes a phase transition at some temperature T_c to a deconfined phase where it is believed that the dominant degrees of freedom are quasiparticles with quantum numbers of quarks and gluons contrary to the low temperature phase $(T < T_c)$ where the dominant degrees of freedom are hadrons. The existence of this phase transition was shown using lattice Monte-Carlo simulations of Quantum Chromodynamics (QCD), the theory describing strongly interacting particles, some 20 years ago [1].

The deconfined high temperature phase of QCD is generally described as the plasma of strongly interacting matter. The dominant degrees of freedom are quarks and gluons, which a priori are massless degrees of freedom. At high temperature, however, they aquire a thermal mass resulting from interactions with the thermal heat bath. More precisely these masses represent a specific limit of the quark and gluon self-energies which have a rather complex energy (ω) and momentum $(\boldsymbol{p})$ dependence. At high temperature the quarks and gluons should thus be considered as quasi-particles propagating in the thermal heat-bath.

Quasiparticles show up as poles in the retarded quark and gluon propagators. The corresponding dispersion relations, the position of the poles, were studied in so-called HTL perturbation theory which is valid if the coupling constant $g(T)$ which controls the interaction among quarks and gluons becomes significantly smaller than unity and therefore a separation of different scales holds, $1/T \ll 1/gT \ll 1/g^2 T$. These length scales characterize the typical scales over which thermal, electric and magnetic excitations propagate and thus also characterize the magnitude of corresponding quark and gluon masses (see e.g. [2] for a review). In the interesting temperature region close to the transition temperature T_c, however, the coupling is large $g \gtrsim 1$. Nevertheless, the corresponding quasiparticle picture finds application in refined perturbative calculations of the bulk thermodynamic properties where it helps to improve the convergence of the perturbative series [3] as well as in more phenomenological approaches [4]. In view of these facts a non-perturbative study of quark and gluon dispersion relations in the deconfined phase is highly desirable.

In the next section we describe in detail the goals of our project as well as some computational details. In section 3 we summarize our numerical results. Finally section 4 contains our conclusions.

2   Simulating finite temperature QCD on lattice

2.1   Goals and parameters of lattice Monte-Carlo simulation

Lattice Monte-Carlo simulations can provide information on the imaginary-time (Matsubara) propagator $D(i\omega_n, p)$ [2] (ω_n being the Matsubara frequencies). This is related to the retarded propagator by analytic continuation $D_R(p_0, p) = -D(p_0 + i\epsilon, p)$. This implies

$$D(i\omega_n, p) = -\int_{-\infty}^{+\infty} d\omega \frac{\rho(\omega, p)}{i\omega_n - \omega}, \qquad (2.1)$$

where $\rho(\omega, p) = \frac{1}{\pi}\mathrm{Im}D_R(\omega + i\epsilon, p)$ is the spectral function. Quasiparticles (quarks and gluons) appear as complex poles in $D_R(p_0, p)$ or, equivalently, as peaks in the spectral function $\rho(\omega, p)$.

As quark and gluon propagators are gauge dependent quantities it is necessary to fix a particular gauge which we chose to be the Coulomb gauge. In this gauge the condition $\partial_i A_i(\tau, x) = 0$ is imposed independently on different time slices which allows to construct a transfer matrix [5] [a]. As the consequence of this the spectral function of quarks and gluons is positive. Though the quark and gluon propagators are gauge dependent quantities the position of the peaks in the spectral function (poles in the retarded propagators) is gauge independent at any order of perturbation theory [6]. Gauge independence of the peak position can be proven also non-perturbatively in a class of gauges which allow the construction of a transfer matrix [5]. At finite temperature there are two kinds of quasiparticle excitations (corresponding to two branches in the dispersion relation) [2]: (i) the real quasiparticles (quarks and transverse gluons), which are the analog of partonic degrees of freedom at zero temperature, and (ii) collective excitations (plasmino and longitudinal gluons), which have exponentially small residues for momenta $p > T$.

We calculated the temporal quark and gluon propagators, i.e. the propagators in the mixed (τ, p)-representation (τ being the imaginary time),

$$D(\tau, p) = \sum_n D(i\omega_n, p) \exp(-i\omega_n \tau) \qquad (2.2)$$

for several values of the spatial momenta p. Using Eq. (2.1) the gluon propagator in the mixed representation can immediately be written in terms of

---

[a] To fix the gauge completely an additional time-dependent gauge transformation is necessary. This, however, depends on temporal link variables $U_0(\tau, x) \equiv \exp(iagA_0(\tau, x))$ only [8].

the spectral function,

$$D^g_{(T,L)}(\tau, p) = \int_0^\infty d\omega \rho_{(T,L)}(\omega, p) \frac{\cosh(\omega(\tau - 1/2T))}{\sinh(\omega/2T)}, \qquad (2.3)$$

where T and L refer to transverse and longitudinal gluons respectively. The most general form of the temporal quark propagator is

$$D^q(\tau, p) = \gamma_0 F(\tau, p) + \boldsymbol{\gamma} \cdot \boldsymbol{n} G(\tau, p) + H(\tau, p), \qquad (2.4)$$

with $\boldsymbol{n} = \boldsymbol{p}/p$. In the chiral limit the last term vanishes. Using the most general form for the retarded quark propagator [7] and Eq.(2.1) one can derive the following representation for the functions F and G:

$$F(\tau, p) = \int_0^\infty d\omega \rho_F(\omega, p) \frac{\cosh \omega(\tau - 1/2T))}{\cosh(\omega/2T)}, \qquad (2.5)$$

$$G(\tau, p) = \int_0^\infty d\omega \rho_G(\omega, p) \frac{\sinh \omega(\tau - 1/2T))}{\cosh(\omega/2T)}. \qquad (2.6)$$

Equations (2.3), (2.5) and (2.6) relate the temporal propagators which we calculate to the spectral function and thus in principle can be used to reconstruct the spectral functions. However, the propagators are calculated for ~ 10 values of τ while for any reasonable discretization of the ω-range the spectral function $\rho(\omega)$ is parameterized by ~ 100 parameters. This clearly is an ill-posed problem and additional input is needed to fix these parameters. In order to do so we use the *Maximum Entropy Method* (MEM) (see [9] for a review). In this approach one searches for a spectral function which maximizes the quantity $\alpha S - L$, where L is the likelihood function corresponding to the standard χ^2-fit, α is a real parameter and S is an entropy functional which incorporates any prior knowledge on the structure of the spectral function through the so-called default model $m(\omega)$ [9]. For a positive definite spectral function $\rho(\omega)$, $S[\rho(\omega), m(\omega)]$ is the Shannon-Jaynes entropy (for other cases see [10])

$$S[\rho(\omega), m(\omega)] = \int_0^\infty d\omega \big(\rho(\omega) - m(\omega) - \rho(\omega) \ln \frac{\rho(\omega)}{m(\omega)} \big) \big). \qquad (2.7)$$

Now we need to specify the default model. From Eqs. (2.3), (2.5) and (2.6) one can see that $\rho_{T,L}$ and $\rho_{F,G}$ have dimension of inverse mass squared and inverse mass respectively. For large ω the spectral functions can be calculated using perturbation theory. On the other hand if ω is large enough it is the only dimensionful scale. Therefore

$$\rho_T(\omega \to \infty) = m_0^T/\omega^2, \qquad (2.8)$$

$$\rho_F(\omega \to \infty) = \rho_G(\omega \to \infty) = m_0^F/\omega, \qquad (2.9)$$

where $m_0^{T,F}$ is a dimensionless real number and $m_0 \sim \mathcal{O}(\alpha_s)$ (in what follows we omit the superscript on m_0).

2.2   Simulation Programme and CPU requirements

We have performed simulations in quenched QCD[b] with Wilson fermions on isotropic (i.e. with the same lattice spacing in time and space like directions) $64^3 \times 16$ lattices. Propagators have been calculated at two values of the temperature, $1.5T_c$ (corresponding to gauge coupling $\beta = 6.972$) using 20 gauge configurations and at $3T_c$ ($\beta = 7.457$) using 40 gauge configurations.

Because of the large memory requirements of our calculations we have separated the various steps into three independent programs: generation of gauge configurations, gauge fixing and calculations of the propagators. The typical memory requirement at each of these steps is about 500Mwords (4GByte).

The programs for gauge field updates, gauge fixing and calculation of the propagators have been written in FORTRAN 90 and use MPI. In the generation of gauge field configurations we use the Kennedy-Pendleton heatbath algorithm [12] combined with an overrelaxation step to simulate the SU(3) gauge theory. In fact, here we use 4 overrelaxation steps followed by one heatbath update. The acceptance rate of the update routine is about 98%. We have performed simulations on 4-dimensional lattices of size $64^3 \times 16$. To perform simulation on such large lattices at least 64 PEs have to be used in order to satisfy the overall memory requirements (assuming 128 MB per node). For one complete update of gauge fields on the T3E-900 (1 heatbath and 4 overrelaxation steps) we need 44.74 sec on 64 PEs. We have performed simulations at two temperatures, $3T_c$ and $1.5T_c$. For these we found that the typical autocorrelation time characterizing the decorrelation of subsequent gauge field configurations is about 100 (200). Thus we need 100 (200) updates to produce one independent configuration. This means that in order to generate one independent configuration for the two temperatures considered we needed 238.6 CPUh (normalized to the time per processor unit).

In our analysis of propagators we use the Coulomb gauge. The gauge fixing routines for Coulomb gauge use the overrelaxation algorithm [13]. We find that this gauge fixing algorithm is very efficient. The typical number of iterations required to fix the Coulomb gauge on a configuration with accuracy $|\partial_i A_i|^2 < 10^{-6}$ is about 1300. As an alternative we have tested an algorithm based on Fast Fourier Transformation which needs approximately 300 iterations to achieve the same accuracy on a $32^3 \times 4$ lattice. It, however, is 5-6 times more expensive in terms of CPU time. Still, the gauge fixing is the most time consuming part of the project. The costs of gauge fixing of one configuration is 305.3 CPUh.

3   Numerical results

In this section we will summarize our numerical results on the temporal quark and gluon propagators and on the quasiparticle properties extracted from them. Some preliminary results have been published in [14,15].

---

[b] for a recent review, further references and introductory articles see [11].

3.1   Temporal quark and gluon propagators

First let us summarize our numerical results for the temporal propagators. The finite lattice volume introduces an infrared cutoff, the smallest non-zero momentum available on our $(64^3 \times 16)$ lattice is $p_{min} = 1.57T$ which is quite large. Because of this we cannot resolve the particular behavior of the collective excitations. In particular the longitudinal gluon propagator is compatible with zero within present statistical accuracy. The temporal gluon propagators are influenced by a zero Matsubara mode contribution $D^g_{(T,L)}(i\omega_n = 0, p)$, which is the static magnetic propagator in momentum space studied in detail in Ref. [8]. At $p = 0$ the magnetic propagators are strongly volume dependent and very large lattices are needed to perform a reliable infinite volume extrapolation [8]. As a consequence the dispersion relation at zero momentum, i.e the plasmon frequency ω_P, cannot be reliably determined from our present calculations. Therefore in what follows the gluon propagators are analyzed only for non-zero momenta.

For high temperatures (considerably larger than T_c) one expects that perturbative calculations based on Hard Thermal Loop (HTL) approximation should work [2]. Therefore we compare our data for temporal quark and gluon propagators with the predictions of perturbation theory in HTL approximation at $T = 3T_c$. In Figure 1 we show our results for the transverse gluon propagator $D^T_g(\tau, p)$ at three values of the spatial momenta calculated on 40 gauge fixed configurations. We also show there the prediction of the HTL approximation using a coupling constant $g \simeq 1.6$ suggested by the short distance behavior of the heavy-quark potential [14]. As one can see from the figure the data deviate substantially from the HTL prediction even at momenta $p = 3.14T$. We note that corrections to the HTL approximation will

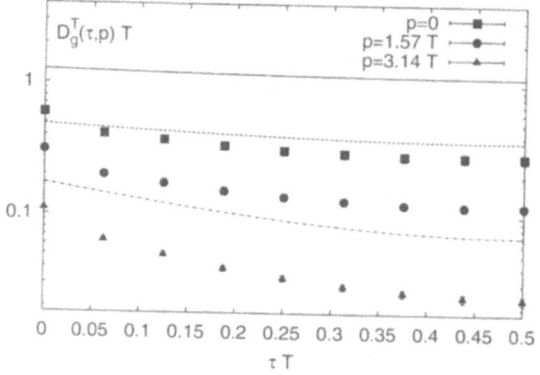

Fig. 1. The transverse gluon propagator for different momenta. The solid, dashed and dashed-dotted lines correspond to the prediction of perturbation theory in HTL approximation for momenta $p/T = 0$, 1.57 and 3.14 respectively.

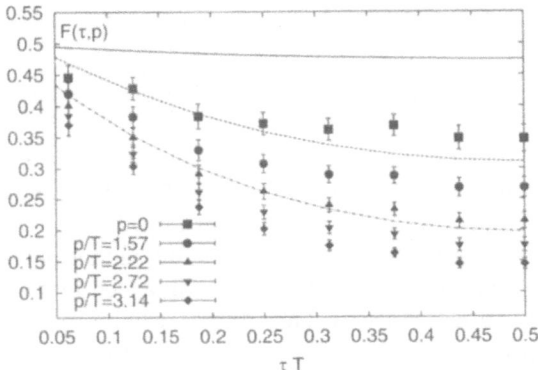

Fig. 2. The quark propagators for different momenta. The solid, dashed and dashed-dotted lines correspond to the prediction of perturbation theory in HTL approximation for momenta $p/T = 0$, 1.57 and 3.14 respectively.

not resolve this discrepancy since they lead to smaller $\omega(p)$ values shifting the propagator to values larger than the HTL result [16].

Let us now discuss the quark propagators. First we note that $H(\tau, p)$ is zero within present statistical accuracy for both temperatures and all values of the spatial momenta considered. In Figure 2 we show our results for $F(\tau, p)$ calculated on 40 configurations and compared with predictions of the HTL approximation. As in the case of the gluon propagator we find large deviations from the HTL predictions. The situation is similar for $G(\tau, p)$. We have found that there is no choice of g which can provide agreement between lattice results and HTL.

3.2   Determination of the spectral functions

As discussed already in section 2.1, the spectral functions are reconstructed from temporal quark and gluon correlators. First of all we are interested in the position of peaks in the spectral functions as these determine the dispersion relations of quasiparticles and as such are expected to be gauge-independent. At zero temperature the peaks in the spectral function can also be determined from exponential fits to the large distance behavior of the different correlators. Alternatively one can determine them from so-called effective masses, which reach a plateau at large distances. At finite temperature both methods are inadequate for the determination of the peak position as the temporal lattice extent is always limited by the inverse temperature. However, if the spectral function is dominated by a single particle state than effective masses may still be used to determine the dispersion relation. The effective masses are introduced by the following formula for $D_T(\tau, p)$ and $F(\tau, p)$

$$\frac{D(\tau, p)}{D(\tau + 1, p)} = \frac{\cosh(m(\tau)(\tau - 1/(2T)))}{\cosh(m(\tau)(\tau + 1 - 1/(2T)))}, \tag{3.10}$$

with D being $D_T(\tau, p)$, $F(\tau, p)$, while for $G(\tau, p)$ it is defined as

$$\frac{G(\tau, p)}{G(\tau + 1, p)} = \frac{\sinh(m(\tau)(\tau - 1/(2T)))}{\sinh(m(\tau)(\tau + 1 - 1/(2T)))}. \qquad (3.11)$$

First let us discuss the gluon spectral function. In Figure 3 we show the gluon spectral function at $T = 3T_c$ and for $p = 1.57T$. One can see a rather broad peak around $\omega/T \sim 2$ and a very broad structure for $\omega > 10T$. The spectral function itself strongly depends on the default model parameterized by m_0. The position of the peak, however, is not influenced by the default model at least within the present statistical accuracy. At finite temperature the gluonic quasiparticle peak can also aquire a finite width due to thermal effects [2]. However, the width observed by us at present seems to be primarily due to insufficient statistics and too few data points used in the analysis. A similar conclusion has been drawn in [9]. In Figure 3 we also show the effective masses from the temporal gluon propagator for $T = 3T_c$ and $p = 1.57T$,

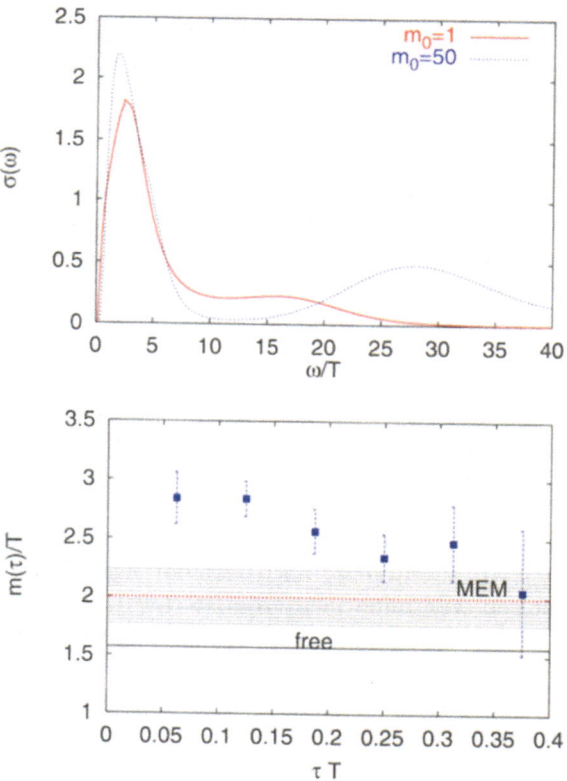

Fig. 3. The gluon spectral function for $3T_c$ and $p = 1.57T$ (top) and effective gluon mass (bottom). The dashed line and the band on the right figure is the position of the peak of the spectral function from the MEM analysis and its uncertainty.

where the peak position from the MEM analysis is also shown. As one can see from the Figure the effective masses do not reach a plateau and stay above the peak position extracted from the MEM analysis. This is in accordance with the observation that the spectral function receives large contributions from $\omega \gg p$. The situation is similar for other values of the spatial momenta and temperature. Now we turn to the discussion of the quark propagators. It turns out that our data on G are too noisy to apply the MEM analysis to them. The local masses extracted from F and G, however, are identical within statistical errors. In fact, if the quark propagators are dominated by a single quasiparticle contribution, the local masses extracted from F and G should be identical. We therefore applied the MEM analysis only for F. The reconstructed spectral function $\rho_F(\omega, p)$ is shown in Figure 4. It has only a single peak indicating the absence of the plasmino branch for the values of the momenta studied by us (at zero momentum there is no distinction between the quasiparticle (quark) and plasmino branch). In contrast to the

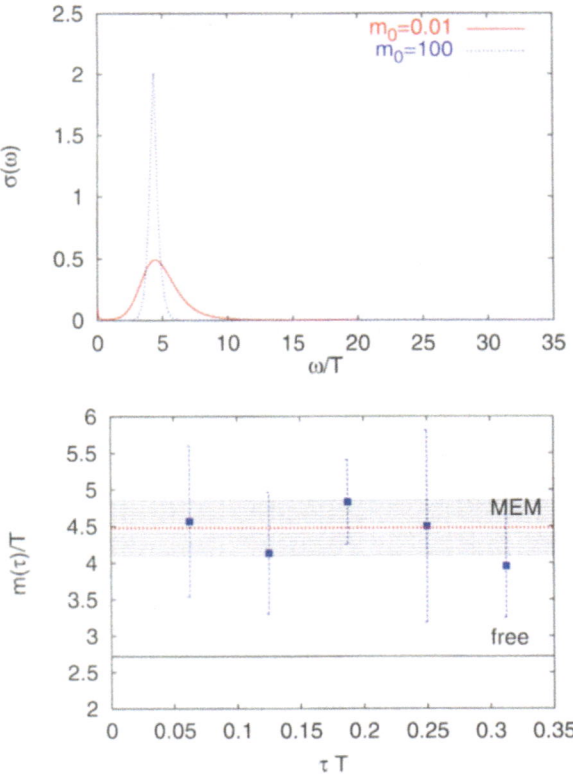

Fig. 4. The quark spectral function for $T = 1.5T_c$ and $p = 2.72T$ (top) and the corresponding local masses (bottom). The dashed line and the band on the right figure are the peak position from the MEM analysis and uncertainty of its value.

gluon spectral function the quark spectral function extracted from F has negligible continuum contribution above the light cone. As a consequence the corresponding effective masses $m_q(\tau, p)$ shown in Figure 4 reach a plateau already at $\tau T = 0.065$. The position of the peak in the spectral function $\omega_q(p)$ agrees with the average value of the local masses $m_q(\tau, p)$.

3.3   Dispersion relation and quasiparticle masses

Performing the analysis described in the previous section for all values of the spatial momenta we can determine the dispersion relations of quarks and gluons, $\omega_q(p)$ and $\omega_g(p)$. Our numerical results for the dispersion relations of quark and gluons $\omega_q(p)$ and $\omega_g(p)$ are summarized in Figure 5. While at $T = 3T_c$ the dispersion relation both for quarks and gluons is close to the free dispersion relation $\omega^2(p) = p^2$ one sees large deviations from it at $T = 1.5T_c$. In order to quantify the deviations from the free propagation we have fitted the dispersion relations to $\omega_{q,g}^2(p) = p^2 + m_{q,g}^2$ and determined the values of quasiparticle masses m_q and m_g. For $T = 3T_c$ we have found $m_q/T = 1.7 \pm 0.1$ and $m_g/T = 1.2 \pm 0.1$. The value of the gluon mass is compatible with the leading order result of HTL perturbation theory $m_g^2 =$

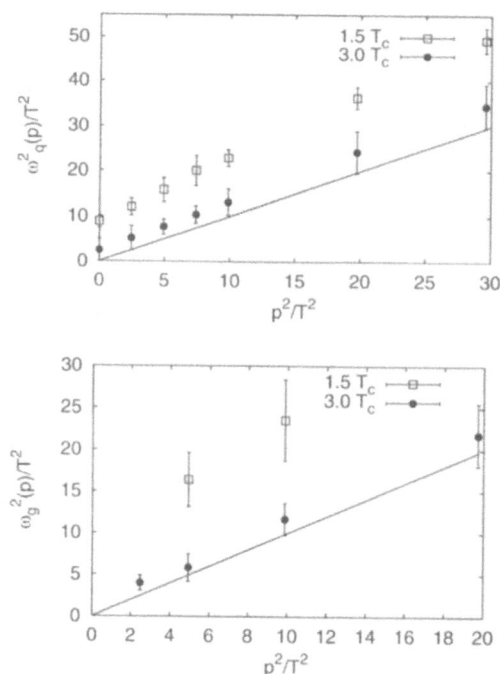

Fig. 5. The dispersion relation of quarks (top) and gluons (bottom) for temperatures $T = 1.5T_c$ and $3T_c$.

$g^2 T^2/2$ if we assume for the gauge coupling $g(3T_c) \sim 1.6$ as suggested by the short distance behavior of the heavy quark potential [14]. With the same value of g the leading HTL result for m_q is roughly a factor 2 smaller than the value found by us. For $T = 1.5T_c$ we have obtained $m_q/T = 3.9 \pm 0.2$, $m_g/T = 3.4 \pm 0.3$. Here the value of m_q was obtained by omitting the first three values of $\omega_q(p)$ from the fit. The values of the quasiparticle masses at $T = 1.5T_c$ are considerably larger than the corresponding ones at $3T_c$ and those expected from perturbation theory. This is consistent with the temperature dependence of quasiparticle masses $m_{q,g}/T$ used in quasiparticle models for the equation of state [4].

4 Conclusions

We have calculated the temporal quark and gluon propagators on large ($64^3 \times 16$) isotropic lattice and extracted the spectral function from them. Calculating the spectral function at several spatial momenta we were able to determine the dispersion relation for quarks and gluons and extract the temperature dependent quasiparticle masses. We have found that the quasiparticle masses measured in units of temperature increase substantially as the temperature is lowered from $3T_c$ to $1.5T_c$.

Our temporal propagators are quite different from those calculated within the HTL perturbation theory. We have found that in general medium effects are stronger than suggested by hard thermal loop perturbation theory. In particular, the quasiparticle masses determined by us are generally larger than those predicted by HTL perturbation theory (except for the gluon mass at $3T_c$). We also see a large continuum contribution in the gluon spectral function which is absent in the HTL approximation.

Due to the finite lattice volume we were not able to resolve collective mode such as the plasmino mode. For a detailed study of these modes one should consider even larger spatial volumes which eventually will become possible by considering anisotropic lattices with spatial lattice spacing larger than the temporal one.

References

1. J. Kuti et al, Phys. Lett. **B98** (1981) 199; L.D. McLerran and B. Svetitsky, Phys. Rev. **D24** (1981) 450; J. Engels, F. Karsch, H. Satz and I. Montvay, Phys. Lett. **B101** (1981) 89
2. M. Le Bellac, *Thermal Field Theory* (Cambridge University Press 1996)
3. F. Karsch et al, Phys. Lett. **B401** (1997) 69; O.J. Andersen et al, Phys. Rev. Lett. **83** (1999) 2139; J.P. Blaizot et al, Phys. Rev. Lett. **83** (1999) 2906
4. U. Heinz and P. Levai, Phys. Rev. **C57** (1998) 1879; A. Peshier et al, Phys.Rev. **C61** (2000) 045203
5. O. Philipsen, Phys. Lett. **B521** (2001) 273; hep-lat/0112047
6. R. Kobes et al, Nucl. Phys. **B355** (1991) 1; A.S. Kronfeld, Phys. Rev. **D58** (1998) 051501

7. A. Weldon, Phys. Rev. **D61** (2000) 036003
8. A. Cucchieri et al, Phys. Lett. **B497** (2001) 80; Phys. Rev. **D64** (2001) 036001
9. M. Asakawa et al, Prog. Part. Nucl. Phys. **46** (2001) 459
10. K. Langfeld et al, Nucl. Phys. **B621** (2002) 131
11. F. Karsch, Lect. Notes Phys. **583** (2002) 209 (hep-lat/0106019)
12. A.D. Kennedy and B.J. Pendleton, Phys. Lett. **B156** (1985) 393
13. F. Karsch and J. Rank, Nucl. Phys. B (Proc. Suppl.) **42** (1995) 508
14. P. Petreczky et al, Nucl. Phys. **A698** (2002) 400
15. P. Petreczky et al, Nucl. Phys. B (Proc. Suppl.) **106** (2002) 514
16. H. Schulz, Nucl. Phys. **B413** (1994) 353; A. Peshier et al, Ann. Phys. **266** (1998) 162

Regional Climate Simulation for Central Europe (RECLICH)

Klaus Keuler

Brandenburg University of Technology Cottbus, Chair for Environmental Meteorology, D-03013 Cottbus, Germany

Abstract. The results of the same regional climate simulation executed on different computer systems are compared to assess the influence of computer architecture and compiler characteristics on the results of long-term numerical simulations. The comparisons of temperature and precipitation fields show significant differences at any time. The differences of the monthly mean values of the same fields are, however, much smaller. This indicates that the influence of the computer system selected for the simulation can be neglected for the interpretation of climate values but may lead to non negligible uncertainties in the weather forecast for a specific date.

1 Introduction

Changes of the global climate in the near future may cause various effects in different parts of the world. The increase of the global mean surface air temperature is only one parameter that indicates a general warming of the earth's atmosphere. The amount of warming, however, will as likely as not vary between different regions of the globe. This is also valid for changes in precipitation. We expect a general increase in global precipitation but the modifications of the annual or seasonal precipitation in individual regions may substantially differ from the mean global tendency.

The major objective of this project is to investigate in which way a probable change of the global climate due to a further increase of greenhouse gas concentrations will affect the climate conditions for Europe. Therefore, two regional climate simulations will be performed representing present-day climate conditions on the one hand and a future climate scenario on the other hand. The model used for these simulations is the regional climate model REMO (Jacob and Podzun 1997) which has been developed at the Max-Planck Institute for Meteorology (MPIM) in Hamburg particularly for the purpose of continuous long-term simulations of atmospheric processes with a high horizontal resolution.

The first year of the project has been used for the implementation of the regional model and its runtime control system. During the first tests of the model on the NEC SX-5 at the High Performance Computer Center Stuttgart (HLRS) a fatal error in the model code was detected by the developers at MPIM. Unfortunately it had taken several months to fix the error and to subject the model to a number of additional and detailed tests at MPIM until the new version was released again. This had caused a large delay of the project so that the intended simulations could be started only a short while ago and are still running.

In Sect. 2 the setup for the intended model experiments is explained. Then the implementation of the model system and some of the related problems are described. A few results of the first test runs are presented in Sect. 4. A brief perspective to the upcoming tasks in Sect. 5 is followed by a summary of some specific performance characteristics of the regional model on the HLRS computer platforms.

2 Model Experiments

The regional climate model REMO is a so called limited area model with open boundaries. The meteorological conditions outside the model domain have to be prescribed as time dependent boundary values. In our case, these boundary values are interpolated in space and time from the results of a global climate simulation with the general circulation model (GCM) ECHAM4 (Roeckner et al. 1996) at a horizontal resolution of T106 (about 1.1° in longitude and latitude). The horizontal resolution of the regional model is 1/6° (about 18 km) and the model domain indicated in Fig. 1 covers an area of about $2200 \times 2600 \, \text{km}^2$.

The global simulations used for the interpolation of the boundary values of the regional model represent two ten year time slices (Wild and Ohmura 2000) for the periods 1971–1980 and 2041–2050. The greenhouse gas concentrations for these periods are specified according to the modified IPCC scenario IS92a (Houghton 1996). The climate conditions for both periods are determined by ten year mean values of selected climatological parameters like air temperature, precipitation, and wind speed and are calculated from the results of the two corresponding regional climate simulations with REMO. The differences in these parameters between both simulations yield the climate change signal for central Europe as a result of an increasing greenhouse gas concentration.

3 Technical Realization

The model configuration for the regional simulations was developed and tested in a number of experiments on a Cray C90 vector computer and on a parallel processor system, an Origin 2000 from SGI, at the Deutsches Klimarechenzentrum (DKRZ) in Hamburg. The results of the two GCM time slice experiments, which shall be used here as input to drive the regional simulations, are archived on a large fileserver at the DKRZ. They claim an overall storage of 345 GB and have to be provided at runtime to the regional simulation.

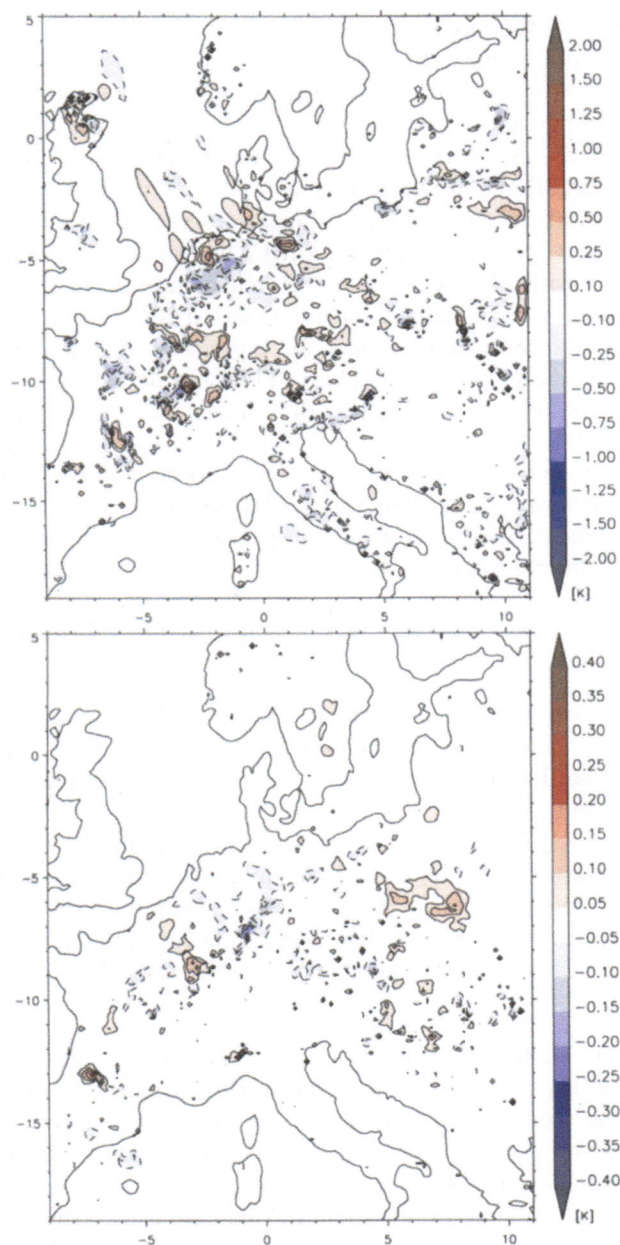

Fig. 1. Differences of the surface air temperature between simulations on different computer systems. Upper panel: differences at an arbitrary date of the simulation. Lower panel: differences of the monthly means.

The transfer of the model system, of the automated job control necessary to ensure the continuous run of the model, and of the input data from the DKRZ to the computer platforms at the HLRS has required several modifications and adaptations.

First of all the entire model code written in FORTRAN 77 with a few elements of FORTRAN 90 had to be recompiled. This required an appropriate exchange of all statements for optimization (vectorization, auto-tasking) within the code. Compiler options for CRAY had to be replaced in the makefiles by the corresponding NEC syntax. In addition, most of the I/O routines had to be modified to process 32 bit unformatted IEEE-data.

After this, the data management for input and output files during the simulation was completely reorganized. This was necessary because of the different policies for temporal limited disk space and data administration at DKRZ and HLRS.

Finally, the control of the model runs had to be adapted to the requirements of the computer infrastructure at HLRS. This concerned the batch-queue management, the job control – in particular the coordination of jobs running on different computer systems –, and the handling of input- and output-files.

A total of four different computers are involved in processing the model simulations.

1. The NEC Server VOLVOX ('crossi') provides the cross compiler to generate the executables of REMO source code for the SX-5 platform.
2. The fileserver hwwfs1 is used as intermediate and permanent storage for input and output data.
3. The NEC SX-5 Cluster runs the time-consuming model simulations.
4. A workstation which has been additionally installed into the HLRS environment serves as an interface between the fileservers at DKRZ and HLRS. It became necessary due to conflicts between the ftp-protocol of the UNITREE file system at DKRZ and the firewall around the HLRS platforms.

The climate simulation runs with a time step of two minutes to ensure numerical stability of the applied discretization schemes. During the whole simulation over a period of ten years, a new set of input data with a volume of 12 MB must be provided every six hours. The results of the regional simulation are stored in the same interval claiming a disk space of 23 MB for each file. The whole simulation is subdivided into a sequence of jobs, each of them covering a period of ten days, to allow a reasonable job and data management.

At the beginning of a ten year simulation the first year of input data is completely transferred from the DKRZ UNITREE system to the interface-workstation at HLRS and the first thirty days are stored on a permanent disk space (WORKIN) of the fileserver hwwfs1. When a ten day model run is started, the required series of input files is transferred from WORKIN to the temporary disk storage (scratch area) at SX-5. The results of the regional simulation are also stored onto this scratch area. At the end of the run the six-hourly output files and the restart files for the continuation of the simulation are moved from the scratch area back to a file system (WORKOUT) on hwwfs1. Due to the quota on the permanent disk space, three service procedures are started now on the fileserver. The first one cleans the WORKIN file system, which means, it removes all input files that have already been used. The second one archives the results of the simulation to the data migration facility (DMF) subsystem of the fileserver hwwfs1 and removes them from WORKOUT. The third one gets the the next ten days of input data from the interface-workstation to WORKIN. Then the next job is submitted to the SX batch queue and the procedures are repeated. While the model is running, the input data for the next year are transferred from UNITREE to the interface-workstation and an already expired year is removed from this workstation.

This network of different jobs running synchronized on different machines requires a very stable performance of the computers involved. The whole job construction would collapse and has to be restarted by hand if one of the systems crashes or if the network breaks down at an unfavorable moment.

4 Results of Test Runs

The job control programs described above were implemented and tested in numerous configurations to find an optimal job-scheduling. Then, several tests with simulations over periods of a few days up to one month were carried out to check the results of the regional model and to estimate its performance. One aspect among other things was to investigate how accurately the results of previous simulations can be reproduced on a different computer. Therefore, the results of a one month simulation performed on the SX-5 at HLRS for an arbitrary January were compared with the results of the same simulation carried out on a SGI Origin 2000 parallel processor system at DKRZ.

As expected, the results were not exactly the same. Figure 1 shows the differences of the surface air temperature between the SX-5 and the SGI run. The upper panel presents the deviations of the temperature fields at a certain time – here at the beginning of day 30 of the simulation – averaged over the previous six hours. Positive and negative deviations occur all over the model domain with a magnitude of up to 1 K. Such differences develop at any time of the simulation, even during the first day, move around over a certain distance and vanish again after several hours. The origins of these cells with positive and negative deviations seem to be arbitrarily distributed over the model domain. Their size and magnitude slightly vary during the whole simulation but do not systematically increase.

These deviations are caused by small differences in the numerical calculation due to different processor architectures. Differences in the computational accuracy can be regarded in a physical sense as an induced small perturbation to the atmospheric system. Because of its general chaotic behavior, the atmosphere responds to this disturbance with a slightly different development. Over a certain time these primarily small differences can increase and may lead to a significantly different further development of weather conditions in some parts of the atmosphere and thus in the model simulation too. For example, a small deviation at a certain location may cause a diverging formation of clouds in this region. As a consequence, the incoming solar radiation at the surface is modified for several hours. This yields again a significant deviation of the air temperature between both runs. However, as both simulations are using exactly the same boundary values, the general development of the weather conditions has to be the same in both cases and the differences between both solutions always remain in a limited range.

The deviations are arbitrarily produced during the computation and cannot be predicted as well as the effects of small disturbances in real weather forecasts. Due to the stochastic nature of these processes, however, the overall effect on long-term mean values, which in particular means the effect on climate conditions for a region, should decrease with increasing length of the averaging period. This is approved by the lower panel of Fig. 1, which shows the differences between the monthly mean values of the air temperature of the two simulations. The differences are considerably smaller and remain below a few tenth of a degree centigrade but are still visible on a one month time-scale.

Similar deviations occur in the precipitation fields presented in Fig. 2. The precipitation values at the beginning of day 30 accumulated over the previous six hours (upper panel) partly differ by more than 2 mm. As for the temperature fields, the relative deviations on a longer time-scale are much smaller. The lower panel in Fig. 2 shows that the absolute differences of the precipitation accumulated over the whole month are mostly smaller than 6 mm. This value corresponds to an averaged six hourly precipitation intensity difference of only 0.05 mm. The variations in the simulated precipitation appear to be more significant than those in the temperature distribution and are related to the regions with the highest amount of precipitation. The reason for the precipitation differences is the same as for the temperature deviations. Small variations in the calculation result in modifications of the atmospheric conditions, which may considerably affect the amount of a single precipitation event.

Therefore, the sensitivity of atmospheric circulation models to the computer used for the simulations has to be taken into account to estimate the uncertainty or fuzziness of daily and seasonal weather prediction. In conclusion, this uncertainty limits the predictability for atmospheric processes in addition to any inaccuracies of the physical model itself.

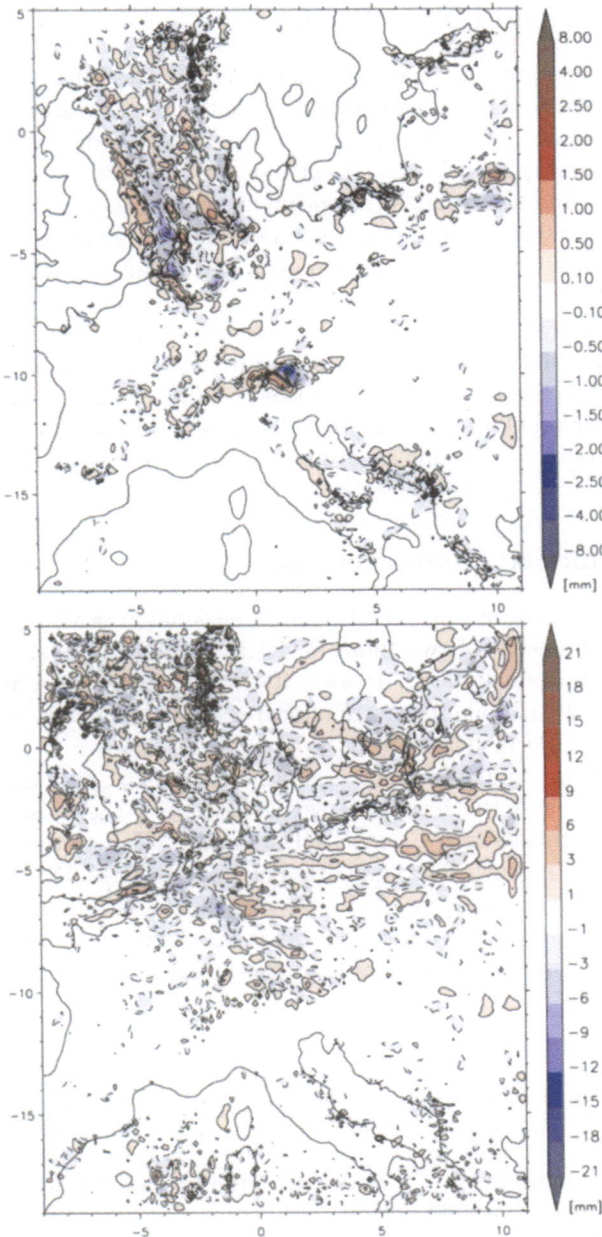

Fig. 2. Differences of the precipitation between simulations on different computer systems. Upper panel: differences of the six hourly accumulated precipitation amount at an arbitrary date of the simulation. Lower panel: differences of the monthly accumulated precipitation.

5   Summary and Next Steps

The regional climate model REMO has been successfully installed on the SX-5 at the HLRS. The implemented job-control enables an automatic continuous simulation of climate conditions over a period of several years as a sequence of ten-day model runs. Input files are provided on demand during the simulation and the results are continuously archived for further analyses. The test-simulations indicate that the model does a correct job but that the numerical results for a single date or month may differ within a certain range from those obtained on an other computer.

The first ten-year simulation for present-day climate conditions (control run) has already been started but is still running. The climate means of this simulation will be compared with the corresponding climate parameters of the second ten-year simulation (scenario run) to assess the possible changes of the climate conditions for Europe according to the prescribed global greenhouse gas scenario.

6   Performance Characteristics

The regional simulation runs on a three-dimensional grid with $121 \times 145 \times 24$ boxes and a time-step of two minutes. Each ten-day time- slice simulation takes 6400 s on a single CPU. It uses 490 MB of input data and produces an output of 920 MB. The program allocates 450 MB memory and performs with 1750 MFLOPS or 64 MIPS.

For a ten-year simulation an overall amount of 510 GB is transferred between the SX-5 and the fileserver hwwfs1, between the fileserver and the interface- workstation, and between HLRS and DKRZ or the university in Cottbus via the Internet. The simulation claims 640 h of CPU time. As a result of the average load of the SX-5, one year of simulation takes a real-time of ten to fifteen days. Therefore, one regional simulation over ten years is suggested to be finished within three to five months.

References

Houghton, J. Th., Intergovernmental Panel on Climate Change (IPCC): Climate change 1995: The science of climate change. Univ. Press, Cambridge (1995) 572 pp

Jacob, D., Podzun, R.: Sensitivity studies with the regional climate model REMO. Meteorol. Atmos. Phys. **63** (1997) 119–129

Roeckner, E., Arpe, K., Bengtsson, L., Christoph, M., Claussen, M., Dümenil, L., Esch, M., Giorgetta, M., Schlese, U., Schulzweida, U.: The atmospheric general circulation model ECHAM4: Model description and simulation of present-day climate. Max-Planck Institute for Meteorology, Report no. **218** (1996)

Wild, M., Ohmura, A.: Change in mass balance of polar ice sheets and sea level from high-resolution GCM simulations of greenhouse warming. Ann. Glaciol. **30** (2000) 197–203

Variation of Non-Dimensional Numbers and a Thermal Evolution Model of the Earth's Mantle

Uwe Walzer[1], Roland Hendel[1], and John Baumgardner[2]

[1] Institut für Geowissenschaften, Friedrich-Schiller-Universität,
   Burgweg 11, 07749 Jena, Germany
[2] Los Alamos National Laboratory, MS B216 T-3, Los Alamos, NM 87545, USA

Abstract. A 3-D compressible spherical-shell model of the thermal convection in the Earth's mantle has been investigated with respect to its long-range behavior. In this way, it is possible to describe the thermal evolution of the Earth more realistically than by parameterized convection models. The model is heated mainly from within by a temporally declining heat generation rate per volume and, to a minor degree, from below. The volumetrically averaged temperature, T_a, diminishes as a function of time, as in the real Earth. Therefore, the temperature at the core-mantle boundary, $T_{CMB,av}$, has not been kept constant but the heat flow, in accord with Stacey (1992). Therefore, $T_{CMB,av}$ decreases like T_a. This procedure seems to be reasonable since evidently nobody is able to propose a comprehensible thermostatic mechanism for CMB. First of all, a radial distribution of the starting viscosity has been derived using PREM and solid-state physics. The time dependence of the viscosity is essential for the evolution of the Earth since the viscosity rises with declining temperature. For numerical reasons, the temperature-dependent factor of the model viscosity is limited to four orders of magnitude.

The focus of this paper is an investigation of the variation of parameters, especially of the non-dimensional numbers as the Rayleigh number, Ra, the Nusselt number, Nu, the reciprocal value of the Urey number, Ror, the viscosity level, r_n, etc. For $0.0 \leq r_n \leq +0.3$, the authors arrived at Earth-like models. This interval contains the starting model. The quantification of the essential features of the model is provided by eight plots. Numerical procedure: The differential equations are solved using a fast multigrid solver and a second-order Runge-Kutta procedure with a FE method. On 128 processors, runs with 10649730 grid points need about 50 hours. Figure 11 shows the scaling degree of our code.

If the temperature dependence of the viscosity, Eq. (4), is replaced by Eq. (10) then, in the interval $0.0 \leq r_n \leq +0.3$, reticulately connected thin cold sheet-like downwellings are found from the surface down to 1350 km depth. However, the movements along the upper surface are not plate-like.

1 Introduction

This paper presents a new thermal evolution model of the Earth's mantle. It is based on the assumption of thermal convection in a compressible spherical shell heated from within, for the most part. The focus is the variation of dimensionless parameters in order to show the reliability of the model.

The construction of this introduction results from the first paragraph. Here, some essential papers of other authors are reviewed. The main sources of the internal heating of the Earth stem from the transformation of kinetic energy of the parential bodies into heat during accretion (Safronov, 1972) and from the decay of radioactive elements with long half-life periods. Spohn and Schubert (1991) demonstrated that, after an early impact of a Mars-sized body, the Earth needs 1-10 Ma to come to equilibrium. Using such a starting point, Schubert et al. (1979, 1980) calculated the thermal evolution in a parameterized model where, from the integration of the energy conservation in the Earth's mantle, it follows that

$$Mc\frac{\partial T_a}{\partial t} = MH_0e^{-\lambda t} - Aq \tag{1}$$

where M is the mass of the mantle, c the specific heat, T_a the volume-averaged temperature of the mantle, H_0 the specific radiogenic heat generation rate at the beginning of mantle evolution, λ a unified rate constant, A the magnitude of the Earth's surface and q the mean mantle heat flow at the surface. According to the presumed heating mode, different Nu-Ra relations are used in these and the subsequently mentioned papers. Christensen (1984, 1985) studied the influence of a strongly temperature-dependent viscosity on thermal convection. Dependent on the assumed degree of temperature dependence of the viscosity, Christensen (1985), Solomatov (1995) and Reese et al. (1999) arrived at three different convection modes, a first regime with low temperature influence and movable upper boundary layer, the sluggish-lid regime and the stagnant-lid regime. For each regime, a different Nu-Ra relation has been derived.

Our evolution model of the mantle is a convective spherical-shell model, since the balance equations of energy, momentum and mass and later on also the balance equations of the sums of the number of the atoms of radioactive parent nuclides and the corresponding radiogenic daughter nuclides can be more realistically formulated for a spherical shell. Furthermore, the compressibility and associated effects such as viscous heating and adiabatic heating are relevant for the larger terrestrial planets (Earth, Venus). Therefore, these effects are taken into account in our model. For that reason, some papers of other authors are discussed here which show similar features. Bercovici et al. (1992) calculated compressible-fluid convection using the anelastic liquid approximation (ALA). The material parameters are supposed to be constants, expect the density $\rho(r)$. So, also the viscosity, η, is constant. The superadiabatic temperature difference, ΔT_{sa}, between the upper and the lower surface of the spherical shell is kept constant, too. The model has no internal heating. T_{bot} is the temperature at the lower surface, Di is the dissipation number, Ra_a is the Rayleigh number.

$$Ra_a = \langle \rho\alpha gh^3 \Delta T_{sa}/\eta\kappa \rangle \tag{2}$$

where α is the coefficient of thermal expansion, g gravity, h the depth of the mantle and κ the thermal diffusivity. $\langle\rangle$ denotes the volume-averaged value. In order to investigate compressibility effects, Ra_a, $\overline{Di}$ and $\overline{T}_{bot}/\Delta T_{sa}$ have been systematically varied. The thickness of the upper stagnant layer grows with rising $\overline{T}_{bot}/\Delta T_{sa}$. The model by Zhang and Yuen (1996) is somewhat more complex and resembles more to the mantle. ALA is used, too, but additionally the assumption

$$\tilde{\alpha}/\tilde{\gamma} = \tilde{\rho}_{ref}^{-2} \tag{3}$$

where the curly overbar denotes non-dimensional quantities; γ is the Grueneisen parameter and ρ_{ref} is a depth-dependent reference density using the Adams-Williamson equation. $\tilde{\alpha}$, $\tilde{\rho}_{ref}$, the thermal conductivity, $\tilde{k}$, and the reference temperature, $\tilde{T}_{ref}$, are weakly depth-dependent, $\tilde{\eta}$ is moderately depth-dependent. The number and the strength of plumes and subducting zones strongly depend on the depth dependence of the viscosity. Using other viscosity-depth relations, we can corroborate this experience. Viscous dissipation is maximum near the upper and lower surface of the spherical shell. Zhang and Yuen (1996) concluded that adiabatic and viscous heating increase each other in downwellings whereas viscous heating retards adiabatic cooling in plumes. For incompressible convection, this asymmetry causes a rising temperature in the interior of the shell.

2 Our model

In this paper, we investigate the effects of a systematic variation of non-dimensional parameters on our model. On the one hand, we used non-dimensional numbers which are fixed for one run, namely the Rayleigh numbers, $Ra(1)$ and $Ra(2)$, the Nusselt numbers, $Nu(1)$ and $Nu(2)$, the reciprocal values of the Urey numbers, $Ror(1)$ and $Ror(2)$, and a non-dimensional number, r_n, which characterizes the level of the viscosity profile. Below, these numbers will be defined. On the other hand, there are also time-dependent non-dimensional quantities, $Ra(t)$, $Nu(t)$ and $Ror(t)$, since the Earth's mantle as well as our model irreversibly evolve.

We calculate the thermal convection in a spherical shell with a ratio of the radii as in the Earth's mantle according to PREM, with an infinite Prandtl number, with impermeable free-slip boundaries, with a spatially homogeneous internal heating, subsiding as a function of time. The abundances of ^{238}U, ^{235}U, ^{232}Th and ^{40}K of the bulk silicate Earth are taken from McCulloch and Bennett (1994). These abundances deviate only slightly from those of the primitive mantle according to Hofmann (1988). The decay of the heat generation is an essential feature of the evolution of the mantle and must not be neglected. The primordial heat of the mantle is introduced by an initial adiabatic temperature distribution which is 3900 K at the core-mantle boundary (CMB) in our model. The upper surface temperature is thought to be constant at 288 K, constant with respect to time and space. Even if only the primordial heat would exist, the thermal history of the mantle would be a decline in temperature. Radioactive decay, heat conduction and viscous dissipation are irreversible processes. Therefore, neither in this relatively simple model nor in the real mantle, steady states will occur during the considered time span of evolution. The non-dimensional numbers mentioned at the beginning make us independent of the details of the dimensional quantities which can be found in Walzer et al. (2002a). Furthermore, the model is compressible and takes into account the adiabatic and viscous heating. There are no continents in the model. So, the boundary conditions at the upper surface are equal everywhere. Moreover, the model does not contain chemical differentiation (Walzer and Hendel, 1999; Ogawa, 2000; DeSmet et al., 2000) and convective mixing of chemical reservoirs (Schmalzl, 1996). Similar to the model of Zhang and Yuen (1996), our method contains some quantities which weakly depend on the radius, r: The Figs. 1 and 2 by Walzer et al. (2002a) show the pressure, P, the reference density, ρ_{ref}, the bulk modulus, K, the Grueneisen parameter, γ, the thermal expansion coefficient, α, the specific heat at constant pressure, c_p, and the specific heat at constant volume, c_v, as a function of the radius, r. The coefficient α stems from Chopelas and Boehler (1992) and some related papers wheras the rest of the quantities have been derived from PREM and some usual solid-state physics formulae. The thermal conductivity, k, of the model is identical with that of Bercovici et al. (1992). A number of viscosity distributions is essential for the model. The P, T-dependence of the viscosity as well as the mineralogical phase boundaries at 410 and 660 km depth are reflected in these distributions. The basic distribution of the viscosity, η, results from

$$\eta(r) = 10^{r_n} \cdot \eta_1(r) \cdot \exp\left[\frac{c_1}{T_{av}/T_{00} + 1} - \frac{c_1}{2}\right] \tag{4}$$

for $r_n = 0$ where $\eta_1(r)$ is shown by Fig.1 and derived by Walzer and Hendel (2002) and Walzer et al. (2002a) using the seismological model PREM and solid-state physics.

The overall features of $\eta_1(r)$ show a certain resemblance with the viscosity profile according to Forte and Mitrovica (2002, their Fig. 2) and Ekström and Dziewonski (1998) although that profile has been derived by a totally different method. The latter method used density anomalies in the mantle and a simplified distribution of observed plate velocities given by the NUVEL–1 model. The mentioned density anomalies are derived from the observed shear-velocity anomalies and the observed free-air gravity anomalies which are calculated using the non-hydrostatic geopotential derived from satellite data.

The quantity r_n in Eq. (4) denotes a non-dimensional real number which characterizes the general level of the viscosity profile. In the third factor of Eq. (4), $c_1 = 9.2103$, $T_{00} = 1500\,\mathrm{K}$ and T_{av} is the laterally averaged temperature (in Kelvin) which is a function of the time, t, too. So, it is guaranteed that the temperature-dependent factor of the viscosity varies only in a range of

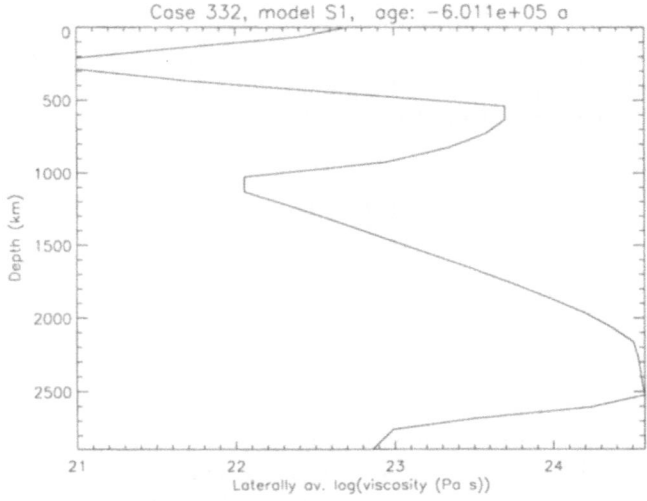

Fig. 1. Basic viscosity, $\eta_1(r)$, of the mantle as a function of depth, according to Walzer et al. (2002a).

four orders of magnitude. This assumption does not essentially distort the physics of the problem but, so, numerical problems are avoided.

Using our model, we want to calculate the last 4990 Ma of the thermal evolution of the mantle. This is the time after the formation of the Earth's core. So, not only some 100 Ma are relevant for our considerations.

a) *It is an essential feature of the evolution of the Earth, that its direction is defined by irreversible processes,* especially that the heat generation rate per volume, Q, decreases with time and that the Earth cools down, i.e. that the volume-averaged temperature, T_a, diminishes. The latter process would happen also if the internal heating would take place only at the beginning by accretion. From investigations of komatiites, it is well known that the temperature of the upper part of the mantle grows less by about 100 K per 10^9 a.

b) From geochemistry, it is well-known that the radioactive elements are concentrated mainly in the silicate shell but not in the metallic Earth's core. From this it can be concluded that the mantle is mainly heated from within and, only to a minor degree, from below.

From a) and b) it follows that it is *totally improbable that the laterally averaged CMB temperature, $T_{CMB,av}$, is a constant with respect to time.* That has nothing to do with the fact that T_{CMB} should not depend on the location vector since the outer core is a metallic liquid. In spite of this consideration, convection investigations assuming a T_{CMB} which is constant with respect to time lead to good results if the considered time span is only some 100 Ma. In Section 6.7.5, entitled Constancy of the Core-to-Mantle Heat Flux, Stacey

(1992) shows that it is a better approximation for evolution models to assume a constant heat flow at CMB. Additional arguments are given by Walzer and Hendel (1999) in Appendix A. In the present model $q_{CMB} = 28.9\,mWm^{-2}$ is assumed using a result by Anderson (1998). Because of the paragraphs a) and b), the Rayleigh number

$$Ra = \left\langle \frac{\rho\alpha g h^3}{\kappa\eta_{al}} \cdot \frac{(Qh + q_{CMB})h}{k} \right\rangle \tag{5}$$

is better adapted to the problem than

$$Ra_B = \left\langle \frac{\rho\alpha g h^3}{\kappa\eta_{al}} \right\rangle \cdot \Delta T \tag{6}$$

The bracket $\langle\rangle$ denotes a volume average but not a temporal average. The temperature difference, $\Delta T = T_{CMB} - T_{ob}$, is no constant for the present problem, T_{ob} being the constant upper-surface temperature. The heat generation rate per volume, Q, is given by

$$Q = \rho \sum_{\nu=1}^{4} a_{\mu\nu}\, a_{if\nu}\, H_{0\nu}\, \exp\left(-t/\tau_\nu\right) \tag{7}$$

where $a_{\mu\nu}$ is the abundance of the heat-producing elements, $a_{if\nu}$ the isotopic abundance factor, $H_{0\nu}$ the specific heat generation per unit time, $4.49 \times 10^9\,a$ ago, τ_ν the $1/e$ life of the corresponding nuclide and ν the consecutive index of the mentioned four radionuclides.

The quantity Q is monotonously declining as a function of time and also η_{al} is a function of the time since the cooling of the Earth (see paragraph a)) is an essential feature of the Earth's evolution and because of Eq. (4). Therefore, Ra is a function of time. In the main part of the time, Ra is growing less. So, steady-state models with constant Ra are not adequate to the problem. The definition

$$\log \eta_{al} = \langle \log \eta \rangle \tag{8}$$

has been used since the high values of η would dominate in a simple volume-proportional average.

3 Quantification of the model results. Non-dimensional numbers. Variation of the parameters. Conclusions.

Figure 2 presents the temperature distribution and the creeping velocities on an equal-area projection, with constant radius, for the exponent $r_n = 0$. Similar distributions have been computed for different radii (or depths in the mantle) and for different ages in the evolution of the system. Only non-dimensional quantities of different runs have been compared in the Figs. 3

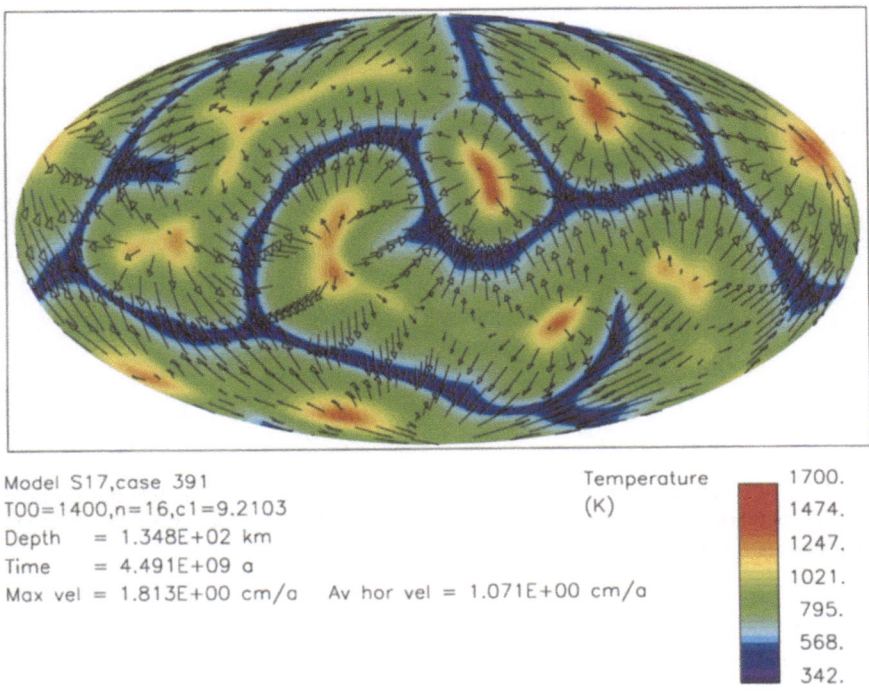

Model S17,case 391
T00=1400,n=16,c1=9.2103
Depth   = 1.348E+02 km
Time   = 4.491E+09 a
Max vel = 1.813E+00 cm/a   Av hor vel = 1.071E+00 cm/a

Temperature
(K)

1700.
1474.
1247.
1021.
795.
568.
342.

Fig. 2. Colors represent the temperature, arrows stand for the creeping velocities on an equal-area projection in 134.8 km depth. This distribution was computed by a run with Eq. (4) and $r_n = 0$ for the geological present.

through 9, where the systematic variation of the level of the viscosity profile was generated by the variation of r_n. Walzer et al. (2002b) investigated the influence of another temperature dependence of the viscosity on the distribution of non-dimensional quantities. If the models are more realistic, it will be difficult to discuss the mechanism since there is no simple superposition of the partial mechanisms because of non-linear relationships. To state the facts oversubtly, sometimes we have to decide whether we want to model nearer to geophysics or nearer to theoretical fluid mechanics. The mentioned papers of the authors try to bridge this gap.

Figure 3 shows the evolution of Ra (cf. Eq. (5)) as a function of r_n. As expected, highly viscous models evolve considerably slowly. Based on PREM and solid-state physics, our first supposition for the Earth was $r_n = 0$ (Walzer et al., 2002a). Another non-dimensional quantity, Ror, is represented as a function of r_n in Fig. 4. Ror is the ratio of the heat output per unit time radiated into space at the Earth's surface which stems from the Earth's interior to the radiogenic heat per unit time generated in the Earth's interior.

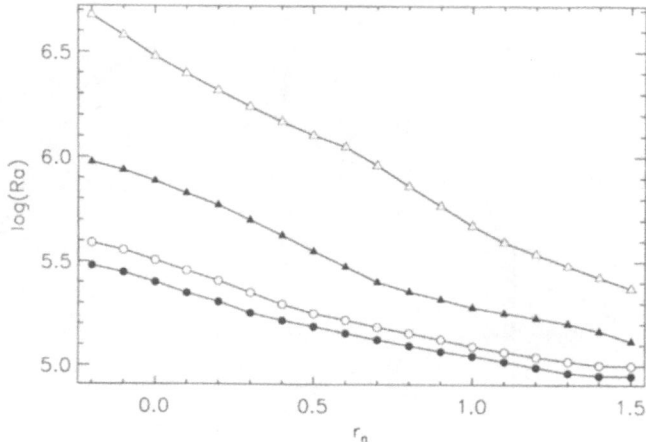

Fig. 3. The Rayleigh number, $Ra(t)$, as a function of the non-dimensional parameter r_n. Open triangles represent an age of 4000 Ma, filled triangles stand for 2000 Ma, open circles for 500 Ma, filled circles for 0 Ma.

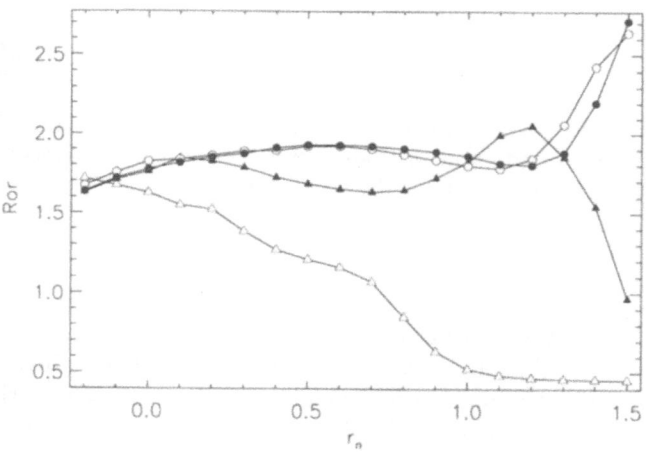

Fig. 4. The reciprocal value of the Urey number, $Ror(t)$, versus r_n. For explanation of symbols see Fig. 3.

The well-known Urey number is the reciprocal value of Ror. We have

$$Ror = Aq/M(Q/\rho) \qquad (9)$$

where all used quantities are defined above. Excluding the initial phase of the Earth's evolution with the differentiation of core and primordial mantle, the time average of Ror is 1.85 according to Stacey and Stacey (1999). According

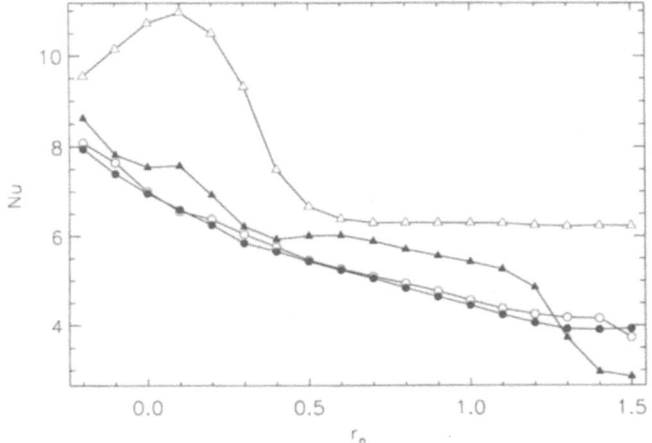

Fig. 5. The Nusselt number, $Nu(t)$, versus r_n. For explanation of symbols see Fig. 3.

to Fig. 4, this requirement is fulfilled for the runs with $r_n = +0.1$ and $r_n = +0.2$, in case of wider error bars for $0.0 \leq r_n \leq +0.3$. The interval $+1.0 \leq r_n \leq +1.3$ does not give an appropriate solution since Fig. 5 shows too low Nusselt numbers for this range. For an age of 4000 Ma, the maximum Nusselt number is at $r_n = +0.1$ according to Fig. 5. This is within the most favored r_n-span and near $r_n = 0.0$ which stems from the derivation of Walzer et al.(2002a). Fig. 7 shows that the maximum of the Nusselt numbers for an age of 4000 Ma and a relative maximum of Nu for an age of 2000 Ma are situated at about $Ra(1) = 0.89 \times 10^6$. The quantity $Ra(1)$ is the Rayleigh number, Ra, defined by Eq. (5), temporally averaged over the last 4000 Ma. A comparison of the Figs. 5 and 7 demonstrates that for the low Nusselt numbers, Nu, which correspond to the interval $+1.0 \leq r_n \leq +1.3$, the relation $Ra(1) < 4 \times 10^5$ applies. Since $Ra(1) = 0.89 \times 10^6$ seems to be more realistic, this is a further reason to exclude $+1.0 \leq r_n \leq +1.3$ and higher r_n for Earth-like models. Taking no account of these strongly viscous models, Fig. 5 shows the greatest decrease in the Nusselt number during the last 4000 Ma.

Figure 6 demonstrates the distribution of Nu versus $Ra(2)$. The latter value is the time average of Ra over the last 2000 Ma. $Ror(2)$ is defined by analogy. $Ror(2)$ versus $Ra(2)$ is shown by Fig. 8. The realistic value of $Ror(2) = 1.85$, realistic according to Stacey and Stacey (1999), is situated somewhere in the flat maximum of Fig. 8. The relation between the temporal averages over the last 4000 Ma of Nusselt numbers and of Rayleigh numbers is depicted in Fig. 9. The exact relations to theoretical values can be checked only for a simplified mechanism. But according to tendency, the $Nu(1)$-

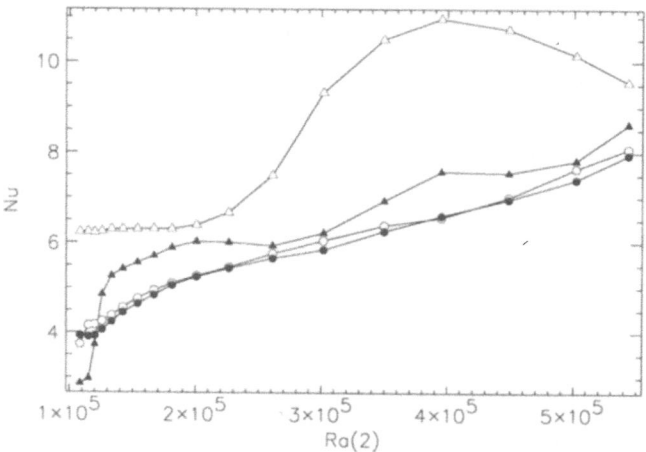

Fig. 6. The Nusselt number, $Nu(t)$, versus $Ra(2)$. The quantity $Ra(2)$ is the Rayleigh number, Ra, averaged over the last 2000 Ma. For explanation of symbols see Fig. 3.

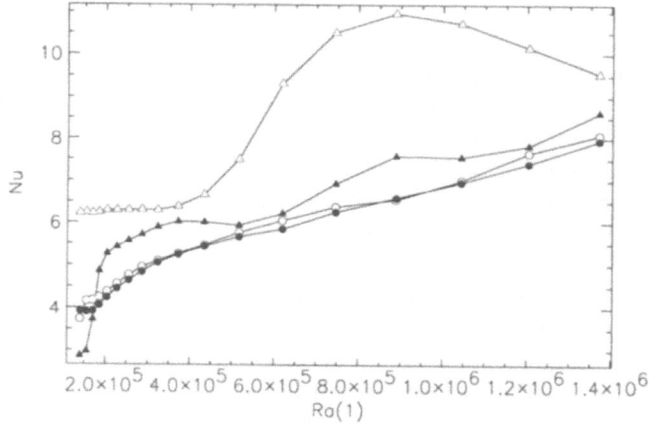

Fig. 7. Nu versus $Ra(1)$. The quantitiy $Ra(1)$ is the Rayleigh number, Ra, averaged over the last 4000 Ma. For explanation of symbols see Fig. 3.

$Ra(1)$ curve corresponds to the usual theoretical expectations. Figure 10 is the only picture with a dimensional quantity, namely qob, the heat flow in mWm^{-2}, averaged over the upper surface of the mantle. The quantity qob is plotted as a function of the non-dimensional r_n. The present-day mean heat flow at the Earth's surface is $87\,mWm^{-2}$, the present day mean heat flow at the upper surface of the mantle is $72\,mWm^{-2}$ according to Schubert et al. (2001). For the latter value, the contribution of the crust has been dropped.

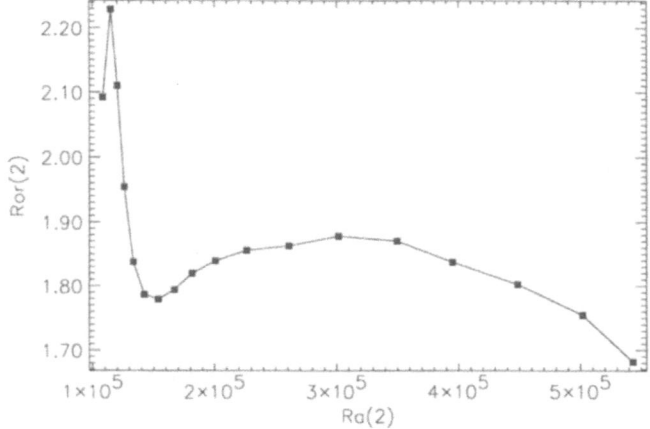

Fig. 8. $Ror(2)$ versus $Ra(2)$. The supplement (2) means the time average over the last 2000 Ma.

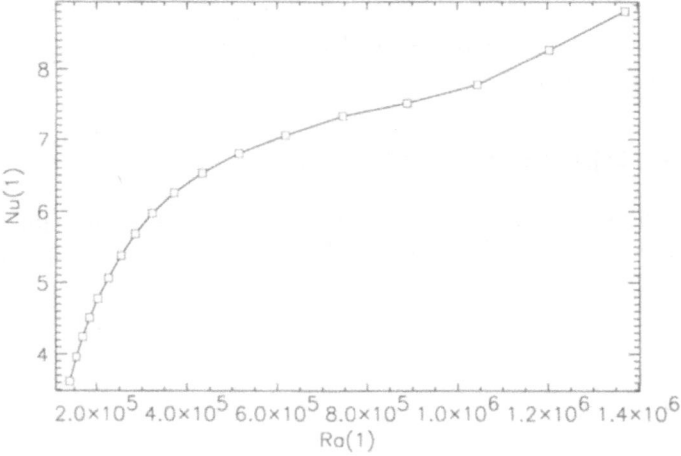

Fig. 9. The Nusselt number, $Nu(1)$, versus the Rayleigh number, $Ra(1)$. The supplement (1) stands for the time average over the last 4000 Ma.

An approximate equality of the computed present-day heat flows, i.e the solid circles of Fig. 10, with the observed $72\,mWm^{-2}$ is given for $r_n = +0.1$ and $r_n = +0.2$, if somewhat larger error bars are accepted, for $0.0 \leq r_n \leq +0.3$.

A second corresponding interval would be $+1.0 \leq r_n \leq +1.3$, but this second interval has to be excluded for the above mentioned reasons. Since the observed qob and Ror values stem from different information sources, this can be regarded as a confirmation that, using the interval $0.0 \leq r_n \leq +0.3$,

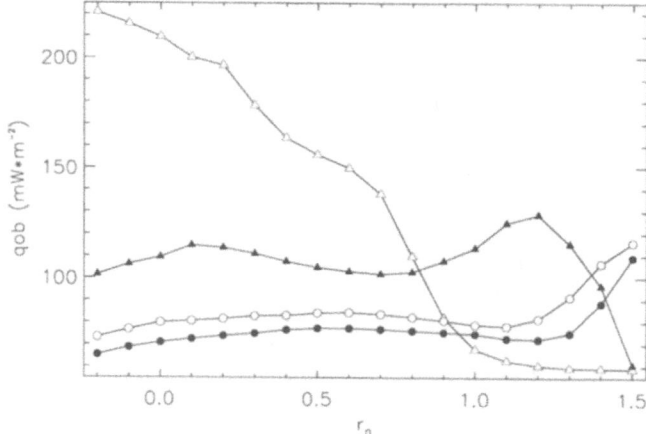

Fig. 10. The laterally averaged heat flow, $qob(t)$, at the upper surface of the spherical shell versus the non-dimensional r_n. For explanation of symbols see Fig. 3.

the model produces the most Earth-like results. Further conclusions are to be found in the abstract.

4   Some aspects of computing

The differential equations of convection in a compressional spherical shell are solved using a three-dimensional FE method, a fast multigrid solver and

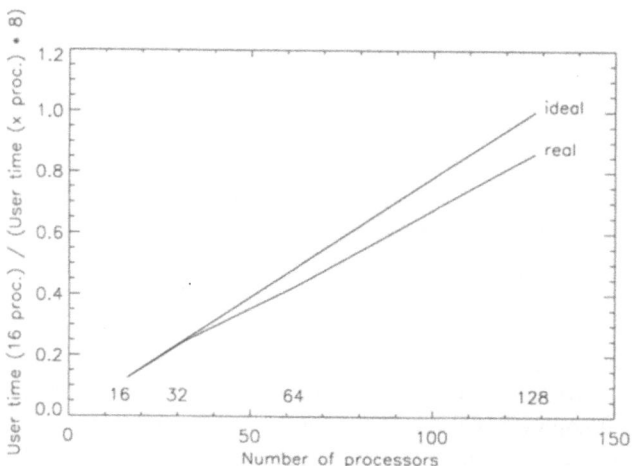

Fig. 11. Scaling degree

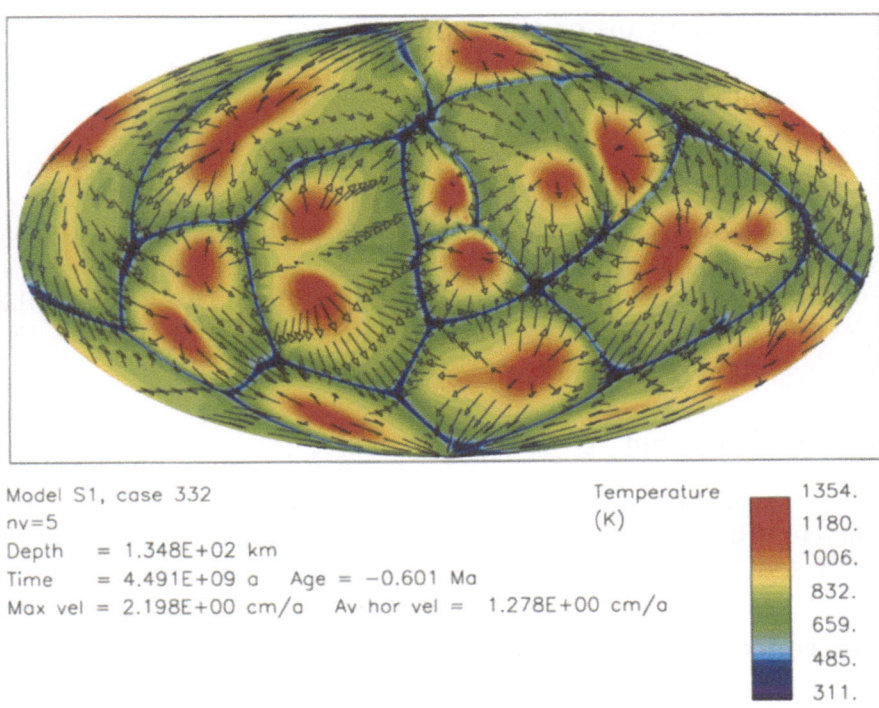

Model S1, case 332   Temperature   1354.
nv=5   (K)   1180.
Depth   = 1.348E+02 km   1006.
Time   = 4.491E+09 a   Age = −0.601 Ma   832.
Max vel = 2.198E+00 cm/a   Av hor vel = 1.278E+00 cm/a   659.
                                                                          485.
                                                                          311.

Fig. 12. Colors represent the temperature, arrows stand for the creeping velocities
on an equal-area projection in 134.8 km depth. This distribution is the result of
a run with Eq. (10) and $r_n = 0$ for the geological present. Although near the
surface there is no plate-like distribution of the velocity arrows, the subducting
downwellings (blue) are nearly slab-like. They are relatively distinct blue lines in
1350 km depth, yet, notwithstanding that a Newtonian rheology was presumed.

the second-order Runge-Kutta procedure. The mesh was generated by projection of a regular icosahedron onto a sphere. So, the spherical surface was
divided into twenty spherical triangles or ten spherical diamonds. The successive dyadic mesh refinement procedure connects the midpoints of each
side of a spherical triangle with a great circle. So, each triangle is subdivided
into four smaller triangles. The radial distribution of the different spherical-
surface triangular networks is so that the volumes of the cells are nearly
equal. More details are given by Baumgardner (1983), Bunge et al. (1997)
and Yang (1997). We used 1351746 or 10649730 grid points per run. $Ra(t)$,
$Nu(t)$, $Ror(t)$ and other functions are nearly identical for runs with the different grid point numbers. For the most runs we used 128 processors on hwwt3e.
The runs need 1 through 5 hours of run time. Figure 11 shows the scaling
degree of the code for runs with 16, 32, 64 and 128 processors using 1351746
grid points.

5   A modification of the model

If we replace Eq. (4) by

$$\eta(r,\theta,\phi,t) = 10^{r_n} \cdot \eta_1(r) \cdot \exp\left[c_t \cdot T_m \left(\frac{1}{T\,(r,\theta,\phi,t)} - \frac{1}{T_{av}\,(r,t)}\right)\right] \quad (10)$$

then we receive extremely thin slab-like downwellings. They are represented by thin blue lines in Fig. 12. The only difference between Fig. 2 and Fig. 12 is caused by the different temperature dependences (4) and (10). θ denotes the colatitude and ϕ the geographical longitude. In the lower half of the mantle, $c_t = 2$. In the upper half of the mantle, $c_t = 0$.

Acknowledgment

The provision of computer resources at the Rechenzentrum der Universität Stuttgart (HLRS) and at the John von Neumann Institute of Computing Juelich (NIC) is gratefully acknowledged.

References

Anderson, O.L., 1998. The Grüneisen parameter for iron at outer core conditions and the resulting conductive heat and power in the core. Phys. Earth Planet. Int. **109**, 179–197.

Baumgardner, J.R., 1983. A three dimensional finite element model for mantle convection. Thesis, Univ. of California, Los Angeles.

Bercovici, D., Schubert, G., Glatzmaier, G.A., 1992. Three-dimensional convection of an infinite-Prandtl-number compressible fluid in a basally heated spherical shell. J. Fluid Mech. **239**, 683–719.

Bunge, H.-P., Richards, M.A., Baumgardner, J.R., 1997. A sensitivity study of three-dimensional spherical mantle convection at 10^8 Rayleigh number: effects of depth-dependent viscosity, heating mode, and an endothermic phase change. J. Geophys. Res. **102**, 11991–12007.

Christensen, U.R., 1984. Heat transport by variable viscosity convection and implications for the Earth's thermal evolution. Phys. Earth Planet. Int. **35**, 264–282.

Christensen, U.R., 1985. Thermal evolution models for the Earth. J. Geophys. Res. **90**, 2995–3007.

Chopelas, A., Boehler, R., 1992. Thermal expansivity in the lower mantle. Geophys. Res. Lett. **19**, 1983–1986.

De Smet, J., Van den Berg, A.P., Vlaar, N.J., 2000. Early formation and long-term stability of continents resulting from decompression melting in a convecting mantle. Tectonophysics **322**, 19–33.

Ekström, G., Dziewonski, A.M., 1998. The unique anisotropy of the Pacific upper mantle. Nature **394**, 168–172.

Forte, A.M., Mitrovica, J.X., 2001. Deep-mantle high-viscosity flow and thermo-chemical structure inferred from seismic and geodynamic data. Nature **410**, 1049–1056.

Hofmann, A.W., 1988. Chemical differentiation of the Earth: the relationship between mantle, continental crust, and oceanic crust, Earth Planet. Sci. Lett., **90**, 297–314.

McCulloch, M.T., Bennett, V.C., 1994. Progressive growth of the Earth's continental crust and depleted mantle: Geochemical constraints. Geochim. Cosmochim. Acta **58**, 4717–4738.

Ogawa, M., 2000. Coupled magmatism-mantle convection system with variable viscosity. Tectonophysics **322**, 1–18.

Reese, C.C., Solomatov, V.S., Moresi, L.-N., 1999. Non-Newtonian stagnant lid convection and magmatic resurfacing on Venus. Icarus **139**, 67–80.

Safronov, V.S., 1972. Evolution of the Protoplanetary Cloud and Formation of the Earth and Planets. Nauka, Moscow. Translation by the Israel Program for Scientific Translation.

Schmalzl, J., 1996. Mixing properties of thermal convection in the Earth's mantle. Geologica Ultraiectina, no. 140

Schubert, G., Cassen, P., Young, R.E., 1979. Subsolidus convective cooling histories of terrestrial planets. Icarus **38**, 192–211.

Schubert, G., Stevenson, D., Cassen, P., 1980. Whole planet cooling and the radiogenic heat source contents of the Earth and Moon. J. Geophys. Res. **85**, 2511–2518.

Schubert, G., Turcotte, D.L., Olson, P., 2001. Mantle Convection in the Earth and Planets. Cambridge Univ. Press, Cambridge, 940 pp.

Solomatov, V.S., 1995. Scaling of temperature- and stress-dependent viscosity convection. Phys. Fluids **7**, 266–274.

Spohn, T., Schubert, G., 1991. Thermal equilibrium of the Earth following a giant impact. Geophys. J. Int. **107**, 163–170.

Stacey, F.D., 1992. Physics of the Earth, 3rd edn., Brookfield Press, Brisbane, 513 pp.

Stacey, F.D., Stacey, C.H.B., 1999. Gravitational energy of core evolution: Implications for thermal history and geodynamo power. Phys. Earth Planet. Int. **110**, 83–93.

Walzer. U., Hendel, R., 1999. A new convection-fractionation model for the evolution of the principal geochemical reservoirs of the Earth's mantle. Phys. Earth Planet. Int. **112**, 211–256.

Walzer. U., Hendel, R., 2002. Chemical differentiation, viscosity and the thermal evolution of the mantle. Pure Appl. Geophysics, in press.

Walzer. U., Hendel, R., Baumgardner, J., 2002a. A 3-D compressible spherical-shell model of the thermal evolution of the Earth's mantle. Tectonophysics, in press.

Walzer. U., Hendel, R., Baumgardner, J., 2002b. A sensivity study of a 3-D compressible spherical-shell model of mantle convection. Earth Planet. Sci. Lett., in press.

Yang, W.-S., 1997. Variable viscosity thermal convection at infinite Prandtl number in a thick spherical shell. Thesis, Univ. of Illinois, Urbana-Champaign.

Zhang, S., Yuen, D.A., 1996. Various influences on plumes and dynamics in time-dependent, compressible mantle convection in 3-D spherical-shell. Phys. Earth Planet. Int. **94**, 241–267.

Solid State Physics

Prof. Dr. Werner Hanke

Institut für Theoretische Physik und Astrophysik
Universität Würzburg
Am Hubland, D-97074 Würzburg

In the area of solid state physics four contributions have been selected by the reviewing committee for oral presentations.

The first topic concerns the importance of intermediate-range order in silicates and is a molecular dynamics simulation study of J. Horbach, A. Winkler, W. Kob and K. Binder from the University of Mainz. This topic is a prime example for large-scale molecular dynamics computer simulations, which require the use of parallel computers. Only then it is possible to simulate these systems on a scale of several ns for system sizes, which are big enough to study the structure and dynamics of intermediate length scales (of up to 8,000 particles in the present study). Using this technique of highly developed computer simulations insight can be gained, which is not obtainable in an experiment, since in the latter one would have to resolve the full microscopic information in form of particle trajectories. In the present study of the silicates, no neutron scatting experiments can be done at very elevated temperatures above 2,000 K. On the other hand, the structure and dynamical properties are already present at these very high temperatures. These facts have been exploited in the present project showing that the silicate systems exhibit so-called intermediate-range order on length scales that are larger than the one given from the tetrahedron network structure in pure silica (SiO_2). The new intermediate length scales were shown in the project to be important for dynamic properties in silicates.

A rather different, but nevertheless very interesting solid state application has been presented by S. Kirchner, J. Kroha and P. Wölfle (University of Karlsruhe) on the topic "Self-consistent auxiliary particle theory for strongly correlated fermion systems". This topic also deals with the effect of disorder, but now on the fermions, namely in the strongly correlated fermionic system. More precisely, the so-called single impurity Anderson model was studied in terms of various schemes employing up-to-date numerical recipes. These are the non-crossing approximation (NCA), the conserving T-matrix approximation (CTMA) and a variance of the two. The paper of the Karlsruhe group explains convincingly why massive parallel computers are necessary when going beyond the much studied non-crossing approximation. The achievements of the project are of general interest in that the problem of magnetic impurities in metals is in fact of interest in its own right: Magnetic impurities in metals can lead to a minimum of the resistivity at finite temperature, the "celebrated" Kondo-effect. Furthermore, the single impurity Anderson model has been employed to understand the electronic properties of so-called heavy-fermion metals.

The project "Exitonic and local-field effects in optical spectra from real-space time-domain calculations" by W. G. Schmidt, P. H. Hahn and F. Bechstedt, University of Jena, presents a novel approach to solve the so-called Bethe-Salpeter equation for the dynamic response function. The knowledge of this dynamic response function is of great use in solid stated physics. The reason for this is that its poles contain the so-called elementary excitons of the solid, such as excitons, plasmons, electron-hole excitations, etc. Therefore, this response function is intimately connected with experiments. The formulation of the Bethe-Salpeter equation in this project is taken from weak-coupling perturbation theory. Therefore, it is applicable to systems like semiconductors, where the typical ratio of the Coulomb interaction energy of the electrons over the kinetic energy (bandwidth) is very small and can be used as a systematic small parameter in the many-body perturbation theory. The specific application deals with the surface of the semiconductor silicon. In this case, it was shown that, by comparing with surface optical spectra, many-body interactions between the electron excited into the conduction band and the hole left behind in the valence band (i.e. excitons) are largely responsible for the observed line shapes.

The last topic presented as a talk by the Würzburg group (Z. B. Huang, W. Hanke and E. Arrigoni) deals with an open question from the microscopic theory of superconductivity. In any solid state system (even in the high-temperature superconductors, where many indications are that the pairing mechanism is mainly driven by electronic correlations), there is, however, still one contribution, which additionally contributes to the electronic pairing, i. e. via the electron-phonon interaction. This standard electron-phonon mechanism is in fact the "glue" for conventional pairing in the low-temperature superconductors, such as aluminum (Al), lead (Pb), etc. Two new numerical techniques have been developed and applied to the strongly correlated three-band Hubbard model (which allows to correctly simulate both the Cu(d) and O(p_x, p_y)-electronic degrees of freedom of the high-T_c superconductors (HTSL)) to extract the so-called electron-phonon vertex function. The vertex function characterizes the strength of the coupling between electrons and phonons in the presence of strong correlations. This problem has never been accessible to a controlled investigation, because of the formidable task to deal simultaneously with strong correlations of the electrons and their interactions with the phonons. In the current project, using high-performance computing, it is demonstrated that the interplay of electron lattice interactions and strong Coulomb effects can lead to a substantial enhancement of this vertex function in the relevant doping regime for the high-T_c cuprates. This finding may have a substantial impact on the pairing theory in the HTSC, in that the vertex function enters quadratically in the expression for the electron-phonon coupling constant λ.

The following contributions, which are now summarized, were presented as posters:

The exact diagonalization study of spin, orbital and lattice correlations in CMR manganites by A. Weiße, G. Wellein and H. Fehske from the University of Bayreuth deals with a microscopic understanding of a technologically very interesting material: The manganites exhibit the so-called colossal magneto-resistance (CMR), which is used in today's magnetic data recording devices. It is more and more becoming evident that this CMR behavior is intimately related to the complex phase diagram of the manganites. On the other hand, it is also becoming evident that the very complex and fascinating phase diagrams of these materials are due to a subtle interplay of almost all degrees of freedom known in solid state physics, namely magnetic, charge and orbital interactions, temperature and doping influence, etc. The advantage of the present study is that it offers an exact i. e. diagonalization result. The disadvantage, on the other hand, is that it is limited to very small, namely four-site, clusters. There it was demonstrated in the project how the coupling of the phononic degrees of freedom, namely the lattice vibrations, influences the order of orbitals and charges, spins and the correlations between them.

Another topic was the subject by B. Nielaba from the University of Konstanz dealing with the project "nanoPIMC". The project is concerned with an exciting numerical undertaking to gain improved insight into "nanostructures in reduced geometry". The detailed theoretical studies of these nano-systems are still in an initial stage, despite the fact that many experimental results concerning the structural, electronic etc. properties of these systems have recently been observed. The main reason for this difficulty is that systems, which are far away from the thermodynamic limit or the so-called infinite-size limit, are clearly very difficult to attack by analytical techniques, which are either structured to deal with the limit of infinitely many particles or just work for, on the other hand, extremely few particles of the order two to five. Therefore, this field has recently gained enormous impetus by the fast development of high-performance computing, which allows the simulations in reduced geometry of about 10 to 10,000 particles. The project has used mainly the Car-Parrinello type of computational approach, which is based on density functional theory. The present report contains new insights into pore condensates and phas transitions in colloid systems in external potentials and reduced geometry.

Another topic which has been selected for a poster is the project "Thermodynamics and dynamics of correlated electron systems" by C. Lavalle, M. Rigol, M. Feldbacher, F. F. Assaad and A. Muramatsu, University of Stuttgart. In this project, which deals with low-dimensional models of strongly correlated fermions, such as the t-J model, Quantum-Monte-Carlo simulations have been performed in order to gain insight into the thermodynamic behavior. Interesting insights emerged for the one-dimensional case, where it is well-known that the Fermi-liquid picture is no longer valid and the elementary excitations are spinons and holons. The latter quantities are fascinating objects carrying separately the charge and spin of the electron and, amazingly, have been observed in a variety of spectral experiments. Another model, which has been studied in detail, is the Kondo lattice model, which describes e. g. heavy-fermion materials. These materials are characterized by effective masses up to three orders of magnitude larger than that of the bare electron mass. New insights have been obtained for the microscopic reason how the heavy-fermion state is coming about.

A last project, which was presented in a poster is the "Self-trapping of the silicon Si(111)-(2x1) surface exciton" work by M. Rohlfing, University of Münster. This project is somewhat related to the work of the group around F. Bechstedt, University of Jena, described above and deals with the localization of the surface exciton at semiconductor surfaces due to self-trapping. Details of optical experiments, which have been recently observed, are characteristic for this surface exciton, such as an unusual linewidth behavior of the optical response. To study this exciton self-trapping, computational techniques are employed, which are designed for excited states in weakly correlated materials, such as density-functional theory and weak-coupling many-body perturbation theory. This scheme is, technically, combined with an appropriate local-orbital representation of the electrons, namely a tight-binding representation that allows to simulate very large super-cells, usually not encountered in solid state physics. These super-cells contain up to several thousand atoms.

In summary, in the field of supercomputing solid state physics, a variety of very interesting applications have been presented. They range from the more standard treatments of weakly correlated electron systems or band electron systems, such as semiconductors, to strongly correlated materials, such as the high-T_c superconductors and the technologically also very relevant manganites. Beautiful other examples dealing with high-performance techniques, such as the Car-Parinello simulation technique have been followed up to study clusters and surface structures of nano-systems. In these studies and in related studies of silicates the aim is to understand system sizes of typically several thousand atoms. It is becoming more and more evident that these system sizes of intermediate length scales already contain salient fingerprints of both the structures and dynamics of the actual experimental systems in the thermodynamic limit or infinite-size limit.

The Importance of Intermediate Range Order in Silicates: Molecular Dynamics Simulation Studies

Jürgen Horbach[1], Anke Winkler[1], Walter Kob[2], and Kurt Binder[1]

[1] Institut für Physik, Johannes Gutenberg–Universität,
   D–55099 Mainz, Staudinger Weg 7, Germany
[2] Laboratoire des Verres, Université Montpellier II,
   Place E. Bataillon, cc69, 34095 Montpellier, France

Abstract. We present the results of large scale computer simulations in which we investigate the structural and dynamic properties of silicate melts with the compositions $(Na_2O)2(SiO_2)$ and $(Al_2O_3)2(SiO_2)$. In order to treat such systems on a time scale of several nanoseconds and for system sizes of several thousand atoms it is necessary to use parallel supercomputers like the CRAY T3E. We show that the silicates under consideration exhibit additional intermediate range order as compared to silica (SiO_2) where the characteristic intermediate length scales stem from the tetrahedral network structure. For the sodium silicate system it is demonstrated that the latter structural features are intimately connected with a surprising dynamics in which the one–particle motion of the sodium ions appears on a much smaller time scale than the correlations between different sodium ions.

1 Introduction

Silicate melts and glasses are an important class of materials in very different fields, e.g. in geosciences (since silicates are geologically the most relevant materials) and in technology (windows, containers, optical fibers etc.). From a physical point of view it is a very challenging task to understand the properties of those materials on a microscopic level, and in the last twenty years many studies on different systems have shown that molecular dynamics computer simulations are a very appropriate tool for this purpose [1–4]. The main advantage of such simulations is that they give access to the whole microscopic information in form of the particle trajectories which of course cannot be determined in real experiments.

In *pure* silica (SiO_2) the structure is that of a disordered tetrahedral network in which SiO_4 tetrahedra are connected via the oxygens at the corners. In recent simulations we have studied in detail various aspects of static and dynamic properties of silica such as structural and thermodynamic properties of the glass state [5,6], the diffusion dynamics and structural relaxation [7–10], the frequency dependent specific heat [11], the vibrational degrees of freedom [12] and free surfaces [13–15]. In this paper we consider silicates that contain additional oxide components. Especially silicates with alkali oxides have been investigated very recently in several molecular dynamics simulations [16–22]. We investigate here the systems $(Na_2O)2(SiO_2)$ and $(Al_2O_3)2(SiO_2)$ (for which we use in the following the abbreviations NS2 and AS2, respectively). Whereas sodium in NS2 plays the role of a network modifier that partially destroys the SiO_4 network, aluminium in AS2 is also a network former in that it prefers a four–fold coordination by oxygen atoms. However, the packing of the AlO_4 tetrahedra is different from that of the SiO_4 tetrahedra which is indicated for instance by a different coordination number distribution of aluminium and silicon by oxygen atoms (mainly two–fold for silicon and two– and three–fold for aluminium) [23]. As we show in the following the insertion of sodium or aluminium atoms does not only modify the structure on local length scales but it introduces also new intermediate length scales that can be identified by means of the partial static structure factors. These length scales are important for the dynamic properties as we will demonstrate for the case of NS2.

The rest of the paper is organized as follows: In the next section we give the main computational details and discuss the efficiency of our simulation code on the T3E at the HLRZ Stuttgart. Then we present the structural properties of AS2 and NS2 on intermediate length scales (Sec. 3) and the dynamics of NS2 (Sec. 4). Eventually we summarize our results (Sec. 5).

2   Model and details of the simulations

In a classical molecular dynamics (MD) computer simulation one solves numerically Newton's equations of motion for a many particle system. If quantum mechanical effects can be neglected such simulations are able to give in principle a realistic description of any molecular system. The determining factor of how well the properties of a real material are reproduced by a MD simulation is given by the potential with which the interaction between the atoms is described. The model potential we use to compute the interaction between the ions in NS2 and AS2 is the one proposed by Kramer *et al.* [24] which is a generalization of the so–called BKS potential [25] for pure silica. It has the following functional form:

$$\phi_{\alpha\beta}(r) = \frac{q_\alpha q_\beta e^2}{r} + A_{\alpha\beta}\exp\left(-B_{\alpha\beta}r\right) - \frac{C_{\alpha\beta}}{r^6} \quad \alpha,\beta \in [\text{Si}, \text{Al}, \text{Na}, \text{O}]. \quad (1)$$

Here r is the distance between an ion of type α and an ion of type β. The values of the parameters $A_{\alpha\beta}, B_{\alpha\beta}$ and $C_{\alpha\beta}$ can be found in the original publication. The potential (1) has been optimized by Kramer *et al.* for zeolites, i.e. for systems that consist of Si, Al, Na, O, and possible other components like phosphor. In that paper the authors used for silicon, aluminium, and oxygen the *partial* charges $q_{Si} = 2.4$, $q_{Al} = 1.9$, and $q_O = -1.2$, respectively, whereas sodium was assigned its real ion charge $q_{Na} = 1.0$. Thus, with this set of charges charge neutrality is fulfilled neither in NS2 nor in AS2. We

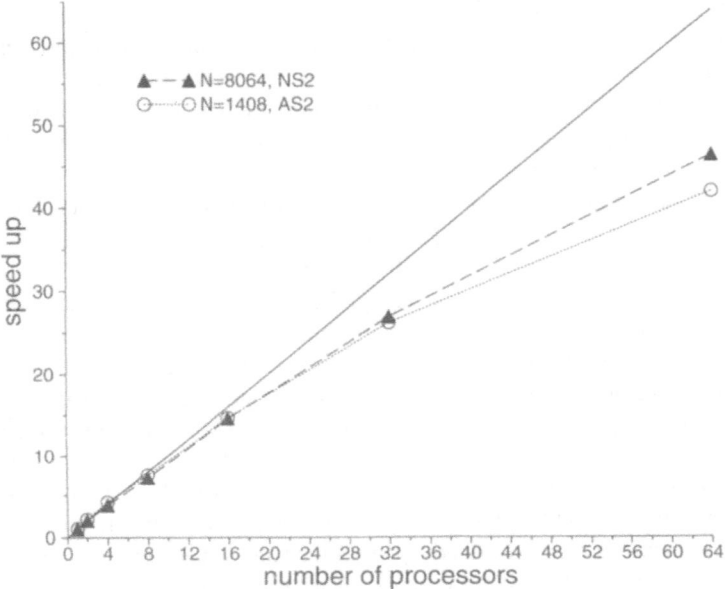

Fig. 1. Speed up factor for the simulations with $N = 8064$ (filled triangles) and $N = 1408$ particles (open circles) as a function of the number of processors n. The bisecting line (straight line) indicates a perfect scaling of the performance with n.

have therefore modified the Kramer potential by setting the partial charge for sodium and aluminium to 0.6 and 1.8, respectively, and by introducing additional short range potentials such that the original functional form of the Kramer potential is approximately recovered on distances of nearest Al–O and Na–O neighbors. More details on the interaction potential can be found in Refs. [16,17,23]. Our models give predictions for structural and dynamic properties of NS2 and AS2 which are in good agreement with experimental findings [17,23]. Furthermore, Ispas *et al.* [26] have shown for $(Na_2O)4(SiO_2)$ that *ab initio* simulations (Car–Parrinello molecular dynamics) yield comparable results regarding the structure to those obtained with molecular dynamics simulations in which our potential model is used.

The simulations have been done at constant volume: For AS2 we fixed the density to $2.6\,g/cm^3$ which is close to the experimental density at $T = 300$ K. In the case of NS2 we did simulations at the two densities $2.37\,g/cm^3$ and $2.5\,g/cm^3$, corresponding to experimental densities in the melt and at room temperatures, respectively. The AS2 system consists of 1480 particles and for the NS2 systems we used system sizes of 8064 particles at the low density and 1152 particles at the high density.

As can be seen from Eq. 1 the interaction potential contains a long–ranged Coulomb term. This part of the interaction is the most time consuming in the calculation of the forces. To do this we made use of the so–called Ewald summation technique [27], a method that scales with the particle number N as $N^{3/2}$. Thus, for systems which contain about 8000 particles a huge numerical effort is required: The longest runs (at the lowest temperatures) had a length of about 10 million time steps for which a time of two weeks was needed on 64 processors thus giving a total CPU time of about 128 weeks of (single) processor time.

The equations of motion were integrated with the velocity form of the Verlet algorithm. The time step of the integration was 1.6 fs. The temperature range investigated was 4000 K$\geq T \geq$ 2100 K in the case of NS2 and 6100 K$\geq T \geq$ 2300 K in the case of AS2. To equilibrate the systems the temperatures were controlled by coupling them to a stochastic heat bath, i.e. by substituting periodically the velocities of the particles with the ones from a Maxwell-Boltzmann distribution with the correct temperature. After the system was equilibrated at the target temperature, we continued the run in the microcanonical ensemble, i.e. the heat bath was switched off. We have done production runs up to several ns real time which corresponds to several million time steps. We have also calculated glass structures at $T = 300$ K. The glass state was produced by cooling the system from equilibrated configurations at our lowest temperatures with a cooling rate of about 10^{12} K/s. Note that we show in the following sections only results for the lowest temperatures, i.e. at $T = 2100$ K for NS2 and at $T = 2300$ K for AS2 as well as at $T = 300$ K for both systems, because the results for the higher temperatures lead essentially to the same conclusions that we will draw below. However, a detailed discussion of the temperature dependence of the systems under consideration can be found in Refs. [17].

The program code was written in FORTRAN. All the parallelization was done by using MPI subroutines. More details on the parallelization can be found in Refs. [9,23]. Of course, the performance of a parallel code never scales perfectly with the number of processors because the communication between the processors requires an additional amount of CPU time. Fig. 1 shows the speed up factor on the Cray T3E of the HLRZ Stuttgart as a function of the number of processors n, i.e., the factor by which the code is faster if one uses n processors instead of one. The bisecting line indicates the limiting case where the communication overhead is not influenced by the speed of the code. We see that the curves for $N = 1408$ and $N = 8064$ scale nearly perfectly for $n \leq 16$. For $n = 64$ we obtain still a speed up factor of about 46.4 for $N = 8064$ particles whereas this factor is 42 for $N = 1408$. In most of our simulations we have used 64 processors for the large systems and 32 processors for the small ones.

3   Intermediate length scales in silicates

An appropriate quantity to investigate the structure of atomic systems on intermediate length scales is the static structure factor. It is essentially the

Fourier transform of the pair correlation function which gives the probability of finding an atom at a distance r from another atom [2]. The structure factor can be directly measured in neutron scattering experiments from the intensity of the radiation observed with a momentum transfer $\hbar\mathbf{q}$ ($\hbar$: Planck's constant, $\mathbf{q}$: wave–vector of the momentum transfer). In a three–component system one can define six partial structure factors as [2]

$$S^{\alpha\beta}(q) = \frac{1}{N} \sum_{k=1}^{N_\alpha} \sum_{l=1}^{N_\beta} \langle \exp\left(i\mathbf{q} \cdot [\mathbf{r}_k - \mathbf{r}_l]\right) \rangle. \tag{2}$$

where the first sum runs over N_α particles of type α and the second one over N_β particles of type β.

Figure 2 shows $S_{\alpha\beta}(q)$ for AS2 at the temperature $T = 2300$ K. For $q > 2.3$ Å^{-1} the partial structure factors reflect length scales of nearest neighbors. In AS2 the smallest distances between atoms are those of Al–O and Si–O bonds that have lengths of about 1.6 to 1.65 Å. The peaks around $q_2 = 1.7$ Å^{-1} in $S_{\alpha\beta}(q)$ (marked by dashed vertical lines in Fig. 2) are due to the order that arises from the tetrahedral network structure. The length

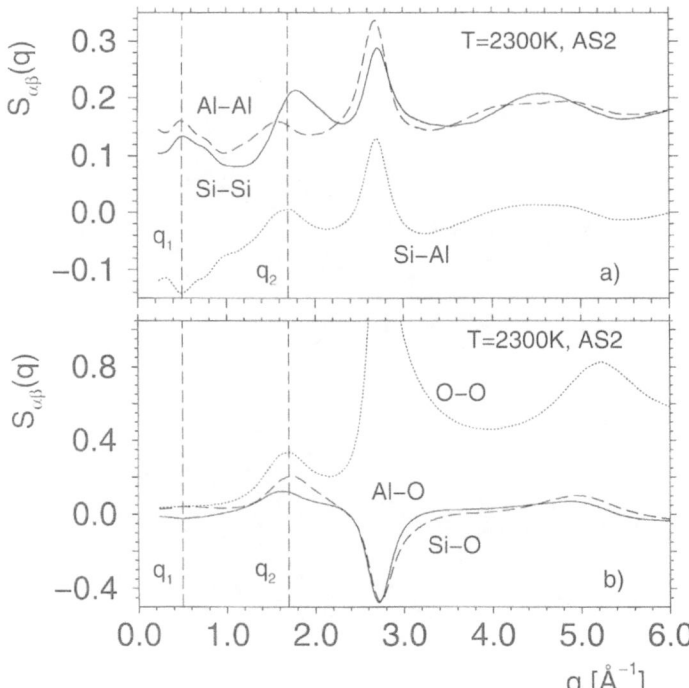

Fig. 2. Partial static structure factors for AS2 at $T = 2300$ K. a) Al–Al, Si–Si, and Si–Al correlations, b) Si–O, Al–O, and O–O correlations. For the meaning of the dashed vertical lines see text.

scale $2\pi/1.7$ Å^{-1} = 3.7 Å that corresponds to this peak is approximately the spatial extent of connected SiO$_4$ and AlO$_4$ tetrahedra. Note that in silica a peak at 1.7 Å^{-1} is also a very prominent feature and is called there first sharp diffraction peak. But in contrast to silica one observes in AS2 an additional peak at $q_1 = 0.5$ Å^{-1} in the Al–Al, Si–Si, and Si–Al correlations and only weakly pronounced also in the remaining correlations in which oxygen is involved. q_1 corresponds to a length scale of about 12.5 Å and has its reason in a slightly different ordering of AlO$_4$ complexes as compared to the SiO$_4$ network (for details see [23]). This relatively large length scale shows that large system sizes are required to analyze the structure of systems like AS2 in a sensible way. The different ordering of AlO$_4$ leads to a structure where an AlO$_4$ tetrahedron prefers to be surrounded on a local scale by other AlO$_4$ tetrahedra. This leads to a structure where AlO$_4$ complexes are connected to each other as string–like objects through the system that form a percolating network. This is illustrated by the snapshhot in Fig. 3 where the aluminium and silicon atoms are shown as the blue and gold spheres, respectively. Note that it does not matter that this snapshot is at $T = 300$ K and not at $T = 2300$ K as the structure factors in Fig. 2 because we find only small differences in structural quantities at both temperatures. Thus, we see

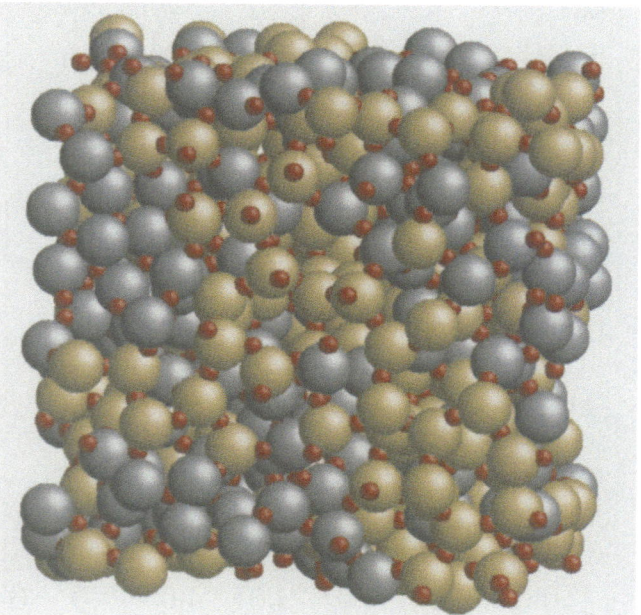

Fig. 3. Snapshot of (Al$_2$O$_3$)2(SiO$_2$) (AS2) at $T = 300$ K. The size of the spheres is chosen such that one can identify aluminium– and silicon–rich regions: The aluminium and oxygen atoms are shown respectively as big blue and gold spheres, whereas the oxygen atoms are shown as small red spheres.

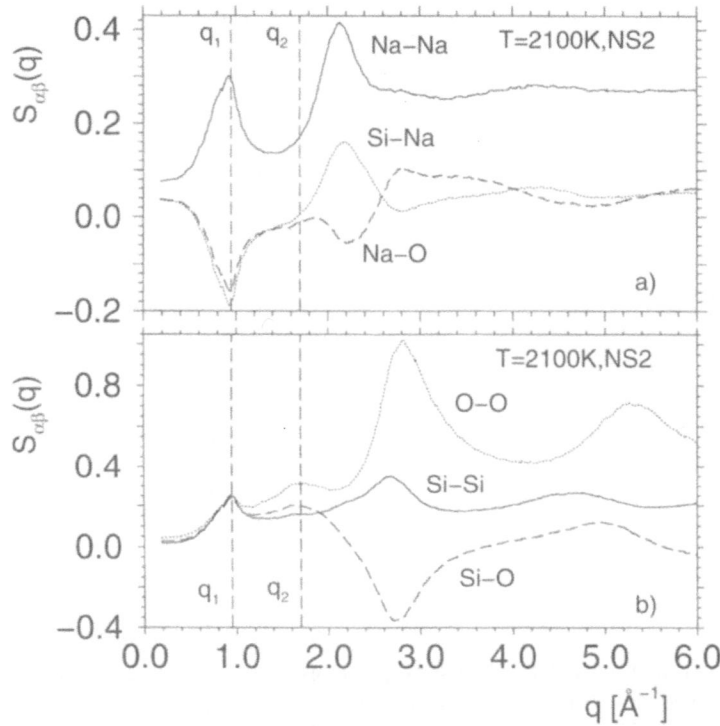

Fig. 4. Partial static structure factors for NS2 at $T = 2100$ K. a) Na–Na, Si–Na, and Na–O correlations, b) Si–O, Si–Si, and O–O correlations. For the meaning of the dashed vertical lines see text.

that the aluminium atoms are not at all homogeneously distributed and if one only considers the Al atoms voids with a size of about $2\pi/q_1$ are found that lead to the peak at q_1 in S_{Al-Al}. It is not surprising that these voids are also reflected in the Si–Si and Si–Al correlations but much less in the correlations containing oxygen (as can be seen in Fig. 2b): The oxygen atoms are essentially homogeneously distributed on the length scale $2\pi/q_1$ since they are nearest neighbors of silicon and aluminium with a similar length of Si–O and Al–O bonds.

The sodium ions in NS2 play a different role from the aluminium atoms in AS2 since they partially destroy the SiO_4 network. This can be directly recognized in the partial structure factors for NS2 which are shown in Fig. 4 at $T = 2100$ K for the density $\rho = 2.37$ g/cm³: The peak at $q_2 = 1.7$ Å^{-1} that reflects the structure of a tetrahedral network is absent in the correlations with sodium (Fig. 4a) and is especially in S_{Si-Si} much weaker pronounced than in AS2 (Fig. 4b). But we observe again a second prepeak at smaller q, now around $q_1 = 0.95$ Å^{-1}. This q value is of the order of the length scale of next nearest Na–Na or Si–Na neighbors (around 6.6 Å). Again, the peak at q_1

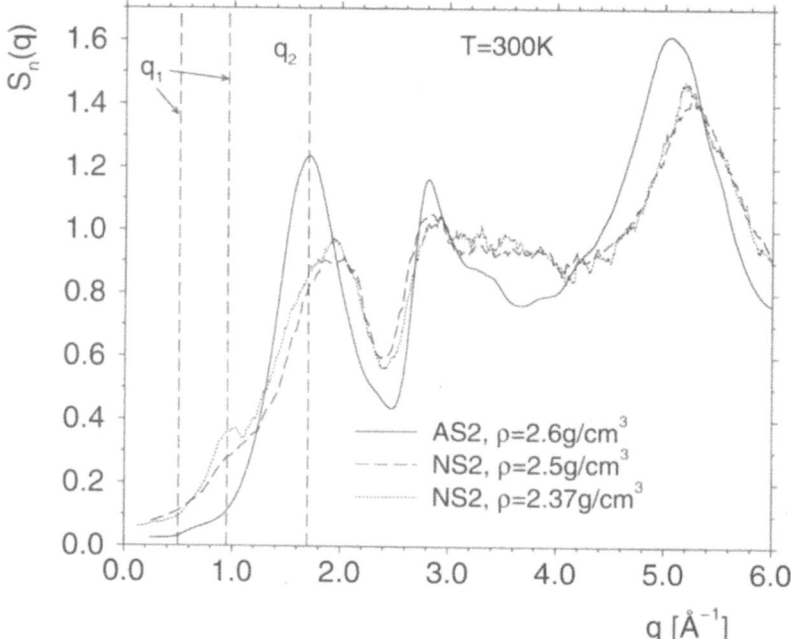

Fig. 5. $S_n(q)$ at $T = 300$ K for AS2 and for NS2 at the indicated densities. The dashed vertical lines mark the position of the peaks at $q_1 = 0.5$ Å^{-1} in AS2, $q_1 = 0.95$ Å^{-1} in NS2, and $q_2 = 1.7$ Å^{-1} in both systems.

is the characteristic wave–vector of a percolating network that is now formed by the sodium atoms. At first glance it seems to be surprising that also S_{O-O} exhibits a peak at q_1. But the role of oxygens is different in NS2 from that in AS2: The nearest neighbor distance for Na–O, 2.2 Å, is larger than for Si–O which is 1.6 Å. And the arrangement of oxygen around sodium is different from the tetrahedral one around silicon (for more details see Ref. [17]). Thus, the distribution of oxygen atoms in NS2 is not homogeneous on the length scale $2\pi/q_1$.

So far we have seen that NS2 and AS2 exhibit intermediate order on a relatively large length scales. This gives rise to a prepeak in $S_{\alpha\beta}(q)$ at q_1 which is 0.5 Å^{-1} for AS2 and 0.95 Å^{-1} for NS2. But does one see these peaks at q_1 also in experiments? In experiments such as neutron scattering one does not have access to the partial structure factors for systems like NS2 or AS2. Here one measures a sum of the partial structure factors whereby the different contributions are weighted by the neutron scattering lengths b_α:

$$S_n(q) = \frac{1}{\sum_\alpha N_\alpha b_\alpha^2} \sum_{kl} b_k b_l \langle \exp\left(i\mathbf{q} \cdot [\mathbf{r}_k - \mathbf{r}_l]\right)\rangle. \tag{3}$$

The values for b_α are $0.4149 \cdot 10^{-12}$ cm, $0.3449 \cdot 10^{-12}$ cm, $0.363 \cdot 10^{-12}$ cm, and $0.5803 \cdot 10^{-12}$ cm for silicon, aluminium, sodium, and oxygen, respectively [28]. By weighting the $S_{\alpha\beta}(q)$ from our simulation with the b_α in accordance with Eq. (3) one can easily calculate the quantity $S_n(q)$. It is shown in Fig. 5 at $T = 300$ K for NS2 at the two densities $\rho = 2.37$ g/cm^3 and 2.5 g/cm^3 and for AS2 at $\rho = 2.6$ g/cm^3. We infer from this figure that the aforementioned prepeaks at q_1 can be seen in AS2 and in NS2 at the higher density only as a weakly pronounced shoulder. Thus it would be difficult to identify them in a neutron scattering experiment. Only in NS2 at the lower density one can clearly see the prepeak at $q = 0.95$ Å^{-1}. But at this density we observe a negative pressure of about -1.6 GPa at $T = 300$ K, a condition that would be difficult to realise in an experiment. However, in an experiment under normal pressure conditions the density decreases if one goes to higher temperatures. And indeed, very recent neutron scattering experiments of Meyer et al. do find the feature at q_1 for NS2 [29]. Meyer et al. have measured for the first time the temperature dependence of the structure factor from $T = 300$ K (where the system is in a glass state) to $T = 1600$ K (where one has a melt). They find that the feature at q_1 becomes more and more pronounced by increasing the temperature and one can clearly identify it at $T = 1600$ K. This behavior is similar to what we see in our simulations and can be understood by an decreasing density in the experiment if the temperature is increased.

4   The dynamics of NS2

In a recent simulation we have demonstrated that the dynamics in in NS2 is much faster than the one in pure silica [17]. Even at a relatively high temperature of $T = 2750$ K the diffusion constants of silicon and oxygen are two orders of magnitude larger in NS2 than in SiO$_2$. Furthermore, in NS2 the sodium diffusion decouples more and more from the silicon and oxygen diffusion such that at temperatures $T \leq 2500$ K the dynamics of the Na atoms is about two orders of magnitude faster than the one of the oxygen and silicon atoms [17]. This is in qualitative agreement with the expermental fact that NS2 is an ion conducting material.

Thus, since essentially the Si and O atoms do not move with respect to the movement of the Na atoms one may expect that sodium diffusion is restricted to a small subspace of the configuration space. The Si and O atoms form a quasi–frozen matrix for the Na atoms and it would be surprizing if the sodium atoms are able to diffuse into this matrix. In order to check this idea we have calculated a (coarse grained) probability of finding no sodium atom at a given location in space. Following the approach of Jund et al. [19] we calculate this probability by dividing the system into 48^3 cubes (of length $L/48 \approx 1.01$ Å). Then we calculate the probability $P(t)$ that a cube which does not contain a sodium ion at time zero is also not visited by a sodium ion until a later time t. The time dependence of $P(t)$ is shown in the inset of

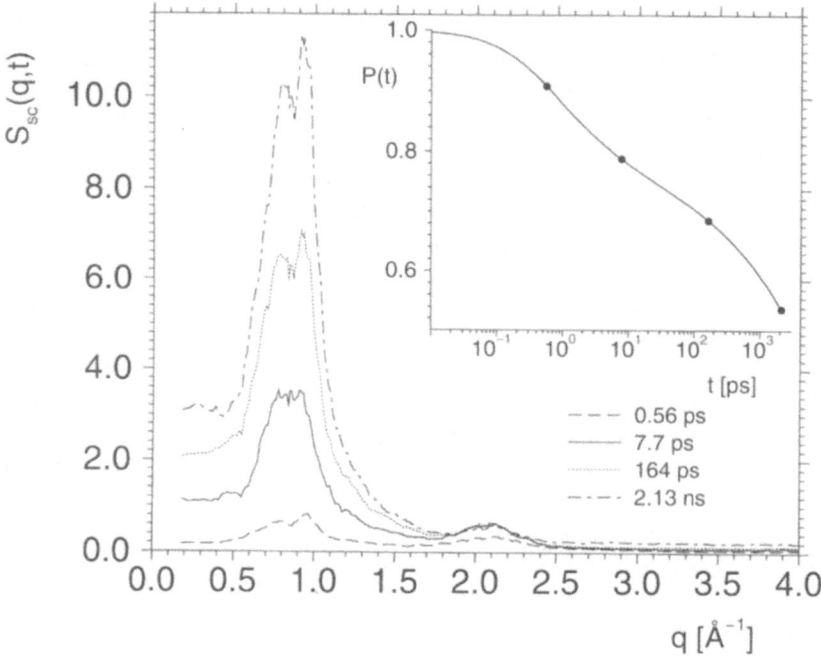

Fig. 6. Swiss cheese structure factor $S_{sc}(q,t)$ at the indicated times. The inset shows the probability $P(t)$ (see text). The circles on the curve for $P(t)$ are at the times at which $S_{sc}(q,t)$ is shown.

Fig. 6. From this graph we recognize that after 2.5 ns, i.e. after more than the α-relaxation time of the matrix [17], more than 50% of the cubes have not yet been visited by a sodium atom. (We mention that after this time the mean squared displacement of the Na atoms is more than $(45 \text{ Å})^2$, which shows that these atoms have moved a large distance. On this time scale also the *local* structure of the Si–O matrix is partially reconstructed [17].) Hence we can conclude that on this time scale the sodium free region forms a percolating structure around a network of channels, i.e. it has somewhat the structure of a Swiss cheese. In order to investigate the structure of this percolating region we define a "Swiss cheese" structure factor $S_{sc}(q,t)$ as follows: We assign to each cube which has not been visited by a sodium atom until time t a point and we compute the static structure factor of $N_{sc}(t) = P(t)(48^3 - N_{Na})$ points:

$$S_{sc}(q,t) = \frac{1}{N_{sc}(t)} \sum_{k,l=1}^{N_{sc}(t)} \langle \exp(i\boldsymbol{q} \cdot (\boldsymbol{r}_k - \boldsymbol{r}_l)) \rangle \ . \tag{4}$$

This quantity is shown in Fig. 6 for four different times: $t = 0.56$ ps, 7.7 ps, 164 ps, and 2.13 ns. We see that $S_{sc}(q,t)$ has a pronounced peak at $q_1 = 0.95 \text{ Å}^{-1}$ which is also a prominent feature in $S_{Na-Na}(q)$, as we have seen

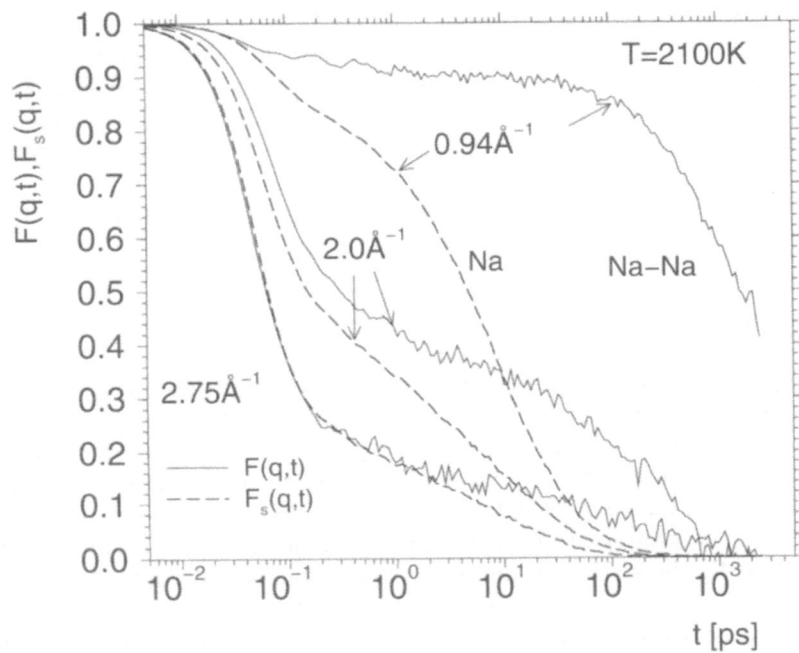

Fig. 7. Coherent intermediate scattering functions $F(q, t)$ for sodium–sodium correlations (bold solid lines) and incoherent intermediate scattering functions $F_s(q, t)$ (dashed lines) at $T = 2100$ K for the indicated values of q.

in the preceding section. Hence we can now conclude that the peak at q_1 in $S_{\text{Na-Na}}(q)$ corresponds to the typical distance between the channels. Note that with increasing time the height of this peak increases quickly. However, it is clear that the peak at q_1 decreases again on the time scale on which the matrix starts to reconstruct itself significantly and thus rearranges the channel structure.

We address now the question how the sodium ions relax inside the channels. An appropriate quantity to investigate this point are time dependent density–density correlation functions, i.e. the coherent intermediate scattering function $F(q, t)$ and its self part, the incoherent intermediate scattering function $F_s(q, t)$ [2]. In Fig. 7 we show $F(q, t)$ for the Na–Na correlations (solid lines) as well as $F_s(q, t)$ for the sodium atoms (dashed lines) for three different wave–vectors: $q = 0.94$ Å^{-1}, 2.0 Å^{-1}, and 2.75 Å^{-1}. From this figure we infer immediately a surprising result: At $q = 0.94$ Å^{-1}, i.e. at the characteristic q value of the sodium channel structure, $F(q, t)$ decays on a time scale which is two orders of magnitude larger than the one for $F_s(q, t)$. Such a strong difference cannot be explained by a de Gennes narrowing argument [2]. Instead this result can be rationalized by the idea that the sodium atoms move quickly between preferential sites, since this type of motion gives rise to a fast decorrelation of the incoherent function whereas it does not affect the coherent one. Only on the time scale of the relaxation of the SiO_2 matrix also the coherent function starts to decay. Note that the slow decay of $F(q,t)$ is found only for wave–vectors around 0.95 Å^{-1}. For different q the function decays significantly faster as can be seen from the other curves shown in Fig. 7.

More details on the issues discussed in this section can be found in Ref. [18].

5   Summary

Large scale molecular dynamics computer simulations as the ones presented in this paper for AS2 and NS2 require the use of parallel computers such as the Cray T3E. Only then it is possible to simulate these systems on a scale of several ns for system sizes which are big enough to study the structure and dynamics on intermediate length scales (up to 8000 particles in our case). Although no neutron scattering experiments can be done yet for temperatures above 2000 K, the structural and dynamical properties are already present at these high temperatures and thus, one can gain insight into features that one observes in experiments. Furthermore, this insight is much more detailed in a MD simulation than in an experiment since one has access to the full microscopic information in form of the particle trajectories.

We have exploited this fact for the case of AS2 and NS2 by showing that these systems exhibit intermediate range order on length scales that are larger than the one given from the tetrahedral network structure in *pure* silica. The reason for this is a different ordering of Al–O and Na–O complexes and leads to a percolating network of these structural elements through the SiO_4 network. We have shown for the example of NS2 that this intermediate range order is also important to understand the dynamics: In NS2 the sodium ions that move through channels in the Si–O matrix and the structure of these channels is connected with the prepeak in the static structure factor at 0.95 Å^{-1}. The presence of these channels leads to a surprising decoupling of the fast (single particle) sodium motion from correlations between different sodium atoms that decay on the time scale of the channel relaxation.

Acknowledgments

We thank the HLRZ Stuttgart for a generous grant of computer time on the CRAY T3E. A. W. is grateful to SCHOTT Glas for partial financial support.

References

1. C. A. Angell, J. H. R. Clarke, and L. V. Woodcock, Adv. Chem. Phys. **48**, 397 (1981).

2. U. Balucani and M. Zoppi, *Dynamics of the Liquid State* (Clarendon Press, Oxford, 1994).

3. W. Kob, J. Phys.: Condens. Matter **11**, R85 (1999).

4. P. H. Poole, P. F. McMillan, and G. H. Wolf Reviews in Mineralogy **32**, 563 (1995).

5. K. Vollmayr, W. Kob, K. Binder, Phys. Rev. B **54**, 15808 (1996).

6. J. Horbach, W. Kob, and K. Binder, J. Phys. Chem. B **103**, 4104 (1999).

7. J. Horbach and W. Kob, Phys. Rev. B **60**, 3169 (1999).

8. J. Horbach and W. Kob, Phys. Rev. E **64**, 041503 (2001).

9. J. Horbach, W. Kob, and K. Binder, p. 186 in *High Performance Computing in Science and Engineering '98*, Eds.: E. Krause and W. Jäger (Springer, Berlin, 1999).

10. K. Binder, J. Non–Cryst. Sol. **274**, 332 (2000).

11. P. Scheidler, W. Kob, A. Latz, J. Horbach, and K. Binder, Phys. Rev. B **63**, 104204 (2001).

12. J. Horbach, W. Kob, and K. Binder, Eur. Phys. J. B **19**, 531-543 (2001).

13. A. Roder, W. Kob, K. Binder, J. Chem. Phys. **114**, 7602 (2001).

14. C. Mischler, W. Kob, and K. Binder, Comp. Phys. Comm. (in press).

15. J. Horbach, C. Mischler, W. Kob, and K. Binder, *Multiscale Computer Simulations in Physics, Chemistry, and Biology: The Example of Silica*, Invited paper at the NATO ARW, Kiev, Ukraine, September 9–12, 2001, "Frontiers in Molecular–Scale Science and Technology of Nanocarbon, NanoSilicon and Biopolymer Multifunctional Nanosystems" (E. Buzaneva, P. Scharff, eds.), Kluwer Academic Press, Dordrecht, 2002, p. 1–15.

16. J. Horbach and W. Kob, Phil. Mag. B **79**, 1981 (1999)

17. J. Horbach, W. Kob, and K. Binder, Chem. Geol. **174**, 87 (2001).

18. J. Horbach, W. Kob, and K. Binder, Phys. Rev. Lett. **88**, 125502 (2002).

19. P. Jund, W. Kob, and R. Jullien, Phys. Rev. B **64**, 134303 (2001).

20. R. D. Banhatti and A. Heuer, Phys. Chem. Chem. Phys. **3**, 5104 (2001).

21. N. Zotov, I. Ebbsjo, D. Timpel, and H. Keppler, Phys. Rev. B **60**, 6383 (1999).

22. J. Oviedo and J. F. Sanz, Phys. Rev. B **58**, 9047 (1998).

23. A. Winkler, Ph. D. Thesis (Mainz University, 2002).

24. G. J. Kramer, A. J. M. de Man, and R. A. van Santen, J. Am. Chem. Soc. **64**, 6435 (1991).

25. B. W. van Beest, G. J. Kramer, and R. A. van Santen, Phys. Rev. Lett. **64** 1955 (1990).

26. S. Ispas, M. Benoit, P. Jund, and R. Jullien, Phys. Rev. B **64**, 214206 (2001).

27. D. Frenkel and B. Smit, *Understanding Molecular Simulation — From Algorithms to Applications* (Academic Press, San Diego, 1996).

28. V. F. Sears, J. Mater. Res. **3**, 29 (1992).

29. A. Meyer, H. Schober, and D. B. Dingwell, Europhys. Lett. (in press).

Selfconsistent Auxiliary Particle Theory for Strongly Correlated Fermion Systems

S. Kirchner, J. Kroha, and P. Wölfle

Institut für Theorie der Kondensierten Materie, Universität Karlsruhe,
76128 Karlsruhe, Germany

Abstract. The single impurity Anderson model (SIAM) in the localized moment regime is used as the generic model to analyze strong correlations in metals. As it turns out the auxiliary particle representation is a convenient way for formulating selfconsistent approximations for SIAMs. We discuss the Non-Crossing Approximation, or NCA, one of the simplest approximations possible and point out why massive parallel machines are necessary when going beyond NCA. The CTMA, an approximation capable of describing the correct groundstate and the SUNCA for impurity systems with finite Coulomb repulsion are discussed and numerical results are presented.

1 Introduction

Over the past two decades, the problem of correlated electrons on a lattice has emerged as a central theme of condensed matter theory. With the exception of one-dimensional systems, there are no systematic analytical methods available for solving models like the Hubbard model. A powerful approximation scheme is the Dynamical Mean Field Theory (DMFT), in which the lattice problem is mapped onto an effective single-impurity Anderson model (SIAM), with self-consistently determined properties of the conduction-electrons [1]. The single-impurity Anderson model describes an impurity level with Coulomb interaction among the electrons on the impurity site hybridizing with a band of free electrons. It is a nontrivial task to solve these quantum impurity models for intermediate to strong Coulomb repulsion. However, several theoretical approaches have been employed to solve these models in certain parameter regimes. The Bethe ansatz method yields an exact solution of models with a flat conduction density of states per spin $N(E)$, and allows to calculate the thermodynamic properties [2]. Bosonization methods and conformal field theory have also been successfully employed. These analytical methods are complemented by numerical methods like Quantum Monte Carlo simulations (for not too low temperatures and moderate Coulomb repulsion) and Wilson's Numerical Renormalization Group which has been very successful for not too large spin degeneracies. Quantum impurity systems are complex enough to incorporate strong correlations among the electrons, but are still at the verge of solvability and therefore have served as a playground for the development of powerful methods to tackle strongly correlated electron problems. The problem of magnetic impurities in metals is nonetheless of interest in its own right: Magnetic impurities in metals can lead to a minimum of the resistivity at finite temperature, termed the "Kondo effect" after J. Kondo [3] who first explained this phenomenon. Ever since the discovery of the first heavy-fermion compound (CeAl$_3$) in 1975 the single-impurity Anderson model has been employed to understand the Fermi-liquid properties of heavy-fermion materials [4].

2   The Single Impurity Anderson Model

In its simplest form the SIAM consists of one localized level with four different states, the *empty, spin-up, spin-down,* and *doubly occupied* state interacting via a hybridization term with a band of conduction-electrons. Its Hamiltonian is

$$H = \sum_{k,\sigma} \epsilon_k c_{k,\sigma}^\dagger c_{k,\sigma} + \sum_\sigma \epsilon_d d_\sigma^\dagger d_\sigma + U\, n_{d,\uparrow} n_{d,\downarrow}$$
$$+ \sum_{k\sigma}(V_k\, d_\sigma^\dagger c_{k\sigma} + V_k^*\, c_{k,\sigma}^\dagger d_\sigma), \tag{1}$$

where $n_{d,\sigma}$ is the number operator on the local level (the 'd-level') with spin projection σ. U is proportional to the Coulomb repulsion between the electrons on the d-level, and $c_{k,\sigma}^\dagger$ creates a conduction-electron of wave number k and spin projection σ. In the absence of a magnetic field, the local spin-up and spin-down state are degenerate. The Coulomb repulsion U among electrons on the impurity induces correlations between the conduction electrons. In many cases U is comparable to or larger than the half bandwidth of the conduction band and perturbation theory in U will not be valid. The Hamiltonian in equation (1) possesses a local SU(2) symmetry. Setting $U = \infty$ the only parameter of the SIAM of equation (1) is given by ϵ_d/Γ, where we have introduced $\Gamma = \pi N(0)V^2$. Due to hybridization processes with conduction states the local level is broadened and acquires a finite lifetime $\tau = 1/\Gamma$. The infinite-U Anderson model possesses three different physical regimes. These are the empty-orbital regime, the mixed valence regime and the localized moment regime. In the empty-orbital regime, where the local level is well above the Fermi energy, we have $\epsilon_d \gg \Gamma$. If the local level has considerable overlap with the Fermi edge, $|\epsilon_d| \approx \Gamma$, charge fluctuations taking place on the scale $|\epsilon_d|$ are of the same order as the inverse lifetime Γ. Thus, in the mixed valence regime both spin and charge fluctuations on the local level determine the physical behavior. Lowering the ratio $|\epsilon_d/\Gamma|$ further with the local level below the Fermi energy such that $-\frac{\epsilon_d}{\Gamma} \gg 1$, the localized moment or Kondo regime is reached, where charge fluctuations are suppressed and the expectation value of the local number operator approaches one. In this regime the SIAM is equivalent to the Kondo model at small frequencies and low temperatures [5]. The fact that the $V = 0$ ground state is spin-degenerate leads in this regime to a many-body or strongly correlated ground state of the system with finite V.

The dominant processes and the local density of states in the Kondo regime are shown in Fig. 1. To study the low-energy properties of the Kondo model one can apply renormalization group ideas and study how the exchange coupling J between the local spin and the conduction electron spin changes upon a change of the band cutoff D. This is known as poor man's scaling [6]. One obtains

$$\frac{dJ}{d\ln D} = -2N(0)J^2,$$

where $N(0)$ is the density of states at the Fermi energy. This shows that there exists a scale, the Kondo scale or Kondo temperature T_K, where the exchange coupling becomes of the order of one and simple perturbation theory must fail. This break-down is associated with the appearance of infrared divergences, that is terms $\sim \ln(T/D)$ on the level of bare perturbation theory. This scale is given by

$$T_K = D\, e^{-1/(2N(0)\,J)}. \tag{2}$$

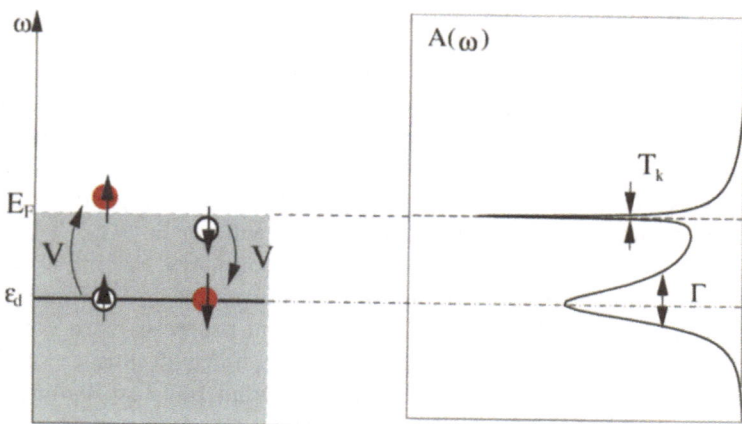

Fig. 1. Left side: Virtual hybridization processes (hybridization strength V) between a localized level at ϵ_d, well below the Fermi energy ϵ_F and excited electron-hole pairs. The fact that the $V = 0$ ground state is spin-degenerate leads to a many-body ground state of the system with finite V.

Right side: The local density of states of the impurity has a broad high energy feature at an energy of about ϵ_d with a width of $N\Gamma$ (in the Kondo regime) with $\Gamma = \pi N(0)V^2$ where $N(0)$ is the density of states per spin at the Fermi energy and a very sharp resonance at (or slightly above) the Fermi energy with a characteristic width $T_k \propto \exp(-\pi|\epsilon_d|/2\Gamma)$, the Kondo temperature. The magnitude of the resonance at the Fermi edge is pinned by Friedel's sum rule at approximately $(\pi\Gamma)^{-1}$ [$n_d \to 1$].

The exchange coupling J is related to the parameters of the SIAM via

$$J_{\text{eff}} = \frac{V^2}{|\epsilon_d|} + \frac{V^2}{|\epsilon_d + U|}.$$

3 Auxiliary Particle Representation

In the auxiliary particle presentation one introduces new creation operators to decompose the impurity electron creation operator as

$$d_\sigma^\dagger = f_\sigma^\dagger b + \sigma a^\dagger f_{-\sigma}, \tag{3}$$

where $f_\sigma^\dagger$ obeys fermion commutation relations and creates the singly occupied level with spin projection σ out of the new vacuum, and $b^\dagger$ and $a^\dagger$ are Bose operators creating the empty and doubly occupied states. In this representation the SU(N)×SU(M) Anderson impurity Hamiltonian is given by

$$H = \sum_{k\sigma\mu} \epsilon_k c_{k\sigma\mu}^\dagger c_{k\sigma\mu} + \epsilon_d \sum_\sigma f_\sigma^\dagger f_\sigma + V \sum_{\sigma\mu} (c_{0\sigma\mu}^\dagger b_{\bar\mu}^\dagger f_\sigma + h.c.) + \lambda Q, \tag{4}$$

where we again considered the limit of infinite Coulomb repulsion, implying that the impurity level (called d-level here) is at most singly occupied, labeled by spin $\sigma = 1, \ldots, N$. The empty level is M-fold degenerate, labeled by $\mu = 1, \cdots, M$, and $b_{\bar\mu}$ transforms according to the conjugate representation of SU(M). The SU(N)×SU(M) SIAM Hamiltonian introduced by P. Nozières and A. Blandin [7] is a generalization of the Hamiltonian of equation (1) to M degenerate conduction bands hybridizing with the impurity level. For $M \neq 1$ the ground state of the model is non-Fermi-liquid like [8]. The decomposition of the electron propagator in equation (3) will be a faithful representation of the original problem only, if an additional constraint is obeyed which limits the number of auxiliary particles in such a way that at any time the impurity site is either empty or singly occupied. Introducing the operator Q as

$$Q = \sum_\sigma f_\sigma^\dagger f_\sigma + \sum_\mu b_\mu^\dagger b_\mu,$$

the constraint is simply given by $Q = 1$. The constraint has been incorporated into the Hamiltonian (4) via the Lagrange parameter λ, which will be taken to infinity at the appropriate stage to effect the projection onto the $Q = 1$ sector of the Hilbert space. The Hamiltonian H possesses an internal gauge symmetry w.r.t. simultaneous U(1) gauge transformations of f_σ and b_μ, which reflects the fact that Q is conserved, $[Q, H] = 0$. In going from the electron operator to auxiliary particles we have transformed an initially nonholonomic constraint ($n_d \leq 1$) into an holonomic one ($Q = n_f + n_b = 1$).

The commutator or anticommutator of the slave operators are simple c-numbers and therefore the slave-particle decomposition permits the application of Wick's theorem in the grand-canonical pseudo-particle space and diagrammatic perturbation theory in its standard form applies. The auxiliary particle self-energies Σ_α, $\alpha = f, b, c$, are defined by $G_f^{-1} = \omega - \epsilon_d - \lambda - \Sigma_f(\omega)$, $G_b^{-1} = \omega - \lambda - \Sigma_b(\omega)$, $G_c^{-1} = G_{c0}^{-1}(\omega) - \Sigma_c(\omega)$, where $G_{c0}^{-1}(\omega) = \sum_k(\omega - \epsilon_k)$. The constraint $Q = 1$ strongly affects the dynamics of G_f and G_b. It can be shown that neither backward propagation in time nor anomalous propagators exist for auxiliary particles [9]. Therefore G_b and G_f are analogous to the core hole propagator in the well known X-ray edge problem [10–12]. The core hole propagator displays a singular threshold behavior as a consequence of the orthogonality catastrophe. Likewise, the constraint $Q = 1$ leads to infrared singular threshold behavior of the pseudo-particle Green's functions $G_{f,b}(\omega) \propto \omega^{-\alpha_{f,b}}$ below a characteristic energy scale which in the empty-orbital regime is given by ϵ_d, in the mixed valence regime by Γ, and in the local moment regime by the Kondo temperature T_K. This low-energy scale T_K of the SU(N)×SU(M) SIAM Hamiltonian is given by

$$T_K = D\Big(\frac{N\Gamma}{\pi|\epsilon_d|}\Big)^{(M/N)} \exp\Big(-\frac{\pi|\epsilon_d|}{N\Gamma}\Big).$$

4   Numerical Evaluation: NCA, CTMA, and SUNCA

In solving the Anderson model, Eq. (4) we use the scheme of conserving approximations, where all physical quantities are derived from functional derivatives of an approximative free energy functional Φ. In the following we will discuss three approximations. The well-known Non-Crossing Approximation (NCA), the Conserving T-Matrix Approximation (CTMA), an approximation devised to overcome the deficiencies of the NCA, and the Symmetric-U Non-Crossing Approximation (SUNCA), the extension of the NCA to finite Coulomb repulsion. The criteria for choosing Φ are at different levels of approximation, (1) the smallness of the parameter $V N(0) \ll 1$ to select the leading diagram (NCA), (2) the inclusion of the dominant spin and charge fluctuations (CTMA), and (3) the symmetric treatment of the particle-hole and particle-particle channel in a Schrieffer-Wolff transformation (SUNCA).

4.1   NCA

The lowest-order (V^2) diagram defines the so-called Non-Crossing-Approximation (NCA) which in the present form was introduced by Y. Kuramoto [13] but can even be traced back to work by H. Keiter and J. Kimball [14]. The NCA is a well studied approximation whose virtues and shortcomings are known [15,16]. The main virtue of the NCA is its simplicity. The auxiliary particle self-energies are simple convolutions of the pseudo-propagator with the conduction-electron propagator. The self-energies Σ_f and Σ_b and the

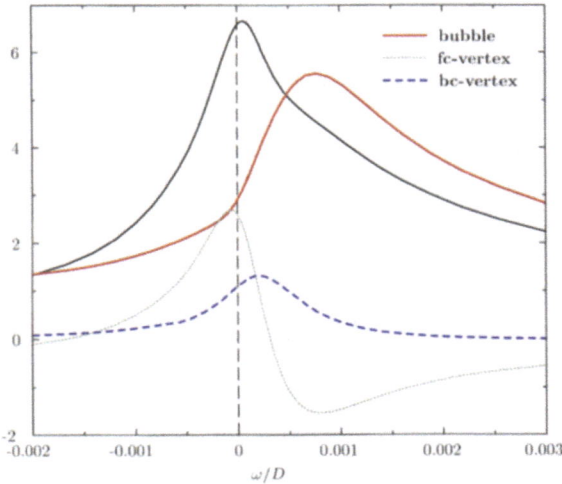

Fig. 2. Decomposition of the CTMA result for the local DOS $\rho(\omega)$ into the bubble and the vertex corrections from $T^{(cf)}$ (fc vertex) and $T^{(cb)}$ (bc vertex) at $T = 0.01 T_K$. It can be shown that the bubble and the vertex contributions $T^{(cf)}$ and $T^{(cb)}$ diverge at $\omega/D = 0$ in such a way that the sum (solid black curve) remains finite.

impurity Green function G_d obey after analytical continuation ($i\omega \rightarrow \omega - i0 \equiv \omega$) and projection onto the physical subspace:

$$\Sigma_{f\sigma}^{(NCA)}(\omega) = \Gamma \sum_\mu \int \frac{d\varepsilon}{\pi} \left[1 - f(\varepsilon)\right] A_{c\mu\sigma}^0(\varepsilon) G_{b\bar{\mu}}(\omega - \varepsilon) \tag{5}$$

$$\Sigma_{b\bar{\mu}}^{(NCA)}(\omega) = \Gamma \sum_\sigma \int \frac{d\varepsilon}{\pi} f(\varepsilon) A_{c\mu\sigma}^0(\varepsilon) G_{f\sigma}(\omega + \varepsilon) \tag{6}$$

$$G_{d\mu\sigma}^{(NCA)}(\omega) = \int d\varepsilon \, e^{-\beta\varepsilon} [G_{f\sigma}(\omega + \varepsilon) A_{b\bar{\mu}}(\varepsilon) - A_{f\sigma}(\varepsilon) G_{b\bar{\mu}}(\varepsilon - \omega)] \tag{7}$$

$$= \int d\varepsilon \, [G_{f\sigma}(\omega + \varepsilon) A_{b\bar{\mu}}^-(\varepsilon) - A_{f\sigma}^-(\varepsilon) G_{b\bar{\mu}}(\varepsilon - \omega)] \,,$$

where $A_{c\mu\sigma}^0 = \frac{1}{\pi} \text{Im} G_{c\mu\sigma}^0 / N(0)$ is the conduction-electron density of states per spin and channel, normalized to the density of states per spin at the Fermi level $N(0)$, and $f(\varepsilon) = 1/(\exp(\beta\varepsilon) + 1)$ denotes the Fermi distribution function. Together with the expressions for the Green's functions, equations (5)–(7) form a set of selfconsistent equations for $\Sigma_{b,f,c}$, comprised of all diagrams without any crossing propagator lines.

4.2   CTMA

It is a nontrivial task to construct an approximation that is capable of correctly describing the ground state properties of the SU(N)×SU(M) Anderson

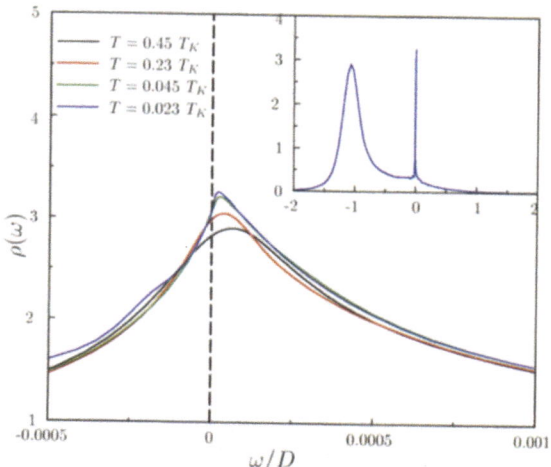

Fig. 3. CTMA result for the local density of states in the two-channel model with $\epsilon_d = -4\,\Gamma$ at various temperatures. The ground state is non-Fermi liquid like which leads to the cusp near the Fermi energy $\omega = 0$. (The cusp is shifted away from the Fermi energy due to the presence of a potential scattering term because of broken particle-hole symmetry.) The resulting crossover scale was $T_K/D = 2.2 \cdot 10^{-4}$. The non-analytical behavior is thus observed on the scale of T_K [19]. The inset shows the complete curve.

model [17]. It follows from a perturbative RG analysis that already the correct description of the high energy dynamics ($T \gg T_K$) requires the inclusion of physical processes which are no longer given by simple convolutions of the auxiliary particle Green functions [18]. Instead, one has to solve a set of Bethe-Salpeter equations in each step of the selfconsistency cycle. A numerically challenging and costly task. The linear Fredholm integral equations of the second kind for the T-matrix $T^{(cf)\,(\pm)\,\mu}_{\sigma,\tau}$ and $T^{(cb)\,(\pm)\,\sigma}_{\mu,\nu}$ read

$$T^{(cf)\,(\pm)\,\mu}_{\sigma,\tau}(\omega,\omega',\Omega) = I^{(cf)\,(\pm)\,\mu}_{\sigma,\tau}(\omega,\omega',\Omega) \mp \Gamma \int \frac{d\varepsilon}{\pi}\, f(\varepsilon - \Omega) \tag{8}$$

$$\times G_{b\bar{\mu}}(\omega + \varepsilon - \Omega)G_{f\sigma}(\varepsilon)A^0_{c\mu\tau}(\Omega - \varepsilon)T^{(cf)\,(\pm)\,\mu}_{\tau,\sigma}(\varepsilon,\omega',\Omega)$$

$$I^{(cf)\,(\pm)\,\mu}_{\sigma,\tau}(\omega,\omega',\Omega) = \frac{\Gamma^2}{\pi N(0)} \int \frac{d\varepsilon}{\pi}\, f(\varepsilon - \Omega) \times$$

$$G_{b\bar{\mu}}(\omega + \varepsilon - \Omega)G_{f\sigma}(\varepsilon)A^0_{c\mu\tau}(\Omega - \varepsilon)G_{b\bar{\mu}}(\omega' + \varepsilon - \Omega).$$

$$T^{(cb)\,(\pm)\,\sigma}_{\mu,\nu}(\omega,\omega',\Omega) = I^{(cb)\,(\pm)\,\sigma}_{\mu,\nu}(\omega,\omega',\Omega) \pm \Gamma \int \frac{d\varepsilon}{\pi}\, f(\varepsilon - \Omega) \tag{9}$$

$$\times G_{f\sigma}(\omega + \varepsilon - \Omega)G_{b\bar{\mu}}(\varepsilon)A^0_{c v\sigma}(\varepsilon - \Omega)\,T^{(cb)\,(\pm)\,\sigma}_{\nu,\mu}(\varepsilon,\omega',\Omega)$$

$$I^{(cb)\,(\pm)\,\sigma}_{\mu,\nu}(\omega,\omega',\Omega) = -\frac{\Gamma^2}{\pi N(0)} \int \frac{d\varepsilon}{\pi}\, f(\varepsilon - \Omega) \times$$

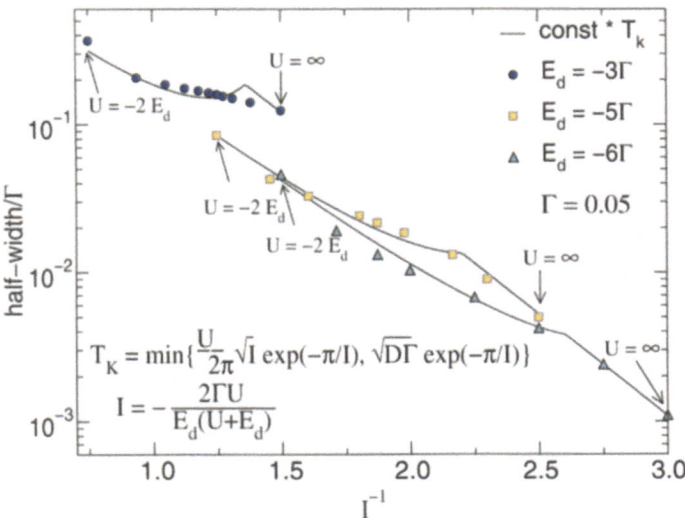

Fig. 4. Kondo temperature T_K for various parameters E_d, U and fixed Γ. Solid lines represent the exact results. Data points are the SUNCA results determined from the width of the Kondo peak in the d-electron spectral function.

$$G_{f\sigma}(\omega + \varepsilon - \Omega)G_{b\bar{\mu}}(\varepsilon)A^0_{c\nu\sigma}(\varepsilon - \Omega) \; G_{f\sigma}(\omega' + \varepsilon - \Omega).$$

These Bethe-Salpeter equations for the irreducible two-particle vertex define a novel approximation, the CTMA [16,20]. The auxiliary particle self-energies are given as simple integrals of the T-matrices. Only after convergence is reached on the level of the auxiliary particle self-energies we calculate the physical quantities, which are generally given as integrals of the slave particle functions. Figure 2 shows our result for the local density of states and its decomposition into leading order and vertex contributions for the one-channel model (N = 2, M = 1). It follows from the threshold behavior of the auxiliary Green functions that the leading order and vertex contributions diverge at $\omega = 0$, $1/\beta = 0$, although the density of states has to obey Fermi liquid sum rules [9]. Figure 3 shows the CTMA result for the local density of states at various temperatures in the two-channel model (N = 2, M = 2), where non-Fermi liquid behavior is found [19].

4.3 SUNCA

In order to be of practical use in, for example, DMFT calculations it is vital that a given approximation is capable of capturing the correct low-energy scale for all values of the Coulomb repulsion U. It is not a trivial task to construct such an approximation and to our knowledge all existing extensions of the NCA to finite U fail do to so [9]. Only the SUNCA, the proper extension of the NCA to finite U, recently proposed in [21], can be shown to correctly

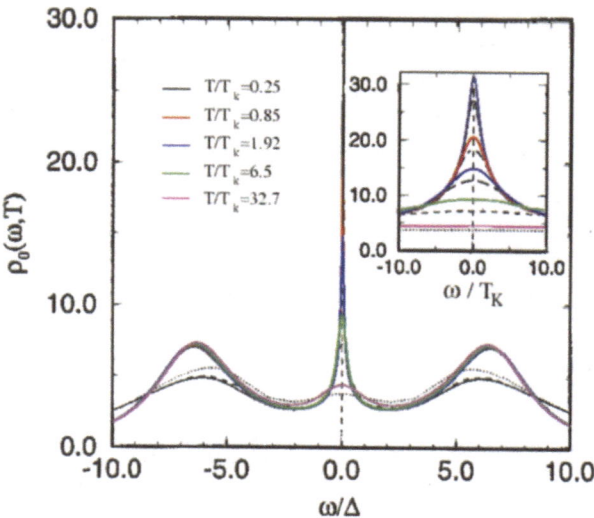

Fig. 5. Comparison of SUNCA results for the local electron spectral function (full lines) with NRG results reported by Costi et al. in [22] (dashed lines).

reproduce the parameter dependence of the low-energy scale T_K. That indeed the correct parameter dependence is obtained is shown in Fig. 4.

In the case of finite U a further auxiliary particle Green function G_a is needed to describe the doubly occupied impurity site. This increases the dimension N_G of the general fixed point problem. Although the SUNCA is conceptionally simpler than the CTMA it is numerically much more involved than the simple NCA. A symmetric treatment of the particle-particle and particle-hole channel requires the use of so-called latter diagrams, which lead again to coupled integral equations for the vertex functions $V_{a\sigma}(\omega, \Omega)$ and $V_{b\sigma}(\omega, \Omega)$:

$$V_{a\sigma}(\omega, \Omega) = \Gamma \int \frac{d\epsilon}{\pi} f(\epsilon - \Omega) A^0_{c-\sigma}(\epsilon - \Omega) G_{f-\sigma}(\epsilon) G_a(\epsilon + \omega - \Omega)$$

$$+ \Gamma \int \frac{d\epsilon}{\pi} f(\epsilon - \Omega) A^0_{c-\sigma}(\epsilon - \Omega) G_{f-\sigma}(\epsilon) G_a(\epsilon + \omega - \Omega) V_{a-\sigma}(\epsilon, \Omega)$$

$$V_{b\sigma}(\omega, \Omega) = \Gamma \int \frac{d\epsilon}{\pi} f(\epsilon - \Omega) A^0_{c\sigma}(\Omega - \epsilon) G_{f-\sigma}(\epsilon) G_b(\epsilon + \omega - \Omega)$$

$$+ \Gamma \int \frac{d\epsilon}{\pi} f(\epsilon - \Omega) A^0_{c\sigma}(\Omega - \epsilon) G_{f-\sigma}(\epsilon) G_b(\epsilon + \omega - \Omega) V_{b-\sigma}(\epsilon, \Omega).$$

A comparison of our results with NRG results is shown in Fig. 5.

4.4   Implementation

The numerical task for the discussed approximations is in general given by a vector fixed point equation for the auxiliary particle Green functions in $N_g = N_f + N_b$ dimensions, where N_f (N_b) is the number of grid points necessary to resolve G_f (G_b). In the case of finite U (SUNCA) or in the case of broken spin degeneracy the number of Green functions and hence the dimension N_g increases accordingly. To solve the fixed point problem we use a relaxation method. The singular behavior of G_f and G_b discussed in section 3 poses considerable difficulties to the numerical evaluation, since a proper resolution of all thresholds is necessary. Whereas NCA requires the compution of simple convolutions the numerical evaluation of the CTMA is very well suited for parallelization. Within each iteration the most time consuming step is to solve equations (8) and (9). Although the integral equations are linear, a simple matrix inversion is complicated by the fact that features of the kernel at different (physical) scales have to be resolved. These features result from:

- the asymptotic behavior of Green functions $\mathrm{Re}\{G_{f,b}(\omega)\} \sim 1/\omega$ for $\omega \rightarrow \infty$,
- the high energy peak at roughly ϵ_d in $G_b(\omega)$ or $G_f(\omega)$ (depending on the gauge),
- the threshold behavior of the auxiliary particle functions whose width is given by the temperature $1/\beta$.

Alternatively, one may use an iterative approach to solve the integral equations. In both cases the center-of-mass frequency Ω enters the integral equations as a conserved parameter. Therefore, equations (8) and (9) are straightforwardly parallelized. N_g is one of the parameters determining the number of nodes N_{node}, in the sense that if N_{node} is a divider of N_g a high degree of vectorization is reached. Each node solves the integral equations for a given subset of Ω. Then, the corresponding self-energy contributions are calculated and the updated auxiliary particle Green functions are determined. We use the MPI standard throughout our program. On the T3E, we usually use $N_{\mathrm{node}} = 62$ nodes for $N_g = 500$ grid points. The degree of vectorization is roughly 79%–84% and for each iteration we need about 0.5×62 CPU minutes. The number of iterations necessary to reach convergence depends on the physical parameters, especially the temperature $1/\beta$, in general the smallest physical scale of the problem in most relevant cases. Typically about 50–70 iterations are needed for one given set of parameters (ϵ_d, Γ, N, M, $1/\beta$). After convergence is reached, physical quantities like the local density of states can be calculated.

Parallelization for the SUNCA proceeds as in the case of the CTMA. Although the dimension N_g is increased, the evaluation within each iteration is much simpler since V_a and V_b depend only on two frequencies. The degree of vectorization reached was in general a few percent lower than for the CTMA.

References

1. A. Georges, G. Kotliar, W. Krauth, and M. Rozenberg. *Rev. Mod. Phys.* **68**, (1996) 13.
2. A. Tsvelick and P. Wiegmann. *Adv. Phys.* **32**, (1983) 453.
3. J. Kondo. *Prog. Theo. Phys.* **32**, (1964) 37.
4. K. Andres, J. Graebner, and H. Ott. *Phys. Rev. Lett.* **35**, (1975) 1779.
5. J. Schrieffer and P. Wolff. *Phys. Rev.* **149**, (1966) 491.
6. P. W. Anderson. *J. Phys.C: Solid State Phys.* **3**, (1970) 2436.
7. P. Nozières and A. Blandin. *J. Phys.* **41**, (1980) 193.
8. D. Cox and A. Ruckenstein. *Phys. Rev. Lett.* **71**, (1993) 1613.
9. S. Kirchner. *Conserving T-Matrix Approach to Quantum Impurities with Application to Quantum Point Contacts.* Ph.D. thesis, University of Karlsruhe (2002). Shaker, Aachen, ISBN 3-8322-0183-1.
10. B. Roulet, J. Gavoret, and P. Nozières. *Phys. Rev.* **178**, (1969) 1072.
11. P. Nozières, J. Gavoret, and B. Roulet. *Phys. Rev.* **178**, (1969) 1084.
12. P. Nozières and C. De Dominicis. *Phys. Rev.* **178**, (1969) 1097.
13. Y. Kuramoto. *Z. Phys. B* **53**, (1983) 37.
14. H. Keiter and J. Kimball. *J. Appl. Phys.* **42**, (1971) 1460.
15. N. Bickers. *Rev. Mod. Phys.* **59**, (1987) 845.
16. T. A. Costi, J. Kroha, and P. Wölfle. *Phys. Rev. B* **53**, (1996) 1850.
17. J. Kroha and P. Wölfle. *Act. Phys. Pol. B* **29**, (1998) 3781.
18. S. Kirchner and J. Kroha. *Journ. Low Temp. Phys.* **126**, (2002) 1233.
19. S. Kirchner, J. Kroha, and P. Wölfle. To be published.
20. J. Kroha, P. Wölfle, and T. A. Costi. *Phys. Rev. Lett.* **79**, (1997) 261.
21. K. Haule, S. Kirchner, J. Kroha, and P. Wölfle. *Phys. Rev. B* **64**, (2001) 155111.
22. T. A. Costi, A. C. Hewson, and V. Zlatić. *J. Phys. C* **6**, (1994) 2519.

Excitonic and Local-Field Effects in Optical Spectra from Real-Space Time-Domain Calculations

W. G. Schmidt, P. H. Hahn, and F. Bechstedt

Computational Materials Science Group
Institut für Festkörpertheorie und Theoretische Optik
Friedrich-Schiller-Universität, Max-Wien-Platz 1, 07743 Jena
(Email: W.G.Schmidt@ifto.physik.uni-jena.de)

Abstract. We present a novel approach to solve the Bethe-Salpeter equation for the polarisation function. Rather than from the usual eigenvalue representation, the macroscopic polarisability is obtained from the solution of an initial-value problem. This allows for the first time to calculate excitonic and local-field effects in optical spectra of large and complex systems such as surfaces. As an example we investigate the optical anisotropy of the hydrogen-passivated Si(110) surface. It is shown that the electron-hole attraction is largely responsible for the peculiar line shape of the surface optical spectrum.

1 Introduction

The correct modelling of optical properties has been a long standing issue of scientific interest. It is significant also from a technological point of view, because methods of optical spectroscopy are rapidly gaining importance for materials characterisation. For example techniques like Reflectance Anisotropy/Difference Spectroscopy (RAS/RDS) have evolved from experimental methods to characterise static surfaces to very powerful *in situ* diagnostic probes which allow for the monitoring and controlling of the surface growth in real time and in challenging environments such as in high pressures or under liquids [1]. However, somewhat in contrast to their frequent use, the present understanding of the physical origin of the observed optical phenomena is still rather limited.

Aspnes and Studna [2] discriminated between two components of surface optical spectra: "intrinsic" contributions arising from optical transitions within the bulk and "extrinsic" contributions directly related to the surface chemistry. The latter can often be traced to specific surface electronic states and serve as fingerprints for surface structural motifs [3–5]. The origin of the intrinsic features, however, is harder to explain. It has been discussed for a long time that these features are likely to be related to many-particle effects [6] and/or surface local fields (LF) [7,8], i.e., the influence of the surface-modified microscopic fluctuations of the electric field on the macroscopic dielectric response. However, no definite assignment has been possible yet.

This is largely due to the numerical expense required for converged calculations of surface optical properties. In particular the intrinsic features of the surface spectra are caused by electronic transitions involving a very large number of surface-modified bulk wave functions [3,4]. Therefore, even calculations assuming a single-particle picture for electronic excitations and neglecting self-energy effects are quite involved for surfaces. The inclusion of many-particle effects such as electron-electron and electron-hole interactions dramatically increases the computational cost. Although exact expressions for the excitonic and LF contributions to the surface optical response based on the Bethe-Salpeter equation (BSE) have been derived decades ago, their application has been limited to very few tight-binding (TB) studies [9–11]. Because of the complexity of the problem, another, more frequently used approach approximates LF effects by modelling the crystal surface by a lattice of polarisable entities, which obey a Clausius-Mossotti-like relation [7,8,12]. Obviously, such models as well as TB calculations necessarily depend on external input parameters and cannot account accurately for the surface induced changes of the electronic structure. Very recently, it has become possible to solve the BSE from *first principles* for bulk semiconductors [13,14] and strongly localised surface states [15]. However, the large numerical effort has restricted such calculations to the interaction of relatively few electron-hole pairs. As yet they have not been applied to the surface optical response in a wide spectral range.

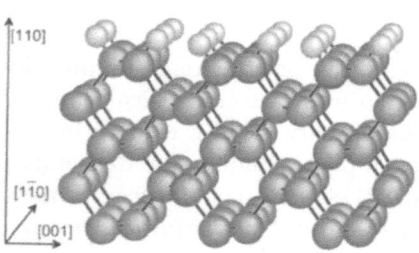

Fig. 1. Upper part of the slab representing the hydrogen-covered Si(110) surface.

Here we use an alternative approach to solve the BSE that allows for the study of large systems such as surfaces. It is shown that many-particle effects on the energies and oscillator strengths of electronic excitations in bulk-like layers are largely responsible for the appearance of the intrinsic features in surface optical spectra. In particular the electron-hole interaction may influence the magnitude and line shape of spectral features.

We use the hydrogen-passivated Si(110) surface (cf. Fig. 1) as a model system. It is one of the first systems studied by RAS [2] and its optical features are mainly intrinsic in character. The passivation of the Si dangling bonds results in there being no surface states in the energy region probed by RAS. The surface spectrum is rather insensitive to the structural and chemical details of the passivation [2,16,17] and has a very characteristic line shape with maxima close to the E_1 and E_2 critical point energies of bulk silicon. Because the RAS spectrum can be easily reproduced, it has become a calibration standard for RAS apparatus and a textbook example for surface optical properties [18].

The physical mechanism leading to the observed line shape, however, is not understood. In their original study [2] Aspnes and Studna argued that the measurements are indicative for the appearance of surface local-fields and/or many-body screening. The strong influence of local fields seems to be supported by model calculations [7,12]. TB studies that neglected LF effects [19] failed to describe the experiment, as did a TB work that included an approximation for LF effects [8]. In the latter study it was concluded that surface defects are responsible for the experimentally observed peaks. While real surface do contain defects, their RAS contributions should be small in that specific case, since the measured spectrum is nearly independent from the surface preparation procedures [2,16,17]. Indeed, a recent *ab initio* calculation [20] that approximated self-energy corrections using a scissors operator, but neglected LF and excitonic effects showed that a hydrogen-terminated Si(110) surface gives rise to optical anisotropies at the bulk critical points without the assumption of surface defects. This work, however, could not account for the peculiar line shape observed experimentally.

We go beyond these previous studies and present a consistent and detailed analysis of how electronic self-energy, LF and excitonic effects manifest themselves in the optical spectrum of Si(110):H. Thereby we proceed in three steps: (i) local density-functional (DFT-LDA) calculations yield the structurally relaxed ground state configuration of the surface, including the Kohn-Sham eigenvalues and eigenfunctions that enter the single- and two-particle Green's functions. (ii) The elec-

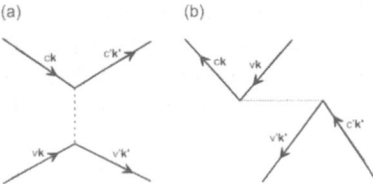

Fig. 2. Schematic representation of electron-hole attraction (a) and electron-hole exchange (b). The screened (unscreened) Coulomb interaction is indicated by a dashed (dotted) line.

tronic quasiparticle spectrum is obtained within the GW approximation [21] to the exchange-correlation self-energy and (iii) the BSE is solved for coupled electron-hole excitations [13–15]. Thereby the screened electron-hole attraction and the unscreened electron-hole exchange (cf. Fig. 2) are taken into account [22–24]. Inclusion of the latter allows for a parameter-free calculation of the LF effects. For surfaces LF effects can be expected from both the microscopic fluctuations of the electric field within the bulk, and from the truncation of the bulk itself. The resulting macroscopic polarisabilities are finally used to compute the reflectance anisotropy for normally incident light polarised parallel to the $[1\bar{1}0]$ and $[001]$ directions [25]. The anisotropy of the reflectivity of light polarised in two perpendicular directions in the surface plane, x and y, is given by [25]

$$\frac{\Delta R}{R}(\omega) := 2\frac{R_{xx} - R_{yy}}{R_{xx} + R_{yy}}(\omega) = \frac{16\pi\omega}{c}\Im\left\{\frac{\alpha_{xx}^{hs}(\omega) - \alpha_{yy}^{hs}(\omega)}{\epsilon_b(\omega) - 1}\right\}. \quad (1)$$

Here $\alpha_{ii}^{hs}(\omega)$ with $i = x, y$ is the diagonal tensor component of the averaged half-slab polarisability and $\epsilon_b(\omega)$ is the bulk dielectric function. Equation 1 contains in principle all contributions to the optical anisotropy such as the ones related to surface electronic states, atomic relaxations, the influence of the surface potential on the bulk wave functions as well as local-field and many-body effects like the electronic self-energy and the electron-hole attraction. Of course, it is not an easy task to include all these contributions in the actual calculations. In the following we discuss our approach to calculate the slab polarisability.

2   Method

2.1   Ground-state calculations within DFT-LDA

We start from *first-principles* pseudopotential calculations, using a massively parallel real-space finite-difference implementation of the DFT-LDA [26], which is characterised by an excellent scaling behaviour on parallel supercomputers such as the Cray T3E [27]. A multigrid technique is used for convergence acceleration. The surface is modeled by periodic supercells containing 12 atomic Si layers (cf. Fig. 1). Silicon dangling bonds at the bottom and top layer are saturated with hydrogen. A vacuum region equivalent to 8 atomic layers in thickness separates the material slabs in [110] direction. Apart from the atoms of the innermost two layers which were kept in their ideal bulk positions, all atomic coordinates are fully relaxed. Four **k** points in the irreducible part of the surface Brillouin zone are used for the self-consistent calculation of the ground-state charge density. For the calculation of the surface optical properties we use 140 uniformly distributed **k** points.

2.2   Self-energy corrections

When compared quantitatively with experiment, optical spectra calculated within DFT-LDA have serious deficiencies. The most serious one is a systematic redshift of the theoretical features compared to the measurements. This is related to the fact that the quasiparticle character of the electrons, i.e. their mutual interaction and screening is insufficiently accounted for. That leads to an underestimation of the excitation energies known as the DFT-LDA band gap problem [28,29]. In order to include electronic self-energy effects one needs to replace the local exchange and correlation potential $V^{XC}(\mathbf{r})$ in LDA by the nonlocal and energy-dependent self-energy operator $\Sigma(\mathbf{r}, \mathbf{r}'; E)$. The so-called GW approximation (GWA) [30,31] where the self-energy operator is expressed as convolution of the dynamically screened Coulomb potential W and the single-particle propagator G, $\Sigma = iGW$, currently forms the basis of nearly all numerical quasiparticle calculations. Since the calculation of surface optical spectra involves a very large number of electronic states, however, we introduce further approximations. It was found that the real quasiparticle wave functions have more than 99.9 % overlap with the original DFT-LDA orbitals used as starting point in the calculations, at least for states close to the fundamental band edges of bulk semiconductors [21]. Therefore we obtain the quasiparticle energies ε_n^{QP} from the DFT-LDA eigenvalues and wave functions in a perturbative manner, by

$$\varepsilon_n^{QP}(\mathbf{k}) = \varepsilon_n(\mathbf{k}) + \langle n\mathbf{k}|\Sigma(\varepsilon_n^{QP}) - V^{XC}|n\mathbf{k}\rangle. \qquad (2)$$

Here, Σ must in principle be computed self-consistently at the energy ε_n^{QP}. In practice, this is approximated by a Taylor expansion of Σ around the DFT-LDA eigenvalue. To simplify the calculation of the self-energy corrections further, we follow the schemes developed by Hybertsen and Louie [32] and Bechstedt et al. [33]: the GW quasiparticle energies are obtained from the DFT-LDA eigenvalues by

$$\varepsilon_n(\mathbf{k})^{QP} = \varepsilon_n(\mathbf{k}) + \frac{1}{1+\beta_{n,\mathbf{k}}} \left[\Sigma_{n,\mathbf{k}}^{st} + \Sigma_{n,\mathbf{k}}^{dyn}\left(\varepsilon_n(\mathbf{k})\right) - V_{n,\mathbf{k}}^{XC} \right], \qquad (3)$$

where the self-energy operator Σ has been divided into static and dynamic contributions. Indices at Σ and V^{XC} indicate diagonal matrix elements with the respective wave functions. $\beta_{n,\mathbf{k}}$ is the linear term in the expansion of Σ^{dyn} around the DFT-LDA eigenvalue $\varepsilon_n(\mathbf{k})$. The static part can be further divided into two parts

$$\Sigma^{st}(\mathbf{r}, \mathbf{r}') = \frac{1}{2} \sum_{n,\mathbf{k}} \psi_{n,\mathbf{k}}(\mathbf{r})\psi_{n,\mathbf{k}}^*(\mathbf{r}') \left[W(\mathbf{r}, \mathbf{r}'; 0) - v(\mathbf{r} - \mathbf{r}') \right] -$$

$$\sum_{v,\mathbf{k}} \psi_{v,\mathbf{k}}(\mathbf{r})\psi_{v,\mathbf{k}}^*(\mathbf{r}') W(\mathbf{r}, \mathbf{r}'; 0) , \qquad (4)$$

representing the Coulomb hole Σ^{COH} and the screened exchange Σ^{SEX}. The $\psi_{n,\mathbf{k}}$ are the DFT-LDA wave functions. The major bottleneck in the GW calculation is the computation of the screened interaction W. An extreme speedup can be achieved by using a model dielectric function, for which several functional forms have been suggested. We use the version suggested by Bechstedt et al. [33]

$$\epsilon(\mathbf{q}, \rho) = 1 + \left[(\epsilon_\infty - 1)^{-1} + \left(\frac{q}{q_{TF}(\rho)} \right)^2 + \frac{3q^4}{4k_F^2(\rho)q_{TF}^2(\rho)} \right]^{-1}, \qquad (5)$$

where k_F and q_{TF} represent the Fermi and Thomas-Fermi wave-vectors, respectively, which depend on the electron density ρ. This expression interpolates between the correct behaviours at high and low $\mathbf{q}$ vectors and, by construction, correctly obtains the static dielectric constant for $\mathbf{q} = 0$. This simple and intuitive model reproduces very well the RPA results for semiconductors [34]. Together with the LDA-like ansatz of Hybertsen and Louie [32] for approximating the spatial dependence of the screening of the inhomogeneous system

$$W(\mathbf{r}, \mathbf{r}'; 0) = \frac{1}{2} \left[W^h \left(\mathbf{r} - \mathbf{r}', \rho(\mathbf{r}) \right) + W^h \left(\mathbf{r} - \mathbf{r}', \rho(\mathbf{r}') \right) \right] \tag{6}$$

by that of a homogeneous electron gas W^h, (5) allows for an analytic solution for Σ_{COH}. The static Coulomb hole contribution to the self-energy takes the form of a local potential

$$\Sigma^{COH}(\mathbf{r}) = -\frac{q_{TF}(\mathbf{r})}{2} \sqrt{1 - \frac{1}{\epsilon_\infty}} \left[1 + \frac{q_{TF}(\mathbf{r})}{k_F(\mathbf{r})} \sqrt{\frac{3\epsilon_\infty}{\epsilon_\infty - 1}} \right]^{-\frac{1}{2}} , \tag{7}$$

where k_F and q_{TF} are computed for the local density $\rho(\mathbf{r})$.

The matrix elements $\Sigma_{n,\mathbf{k}}^{SEX}$ are calculated in Fourier space. In order to speed up the calculations only the diagonal elements in the Fourier transform of W [32] are retained. The effect of local fields on the screening are approximated by using state-averaged electron densities

$$\rho_{n,\mathbf{k}} = \int d\mathbf{r}^3 \rho(\mathbf{r}) |\psi_{n,\mathbf{k}}(\mathbf{r})|^2 , \tag{8}$$

in the calculation of k_F and q_{TF}. Tests made for bulk Si indicate that rather small deviations, of the order of 0.05 eV, are induced by this approximation. Finally, the dynamic terms $\beta_{n,\mathbf{k}}$ and Σ^{dyn} in (3), are approximated by simple integrals of the dielectric function [33]. For the actual calculations we use (5) together with a single-plasmon-pole approximation to describe the frequency dependence. Local-field effects are again included using the mean-density approximation (8). The integrals are numerically evaluated for a dense sampling of ρ and the results for $\beta_{n,\mathbf{k}}(\rho)$ and $\Sigma^{dyn}(\rho)$ are fitted to polynomials. These are then used for a fast computation of the dynamic contributions to the self-energy during the actual GW calculations.

3   Local fields and electron-hole attraction

The excitation energies obtained within the quasiparticle formalism correctly describe one-particle excitations, such as those involved in (inverse) photoemission experiments. For the description of the optical absorption process, however, one needs to go beyond this single-quasiparticle level. In general, the energy of the absorbed photon differs from the bare, algebraic sum of the hole and electron energies, since the electron-hole interaction also needs to be taken into account. The latter can not only produce absorption below the gap, due to bound exciton states, but can also induce appreciable distortions on the line shape above the continuous absorption edge. The calculation of excitonic effects from *first principles* is only a recent achievement [13,14,36],

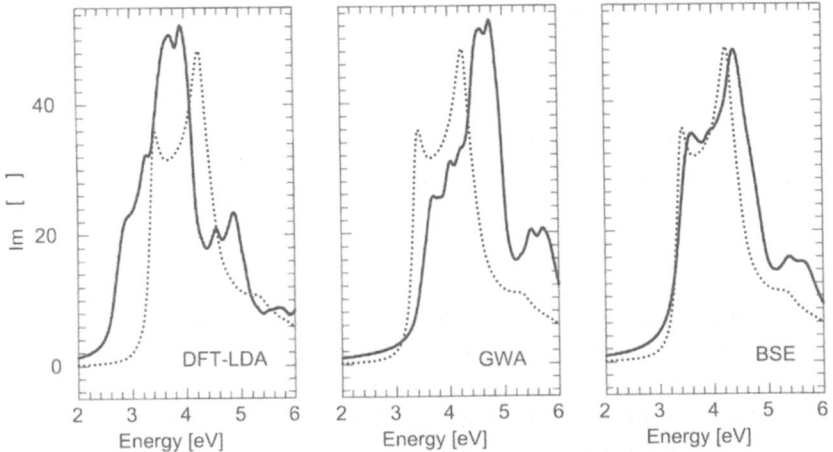

Fig. 3. Dielectric function for bulk Si calculated within DFT-LDA, in GWA and from the BSE (GWA + excitonic and LF effects) in comparison with exp. data from Ref. [35] (dotted line).

and has been restricted to very small systems prior to our work [37]. The polarisation function P including electron-hole attraction and local-field effects can be obtained from the solution of the Bethe-Salpeter equation [22,24],

$$P = P_0 + P_0(\bar{v} - W)P ,\qquad(9)$$

where $\bar{v}$ is the bare Coulomb potential without its long-range part. From the Fourier transform of the diagonal part of P one obtains the macroscopic dielectric function

$$\epsilon^M(\omega) = 1 - \lim_{\mathbf{q}\to 0}\left[v_{00}(\mathbf{q})\int d\mathbf{r}d\mathbf{r}' e^{-i(\mathbf{r}-\mathbf{r}')\mathbf{q}} P(\mathbf{r},\mathbf{r};\mathbf{r}',\mathbf{r}';\omega)\right].\qquad(10)$$

A convenient and natural basis for solving Eq. 9 is given by the orthonormal and complete set of Bloch functions defined, e.g., by the Kohn-Sham problem. Omitting the frequency dependence, the transform of the polarisation into the Bloch-function representation reads

$$P(\mathbf{r}_1,\mathbf{r}_2;\mathbf{r}_3,\mathbf{r}_4) = \sum_{n_1,n_2,n_3,n_4} \psi_{n_1}^*(\mathbf{r}_1)\psi_{n_2}(\mathbf{r}_2)\psi_{n_3}(\mathbf{r}_3)\psi_{n_4}^*(\mathbf{r}_4)P_{(n_1,n_2)(n_3,n_4)},$$

$$\qquad(11)$$

where n denotes both band index and wave vector. If P_0 is explicitly expressed in terms of Bloch functions and quasiparticle energies, using the Lehmann representation of the single-particle Green's function, and transformed into Bloch space by applying $\int d\mathbf{r}_1 d\mathbf{r}_2 d\mathbf{r}_3 d\mathbf{r}_4 \psi_{n_1}(\mathbf{r}_1)\psi_{n_2}^*(\mathbf{r}_2)\psi_{n_3}^*(\mathbf{r}_3)\psi_{n_4}(\mathbf{r}_4)$, the solution of the BSE (9) can be written as

$$P_{(n_1,n_2)(n_3,n_4)} = \left[\hat{H} - I\omega\right]^{-1}_{(n_1,n_2)(n_3,n_4)}(f_{n_4} - f_{n_3}),\qquad(12)$$

where the two-particle Hamiltonian

$$\hat{H}_{(n_1,n_2)(n_3,n_4)} \equiv (\varepsilon_{n_1}^{QP} - \varepsilon_{n_2}^{QP})\delta_{(n_1,n_3)}\delta_{(n_2,n_4)} + (f_{n_2} - f_{n_1}) \times$$

$$\int d\mathbf{r}_1 d\mathbf{r}_2 d\mathbf{r}_3 d\mathbf{r}_4 \psi_{n_1}(\mathbf{r}_1)\psi_{n_2}^*(\mathbf{r}_2)\psi_{n_3}^*(\mathbf{r}_3)\psi_{n_4}(\mathbf{r}_4) \times$$

$$[\delta(\mathbf{r}_1 - \mathbf{r}_2)\delta(\mathbf{r}_3 - \mathbf{r}_4)\bar{v}(\mathbf{r}_1 - \mathbf{r}_3) - \delta(\mathbf{r}_1 - \mathbf{r}_3)\delta(\mathbf{r}_2 - \mathbf{r}_4)W(\mathbf{r}_1, \mathbf{r}_2)]$$

(13)

has been introduced. The $f_n = 0, 1$ is the occupation number of the state n. By performing a matrix inversion for a given frequency ω, the corresponding polarisation is given by Eq. 12. However, for any practical calculation this would be computationally far too expensive, due to the large dimension of $\hat{H}$. This dimension can be reduced by a factor of two, however, if one observes that due to the factors $(f_{n_4} - f_{n_3})$ in (12) and $(f_{n_2} - f_{n_1})$ in Eq. 13, only pairs containing one filled and one empty Bloch state contribute to the macroscopic polarisation. A further reduction of the dimension by a factor of two can be achieved when the off-diagonal blocks coupling the resonant part of $\hat{H}$,

$$\hat{H}_{vck,v'c'k'}^{\mathrm{res}} = (\varepsilon_{ck}^{QP} - \varepsilon_{vk}^{QP})\delta_{vv'}\delta_{cc'}\delta_{\mathbf{k},\mathbf{k}'} +$$

$$2\int d\mathbf{r}_1 d\mathbf{r}_2 \psi_{ck}^*(\mathbf{r}_1)\psi_{vk}(\mathbf{r}_1)\bar{v}(\mathbf{r} - \mathbf{r}_2)\psi_{c'k'}(\mathbf{r}_2)\psi_{v'k'}^*(\mathbf{r}_2) -$$

$$\int d\mathbf{r}_1 d\mathbf{r}_2 \psi_{ck}^*(\mathbf{r}_1)\psi_{c'k'}(\mathbf{r}_1)W(\mathbf{r}_1, \mathbf{r}_2)\psi_{vk}(\mathbf{r}_2)\psi_{v'k'}^*(\mathbf{r}_2),$$

(14)

and the anti-resonant part, $-\left[\hat{H}^{\mathrm{res}}\right]^*$, are neglected. The coupling blocks with contributions only from the interaction terms involving W and $\bar{v}$ are small compared to the (anti-) resonant diagonal blocks containing in addition the quasiparticle transition energies. Apart from special cases, e.g. the calculation of plasmon resonances where the mixing of interband transitions of both positive and negative frequencies must be included in the calculations [38], the coupling can be neglected in the calculation of optical properties [13,39]. Further approximations contained in (14) are the restriction to spin-singlets, static screening and direct transitions, i.e., the neglect of momentum transfer by photons. If furthermore umklapp processes are neglected, the exciton Hamiltonian can be calculated in reciprocal space according to

$$\hat{H}_{vck,v'c'k'}^{\mathrm{res}} = (\varepsilon_{ck}^{QP} - \varepsilon_{vk}^{QP})\delta_{vv'}\delta_{cc'}\delta_{\mathbf{kk}'} -$$

$$\frac{4\pi}{\Omega} \sum_{\mathbf{G},\mathbf{G}'} \left\{ 2\frac{\delta_{\mathbf{GG}'}(1 - \delta_{\mathbf{G}0})}{|\mathbf{G}|^2} B_{cv}^{\mathbf{kk}}(\mathbf{G}) B_{c'v'}^{\mathbf{k}'\mathbf{k}'*}(\mathbf{G}) - \right.$$

$$\left. \frac{\epsilon^{-1}(\mathbf{k} - \mathbf{k}' + \mathbf{G}, \mathbf{k} - \mathbf{k}' + \mathbf{G}', 0)}{|\mathbf{k} - \mathbf{k}' + \mathbf{G}|^2} B_{cc'}^{\mathbf{kk}'}(\mathbf{G}) B_{vv'}^{\mathbf{kk}'*}(\mathbf{G}') \right\},$$

(15)

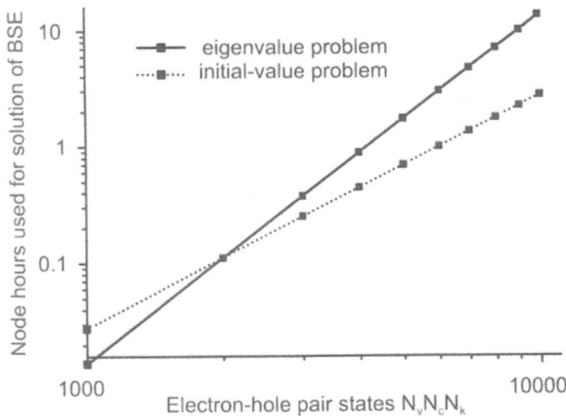

Fig. 4. CPU time needed to solve the BSE on a Pentium PC for bulk Si via an eigenvalue and initial-value problem in dependence on the dimension of the exciton Hamiltonian.

where the Bloch integral

$$B_{nn'}^{\mathbf{kk'}}(\mathbf{G}) = \frac{1}{\Omega} \int d\mathbf{r} u_{n\mathbf{k}}^*(\mathbf{r}) e^{i\mathbf{Gr}} u_{n'\mathbf{k'}}(\mathbf{r}) \tag{16}$$

over the periodic parts u of the Bloch wave functions has been introduced. The calculation of the Hamiltonian according to (15) is computationally very demanding even for bulk systems, due to the rank of the Hamiltonian itself, $N = N_v \cdot N_c \cdot N_{\mathbf{k}}$, as well as due to the double sum over $\mathbf{G}$ and $\mathbf{G'}$, which needs to be performed for each single matrix element. In order to speed up the calculations we therefore replace the inverse dielectric matrix by the same diagonal model dielectric function due to Bechstedt (5), which has been used in the calculation of the screened exchange contribution to the self-energy operator. The local-field effects are again approximated by using state-dependent electron densities in (5), which were calculated using the mean-density approximation (8). Despite these simplifications we obtain for bulk Si an optical spectrum in nearly perfect agreement with experiment, as can be seen in Fig. 3. The experimental results are seemingly even better reproduced than in previous calculations [13], where the screening has been fully included. That is simply related to the fact that the approximations included here allow for a better convergence of the technical parameters. The discrepancies between experiment and theory found in Ref. [13,40], for example, are related to the insufficient **k**-point sampling [41].

After the Hamiltonian has been calculated, one needs to determine the polarisation. The rank of $\hat{H}$ is typically of the order of 10^4 even for small bulk unit cells. This excludes the straightforward evaluation of Eq. 12. The usual approach therefore consists in transforming the inversion into an effective eigenvalue problem, which is then solved by diagonalisation [13,36]. In detail, using the spectral representation

$$
\left[\hat{H} - I\omega\right]^{-1} = \sum_{\lambda,\lambda'} \frac{|A^\lambda\rangle S_{\lambda,\lambda'}^{-1} \langle A^{\lambda'}|}{E_\lambda - w},
\tag{17}
$$

where $|A^\lambda\rangle$ and E_λ are the eigenvectors and eigenvalues of the exciton Hamiltonian

$$
\hat{H}|A^\lambda\rangle = E_\lambda |A^\lambda\rangle \quad \text{and} \quad S_{\lambda,\lambda'} = \langle A^{\lambda'}|A^\lambda\rangle,
\tag{18}
$$

the ii $(i = x, y, z)$ component of the macroscopic polarisability becomes

$$
\alpha_{ii}^M(\omega) = \frac{4e^2\hbar^2}{\Omega} \sum_\lambda \left|\sum_{\mathbf{k}} \sum_{c,v} \frac{\langle c\mathbf{k}|v_i|v\mathbf{k}\rangle}{\varepsilon_c(\mathbf{k}) - \varepsilon_v(\mathbf{k})} A_{vc\mathbf{k}}^\lambda\right|^2 \times
$$
$$
\left\{\frac{1}{E_\lambda - \hbar(\omega + i\eta)} + \frac{1}{E_\lambda + \hbar(\omega + i\eta)}\right\},
\tag{19}
$$

where v_i is the corresponding Cartesian component of the single-particle velocity operator. Here, the contributions of the anti-resonant part of the exciton Hamiltonian have been formally included, while the coupling parts are neglected.

The calculation of the dielectric function using (19) is straightforward, but requires the solution of the eigenvalue problem (18). For small bulk unit cells the diagonalisation of $\hat{H}$ can typically be performed within a couple of hours. However, our work aims at the determination of optical properties for large and complex systems such as surfaces. The minimum slab thickness for the calculation of surface optical properties is about 12 layers. Typically more than 100 $\mathbf{k}$ points are needed to sample the surface Brillouin zone. The dimension of the exciton Hamiltonian $N = N_v \cdot N_c \cdot N_{\mathbf{k}}$ is therefore about $10^6 - 10^7$, even for the relatively small unit cell of an unreconstructed surface. Even with today's powerful supercomputers, the diagonalisation of matrices of this size is prohibitively slow. Therefore an approach different from the one outlined above is needed.

To avoid the diagonalisation bottleneck we exploit an idea by Glutsch *et al.* [42], who traced back a similar eigenvalue problem to an initial-value problem. If a vector $|\mu^i\rangle$ with elements

$$
\mu_{vc\mathbf{k}}^i = \frac{\langle c\mathbf{k}|v_i|v\mathbf{k}\rangle}{\varepsilon_c(\mathbf{k}) - \varepsilon_v(\mathbf{k})}
\tag{20}
$$

is introduced, Eq. 19 takes the form

$$
\alpha_{ii}^M(\omega) = \frac{4e^2\hbar^2}{\Omega} \sum_\lambda |\langle \mu^i|A^\lambda\rangle|^2 \left\{\frac{1}{E_\lambda - \hbar(\omega + i\eta)} + \frac{1}{E_\lambda + \hbar(\omega + i\eta)}\right\},
\tag{21}
$$

which is equivalent to

$$
\alpha_{ii}^M(\omega) = \frac{4e^2\hbar^2}{\Omega} i \int_0^\infty dt e^{i(\omega+i\eta)t} \left\{\langle \mu^i|\xi(t)\rangle - \langle \mu^i|\xi(t)\rangle^*\right\}
\tag{22}
$$

with

$$ i\hbar \frac{d}{dt}|\xi(t)\rangle = \hat{H}|\xi(t)\rangle \quad \text{and} \quad |\xi(0)\rangle = |\mu^i\rangle. \tag{23} $$

The equivalence can be shown by integrating $|\xi(t)\rangle = e^{\frac{\hat{H}t}{i\hbar}}|\mu\rangle$ and exploiting the spectral representation as in (17). We have also verified numerically that Eqs. 21 and 22 lead to exactly the same spectrum [43]. The latter formula, however, requires much less computational resources. We solve the initial value problem (23) using the central-difference method

$$ \hat{H}|\xi(t_{i+1})\rangle = i\hbar \frac{|\xi(t_{i+2})\rangle - |\xi(t_i)\rangle}{2\Delta t}, \tag{24} $$

i.e., by means of a sequence of matrix-vector multiplications. The upper limit of the Fourier integral (22) can be truncated, due to the exponential $e^{-\eta t}$. Therefore, the number of time steps, i.e. matrix-vector multiplications, is nearly independent from the dimension of the system. The operation count for this method scales thus as $\mathcal{O}(N^2)$, compared to the $\mathcal{O}(N^3)$ for the matrix diagonalisation. Moreover, the matrix-vector multiplications can be easily distributed on several processors of a parallel computer, whereas the parallelisation of matrix diagonalisation is less effective, due to the large amount of data transfer across processors. The cross-over point for the CPU time usage of both methods is reached for a number of electron-hole pair states as low as about 2000, as shown in Fig. 4. With the method outline above it is thus possible to include all spectroscopically relevant many-body interactions into calculations of optical properties for systems of an unprecedented size. In the following section we demonstrate the influence of many-body effects on the optical spectrum of the prototypical Si(110):H surface.

4   Results and Discussion

Figure 5 contains the calculated RAS spectra for the Si(110):H surface represented by a 12-layer slab. The DFT-LDA spectrum shows two strong positive RAS features near the E_1 and E_2 bulk critical point energies. However, the features are far too broad. This is partially due to the coarse k-point sampling and a slab which is too thin to allow for a complete description of the surface-perturbed bulk wave functions responsible for the observed optical anisotropies. Denser k-point meshes and thicker slabs lead to a much better description of the optical anisotropy at the E_2 energy (see Fig. 3 in Ref. [20]). They do not improve the poor representation of the line shape and strength of the anisotropy at the E_1 energy, however. Inclusion of quasiparticle effects in the GWA leads to a blue-shift of the spectrum by about 0.6–0.7 eV and changes the line shape. In particular the anisotropy at the E_1 energy is enhanced compared to the DFT-LDA spectrum. The E_1 peak height relative to the E_2 anisotropy is still much smaller than measured, however.

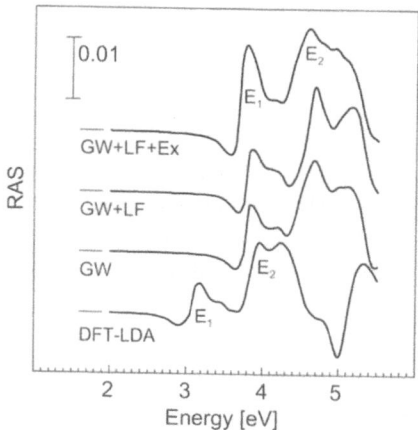

Fig. 5. RAS spectra calculated (within DFT-LDA, in GWA, in GWA with the effects of local fields included, and in GWA with the effects of local fields and the electron-hole attraction included) for Si(110):H described by a 12-layer slab.

In order to determine the influence of excitonic and LF effects on the RAS we calculate the exciton Hamiltonian for the Si(110):H surface according to Eq. 15. In the slab calculation 50 valence and 50 conduction bands at 140 **k** points are taken into account. This would give rise to a rank $N = N_v \cdot N_c \cdot N_{\mathbf{k}} = 350,000$ of the exciton Hamiltonian. Fortunately, however, it turns out that not all matrix elements are needed to calculate a numerically converged optical spectrum for the photon energies considered here. By means of a suitable chosen cutoff function we could reduce the rank of the Hamiltonian to about $100,000$. Still, the calculation of 10^{10} matrix elements requires an appreciable amount of computer time, about 15,000 node hours. The task was solved by distributing the calculation on 128 T3E processors. The calculation of the polarisation itself, using the central-difference method (24), is highly efficient. It requires only about $1,500$ node hours.

We find that LF effects lead to surprisingly small changes of the spectrum. A reduction of the calculated slab polarisabilities upon inclusion of LF effects comparable to the one calculated for bulk Si is observed. The reduction acts both on the $\alpha_{[001]}$ and the $\alpha_{[1\bar{1}0]}$ tensor components. It is therefore largely cancelled in the optical anisotropy. Rather than increasing the ratio of the E_1/E_2 peak heights, LF effects even lead to a small decrease. A drastic enhancement of the optical anisotropy at the E_1 energy and a red-shift of the entire spectrum by about $0.1 - 0.2$ eV result, however, from the inclusion of the attractive electron-hole interaction. This is shown by the uppermost spectrum in Fig. 5. Also the characteristic negative anisotropy below the E_1 energy is enhanced by excitonic effects.

Figure 6 shows the calculated slab polarisability. The $[1\bar{1}0]$ component, i.e., the component probed by light with a polarisation direction parallel to

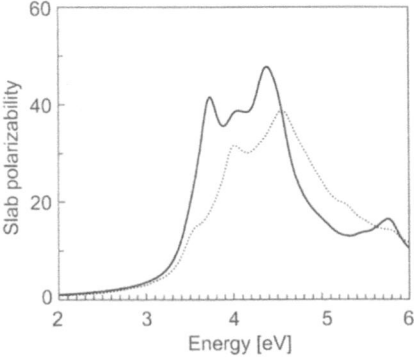

Fig. 6. Imaginary parts of the $\alpha_{[1\bar{1}0]}$ (solid line) and $\alpha_{[001]}$ components (dotted line) of the slab polarisability calculated for Si(110):H in GWA with the effects of local fields and the electron-hole attraction included.

the Si-Si zig-zag chains shows a strong peak close to the E_1 energy. On the other hand, the $\alpha_{[001]}$ component has only a weak shoulder at the E_1 energy. This difference is responsible for the strong optical anisotropies measured for passivated Si(110) surfaces.

The stepwise inclusion of many-particle effects in the calculation leads to a considerable and systematic improvement of the agreement with the experiment. However, even the uppermost curve in Fig. 5 still deviates from the measured data. On the one hand, this concerns the line shape around the E_2 peak. This is mainly due to the insufficient thickness of our slab. That can be seen from DFT-LDA calculations [20] which are computationally far less expensive and can thus be extended to full numerical convergence. For the comparison with the experimental data [16] shown in Fig. 7 we have therefore extrapolated our calculated spectrum to a thicker slab by adding the difference between the 'GW+LF+Ex' and 'GW' curves from Fig. 5 to the RAS spectrum of Si(110):H calculated in GWA for a 24-layer slab. This simple procedure leads to a rather good description of the measured line shape. However, there remains still another discrepancy between calculation and experiment: the calculated peak positions occur at energies that are about 0.3 eV too high. Our calculations were performed at the theoretical equilibrium lattice constant of 5.378 Å. That leads to an increase of the energy splitting between occupied and empty states by about 0.1 eV compared to calculations at the experimental lattice constant. Temperature effects in the measured spectra which are neglected in our calculations result in a red-shift of the optical spectra by a similar amount [44]. The remaining difference to the experiment is related to numerical insufficiencies as the relatively small number of **k** points and our approximations in calculating the screened Coulomb potential W. The latter tend to lead to slightly overestimated excitation energies [33].

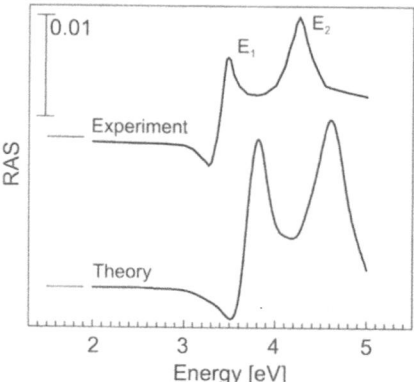

Fig. 7. Measured data for Si(110):H from Ref. [16] are compared with the extrapolation of the 'GW+LF+Ex' curve from Fig. 5 to calculations for a 24-layer slab (see text).

5 Summary

In summary, we have demonstrated that by using a time-evolution rather than a matrix-diagonalisation technique for the solution of the Bethe-Salpeter equation, it is now possible to include excitonic and local-field effects in the calculation of optical properties of complex systems consisting of many atoms. We calculated the optical anisotropy of the prototypical Si(110):H surface, the origin of which has been the subject of a long-standing controversy. It is shown that excitonic effects via strong modifications of the optical response of surface-modified bulk wave functions determine largely the line shape of the optical features. Local-field effects are found to play a much smaller role than previously thought. We expect the method presented here to be extremely useful for the accurate calculation of optical properties for many more systems characterised by a large number of electron-hole pairs.

Acknowledgements

We thank S. Glutsch, L. Reining and V. Olevano for very helpful discussions. Grants of computer time from the Höchstleistungsrechenzentrum Stuttgart are gratefully acknowledged.

References

1. D. E. Aspnes, Solid State Commun. **101**, 85 (1997).
2. D. E. Aspnes and A. A. Studna, Phys. Rev. Lett. **54**, 1956 (1985).
3. W. G. Schmidt, N. Esser, A. M. Frisch, P. Vogt, J. Bernholc, F. Bechstedt, M. Zorn, T. Hannappel, S. Visbeck, F. Willig, and W. Richter, Phys. Rev. B. **61**, R16335 (2000).

4. W. G. Schmidt, F. Bechstedt, and J. Bernholc, J. Vac. Sci. Technol. B **18**, 2215 (2000).
5. W. Lu, W. G. Schmidt, E. L. Briggs, and J. Bernholc, Phys. Rev. Lett. **85**, 4381 (2000).
6. L. Mantese, K. A. Bell, D. E. Aspnes, and U. Rossow, Phys. Lett. A **253**, 93 (1999).
7. W. L. Mochán and R. G. Barrera, Phys. Rev. Lett. **55**, 1192 (1985).
8. B. S. Mendoza, R. Del Sole, and A. I. Shkrebtii, Phys. Rev. B **57**, R12709 (1998).
9. C. H. Wu and W. Hanke, Solid State Commun. **23**, 829 (1977).
10. R. Del Sole and E. Fiorino, Phys. Rev. B **29**, 4631 (1984).
11. R. Del Sole and A. Selloni, Phys. Rev. B **30**, 883 (1984).
12. B. S. Mendoza and W. L. Mochán, Phys. Rev. B **55**, 2489 (1997).
13. S. Albrecht, L. Reining, R. Del Sole, and G. Onida, Phys. Rev. Lett. **80**, 4510 (1998).
14. L. X. Benedict, E. L. Shirley, and R. B. Bohn, Phys. Rev. Lett. **80**, 4514 (1998).
15. M. Rohlfing and S. G. Louie, Phys. Rev. Lett. **83**, 856 (1999).
16. T. Yasuda, D. E. Aspnes, D. R. Lee, C. H. Bjorkman, and G. Lucovsky, J. Vac. Sci. Technol. A **12**, 1152 (1994).
17. K. Hingerl, R. E. Balderas-Navarro, A. Bonanni, P. Tichopadek, and W. Schmidt, Appl. Surf. Sci. **175/176**, 769 (2001).
18. P. Y. Yu and M. Cardona, *Fundamentals of Semiconductors* (Springer-Verlag, Berlin, 1999).
19. A. Selloni, P. Marsella, and R. Del Sole, Phys. Rev. B **33**, 8885 (1986).
20. W. G. Schmidt and J. Bernholc, Phys. Rev. B **61**, 7604 (2000).
21. M. S. Hybertsen and S. G. Louie, Phys. Rev. B **34**, 5390 (1986).
22. L. J. Sham and T. M. Rice, Phys. Rev. **144**, 708 (1966).
23. W. Hanke and L. J. Sham, Phys. Rev. B **12**, 4501 (1975).
24. W. Hanke and L. J. Sham, Phys. Rev. B **21**, 4656 (1980).
25. R. Del Sole, Solid State Commun. **37**, 537 (1981).
26. J. Bernholc, E. L. Briggs, C. Bungaro, M. B. Nardelli, J. L. Fattebert, K. Rapcewicz, C. Roland, W. G. Schmidt, and Q. Zhao, phys. stat. sol. (b) **217**, 685 (2000).
27. W. G. Schmidt, P. H. Hahn, and F. Bechstedt, in *High Performance Computing in Science and Engineering 2001*, p. 178 (Springer-Verlag, Berlin, 2002).
28. F. Aryasetiawan and O. Gunnarsson, Rep. Prog. Phys. **61**, 237 (1998).
29. F. Bechstedt, in *Festköperprobleme / Advances in Solid State Physics*, edited by U. Rössler (Vieweg, Braunschweig/Wiesbaden, 1992), Vol. 32, p. 161.
30. L. Hedin, Phys. Rev. **139**, A769 (1965).
31. L. Hedin and S. Lundqvist, in *Solid State Physics*, edited by H. Ehrenreich, F. Seitz, and D. Turnball (Academic, New York, 1969), Vol. 23, p. 1.
32. M. S. Hybertsen and S. G. Louie, Phys. Rev. B **37**, 2733 (1988).
33. F. Bechstedt, R. Del Sole, G. Cappellini, and L. Reining, Solid State Commun. **84**, 765 (1992).
34. G. Cappellini, R. Del Sole, L. Reining, and F. Bechstedt, Phys. Rev. B **47**, 9892 (1993).
35. D. E. Aspnes and A. A. Studna, Phys. Rev. B **27**, 985 (1983).
36. M. Rohlfing and S. G. Louie, Phys. Rev. Lett. **80**, 3320 (1998).
37. P. H. Hahn, W. G. Schmidt, and F. Bechstedt, Phys. Rev. Lett. **88**, 016402 (2002).

38. V. Olevano and L. Reining, Phys. Rev. Lett. **86**, 5962 (2001).
39. S. Albrecht, Ph.D. thesis, École Polytechnique, Paris, 1999.
40. M. Cardona, L. F. Lastras-Martinez, and D. E. Aspnes, Phys. Rev. Lett. **83**, 3970 (1999).
41. S. Albrecht, L. Reining, G. Onida, V. Olevano, and R. Del Sole, Phys. Rev. Lett. **83**, 3971 (1999).
42. S. Glutsch, D. S. Chemla, and F. Bechstedt, Phys. Rev. B **54**, 11592 (1996).
43. P. H. Hahn, Master's thesis, Friedrich-Schiller-Universität Jena, 2001.
44. P. Lautenschlager, M. Garriga, L. Viña, and M. Cardona, Phys. Rev. B **36**, 4821 (1987).

Frozen Phonon Calculations in the Three-Band Hubbard Model for High-Temperature Superconductors

Zhongbing Huang, Werner Hanke, and Enrico Arrigoni

Institut für Theoretische Physik, Universität Würzburg, Am Hubland,
97074 Würzburg, Germany

Abstract. On the basis of Quantum Monte Carlo (QMC) simulations, we study the influence of specific phonon, i.e., the oxygen half-breathing $(\pi, 0)$ mode in the three-band Hubbard model as a relevant model for high-temperature superconductors. In the hole-doped case, both the diagonal and the off-diagonal couplings dramatically change the total energy and local spin correlations. However, in the electron-doped case, the diagonal electron-phonon coupling changes the total energy much more than the off-diagonal coupling, and an essential difference from hole doping is that the electron-phonon coupling has negegible effects on local spin-spin correlations.

1   Introduction

Recently, a large number of experimental observations in the high-temperature superconductors (HTSC) suggest that the electron-phonon coupling in these systems is strong and -in conjunction with the strong electronic correlations, may play a role for the superconductivity mechanism. In particular, photoemission data indicate a sudden change in the dispersion near a characteristic energy scale [1], which is possibly caused by coupling of electronic quasiparticles to phonon modes. Morever, photoemission also shows that the electron-phonon coupling in p-type (hole-doped actual HTSC) cuprates is stronger than in n-type (electron-doped) ones, which is basically due to the fact that the doped holes predominately go to oxygen sites, whereas the doped electrons prefer occupying the copper sites [1]. In addition, neutron scattering data suggest that the oxygen half-breathing $(\pi, 0)$ phonon couples strongly to doped charge carriers and correspondingly, softens significantly with doping [2,3]. In view of the strong electron-phonon coupling and the coupling of charge ordering to lattice distortions, specific phonon modes are likely to play a decisive role in the structural stability.

In order to understand this asymmetry of electron-phonon coupling in p- and n-type cuprates, and also extract the relevant electron-phonon coupling in the renormalized one-band Hubbard model, we have studied the three-band Peierls-Hubbard model in the physically interested parameters regime, by employing the Constrained Path Monte Carlo (CPMC) method [4].

2 Model and Technique

In the hole representation the three-band Peierls-Hubbard model [5] has the Hamiltonian,

$$H = \sum_{\langle i,j\rangle\sigma} t_{dp}^{ij}(\{u_{ij}\})(d_{i\sigma}^\dagger p_{j\sigma} + h.c.) + \sum_{\langle j,k\rangle\sigma} t_{pp}^{jk}(p_{j\sigma}^\dagger p_{k\sigma} + h.c.) + \epsilon_p \sum_{j\sigma} n_{j\sigma}^p$$

$$+ \sum_{i\sigma} \epsilon_d(\{u_{ij}\})n_{i\sigma}^d + U_d \sum_i n_{i\uparrow}^d n_{i\downarrow}^d + U_p \sum_j n_{j\uparrow}^p n_{j\downarrow}^p + V_{pd} \sum_{\langle i,j\rangle} n_i^d n_j^p. \quad (1)$$

Here the operator $d_{i\sigma}^\dagger$ creates a hole at a Cu $3d_{x^2-y^2}$ orbital and $p_{j\sigma}^\dagger$ creates a corresponding hole in an O $2p_x$ or $2p_y$ orbital. U_d and U_p are the coulomb energies at the Cu and O sites, respectively, $\epsilon_d(\{u_{ij}\}) = \epsilon_d + g\sum_j \pm u_{ij}$ and ϵ_p correspond to orbital energies. V_{pd} denotes the nearest-neighbor Coulomb repulsion. The charge-transfer energy is defined as $\epsilon = \epsilon_p - \epsilon_d$. As written, the model has a Cu-O hybridization $t_{pd}^{ij} = \phi_{ij}(t_{pd} \pm \alpha u_{ij})$, where the $+(-)$ applies if the bond shrinks (stretches) with positive u_{ij}. The phase factor ϕ_{ij} takes a minus sign for $j = i + \hat{x}/2$ and $j = i - \hat{y}/2$. The hybridization is $t_{pp}^{jk} = \pm t_{pp}$ between oxygen sites with the minus sign occurring for $k = j - \hat{x}/2 - \hat{y}/2$ and $k = j + \hat{x}/2 + \hat{y}/2$.

For simplicity, we introduce dimensionless parameters: $u = \alpha u_{ij}/t_{dp}$, or $u = g u_{ij}/t_{dp}$. In units of t_{dp}, the parameters are: $U_d = 6.0$, $\epsilon = 3.0$, $U_p = V_{dp} = 0.0$, and $t_{pp} = 0.3$ and 0.5 in the 6x4 and 6x6 lattices, respectively. For hole doping, the doping density δ is positive. While for electron doping, δ is negative.

Figure 1 shows the displacement pattern of the zone-boundary half-breathing phonon mode. C1,C2,C3 and C4 label the copper sites, and O1,O2 and

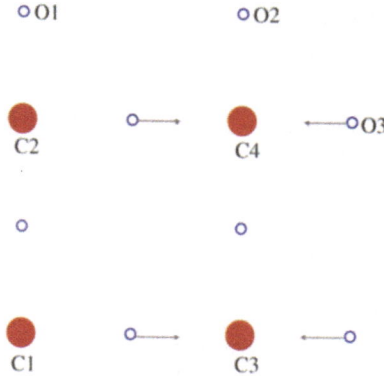

Fig. 1. Displacement pattern in the three-band Hubbard model. Closed circles represent copper, open circles oxygen. The arrowed lines from oxygen sites indicate the lattice displacements.

O3 stand for the different oxygen sites. In the off-diagonal coupling, the hopping term t_{pd} is modified, whereas the on-site energy ϵ_d is changed in the diagonal coupling. Shen *et. al.* [1] have recently discussed the electron-phonon coupling in the cuprate superconductors, mainly in the framework of a single-band Hubbard or t-J model in detail. In p-type materials, because the doped charge carriers mainly occupy oxygen sites, the off-diagonal coupling is stronger than the diagonal coupling: In p-type materials which are hole-doped and constitute the actual high-Tc materials, it is well known that the holes on oxygen form a so-called Zhang-Rice singlet with the holes on copper. This singlet formation is driven by basically the same mechanism responsible for antiferromagnetic exchange J, namely the actual hopping of holes between Cu and O sites. It is, therefore, clear that changing t_{pd} as in the off-diagonal electron-phonon coupling case significantly affects p-type cuprates energetics. However, in n-type materials, the doping charges mainly stay on copper sites, hence it is expected that the diagonal coupling is more relevant. Our first step here is to check the importance of a variety of electron-phonon couplings for the hole- and electron-doped three-band Hubbad model. Then, we aim at deriving from this a sound basis for discussing these couplings in the one-band models, for which we calculate the effective vertex function Γ.

The numerical algorithm is the constrained path Monte Carlo (CPMC) method [4]. In the CPMC method, the ground-state wave function $|\psi_0\rangle$ is projected from a known initial wave function $|\psi_T\rangle$ by a branching random walk in an over-complete space of Slater determinants $|\phi\rangle$.

To completely specify the ground-state wave function for a system of interacting electrons, only determinants satisfying $\langle\psi_0|\phi\rangle > 0$ are needed, because $|\psi_0\rangle$ resides in either of two degenerate halves of the Slater determinant space, separated by a nodal surface $\mathbf{N}$ that is defined by $\langle\psi_0|\phi\rangle = 0$. Without *a priori* knowledge of $\mathbf{N}$, we use a trial wave function $|\psi_T\rangle$ and require $\langle\psi_T|\phi\rangle > 0$. This is what is called the constrained-path approximation.

3 Hole Doping

In Fig. 2, we plot the total energy change as a function of displacement u. On the 6×4 and 6×6 lattices, the energy changes produced by the frozen phonon displacements are similar. For both the diagonal and off-diagonal couplings, the energy decreases with similar values. This suggests that both electron-phonon interactions couple strongly to the electronic degree of freedom. In contrast to Shen *et al's* [1] argument, i.e., that the direct electron-phonon coupling originating from the shift of the d-level due to the oxygen displacement is missing, we observe that the diagonal coupling also influences the system strongly. Especially, the charge distributions on the copper sites are changed by this energy shift. In fact, in the physically relevant parameter regime, the charge density on the copper site is significantly less than one,

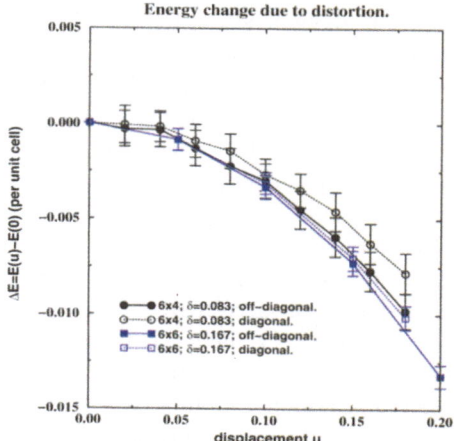

Fig. 2. Energy change vs the lattice displacement for two hole doping densities on the 6×4 and 6×6 lattices.

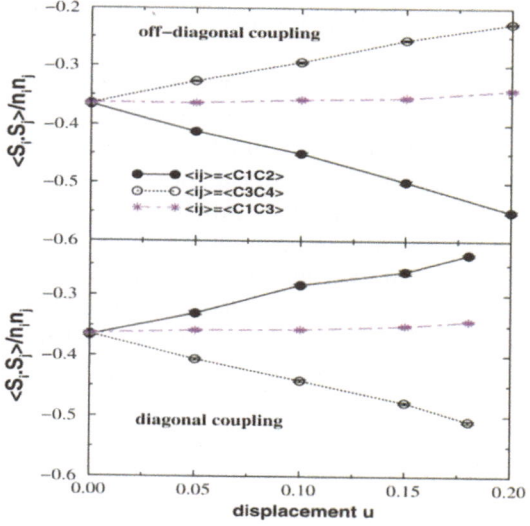

Fig. 3. Spin-spin correlation vs the displacement at the hole doping $\delta = 0.167$ on the 6×6 lattice. $< ij >$ denote the nearest neighbor copper sites i and j (see Fig. 1).

hence also the diagonal coupling should be effective in the hole doping region.

Figure 3 displays the spin-spin correlations between copper sites. For both couplings, we observe that the spin-spin correlation increases or decreases alternatively in the y direction (vertical to the displacement). This pattern is similar to the off-diagonal coupling in the one-band Hubbard model, which changes the hopping matrix in the y direction. We note here that both couplings give rise to similar changes of local spin correlations. In general, when we transform the three-band Hubbard model into the t-J model, the change of copper energy levels caused by the diagonal electron-phonon coupling will induce off-diagonal couplings (changes of t and J) in the t-J model [1,6]. Surprisingly, we find that in the three-band Hubbard model, the off-diagonal coupling has an important effect on the spin correlations.

When we transform the three-band Hubbard model to effective one-band models, i.e., Hubbard and t-J models, the off-diagonal coupling (in the three-band Hubbard model) is transformed into two important components (in the one-band models). On the one hand, due to changes of local energies of the Zhang-Rice singlet and the pure copper d-hole, this off-diagonal coupling is transformed into a diagonal one: we can write it as $H_{\mathrm{diag}} = - \sum_{i\sigma} g(Q)(b_Q + b_Q^\dagger)n_{i\sigma}e^{iQR_i}$ [7]. On the other hand, it can also contribute to the spin-exchange interactions. As shown in Fig. 4, different order perturbative expansions induce different contributions in J [8]. Only when we consider higher spin exchange processes, do we get an off-diagonal coupling in the t-J model. As seen in Fig. 3, this off-diagonal coupling affects the spin-spin correlations, dramatically.

In our simulations, we add the frozen phonon to the electron system, which represents a static lattice deformation pattern, i.e., the adiabatic approximation. As a result, the charge distributions become nonuniform. As is well known, the doped holes mainly reside on the oxygen sites. Therefore, if the electron-phonon coupling dramatically influences the charge distributions

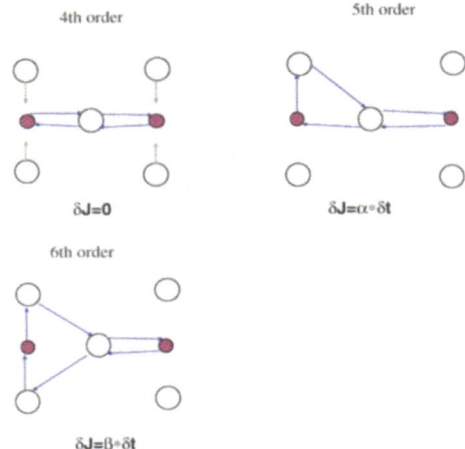

Fig. 4. Cu-Cu Superexchange processes. Closed circles label copper, open circles oxygen. The arrowed solid lines connecting copper and oxygen sites represent hopping processes.

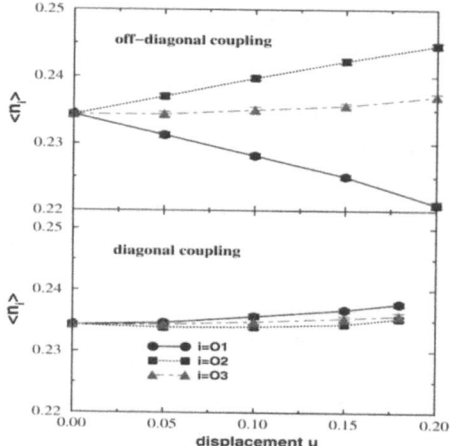

Fig. 5. Charge distributions at oxygen sites vs the displacement at the hole doping density $\delta = 0.167$ on the 6×6 lattice. $i = O1, O2, O3$ denote the oxygen sites (see Fig. 1).

at oxygen sites, it must play an important role in the charge dynamics and possibly superconductivity. From Fig. 5, we see that the diagonal coupling has a rather small effect on the charge distributions, whereas the off-diagonal coupling induces a pronounced CDW pattern in the x direction.

4   Electron Doping

Since experiments found that the electron-phonon coupling is substantial for hole doping, but rather weak for electron doping into the cuprate superconductors, we should observe such an asymmetric effect in the above three-band Hubbard model. In what follows, we therefore consider in detail the electron-doped case.

In Fig. 6 the energy change is plotted as a function of lattice displacement. We see clearly that the diagonal coupling decreases the energy much more effectively than the off-diagonal coupling. This suggests that in electron doping systems, the diagonal coupling is substantially stronger than the off-diagonal coupling.

Figure 7 shows the spin-spin correlations between copper sites. For both couplings, the frozen lattice displacement has a very small effect on the spin correlation. Compared with hole doping (Fig. 3), the off-diagonal spin-phonon coupling induced by the frozen phonon is much weaker. This dramatic difference of the spin-phonon coupling can possibly account for the different electron-phonon coupling constant λ, estimated from photoemission [1]. For p-type superconductors, λ is around 1, while it is close to zero for n-type superconductors. Especially, in the underdoped region, the diagonal couplings

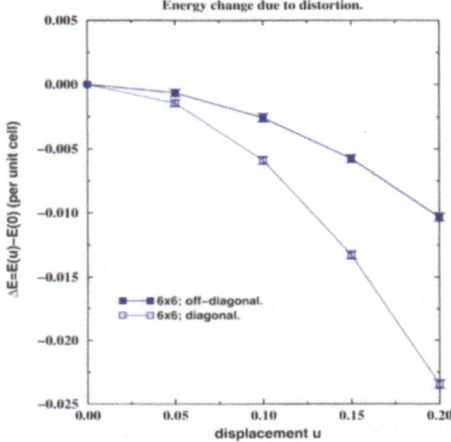

Fig. 6. Energy change vs the displacement at the electron doping $\delta = -0.278$ on the 6×6 lattice.

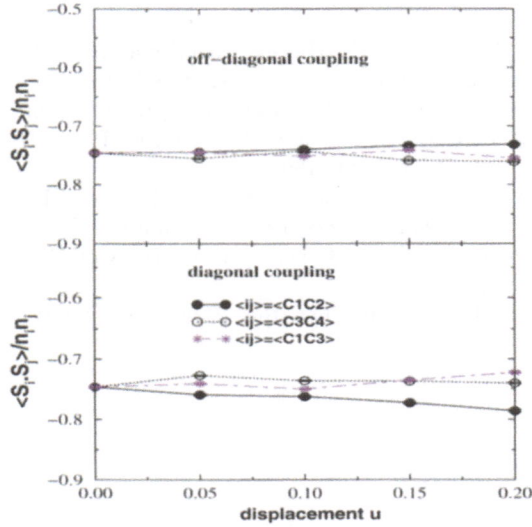

Fig. 7. Spin-spin correlation vs the displacement at the electron doping $\delta = -0.278$ on the 6×6 lattice. $< ij >$ denote the nearest neighbor copper sites i and j (see Fig. 1).

are strongly reduced by the suppression of charge fluctuations, hence the only relevant electron-phonon coupling should be due to the off-diagonal electron-phonon couplings (spin-phonon coupling).

5 Conclusion

The interplay of strong electronic correlations and electron-phonon interactions is an extremely difficult subject to resolve but simultaneously a very important topic in the theory of HTSC. In order to get a first numerically accurate insight into this issue, we have studied the 2D Peierls three-band Hubbard model with frozen-in lattice distortions. We found that in the physically interesting parameters range, both the diagonal and off-diagonal electron-phonon couplings are important for the hole-doped systems, whereas the diagonal coupling contributes more than the off-diagonal coupling in the electron-doped systems. Our present work is done on the unit cells 6×4 and 6×6, and for a few limited dopings. To extend our understanding of the interplay of strong electronic correlations with electron-phonon interactions for more arbitrary displacement patterns and corresponding wave vectors, we plan to extend our calculations to larger lattices.

References

1. Z. X. Shen, A. Lanzara and N. Nagaosa, cond-mat/0108381.
2. R. J. Mcqueeney *et al.*, Phy. Rev. Lett. **82**, 628 (1999).
3. L. Pintschovius and Braden, Phys. Rev. B **60**, R15039 (1999).
4. Shiwei Zhang, J. Carlson, and J. E. Gubernatis, Phys. Rev. Lett. **74**, 3652 (1995); Phys. Rev. B **55**, 7464 (1997).
5. K. Yonemitsu, A. Bishop, and J. Lorenzana, Phys. Rev. Lett. **69**, 965 (1992).
6. J. Lorenzana and G. A. Sawatzky, Phys. Rev. Lett. **74**, 1867 (1995).
7. K. J. von Szczepanski, K. W. Becker, Z. Phys. B – Condensed Matter **89**, 327–334 (1992).
8. H. Eskes and J. H. Jefferson, Phys. Rev. **B48**, 9788 (1993).

Exact Diagonalization Study of Spin, Orbital, and Lattice Correlations in CMR Manganites

Alexander Weiße[1], Gerhard Wellein[2], and Holger Fehske[3,1]

[1] Physikalisches Institut, Universität Bayreuth, D-95440 Bayreuth
[2] Regionales Rechenzentrum Erlangen, Universität Erlangen, D-91058 Erlangen
[3] Institut für Physik, Universität Greifswald, D-17487 Greifswald

Abstract. To understand the interplay of spin, orbital and lattice degrees of freedom in colossal magneto-resistance manganites we numerically diagonalize an $SU(2)$ symmetric spin-orbital model coupled to dynamic Jahn-Teller and Holstein-type phonons. For a four site cluster we demonstrate how the coupling to the lattice changes the order of spins, orbitals and charges, and the correlations between them.

1 Introduction

The transition from a metallic ferromagnetic low-temperature phase to an insulating paramagnetic high-temperature phase observed in some hole-doped manganese oxides (e.g. in $La_{1-x}[Sr, Ca]_x MnO_3$) is associated with an unusual dramatic change in their electronic and magnetic properties, including a spectacularly large negative magneto-resistive response to an applied magnetic field, which might have important technological applications [1].

Apart from this so-called colossal magneto-resistance (CMR) transition hole-doped manganites exhibit a very complex and fascinating phase diagram (see Fig. 1). As a result of the subtle interplay of almost all degrees of freedom known in solid state physics different crystal structures and magnetic, charge and orbital ordered states are observed experimentally in dependence on temperature and doping level. Although such a striking behaviour has stimulated a considerable amount of both experimental and theoretical work [2] in the last decades, much of the basic physics of the CMR still remains controversial.

At present, there has been a renewed interest in simplified model Hamiltonians, capable of describing both the electronic structure of CMR manganites as well as the many-body correlations due to the interaction of charge, spin, orbital, and lattice degrees of freedom. Computational techniques then provide a useful tool to analyze the properties of these microscopic Hamiltonians, at least, in the difficult regime, where the correlations are strong and all interactions have to be treated on an equal footing. Within our HLRS project, reported on in this paper, we went a step in this direction by exactly diagonalizing a rather general low-energy model for the CMR manganites on a small cluster.

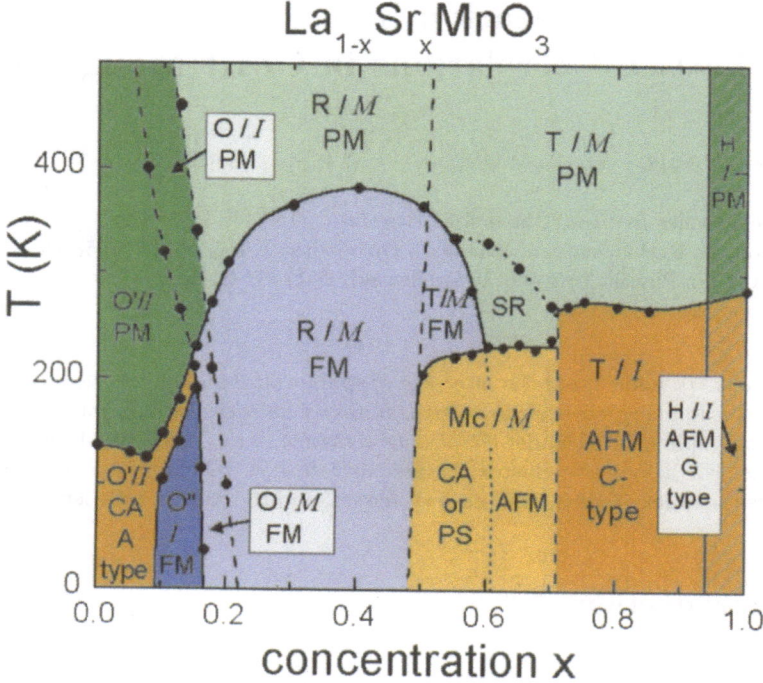

Fig. 1. Phase diagram of $La_{1-x}Sr_xMnO_3$ taken from Ref. [3]. The crystal structures (Jahn-Teller distorted orthorhombic: O', orthorhombic: O, orbital-ordered orthorhombic: O", rhombohedral: R, tetragonal: T, monoclinic: Mc, and hexagonal: H) are indicated as well as the magnetic structures (paramagnetic: PM (green), short-range order (SR), canted (CA), A-type antiferromagnetic structure: AFM (yellow); ferromagnetic: FM (blue), phase separated (PS), and AFM C-type structure) and the electronic state (insulating: I (dark); metallic: M (light)).

2   Theoretical model

The key elements of the electronic structure of the manganites are the partially filled $3d$ states. The cubic environment of the Mn sites within the perovskite lattice results in a crystal field splitting of Mn d-orbitals into e_g and t_{2g} (cf. Fig. 2). In the case of zero doping ($x = 0$) there are four electrons per Mn site which fill up the three t_{2g} levels and one e_g level, and by Hund's rule coupling (J_h), form a $S = 2$ spin state. Doping will remove the electron from the e_g level, and by hopping via bridging oxygen sites the resulting holes acquire mobility.

Due to the specific symmetry of the manganese d and oxygen p orbitals, the transfer of the e_g-electrons shows a pronounced (orbital) anisotropy (see Fig. 3). In the limit of large on-site Coulomb interaction U and Hund's rule coupling J_h the electron transfer is strongly affected by the spin of the core electrons as well. Concentrating on the link between magnetic correlations

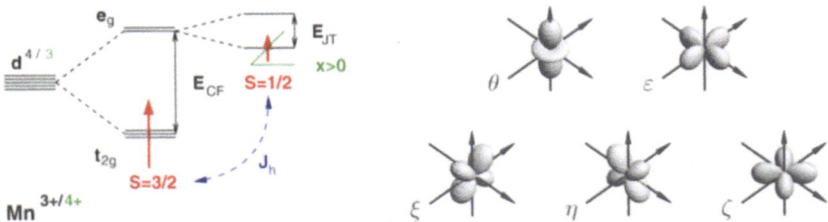

Fig. 2. Left: Crystal-field and Jahn-Teller splitting of the five-fold degenerate atomic Mn $3d$ levels (half-filled t_{2g} triplets form local spins $S = 3/2$ interacting ferromagnetically with electrons in single occupied e_g levels); right: e_g (θ, ε) and t_{2g} (ξ, η, ζ) orbitals.

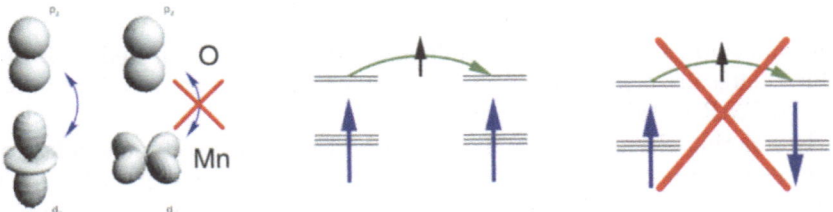

Fig. 3. Left: Along z-direction, electrons can hop only from $\theta = |3z^2 - r^2\rangle$ orbitals; transfer processes involving $\varepsilon = |x^2 - y^2\rangle$ orbitals are forbidden due to vanishing overlap with the in-between O $2p$ states. Right: Double-exchange model: If on-site Coulomb interaction U and Hund's coupling J_h are strong, hopping is allowed if the core spins are aligned and vanishes in the case of antiparallel orientation, i.e., itinerant e_g electrons cause a ferromagnetic interaction of localized t_{2g} spins.

and transport, early studies on lanthanum manganites attributed the low-T metallic behaviour to Zener's double-exchange mechanism, which maximizes the hopping of a strongly Hund's rule coupled e_g-electron in a polarized background of the Mn t_{2g}-electron spins (see Fig. 3).

Recently it has been realized that physics beyond double-exchange is important not only to explain the phase diagram of the manganites but also the CMR transition itself. In particular, orbital and lattice effects seem to be crucial in explaining the CMR phenomenon. More specifically, the orbital degeneracy in the ground state of Mn^{3+} ions connects the system to the lattice, making it sensible to Jahn-Teller distortion and polaronic effects. There are two types of lattice distortions which are important in manganites (see Fig. 4). First the partially filled e_g states of the Mn^{3+} ion are Jahn-Teller active, i.e., the system can gain energy from a quadrupolar symmetric elongation of the oxygen octahedra which lifts the e_g degeneracy. A second possible deformation is an isotropic shrinking of a MnO$_6$ octahedron. This "breathing"-type distortion couples to changes in the e_g charge density, i.e., is always associated with the presence of an Mn^{4+} ion. In the heavily doped

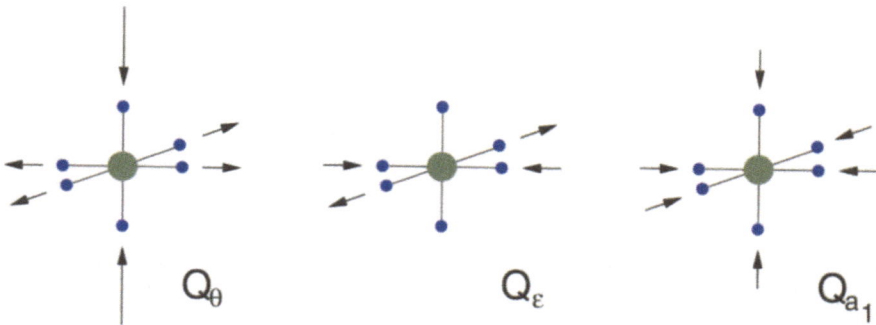

Fig. 4. Jahn-Teller and breathing-type phonon modes.

material, both, breathing-mode collapsed (Mn^{4+}) and Jahn-Teller distorted (Mn^{3+}) sites are created simultaneously when the holes are localized in passing the CMR metal insulator transition.

Restricting the electronic Hilbert space to the large Hund's rule states given by the spin-2 orbital doublet state ^5E [$t_2^3(^4A_2)e$] for Mn^{3+} (d^4) and the spin-$\frac{3}{2}$ orbital singlet state $^4A_2$ [t_2^3] for Mn^{4+} (d^3), within 2nd order perturbation theory the following Hamiltonian results (for details see Ref. [4]):

$$\mathcal{H} = \mathcal{H}_{\text{double-exchange}} + \mathcal{H}_{\text{spin-orbital}}^{\text{2nd order}} + \mathcal{H}_{\text{electron-JT}}$$
$$+ \mathcal{H}_{\text{electron-breathing}} + \mathcal{H}_{\text{phonon}}$$

$$= \sum_{i,\delta,\alpha,\beta} (a_{i,\uparrow} a_{i+\delta,\uparrow}^\dagger + a_{i,\downarrow} a_{i+\delta,\downarrow}^\dagger)\, t_{\alpha\beta}^\delta\, c_{i,\alpha}^\dagger n_{i,\bar\alpha} n_{i+\delta,\bar\beta} c_{i+\delta,\beta}$$

$$+ \sum_{i,\delta,\kappa,\lambda} (J_{\kappa\lambda}^\delta\, \mathbf{S}_i \mathbf{S}_{i+\delta} + \Delta_{\kappa\lambda}^\delta)\, P_i^\kappa P_{i+\delta}^\lambda$$

$$+ g \sum_i \left[(n_{i,\varepsilon} - n_{i,\theta})(b_{i,\theta}^\dagger + b_{i,\theta}) + (d_{i,\theta}^\dagger d_{i,\varepsilon} + d_{i,\varepsilon}^\dagger d_{i,\theta})(b_{i,\varepsilon}^\dagger + b_{i,\varepsilon}) \right]$$

$$+ \tilde{g} \sum_i (n_{i,\theta} + n_{i,\varepsilon} - 2n_{i,\theta}n_{i,\varepsilon})(b_{i,a_1}^\dagger + b_{i,a_1})$$

$$+ \omega \sum_i \left[b_{i,\theta}^\dagger b_{i,\theta} + b_{i,\varepsilon}^\dagger b_{i,\varepsilon} \right] + \tilde{\omega} \sum_i b_{i,a_1}^\dagger b_{i,a_1}. \tag{1}$$

The effective low-energy Hamiltonian $\mathcal{H}$ contains Schwinger bosons $a_{i,\mu}^{(\dagger)}$, i.e. $2\mathbf{S}_i = a_{i,\mu}^\dagger \boldsymbol{\sigma}_{\mu\nu} a_{i,\nu}$ ($\mu, \nu \in \{\uparrow, \downarrow\}$), fermionic holes $c_{i,\alpha}^{(\dagger)}$, phonons $b_{i,\alpha}^{(\dagger)}$ ($\alpha \in \{\theta, \epsilon\}$), and orbital projectors $P_i^{\kappa(\lambda)}$ ($\kappa, \lambda \in \{\xi, \eta, \zeta\}$). In Eq. (1), the first term, being proportional to t, corresponds to the well known double exchange interaction [5]. The second term appears to be a bit more involved, since a rather large number of accessible virtual excitations (proportional to t^2 and t_π^2) contribute (cf. Fig. 5). However, in all cases it is basically the product of a Heisenberg-type spin interaction and two orbital projectors. The coupling be-

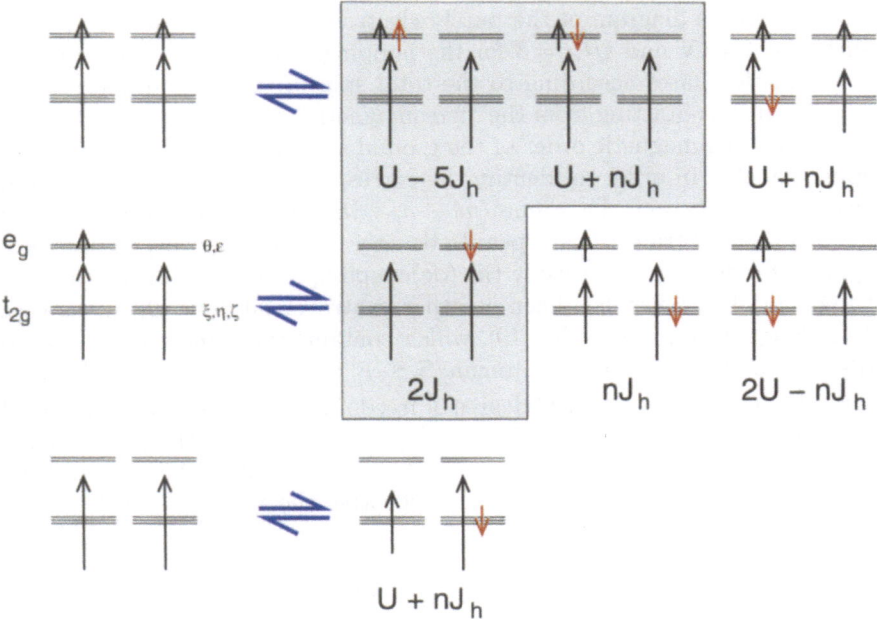

Fig. 5. Virtual excitations accounted for within 2nd order perturbation theory. The shaded region corresponds to t^2-terms, the other terms are proportional to t_π^2.

tween the orbital degree of freedom of the e_g electrons and the optical phonon modes to lowest order in Q is modeled by the $E \otimes e$ Jahn-Teller Hamiltonian (third term) and a Holstein-type interaction (fourth term). The energy of the dispersionless optical phonons are given within harmonic approximation (fifth term). Using a density matrix based optimization procedure [6], we are able to retain the full quantum dynamic of these phonon modes within our numerical solution of the model on small clusters.

3   Numerical results

3.1   Undoped case

Undoped manganites ($LaMnO_3$, $PrMnO_3$) usually exhibit A-type anti-ferromagnetic order and strong Jahn-Teller distortion of the ideal perovskite structure. The origin of the observed magnetic order has been subject to discussions. While different band structure calculations [7] emphasize the importance of lattice distortions for the stability of anti-ferromagnetism, Feiner and Oleś [8] favoured a purely electronic mechanism.

Our calculation points out that both parameters, U/J_h and g, can drive a ferromagnetic to antiferromagnetic transition. The lower right panel of Fig. 6

shows the phase diagram of the purely electronic model, i.e., $g = 0$. We assumed $t = 0.4$ eV and $t/t_\pi = 3$ for the hopping integrals and characterized the magnetic phases according to the total spin of the ground state of the four site cluster. Starting from the "ferromagnetic" phase both, increasing U or g change the magnetic order of the ground state to "antiferromagnetism" (upper panels). In order to identify the corresponding orbital order we consider the local expectation value $\langle n_\theta - n_\varepsilon \rangle$. In view of the distinct driving interactions both transitions appear to be very similar. However, we observe a significant difference, if we study the (de)coupling of spin and orbital degrees of freedom. The latter has been a rather controversial issue [9] in the case of the Kugel-Khomskii model [10], which contains the same kind of second order interactions, as our Hamiltonian, $\mathbf{S}_i\mathbf{S}_j\,\tau_i^\delta\tau_j^\delta$ (here the pseudo-spin operators τ_i^δ operate on the orbital degree of freedom). The lower left-hand panel indicates that the correlation $\langle \mathbf{S}_i\mathbf{S}_{i+\delta}\,\tau_i^\delta\tau_{i+\delta}^\delta \rangle - \langle \mathbf{S}_i\mathbf{S}_{i+\delta} \rangle\langle \tau_i^\delta\tau_{i+\delta}^\delta \rangle$ is a factor of $3 - 5$ smaller, if phonons are responsible for the FM to AFM transition. This behaviour is of course crucial for effective theories that are based on such decoupling schemes.

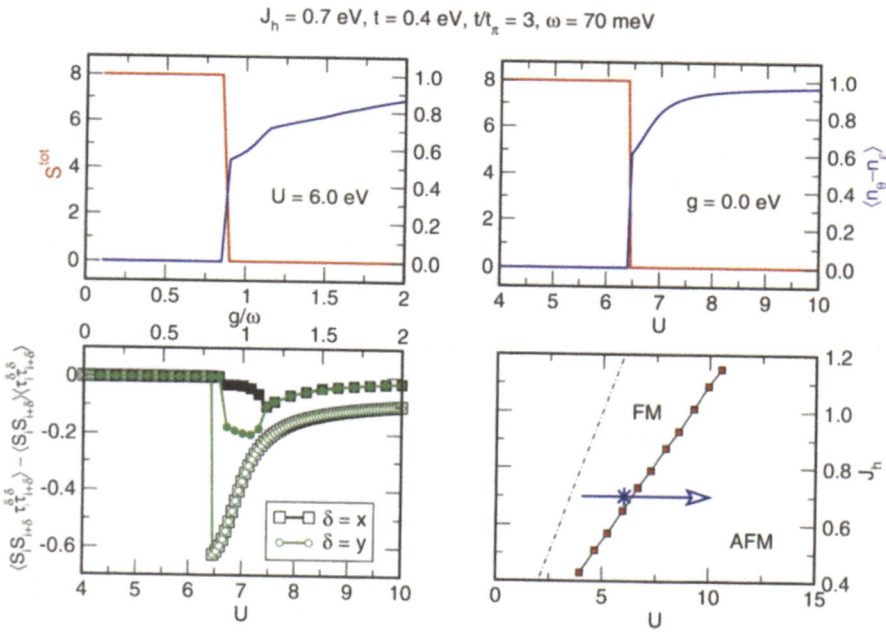

Fig. 6. Upper panels: Total spin S^{tot} (red line) and orbital order (blue line) of the ground state at variable electron-phonon coupling g and Coulomb interaction U. Lower left panel: Dependence of the spin-orbital correlations on electron-phonon ($U = 6$; filled symbols) and Coulomb ($g = 0$; open symbols) interactions. Lower right panel: phase diagram of the electronic model without phonons.

The change of orbital and phonon correlations is illustrated graphically in Fig. 7. Note however, that this is a rather suggestive picture. Studying the eigenstates of the orbital density matrix on a bond $\langle ij \rangle$, we can classify the states according to their behaviour under site exchange. Anti-symmetric states $|a\rangle_{ij} = \frac{1}{\sqrt{2}} \left(|\theta\rangle_i \otimes |\varepsilon\rangle_j - |\varepsilon\rangle_i \otimes |\theta\rangle_j \right)$ are unique, whereas symmetric states $|s(\varphi, \psi)\rangle_{ij} = \frac{1}{\|.\|} \left(|\varphi\rangle_i \otimes |\psi\rangle_j + |\psi\rangle_i \otimes |\varphi\rangle_j \right)$ can be written as a product of two rotated orbitals $|\varphi\rangle_i = \cos(\varphi)|\theta\rangle_i + \sin(\varphi)|\varepsilon\rangle_i$ [4]. Since for small g the orbital configuration is antisymmetric for each bond, we can not draw the correct pattern, but choose only an artificial sketch of it.

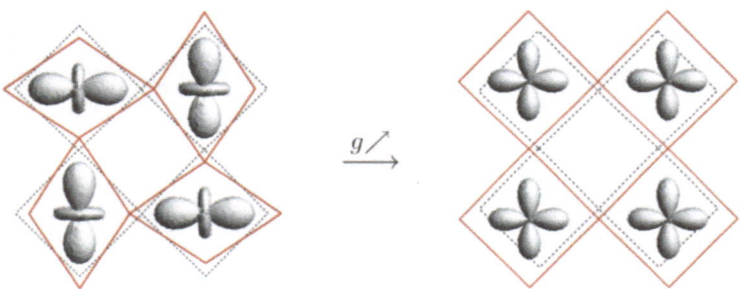

Fig. 7. Evolution of lattice and orbital correlations with increasing electron-phonon coupling g at doping $x = 0$ (schematic view).

3.2   Finite doping

As can be seen from the phase diagram of $La_{1-x}Sr_xMnO_3$ (Fig. 1), in the CMR regime ($0.15 < x < 0.5$), the metallic low-temperature phase is related to ferromagnetic long-range order stabilized by the double exchange interaction. Our numerical calculations for the weakly doped case ($x = \frac{1}{4}$) corroborate the enhancement of ferromagnetic correlations. However, if strong electron-phonon coupling causes localization of the carriers the spin order switches to antiferromagnetism. This coincidence is illustrated in Fig. 8 (upper panels) showing the total spin of the cluster and the kinetic energy in the ground state. The change in the magnetic order is accompanied by the appearance of a lattice distortion and a signature in the fluctuation of the bond length ($\propto \langle q_{x/y}^2 \rangle - \langle q_{x/y} \rangle^2$), which reminds of the data measured close to the critical temperature by Booth et al. [11] (lower left panel). The orbital orientation at the sites which surround the hole is sketched in Fig. 9. Obviously increasing g isolates the lattice sites, each optimizing electron-phonon interaction individually and uncorrelated with the neighbours.

At doping $x = \frac{1}{2}$ the picture is more involved. Strong Coulomb and electron-phonon interactions tend to order the charges in diagonal direction,

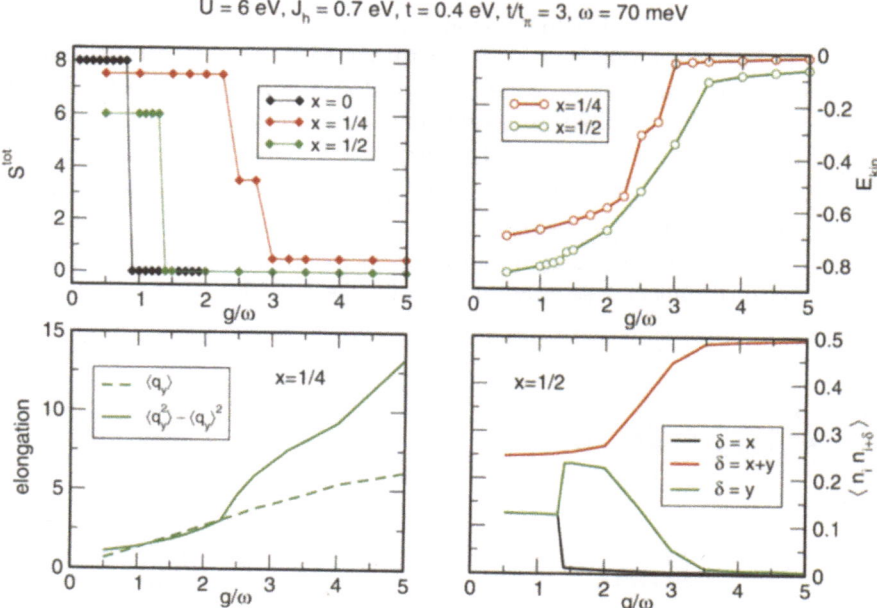

Fig. 8. Upper panels: Total spin S^{tot} and kinetic energy E_{kin} as a function of electron-phonon coupling strength g/ω at various doping levels x. Lower panel: Expectation values $\langle q_y \rangle$ and $\langle q_y^2 \rangle - \langle q_y \rangle^2$ of the bond length in y direction at $x = 1/4$ (left) and density-density correlations at $x = 1/2$.

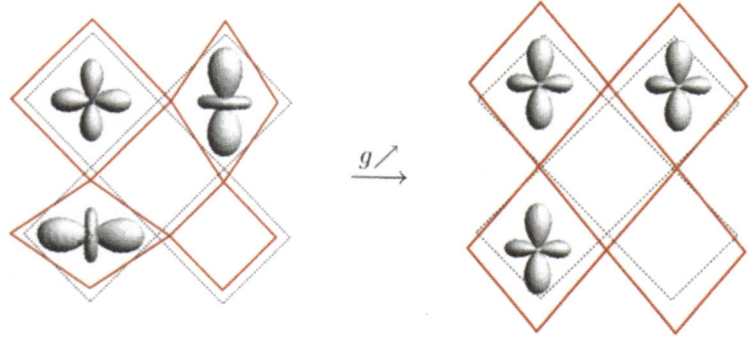

Fig. 9. Evolution of lattice and orbital correlations with increasing electron-phonon coupling g at doping $x = 1/4$.

i.e., in an AB-structure (compare Fig. 8, lower right panel). This allows for a rather large anti-ferromagnetic spin exchange $\propto t^2/J_h$. Consequently ferromagnetic order is unstable at much lower values of g. The ferromagnetic to antiferromagnetic transition is not connected to charge localization and causes only a tiny jump of the kinetic energy. Considering the most relevant eigenstate of the bond orbital density matrix, we observe a symmetric order of complex orbitals along the diagonal [4]. After charge localization is achieved at large g, neighbouring sites are again uncorrelated with respect to orbital ordering and are in some real mixed-orbital state.

4   Performance analysis on supercomputers

Exact diagonalization studies of microscopic electron-phonon models involve very large sparse matrices, even for small clusters. Since the matrix size accessible for diagonalization determines the quality of our results, both continuous access to the most powerful supercomputers and steady improvements of algorithms and implementations are the technical basics of our project. Over the last years Lanczos, Jacobi-Davidson, density-matrix, kernel-polynomial expansion, and maximum-entropy algorithms have been successfully implemented on numerous architectures including CRAY T3E, NEC SX-4/5, IBM SP, Fujitsu VPP700 and Hitachi SR8000 supercomputers. The numerical core of these algorithms is a matrix-vector multiplication (MVM), involving mega-dimensional matrices. Although the matrices are extremely sparse ($\approx 10 - 50$ non-zero entries per row) a memory saving, parallel MVM implementation has been developed, where the non-zero matrix entries are recomputed in each MVM step. The parallel MVM is implemented in FORTRAN and uses the MPI library for data exchange between processors. For CRAY T3E systems MPI calls have been replaced by calls to CRAY shmem library in selected performance critical routines. With a total memory requirement of approximately four vectors of matrix-dimension we are able to perform exact diagonalization studies up to matrices dimensions of 30 billion on present-day german supercomputers.

To demonstrate the scalability of our implementation we consider the more simplified Holstein Hubbard Hamiltonian

$$H = -t \sum_{i,\sigma}(c_{i\sigma}^{\dagger}c_{i+1\sigma} + \text{H.c.}) + U \sum_i n_{i\uparrow}n_{i\downarrow} + g\omega_0 \sum_{i,\sigma}(b_i^{\dagger} + b_i)n_{i\sigma} + \omega_0 \sum_i b_i^{\dagger}b_i,$$

(2)

which, nevertheless, can be taken as a generic model for the interaction of electron and lattice degrees of freedom in solids. Here $c_{i\sigma}^{\dagger}$ creates a spin-σ electron at Wannier site i ($n_{i,\sigma} = c_{i\sigma}^{\dagger}c_{i\sigma}$), $b_i^{\dagger}$ creates a local phonon of frequency ω, t denotes the hopping integral, U is the on-site Hubbard repulsion, g is a measure of the electron-phonon coupling strength. The Holstein Hubbard Hamiltonian allows a scaling of the corresponding matrix size by increasing the number of electrons without changing the principle matrix structure. Since available memory is the limiting factor in our calculations we have fixed a matrix dimension of 2 million per processor in the scalability study presented in Fig. 10. It is well known that the single processor performance of sparse MVM algorithms is bounded by the quality of the memory access and

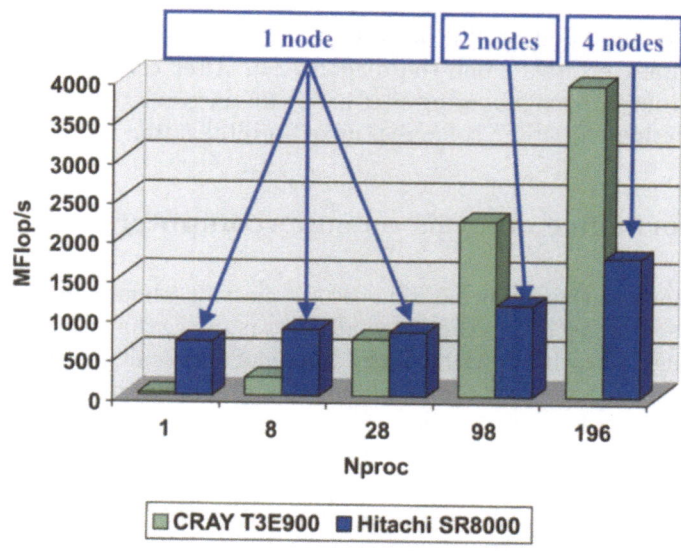

Fig. 10. Scaling study with fixed matrix size of 2 million per CRAY T3E processor. The total performance is depicted as a function of CRAY T3E processors (`Nproc`) used in the calculations. For comparison the corresponding performance numbers of the Hitachi SR8000-F1 at LRZ Munich using one (`Nproc=1,8,28`), two (`Nproc=98`) and four (`Nproc=196`) nodes are given.

thus a performance of roughly 33 MFlop/s has been measured on one processor. For the parallel runs, however, our implementation benefits from the high scalability of CRAY T3E systems and we find a parallel speed-up of approximately 130 on 196 processors, even for complex communication patterns as used in the benchmarks. Nonetheless about 30 (60) CRAY T3E processors are required to get the same performance (amount of memory) as one Hitachi SR8000-F1 node at LRZ Munich using a hybrid parallel programming approach.

The leading edge simulations had to be done on the LRZ system because of the larger main memory which is not available at the HLRS at present. However, even the size of the memory in Munich will not be enough for the problems that we intend to solve in near future. Therefore the installation of a more powerful system with a large aggregate amount of memory at the HLRS would be highly desirable.

5   Concluding Remarks

The work presented in this report is an example for the reliability and predictive power of many-body calculations performed on modern supercomputers.

The implementation of the various optimized program packages on the CRAY T3E at the HLRS Stuttgart provided new and exciting insights into the complex interplay of charge, spin, orbital and lattice degrees of freedom in the currently most intensive studied novel materials: the quasi-1D metals, spin chains and charge-density-wave systems, high-T_c cuprates, polaronic nickelates and colossal magneto-resistance manganites. For doped CMR manganites we showed explicitly how the electron-phonon interaction effectively controls spin and orbital order by affecting charge mobility and orbital degrees of freedom. Our exact diagonalization study of even a small system provides detailed information about correlations and driving interactions behind the rich phase diagram of the manganites. This may support the development of approximate theories.

Acknowledgements

We are indebted to the to the HLR Stuttgart, NIC Jülich, LRZ München and NIC Jülich for the generous granting of their parallel computer facilities. This work was supported by the Deutsche Forschungsgemeinschaft under project Fe 398-1/2.

References

1. S. Jin, T. H. Tiefel, M. McCormack, R. A. Fastnach, R. Ramesh, and L. H. Chen, Science **264**, 413 (1994).
2. Y. Tokura and Y. Tomioka, J. Magn. Magn. Mater. **200**, 1 (1999); J. M. D. Coey, M. Viret, and S. von Molnar, Adv. Phys. **48**, 167 (1999); E. Dagotto, T. Hotta, and A. Moreo, Physics Reports **344**, 1 (2001).
3. J. Hemberger et al., arXiv:cond-mat/0204269; M. Paraskevopoulos et al., J. Phys.: Condens. Matter **12**, 3993 (2000); J. Magn. Magn. Mater. **211**, 118 (2000).
4. A. Weiße and H. Fehske, in preparation.
5. C. Zener, Phys. Rev. **82**, 403 (1951); P. W. Anderson and H. Hasegawa, Phys. Rev. **100**, 675 (1955); K. Kubo and N. Ohata, J. Phys. Soc. Jpn. **33**, 21 (1972); A. Weiße, J. Loos, and H. Fehske, Phys. Rev. B **64**, 054406 (2001).
6. A. Weiße, H. Fehske, G. Wellein, and A. R. Bishop, Phys. Rev. B **62**, R747 (2000); A. Weiße, G. Wellein, and H. Fehske, *High Performance Computing in Science and Engineering '01* edited by E. Krause and W. Jäger, Springer-Verlag, Heidelberg (2002), pp 131-144.
7. D. D. Sarma et al., Phys. Rev. Lett. **75**, 1126 (1995); W. E. Pickett and D. J. Singh, Phys. Rev. B **53**, 1146 (1996); S. Satpathy, Z. S. Popović, and F. R. Vukajlović, Phys. Rev. Lett. **76**, 960 (1996); I. Solovyev, N. Hamada, and K. Terakura, Phys. Rev. Lett. **76**, 4825 (1996).
8. L. F. Feiner and A. M. Oleś, Phys. Rev. B **59**, 3295 (1999).
9. G. Khaliullin and V. Oudovenko, Phys. Rev. B **56**, R14243 (1997); L. F. Feiner, A. M. Oleś, and J. Zaanen, J. Phys. Condens. Matter **10**, L555 (1998).
10. K. I. Kugel and D. I. Khomskii, Sov. Phys. JETP **37**, 725 (1973).
11. C. H. Booth et al., Phys. Rev. Lett. **80**, 853 (1998).

Phase Transitions, Structures and Quantum Effects in Nanosystems

M. Dreher, D. Fischer, K. Franzrahe, P. Henseler, J. Hoffmann,
W. Strepp, and P. Nielaba

Physics Department (Theory), University of Konstanz, 78457 Konstanz, Germany

Abstract. Phase transitions, structures and quantum effects in Nanostructures
are studied by path integral Monte Carlo-, Molecular Dynamics- and Car-Parrinello
methods. We present results of our computations on pore condensates, reentrance
phenomena of systems in external fields, atomic wires and quantum-spin-fluids.

1   Introduction and parallel processing

Many of the interesting effects recently studied in systems on nanometer
length scales take place at low temperatures so that the consideration of quantum mechanics is important. Despite the fact that by experimental techniques
many structural-, elastical-, electronic-, and phase-properties of systems in
the size of a few nanometers have been obtained, the theoretical investigations and analyses are still in an initial stage. This is partly because of the
fact that systems which are far away from the thermodynamic limit (with
infinitely many particles) due to their finite size are difficult to handle by
analytical methods which are suitable for systems with either few particles
(2–5) or in the limit of infinitely many particles. In this field computer simulations have become more and more important, because nano-systems in
reduced geometry contain about 10–10.000 particles, which is nearly ideal
for the application of computer simulation methods. In the research domain
of nanostructures in reduced geometry a relatively small amount of theoretical research has been done. We plan to contribute to bridge this gap by our
project at the HLRS. In particlar in the field of quantum simulations our
group has been able to achieve many interesting contributions [1], which to
a great deal have been obtained by computer time support by the HLRS. By
computer simulation methods we plan to study the structures and the elastic
and electronic properties as well as phase "transitions" in systems of a the
size of a few nanometers located at surfaces. In order to quantify quantum
effects we perform path integral Monte Carlo (PIMC) simulations:

Canonical averages $< A >$ of an observable A in a system defined by the
Hamiltonian $\mathcal{H} = E_{\text{kin}} + V_{\text{pot}}$ of N particles in a volume V are given by: $\langle A \rangle =$
Z^{-1} tr $[A \exp(-\beta \mathcal{H})]$. Here $Z = \text{tr } [\exp(-\beta \mathcal{H})]$ is the partition function
and $\beta = 1/k_B T$ is the inverse temperature. Utilizing the Trotter-product
formula, $\exp(\beta \mathcal{H}) = \lim_{P \to \infty} (\exp(-\beta E_{\text{kin}}/P) \exp(-\beta V_{\text{pot}}/P))^P$, we obtain the

path integral expression for the partition function:

$$Z(N, V, T) = \lim_{P \to \infty} \left(\frac{mP}{2\pi\beta\hbar^2} \right)^{3NP/2} \prod_{s=1}^{P} \int d\{\mathbf{r}^{(s)}\} \tag{1}$$

$$\times \exp\left\{ -\frac{\beta}{P} \left[\sum_{k=1}^{N} \frac{mP^2}{2\hbar^2\beta^2} (\mathbf{r}_k^{(s)} - \mathbf{r}_k^{(s+1)})^2 + V_{\text{pot}}(\{\mathbf{r}^{(s)}\}) \right] \right\}$$

Here, m is the particle mass, integer P is the Trotter number and $\mathbf{r}_k^{(s)}$ denotes the coordinate of particle k at Trotter-index s, and periodic boundary conditions apply, $P + 1 = 1$. This formulation of the partition function allows us to perform Monte Carlo simulations for increasing values of P approaching the true quantum limit for $P \to \infty$.

Thermal averages in the ensemble with constant pressure p are given via the corresponding partition function $\Delta(N, p, T) = \int_0^\infty dV \exp[-\beta pV] Z(N, V, T)$. In Eq. (1) we see that in the path integral formalism each quantum particle k can (for finite P-values) be represented by a closed quantum chain of length P in position space where the classical coordinate of the point $\mathbf{r}_k^{(s)}$ on this chain at the Trotter index s has a harmonic interaction to its nearest neighbors at $\mathbf{r}_k^{(s+1)}$ and $\mathbf{r}_k^{(s-1)}$. An interaction between different quantum particles takes places only between particles $\{\mathbf{r}^{(s)}\}$ with the same Trotter index s. Due to this property the entire system with Trotter index s can be placed efficiently in one processor of a parallel computer with P processors, where the potential energy of all N particles can be computed for this Trotter index s (with an effort $\propto N(N-1)/2$). The different processors then have a physical coupling due to the kinetic energy term resulting in the harmonic interaction between nearest neighbor Trotter indices in the PIMC formalism. Thus only N interactions between the harmonically interacting particles have to be computed due to the harmonic interactions between nearest neighbor Trotter indices s and $s+1$. It is even more efficient to place the system with two neighboring Trotter indices into one processor, since then only two neighboring processors communicate when a local Monte Carlo move (for one quantum particle k) is done. With a Trotter order of $P = 64$ (that means 32 processors) the running time of the simulation is increasing only by about 5 % when doubling the Trotter-order. In contrast to this very good scaling behaviour, in a scalar algorithm with a linear-P dependency the running time would increase by 100%. This shows that only computations on the parallel computer T3E make it possible to investigate systems at temperatures at which the proper approximation of the quantum limit requires large P-values ($P/2$ processors). Many of the studies presented in this report thus in practice are only possible at the T3E.

Besides this inherent advantage of the parallel algorithm for the PIMC simulations we utilize the parallel machine as well most efficiently in running different replicas of the system in parallel in order to increase the statistics in the statistical averages.

2   Car-Parrinello Simulations of Clusters and Surface Structures (SSC Karlsruhe)

Since the discovery of the "supermagic" cluster C_{60} [2] the possibility of the synthesis of new materials consisting of highly stable clusters fascinates many researchers. In case of C_{60} and similar fullerenes like C_{70} and La@C_{82} such materials exist and, e.g., fullerite – the bulk material formed by weakly inter-acting C_{60} "soccer balls" – represents a new form of carbon beside diamond and graphite. The question arises, whether "magic" clusters of other elements like Si or Al might be suitable as building blocks of new cluster materials [3,4].

In combination with experimental studies in the group Ganteför in Kon-stanz we analyzed by Car-Parrinello methods the electronic and structural properties of Si_4-clusters [5]. Experimentally, the clusters are mass-selected and soft landed on an inert van-der-Waals surface [6]. They are probably highly mobile on this surface at room temperature and will immediately form large islands of bulk Si if there wouldn't be a barrier against fusion. The samples are studied using XPS and, in contrast to an earlier study of Si_{10} on amorphous carbon [7], the spectra contradict the formation of large islands supporting the existence of a barrier.

Our density functional (DFT) calculations [5] for the Si_4 clusters have been performed with the approximative gradient-corrected exchange-correla-tion (xc) functionals of Perdew, Burke and Ernzerhof (PBE) [8]. This choice for the xc-functional should give reliable results whenever both localized and extended electron states appear. The computational details are similar as in the studies of Au [9–11], modified to the case of Si [12].

In a first step the ground state structure of an isolated Si_4 cluster was determined. The isolated Si_4 clusters form planar rhomboedric structures with two sharp and two flat corners. Fixing the distance R between two Si-atomic centers on the x-axis in two different Si_4-clusters the potential energy surface was calculated. Two different "reaction channels" have been considered (see Fig.1) starting with structures with symmetry about the x-axis. In Fig.1(a) the distance between the two Si-atoms at the flat angles is fixed, in Fig.1(b) the distance of the atoms at the sharp angles. The total energy is displayed in Fig.1 as a function of R. These calculations correspond to the situation of the clusters in the gas phase. However, since the interaction to the Van-der-Waals surface is small the results obtained can be considered a good approximation to the case of deposited clusters.

Figure 1 displays the calculated dependencies of potential energy of two interacting Si_4 clusters as a function of distance. Neutral Si_4 in its electronic ground state is a planar rhombus and there are several geometries possible for two tetramers approaching each other. We assume the Si_4 clusters lie flat on the surface and, therefore, we restricted to planar geometry of two Si_4 approaching each other with the two obtuse (a) or sharp (b) corners encoun-tering. For the geometry displayed in Fig. 1a the potential energy increases monotonously with decreasing distance corresponding to a repulsive interac-

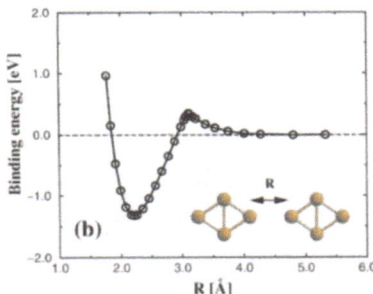
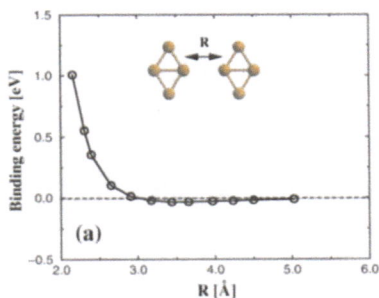

Fig. 1. Calculated potential energy curves for two neutral interacting Si_4 clusters. Two different reaction channels have been calculated: the two tetramers approaching each other with the flat (a) and sharp (b) corners ahead. Case (a) is repulsive, while in case (b) a bond is formed. In (b) an energy barrier is observed at a distance of 3.1A and a height of 0.3eV.

tion. The two clusters do not fuse. If the two Si_4 approach with the sharp corners ahead a bond is formed (Fig. 1b). A minimum with a binding energy of 1.3 eV is calculated corresponding to the formation of a Si_8 cluster. Important in Fig.1b is the small increase of the potential energy at a distance of 3.1A. This barrier is 0.3eV high and, therefore, it might not be overcome at kinetic energies corresponding to room temperature. Accordingly, for both reaction channels the calculation predict a repulsive interaction at low temperatures.

These theoretical findings support the results of an experimental [5] study of Si_4 clusters deposited on HOPG at room temperature (AG Ganteför). Both findings support the idea that this magic silicon cluster is suitable as a building block for a new clusters material consisting of pure silicon. If it will finally turn out to be really possible to synthize such a material this will open a door to a whole new world of material science based on the many magic clusters already found in the gas phase.

3 Computations of Structures and Conductance in Nano-wires

The stability and electronic properties of wires on nanometer length scales have been studied by experimental and theoretical methods recently [13]. Atomic size Au-wires are studied by experimental methods in Konstanz in the group of E. Scheer [14]. In such studies the wires are stretched down to single-atom contacts.

The electronic and structural behavior of these systems is not yet understood entirely on a microscopic level. In our group we have studied [15] the stretching of Au-wires by molecular dynamics methods applying stretching

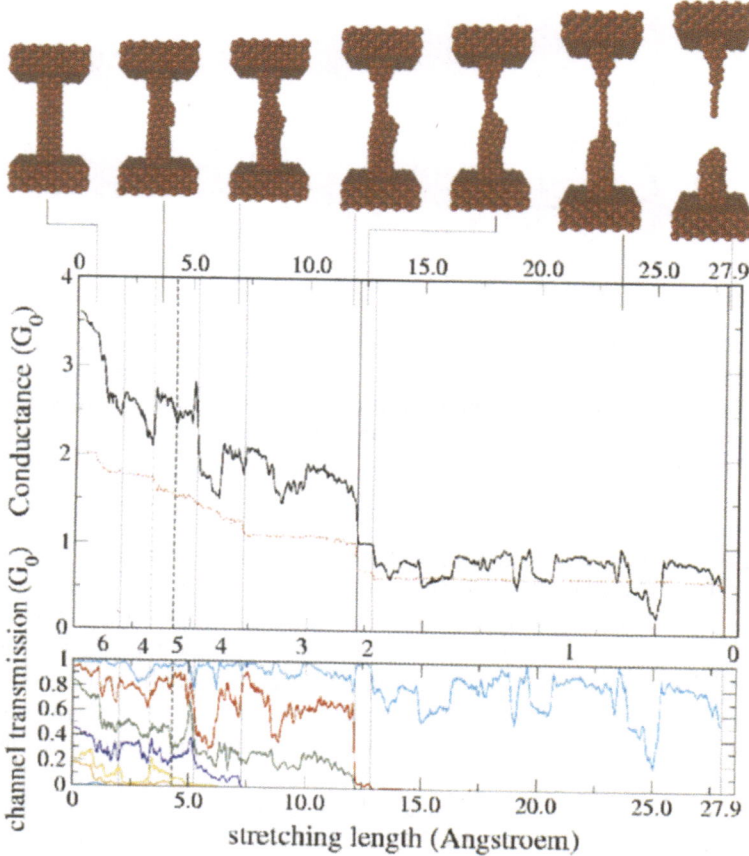

Fig. 2. Structures in Au-wires upon stretching in z-direction and resulting conductance [15].

forces in one spatial direction. In computations with Lennard-Jones particle-interactions single atom contacts and shifts of atomic layers were found, but no formation of atomic chains. Stretching wires with "effective medium"-particle interactions (EMT) however resulted in structures with single atom contacts as well as atomic chains, see Fig. 2. This feature of the EMT-potentials in the literature is sometimes attributed to the fact that the binding energy per particle is dependent on the number of neighboring atoms. In the molecular dynamics simulations a Nose-Hoover thermostat was used to avoid the heating of the wire during the stretching process.

The conductance in these systems was then studied by a combination with "tight-binding"-methods [16]. In the computations the electric current through a nano-contact can be decomposed in single channels and conductance curves can be found for different stretching distances of the electrodes.

During the stretching process the wire is thinning and the smallest effective diameter along the wire can be computed. The conductance in units of the "conductance quantum" $G_0 = 2e^2/h$, the contributions of different channels to the conductance and the smallest diameter (dotted red line) are shown in Fig. 2 for different stretching lengths of the wire. The conductance "plateaus" at $G/G_0 \approx 1, 1.6, 2.6, 3.6$ are close to the experimental values [14]. The conductance fluctuations however seem to be slightly larger compared to the experiment.

The current through the contact depends mainly on the wire atoms and the atomic configurations at the smallest diameter. Interestingly however the atomic configurations in the vicinity of the location of the smallest diameter have a significant contribution to the conductance as well. Further studies are planned.

4 Phase transitions and quantum effects in pore condensates

Materials properties of systems condensed into pores of nanometer length scales (i.e. Vycor, Gelsil) are different from those of bulk systems. Important reasons for these effects are the geometrical finite size effects (the atomistic material structure cannot be neglected), the interaction of the material with the surrounding "glass"-matrix and the large interface contributions relative to the volume. Phase transitions of pore condensates in nano-pores have been investigated by experimental methods recently [17,18].

With computer simulations (CRAY-T3E) we have analyzed [19,20] many interesting properties of "Ar"- and "Ne"-pore condensates recently (modeled as Lennard-Jones systems with particle diameter σ and interaction energy ϵ, in our computations we use particle masses m*$=m\sigma^2\varepsilon/\hbar^2=$ 100 and m*$=1000$ for simplicity well approximating the particle masses of Ne and Ar (m*$=112$ and m*$=1160$)). These systems have – like the "bulk"-systems a gas–liquid phase transition at low temperatures, the precise shape of the phase diagram is strongly influenced by the system geometry (pore radius).

In turns out that with increasing attractive wall interaction the critical density increases, the adsorbate density increases strongly, and the condensate density increases weakly. A meniscus is formed with increasing curvature, the configurations become less stable and the critical temperature decreases. The critical temperature is reduced with decreasing pore diameter. Beginning from the wall a formation of layered shell structures is found which may favor or disfavor the occupancy of sites at the pore axis due to packing effects.

The triple point temperature is influenced by the geometrical finite size effects (pore radius) as well as by the wall–particle interaction. At sufficiently strong wall–particle interaction ($\epsilon^{WP}n^W = 1.509\epsilon$, $\sigma^{WP} = 1.094\sigma$) we find a two step fluid–solid phase transition. In Fig. 3 we show typical configurations for a pore with radius $R = 4\sigma$ at three temperatures. At $T^* = 0.6$ the system is fluid, at $T^* = 0.46$ the central condensate region is frozen and the adsorbate region is fluid, and at $T^* = 0.2$ the adsorbate is frozen as well. At $T^* = 0.575$ a solidification of the condensate phase is found by a jump in the energy. The condensate freezes in one piece from the cylinder wall to the axis. The remaining fluid adsorbate phase solidifies only at $T^* = 0.375$. The temperature dependency of the bond orientational order parameter of the outer layer supports this two step "freezing" scenario as well.

A different scenario was found for even stronger interactions between the wall and the particles: The system freezes at the cylinder wall due to the strong binding energy to the wall. Between the layer at the cylinder wall and the next layer only a very small particle exchange is found. The region close to the cylinder axis however is still fluid and freezes only at lower temperatures. The different behavior of layer-wise freezing in systems with strong wall-particle interaction and the block-wise freezing for small particle-pore-interactions can be understood as follows: particles in the second layer can be treated as particles in a pore with Radius $R^{\text{eff}} = R - \sigma$ with an potential consisting of the particle-wall potential and the particle-particle potential to the particles in the first layer. Latter is in the order of 3ϵ since there are 3 nearest neighbors in the first layer. The effective outer potential for the second layer is in the same order as the outer potential for particles in the first layer for systems with small wall-particle interactions. For even stronger wall particle potentials the effective outer potential for the inner layers differs from the potential of the outer layer. Such two stage freezing phenomena were observed in experiments of melting and freezing of Ar in Vycor pores recently [21].

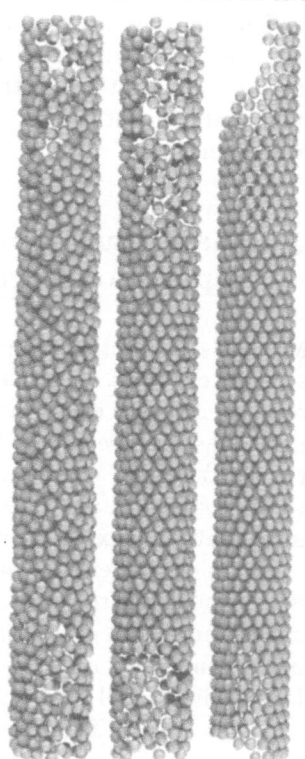

Fig. 3. Configurations of a classical LJ-condensates (N=1500) in a cylindrical pore of radius $R = 4\sigma$, and length $L = 60\sigma$ at $T^* = 0.6$, 0.46, 0.2 (from left to right). The potential parameters for the wall–particle interactions are: $\sigma^{WP} = 1.094\sigma$, $\varepsilon^{WP} n^W = 1.5092\varepsilon\sigma^{-3}$.

For small pore radii and not too small wall–particle interactions we observe a layering structure of the condensate. For a bulk system instead one would expect a crystalline FCC or HCP structure. In agreement with this for large pore diameters and not too strong wall–particle interactions no layering structure is found. In this case structures are formed with local FCC- or HCP-order.

At low temperatures quantum effects become important which have been ignored in most of the existing theoretical studies of pore condensates. By path integral Monte Carlo simulations [1,22–24] the effect of the quantum mechanics on the potential energy as a function of the temperature has been quantified [19,20]. In contrast to classical simulations we obtain by PIMC simulations for Ar- and Ne-condensates an horizontal temperature dependency of the energy resulting in a decrease of the specific heat to zero at small temperatures in agreement with the third law of thermodynamics. The resulting phase diagram for Ar- and Ne-condensates and a comparison with classical computations shows important quantum effects [20]. In the Ne-system (containing the lighter particles) a significant reduction (by about 5-10%) of the critical temperature is found due to quantum delocalizations as well as a strong reduction of the solid density and a crystal structure modification in comparison with the classical case.

5   Phase transitions in systems with repulsive interactions

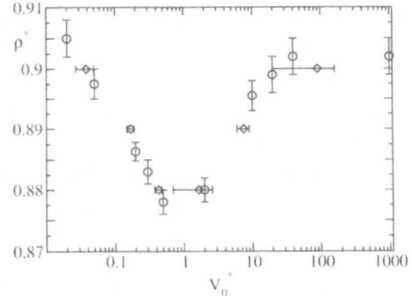

One is often interested in long length scale and long time scale phenomena in solids (eg. late stage kinetics of solid state phase transformations; motion of domain walls interfaces; fracture; friction etc.). Such phenomena are usually described by continuum theories. Microscopic simulations[25] of finite systems, on the other hand deal with microscopic variables.

Fig. 4. Phase diagram [38,39] for the hard-disk system in the ρ^* / V_0^*-plane. Transition points for transitions from the solid to the modulated liquid have been obtained by the order parameter cumulant intersection method [40]. In order to map the phase diagram we scanned in ρ^* for every V_0^*, starting from the high density (solid) region. The system size is $N = 1024$.

We presented [26] a new coarse graining method for the simplest nontrivial case, namely, a crystalline solid in equilibrium, at a non zero temperature far away from phase transitions (for a literature overview see [26]).

Using our new method, we investigated in Ref. [28,42] the melting transition of the solid phase and showed that the hard disk solid is unstable to perturbations which attempt to produce free dislocations leading to a solid → hexatic transition in accordance with KTHNY theory[27]. This transition lies close to a first order solid to liquid melting line. We calculate quantitatively the relative positions of the first order and the KTHNY transitions in the parameter space for this system and explain why earlier simulations failed to arrive at a consensus.

The liquid-solid transition in systems of particles under the influence of external modulating potentials has recently attracted a fair amount of attention from experiments [29,30] (see also references therein), theory and computer simulations [31–36]. We have investigated [38,39] the phase behavior of a two dimensional hard disk system in an external potential. In addition to the pair interaction a particle with coordinates (x, y) is exposed to an external periodic potential of the form: $V(x, y) = V_0 \sin(2\pi x/d_0)$. The constant d_0 is chosen such that, for a density $\rho = N/S_x S_y$, the modulation is commensurate to a triangular lattice of hard disks with nearest neighbor distance a_s: $d_0 = a_s \sqrt{3}/2$. The only parameters which define our system are the reduced density $\rho \sigma^2 = \rho^*$ and the reduced potential strength $V_0/k_B T = V_0^*$, where k_B is the Boltzmann constant and T is the temperature.

In order to analyze the influence of the pair potential range on the reentrance MC-studies [43,44] have been performed for the hard disk potential [38,39], the $1/r^{12}$-, $1/r^6$- and the DLVO-potential [37]. For the short ranged $1/r^{12}$- potential [41] and the DLVO-potential [38] good evidence for a reentrance has been found. With increasing interaction range the effect of "one-dimensional" reentrant melting is less pronounced since more and more neighbors have to be passed by a particle in order to leave its lattice site. This results in a stabilization of the solid relative to the modulated liquid at large V_0^*-values and in a much less pro-

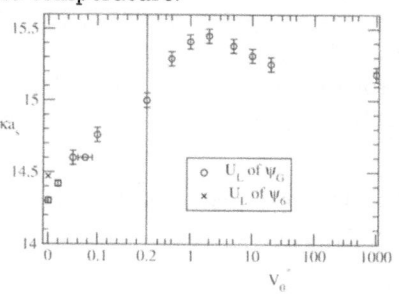

Fig. 5. Phase diagram in the DLVO-system in an external periodic potential with amplitude $V^* = V_0/k_B T$. Symbols: MC-data of the ψ_{G_1} order parameter cumulant intersection points at the transition from the solid to the fluid phase, κ: inverse Debye-screening length.

nounced increase of the transition density for large V_0^*- values in case of the $1/r^6$-potential [41], the data however are still compatible with a reentrance scenario. The phase diagram for the hard disk system is shown in Fig. 4. Within our range of densities, one has a clear signature of a re -entrant liquid phase showing that this phenomenon is indeed a general one as indicated in Ref. [32]. The phase diagram for the DLVO-system [38] is shown in Fig. 5, a pronounced reentrance region can be identified again. Our simulation thus have shown, that systems with DLVO- or "soft disk" interaction behave similar to charge stabilized colloids [30].

Besides this purely classical studies we have analyzed the behavior of quantum hard disks in these systems by PIMC (method: see above) in order to clarify if the reentrance scenario can be transferred to systems with atomic size length scales. It turns out [38] that at small V^*-values the quantum disks are effectively larger than the classical disks due to the quantum delocalization. As a result the transition density is reduced compared to the classical case. At high V^*-values however the delocalization is asymmetrical, the quantum disks are much sharper localized perpendicular to the potential minima than parallel to the minima. This results in an increase in the transition density, approaching the classical values asymptotically.

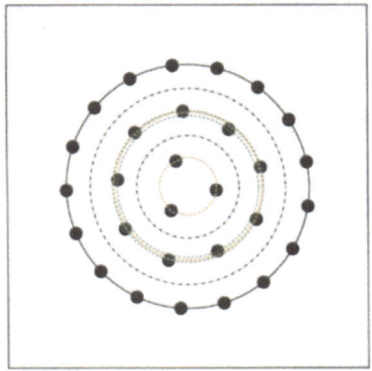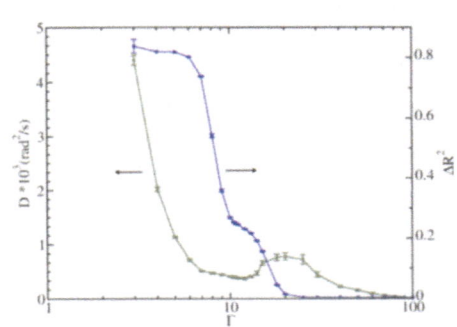

Fig. 6. Left: Ordered shell-structure in $1/r^3$-system with 29 particles. Right: Angular diffusion constant as function of inverse temperature ($\Gamma \propto 1/T$).

By experimental methods at the LS Leiderer (Konstanz) the phase- and diffusion-behavior of colloidal systems in circular containers has been studied [45]. In a corresponding MC-study [46] it turned out that particles with $1/r^3$-interaction potential order in circular layers (see Fig. 6) and the angular diffusion constant shows a reentrance behavior as function of temperature due to radial fluctuations (Fig. 6), in agreement with experimental findings.

6   Fluids with internal quantum states

As a simplified model for adsorbates with internal degrees of freedom we have studied the phase diagram of a fluid with classical translational degrees of freedom and internal quantum mechanical degrees of freedom by PIMC, finite size scaling, Gibbs ensemble- and grand canonical simulation techniques [47,48]. The Hamiltonian of the fluids is:

$$H = \sum_{i=1}^{N} \frac{\mathbf{P}_i^2}{2M} + \sum_{i<j} U(\mathbf{r}_i - \mathbf{r}_j) - \frac{\omega_0}{2} \sum_{i=1}^{N} \sigma_i^x - \sum_{i<j} J(\mathbf{r}_i - \mathbf{r}_j)\sigma_i^z \sigma_j^z \quad (2)$$

The particle masses M are assumed to be sufficiently large to justify the classical treatment of the translational degrees of freedom. The two internal quantum states of a molecule are modeled by the term $-\frac{\omega_0}{2}\sigma_i^x$ (σ^x, σ^z are the Pauli-matrices, $\omega_0/J = 4$). The particles interact as follows: ($U(r) = \infty$: $r < R, U(r) = 0 : r > R$, $J(r) = J : R < r < 1.5R$, $J(r) = 0 : r > 1.5R$).

In case of an antiferromagnetic interaction between the particles of our fluid up to an interparticle distance of 1.4 R we obtained by density functional computations [49], that at high densities and low temperatures the triangular lattice structure as well as the square lattice structure gets destabilized. By Monte Carlo simulations [50] we found, that indeed at high densities and high temperatures the system is in the triangular lattice structure, at low temperatures however a disordered structure appears. The question arises on the stability of the frustrated triangular lattice in a quantum mechanical treatment. By PIMC we have determined the phase diagram at high densities and low temperatures [51]. Since in this case spin-values between -1 and $+1$ are possible, the triangular lattice was stabilized at low temperatures. As phase transition density we obtained approximately the value of the hard disk system without spin interaction, see Fig. 7.

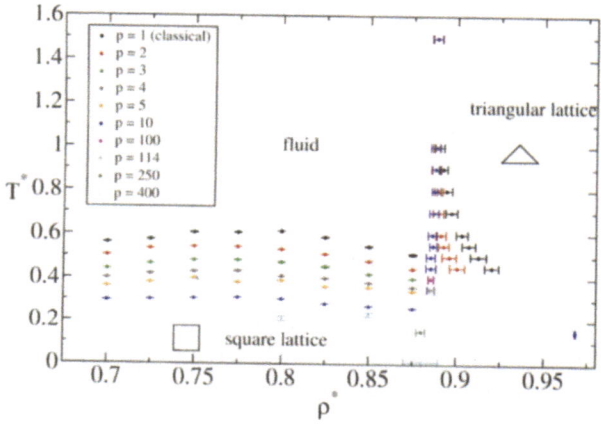

Fig. 7. Phase diagram of the antiferro-model-fluid for different P-values.

At medium densities and low temperatures a quadratical lattice was found without spin-frustration. PIMC-studies [51] showed that the stability region of this phase is shifted to much lower temperatures, see Fig. 7. In total the interesting picture emerges that within a quantum mechanical computation the system with quantum spins can be approximated as system of hard disks without spin interaction.

This very interesting results gives us confidence that the region of very high densities and low temperatures in the phase diagram is indeed accessible and can be characterized by PIMC-studies. We plan to investigate this topic in future studies.

Acknowledgments

We thank W. Andreoni, C. Bechinger, K. Binder, C. Cuevas, A. Curioni, G. Ganteför, M. Grass, J. Heurich, M. Lohrer, M. Rao, E. Scheer and S. Sengupta for useful discussions and cooperations, the HLRS, the SSC Karlsruhe and the NIC for computer time and the DFG for support (SFB 513 and Ni 259/8-2).

References

1. P. Nielaba, in: *Computational Methods in Surface and Colloid Science*, M. Borowko (Ed.), Marcel Dekker Inc., New York (2000), pp. 77–134.
2. H.W. Kroto et al., Science **242**, 1139 (1988).
3. O. Cheshnovsky et al., Chem. Phys. Lett. **138**, 119 (1987); J. Müller et al., Phys. Rev. Lett. **85**, 1666 (2000).
4. U. Röthlisberger, W. Andreoni, and M. Parrinello, Phys. Rev. Lett. **72**, 665 (1994).
5. M. Grass, D. Fischer, M. Mathes, G. Ganteför, P. Nielaba, preprint.
6. B. Klipp et al., Appl. Phys. A **73**, 547 (2001).
7. J.E. Bower, and M. Jarrold, J. Chem. Phys. **97**, 8312 (1992).
8. J.P. Perdew, K. Burke, M. Ernzerhof, Phys. Rev. Lett. **77**, 3865 (1996).
9. H. Grönbeck, W. Andreoni, Chem. Phys. **262**, 1 (2000).
10. R. Car, M. Parrinello, Phys. Rev. Lett. **55**, 2471 (1985).
11. Calculations used the CPMD code by J. Hutter: CPMD 3.0 Copyright IBM Corporation (1990-1997) and MPI Festkörperforschung Stuttgart, 1997.
12. Si PBE pseudopots. and cluster structures from W. Andreoni (unpublished).
13. M. Sorensen et al., PR **B57**, 3283 (1997); H. Ohnishi et al., Nature **395**, 780 (1998); V. Rodrigues et al., PRL **85**, 4124 (2000).
14. E. Scheer et al., PRL **78**, 3535 (1997); E. Scheer et al., Nature **394**, 154 (1998); E. Scheer et al., PRL **86**, 284 (2000).
15. M. Dreher, Diplomarbeit, Konstanz (2002).
16. J.C. Cuevas et al., Phys. Rev. Lett. **81**, 2990 (1998).
17. A.J. Liu et al., PRL **65**, 1897 (1990); Z. Zhang et al., PR **E 52**, 2736 (1995).
18. M.W. Maddox et al., Mol. Simulat. **17**, 333 (1997); L.D. Gelb, K.E. Gubbins, PR **E 56**, 3185 (1997); P. Huber, K. Knorr, Phys. Rev.B **60**, 12657 (1999).
19. J. Hoffmann, Ph.D.-thesis, U. Konstanz (2002).
20. J. Hoffmann, P. Nielaba, preprint.
21. D. Wallacher, K. Knorr, Phys. Rev. **B63**, 104202 (2001).
22. M. Presber, D. Löding, R. Martonak, P. Nielaba, Phys. Rev. **B 58**, 11937 (1998).
23. M. Reber, D. Löding, M. Presber, Chr. Rickward, P. Nielaba, Comp. Phys. Commun. **121-122**, 524 (1999).
24. C. Rickwardt, P. Nielaba, M.H. Müser, K. Binder, Phys. Rev. **B63**, 045204 (2001).
25. D.P. Landau and K. Binder, *A Guide to Monte Carlo Simulations in Statistical Physics*, Cambridge University Press (2000).
26. S. Sengupta, P. Nielaba, M. Rao, K. Binder, Phys. Rev. **E 61**, 1072 (2000).
27. J.M. Kosterlitz, D.J. Thouless, J. Phys. **C 6**, 1181 (1973); B.I. Halperin and D.R. Nelson, PRL **41**, 121 (1978); D.R. Nelson and B.I. Halperin, PR **B19**, 2457 (1979); A.P. Young, PR **B 19**, 1855 (1979).
28. S. Sengupta, P. Nielaba, K. Binder, PR **E61**, 6294 (2000).
29. N.A. Clark, B.J. Ackerson, A.J. Hurd, Phys. Rev. Lett. **50**, 1459 (1983).
30. C. Bechinger, M. Brunner, P. Leiderer, Phys. Rev. Lett. **86**, 930 (2001).
31. J. Chakrabarti, H.R. Krishnamurthy, A.K. Sood, Phys. Rev. Lett. **73**, 2923 (1994).
32. E. Frey, D.R. Nelson, L. Radzihovsky, PRL **83**, 2977 (1999).
33. J. Chakrabarti, H.R. Krishnamurthy, A.K. Sood, S. Sengupta, Phys. Rev. Lett. **75**, 2232 (1995).

34. C. Das, H.R. Krishnamurthy, PR **B58**, R5889 (1998).
35. C. Das, A.K. Sood, H.R. Krishnamurthy, Physica **A 270**, 237 (1999).
36. C. Das, A.K. Sood, H.R. Krishnamurthy, preprint.
37. For an introduction to phase transitions in colloids see, A. K. Sood in *Solid State Physics*, E. Ehrenfest, D. Turnbull Eds. (Academic Press, New York, 1991) **45**,1.
38. W. Strepp, Ph.D-thesis, Univ. of Konstanz (in prep.).
39. W. Strepp, S. Sengupta, P. Nielaba, Phys. Rev. **E63**, 046106 (2001).
40. K. Binder, Z. Phys. **B43**, 119 (1981); K. Binder, Phys. Rev. Lett. **47**, 693 (1981).
41. M. Lohrer, Diplom-thesis, Univ. of Konstanz (2001).
42. K. Binder, S. Sengupta, P. Nielaba, J. Phys.: Cond. Mat. **14**, 2323 (2002).
43. W. Strepp, S. Sengupta, M. Lohrer, P. Nielaba, in "Mathematics and Computers in Simulation" (2002), im Druck.
44. W. Strepp, S. Sengupta, M. Lohrer, P. Nielaba, Comp. Phys. Com. (2002), im Druck.
45. R. Bubeck, C. Bechinger, S. Neser, P. Leiderer, Phys. Rev. Lett. **82**, 3364 (1999).
46. P. Henseler, Diplomarbeit, Konstanz (2002).
47. F. Schneider, D. Marx, P. Nielaba, Phys. Rev. **E 51**, 5162 (1995).
48. N.B. Wilding, P. Nielaba, Phys. Rev. **E53**, 926 (1996)
49. P. Nielaba, S. Sengupta, Phys. Rev. **E 55** (1997).
50. R. Stadelhofer, Diplomarbeit, Konstanz (2001).
51. K. Franzrahe, Diplomarbeit, Konstanz (2002).

Thermodynamics and Dynamics of Correlated Electron Systems

Catia Lavalle[1], M. Rigol[1], M. Feldbacher[1], Fakher F. Assaad[1,2], and Alejandro Muramatsu[1]

[1] Institut für Theoretische Physik III, Universität Stuttgart, Pfaffenwaldring 57, D-70550 Stuttgart, Germany
[2] Max Planck institute for solid state research, Heisenbergstr. 1, D-70569, Stuttgart.

Abstract. Based on state of the art Quantum Monte Carlo simulations, we investigate the metallic states of the one-dimensional t-J model and of a depleted Kondo lattice model in two dimensions. In the one-dimensional case, it is known that correlation effects invalidate the Fermi liquid picture and that the elementary excitations are spinons and holons carrying separately the charge and spin of the electron. In this dimension we will present new results on the single particle spectral function and discuss the implications of spin-charge separation on this quantity. The Kondo lattice model describes heavy fermion materials which generically have Fermi liquid ground states but with effective masses up to three orders of magnitude larger that the bare electron mass. On the basis of numerical simulations, we will show how this heavy fermion state comes about and set the emphasis on the coherence temperature.

1 Introduction

In Fermi liquid theory, the metallic state may be viewed as a gas of non-interacting electrons where correlation effects are *hidden* in a renormalized mass. The elementary excitations are carrying the quantum numbers of the bare electron: unit charge and spin 1/2. The success of the Fermi liquid theory is based on two facts: (i) due to Fermi statistics, the phase space available to the Coulomb repulsion is vanishingly small at low energies and (ii) the Coulomb repulsion is screened. One of the crucial issues in the theory of correlated electron systems is to understand under which circumstances a Fermi liquid ground state will or will not be realized. From the experimental point of view materials such as high temperature superconductors point to a failure of the Fermi liquid theory. On the other hand heavy fermion materials such as $CeCu_6$ generically have Fermi liquid ground states but with effective masses up to three orders of magnitude larger that the bare electron.

In this article, we will concentrate on two models of correlated electron systems: the one-dimensional t-J model as well as a depleted Kondo lattice model. The t-J model reads:

$$H_{t-J} = -t \sum_{<i,j>,\sigma} \tilde{c}^\dagger_{i,\sigma}\tilde{c}_{j,\sigma} + J \sum_{<i,j>} \left(\boldsymbol{S}_i \cdot \boldsymbol{S}_j - \frac{1}{4}\tilde{n}_i\tilde{n}_j \right). \quad (1)$$

Here, $\tilde{c}_{i,\sigma}^{\dagger}$ are projected fermion operators $\tilde{c}_{i,\sigma}^{\dagger} = (1 - c_{i,-\sigma}^{\dagger}c_{i,-\sigma})c_{i,\sigma}^{\dagger}$, $\tilde{n}_i = \sum_{\alpha} \tilde{c}_{i,\alpha}^{\dagger}\tilde{c}_{i,\alpha}$, $\boldsymbol{S}_i = (1/2) \sum_{\alpha,\beta} c_{i,\alpha}^{\dagger}\boldsymbol{\sigma}_{\alpha,\beta}c_{i,\beta}$, and the sum runs over nearest neighbors only. In two dimensions, on a square lattice, this model is believed to capture the physics of high temperature superconductors. Here we will consider the one-dimensional version of the model. In this dimension and irrespective of the correlation strength, it is known that Fermi liquid theory fails. The elementary excitations are not quasiparticles carrying unit charge and spin 1/2 but spinons with spin 1/2 and zero charge and holons with unit charge and zero spin. In other words an injected electron in a one-dimensional solid will split into elementary excitations carrying separately its spin and charge. In the next section, we will describe in detail a new algorithm, the hybrid loop QMC algorithm, that we developed, and that allows us to study static and dynamic properties of the $t - J$ model for arbitrary high doping. We will then show some recent results on the spectral function on the one-dimensional system with finite doping.

The depleted Kondo lattice model (DKLM) we will consider reads:

$$H_{DKLM} = \sum_{k,\sigma} \varepsilon(k)c_{k,\sigma}^{\dagger}c_{k,\sigma} + J \sum_{R} \boldsymbol{S}_{\boldsymbol{R}}^c \cdot \boldsymbol{S}_{\boldsymbol{R}}^f. \qquad (2)$$

The physics under consideration is that of a lattice of magnetic spin-1/2 impurities ($\boldsymbol{S}_{\boldsymbol{R}}^f$) embedded in a metallic host at lattice sites $\boldsymbol{R}$. The metallic state is described by the conduction band ($\varepsilon(k)$). In the above equation, $\boldsymbol{S}_{\boldsymbol{R}}^c = (1/2) \sum_{\alpha,\beta} c_{R,\alpha}^{\dagger}\boldsymbol{\sigma}_{\alpha,\beta}c_{R,\beta}$ and $c_{k,\sigma}^{\dagger}$ creates an electron with z-component of spin σ and crystal momentum k. The above model is believed to capture the physics of heavy fermion materials such as $CeCu_6$. The magnetic moments stem from the partially filled f-shells of the Ce atoms. As mentioned above, the heavy fermion ground state is generically a Fermi liquid with large effective mass or equivalently a low coherence scale. It is the aim of the simulations presented in section 3 to discuss the formation of this heavy fermion ground state and to determine the characteristic energy scale at which it starts forming: the coherence scale.

2   The one-dimensional $t - J$ model at finite doping

In order to have a better understanding of the $t - J$ model it is useful to perform a canonical transformation instead of studying it in its standard formulation (Eq. 1). In this way without introducing approximations and without enlarging the Hilbert space of the system we obtain an Hamiltonian bilinear in the fermion fields that is very suitable to be investigated via a new QMC algorithm, the hybrid loop algorithm, that we developed and that will be described in detail in the following section.

The canonical transformation is as follows [1]

$$c_{i\uparrow}^{\dagger} = \gamma_i^+ f_i - \gamma_i^- f_i^{\dagger}, \quad c_{i\downarrow}^{\dagger} = \sigma_i^- (f_i + f_i^{\dagger}), \qquad (3)$$

where $\gamma_i^\pm = (1 \pm \sigma_i^z)/2$ and $\sigma_i^\pm = (\sigma_i^x \pm i\sigma_i^y)/2$. The spinless fermion operators fulfill the canonical anticommutation relations $\{f_i^\dagger, f_j\} = \delta_{i,j}$, and σ_i^a, $a = x, y$, or z, are the Pauli matrices. The Hamiltonian becomes

$$\tilde{H}_{t-J} = +t \sum_{<i,j>} P_{ij} f_i^\dagger f_j + \frac{J}{2} \sum_{<i,j>} \Delta_{ij}(P_{ij} - 1), \qquad (4)$$

where $P_{ij} = (1 + \boldsymbol{\sigma}_i \cdot \boldsymbol{\sigma}_j)/2$, $\Delta_{ij} = (1 - n_i - n_j)$, and $n_i = f_i^\dagger f_i$.

Starting from a system of interacting spinfull fermions, after the canonical transformation, we obtain a system describing free spinless fermion interacting with a quantum pseudospin background. The constraint to avoid doubly occupied states transforms to the conserved and holonomic constraint $\sum_i \gamma_{i,-} f_i^\dagger f_i = 0$. This constraint simply means, that a spinless fermion and a pseudospin $\downarrow$ are not allowed to sit on the same site.

2.1 The hybrid loop algorithm

Once the canonical transformation is performed, we obtain a Hamiltonian bilinear in the fermion field (Eq. 4). Our aim is now to use the separation of degrees of freedom of the model while performing a $T = 0$ projection that filters the ground state $|\Psi_0\rangle$ out of a trial wave function $|\Psi_T\rangle$

$$|\Psi_0\rangle = \lim_{\Theta \to \infty} e^{-\Theta/2 \, \mathcal{H}} |\Psi_T\rangle. \qquad (5)$$

The projection is performed via a QMC evolution treating the fermionic degrees of freedom with the determinantal algorithm [2],[3] and the spin degrees of freedom with the loop algorithm[4]. The combination of those two algorithms is the hybrid loop algorithm.

The first step is a so called checkerboard decomposition, where the Hamiltonian is split in commuting terms $H_1, H_2, \ldots H_p$, and a Trotter decomposition of the partition sum

$$\mathbf{Z} = \mathrm{Tr} e^{-\beta H} = \mathrm{Tr} \left[\left(e^{-\frac{\beta}{M} H_1} e^{-\frac{\beta}{M} H_2} \ldots e^{-\frac{\beta}{M} H_p} \right)^M \right] + \mathcal{O}\left(\left(\frac{\beta}{M} \right)^2 \right). \quad (6)$$

This maps the d-dimensional quantum system to a $d+1$-dimensional classical system.

Let's now define the trial wave function as a direct product of a fermionic part $(\mathcal{P}|\psi_{Tf}\rangle \equiv |\Psi_{Tf}\rangle$ where $\mathcal{P}$ restricts the Hilbert space to the non double occupied one and it is conserved during the evolution) and a pseudospin part $(\sum_\sigma |\sigma\rangle)$. Let $|\Psi_{Tf}\rangle$ be a product of one-particle states

$$|\Psi_{Tf}\rangle = \prod_{l=1}^{N_p} [\sum_{j=1}^{L} P_{jl} f_j^\dagger] |v\rangle \qquad (7)$$

where $|v\rangle$ is the vacuum state for holes, N_p is the number of holes and L the number of sites in the system.

Since the Hamiltonian is already quadratic in the fermionic degrees of freedom we can directly apply the determinantal algorithm without the need of introducing auxiliary fields. In order to integrate the fermions a complete set of pseudospin states has to be inserted for every timeslice. We then obtain for every realization of the spin field σ_n:

$$\langle \Psi\{\sigma_n\} \mid \Psi\{\sigma_n\}\rangle = \langle v\mid [\prod_{k,l=1}^{N_p} \sum_{i,j=1}^{L} P_{ik}^T f_i P_{jl}^e f_j^\dagger]\mid v\rangle$$

$$= \det[P^T B_M^\sigma ....B_{M/2}^\sigma ....B_1^\sigma P] \tag{8}$$

where the notation stands for $P_{jl}^e \equiv \prod_{n=1}^{M/2} e^{-\Delta\tau H(\sigma_n)} P_{jl} \equiv \prod_{n=1}^{M/2} B_n^\sigma P_{jl}$.

The fermionic dynamics is fixed by the spin background and the fermions contribute to the weight of the spin field realization with Eq. 8. On the other hand the pseudospins are quantum degrees of freedom with their own dynamics and they are simulated via a loop algorithm that leads to a weight for every spin field realization $W(\sigma)$ corresponding to the probability distribution of a Heisenberg antiferromagnet for the configuration σ, where σ is a vector containing all intermediate states $(\sigma_1, \ldots \sigma_n, \ldots, \sigma_1)$.

From this follows that the weight of each configuration is given by

$$\langle \Psi \mid \Psi\rangle = \sum_{\{\sigma\}} W(\sigma)\, \det[P^T B_M^\sigma ....B_{M/2}^\sigma ....B_1^\sigma P] \tag{9}$$

The simulation is therefore divided in two steps. First, a pseudospin configuration is proposed via the loop algorithm. This fixes the fermion evolution that is performed via the determinantal algorithm in the frame of which all the observables can be calculated.

The hybrid loop MC incorporates the advantages of the projector MC (no limitation in the measurable observable) and of the loop MC (absence of critical slowing down). Using this new technique we are able to study static and dynamical properties of the single hole t-J model in all dimension (1D, ladders, 2D) [5], [6], [7] and the t-J model in 1D for all doping in the whole range of parameters (J/t) without sign problem.

In order to study the dynamics of the holes we calculate the time displaced one-particle Green's function for spin up,

$$G(i-j, \tau) = -\langle T\tilde{c}_{i,\uparrow}(\tau)\tilde{c}_{j,\uparrow}^\dagger\rangle, = -\langle T f_i^\dagger(\tau) f_j\rangle \tag{10}$$

where T corresponds to the time ordering operator. Any other observables can be calculated starting from $G(i-j, \tau)$ using the Wick theorem.

The one-particle spectral function $A(\boldsymbol{k}, \omega)$ is connected with the Green's function in imaginary time at $T = 0$, by the spectral theorem

$$G(\boldsymbol{k}, \tau) = \int_{-\infty}^{\infty} d\omega \frac{\exp(-\tau\omega)}{\pi} A(\boldsymbol{k}, \omega). \qquad (11)$$

and the analytic continuation of the data is achieved via MaxEnt [8].

2.2   The spectral function

The one-dimensional $t - J$ model is highly interesting for the understanding of correlated electron systems. In fact due to its rich phase diagram (see Fig. 1), where it is possible to find a Luttinger liquid phase, a superconducting phase, a spin gap phase and a phase separation regime, it is possible to study in a lower dimension all the phases that are suspected in the 2D high temperature superconductors. The issue is still largely unsolved especially for what concerns the dynamical properties of the system for finite values of the parameters and of doping.

Using the hybrid loop MC we are able for the first time to study in detail the spectral function of the model for very large systems ($L = 80,100$) at finite doping and without limitations in parameter space.

Figures 2 and 3 show the spectral function (lower part) and dispersion (upper part) for different doping ($\delta = 0.95$, $\delta = 0.8$) for $J/t = 0.5$ i.e. in the fully Luttinger phase regime.

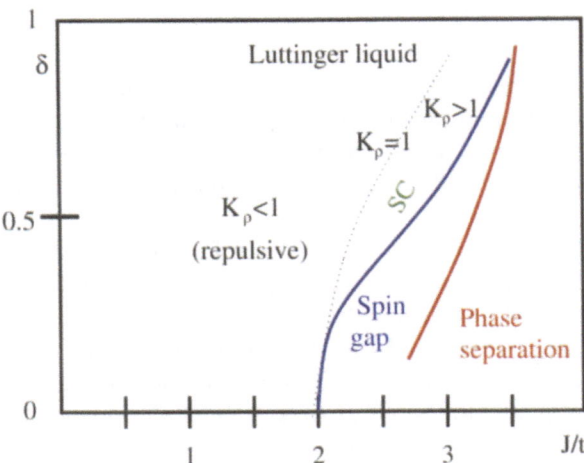

Fig. 1. Phase diagram of the 1D $t - J$ model: one can see the presence of a Luttinger liquid phase, a superconducting phase (SC), a spin gap phase and a phase separation regime .

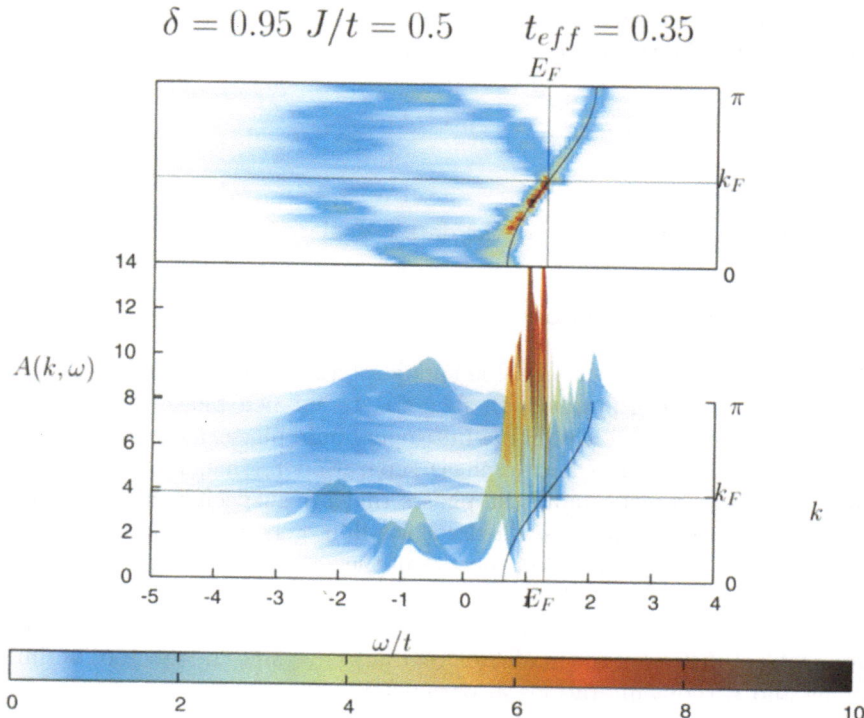

Fig. 2. Spectral function (lower part) and dispersion (upper part) of 1D t-J model for doping $\delta = 0.95$ at $J/t = 0.5$. The region $\omega \leq E_F$ corresponds to the photoemission spectra, the region $\omega > E_F$ corresponds to the inverse photoemission spectra. The fit is done via free cosine band with renormalized amplitude $t_{\mathrm{eff}} = 0.35$.

We can notice that in the finite doping case the high-energy spinon band is almost completely depressed leading to a much broader structure with respect of the one present in the single hole [5] case and that there is a transfer of spectral weight from electron removal to electron addition states as δ increases.

It is possible to see clearly the presence for all $J/t \leq 2$ of shadow bands due to diverging spin fluctuations at $2k_F$. In the $J/t = 0.5$ case their weight near the Fermi level diminishes rapidly with doping as for $J/t \rightarrow 0$ [9], [10].

The low-energy edge band (i.e. the structure close to the Fermi energy) is very well reproduced with a fit by a free renormalized cosine band $-2 * t_{\mathrm{eff}} * \cos(k)$ for small values of J/t ($J/t \leq 2$) at all dopings. It is possible to understand it because of the nature of the elementary excitation. The electron or hole-excitations being a convolution of two free excitation (a holon and a spinon) with two different energy scales, the resulting band at finite energy is still cosine like but gets renormalized by the spin (J) energy scale: this is confirmed by the fact that the t_{eff} depends almost only on J but not on δ.

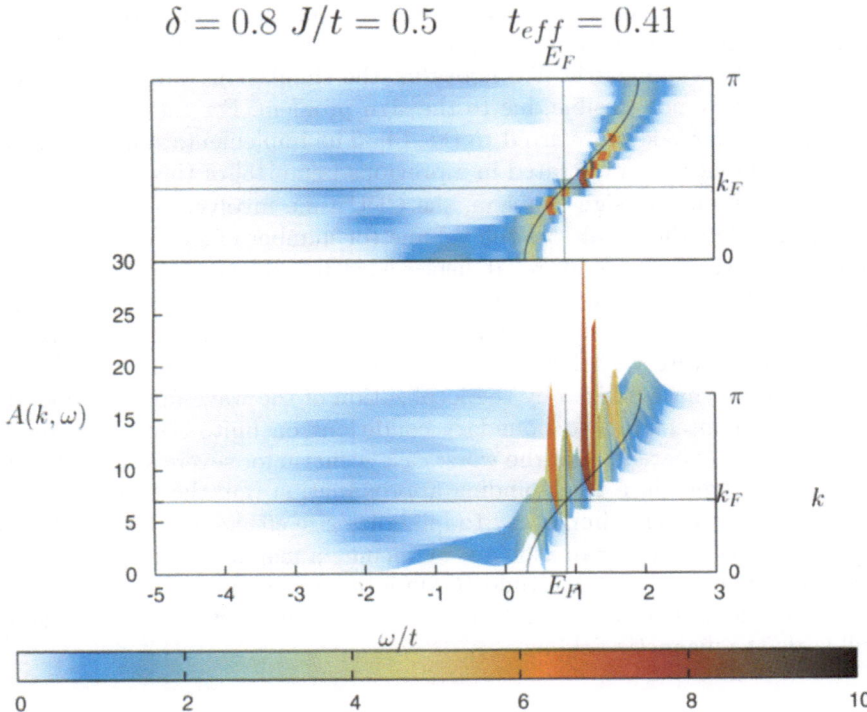

Fig. 3. Spectral function (lower part) and dispersion (upper part) of 1D t-J model for doping $\delta = 0.8$ at $J/t = 0.5$. The region $\omega \leq E_F$ corresponds to the photoemission spectra, the region $\omega > E_F$ corresponds to the inverse photoemission spectra. The fit is done via free cosine band with renormalized amplitude $t_{\text{eff}} = 0.41$.

3   The depleted Kondo lattice model

In this section we will first describe some of the technical aspects involved in the simulations of the DKLM defined in by Eq. 2. By computing the transport and thermodynamic properties of the model, we will pin down the relevant energy scales of the model and in particular concentrate on the *coherence* scale. We will see that this scale compares remarkably well with the single impurity Kondo temperature.

3.1   Technical aspects – reducing size effects

For the simulations we use the determinantal QMC algorithm [3]. In this approach, the many body interaction is decoupled with a Hubbard-Stratonovitch field and the summation over the fields is carried out with Monte-Carlo methods. We have been able to formulate this algorithm without generating a so-called sign problem which leads to an exponential growth of the noise to signal ratio as a function of decreasing temperature and lattice size [11]. This achievment comes constraint, namely that the conduction band has to be particle-hole symmetric. We note that the simulations we present below were previously not feasible due to the sign problem. For the details of the algorithm, the reader is referred to [11–14]. The implementation on massive parallel computers is illustrated in a previous issue [15] of this series.

In the absence of sign problem, the CPU time involved in the determinantal QMC method scales as the volume (or number of sites) to the cubed times the inverse temperature. It hence hard to achieve very large lattices sizes and strategies to reduce finite size effects will become important. Size effects turn out to be particularly severe when the ground state turns out be a metallic state with large coherence temperature. On the other hand, insulators are characterized by the localization of the wave function and are hence rather insensitive to boundary conditions on finite sized systems. It thus becomes apparent, that the worst case scenario for severe size effects are just free electrons in a tight binding approximation (i.e. the first of Eq. 2). We have found a very simple way to minimize size effects for this simple but important case. In the Hamiltonian, we include a magnetic field perpendicular to the plane and of magnitude B. On a torus the minimal magnetic field traversing the lattice corresponds to a single flux quantum Φ_0 [16]. Hence we will scale the magnetic field as

$$\frac{BL^2}{\Phi_0} = 1 \tag{12}$$

where L is the linear size of the lattice. Thus in the thermodynamic limit B scales to zero and we are left with our original problem. However as shown below the approach to the thermodynamic limit is much quicker.

We illustrate the reduction of size effects caused by the inclusion of the magnetic field, we consider the specific heat coefficient: $\gamma = C_v/T$ (See Fig. 4). Upon inspection, one sees that the inclusion of the magnetic field buys us an order of magnitude in temperature before finite size effects set in. The interested reader is referred to Ref. [14] for further details on the methods.

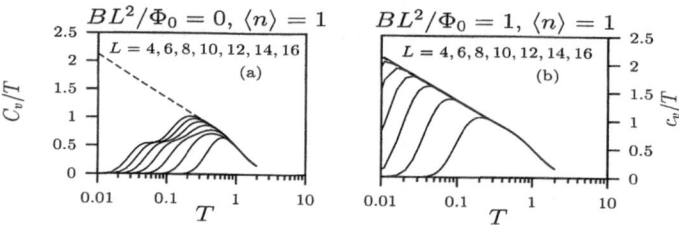

Fig. 4. Specific heat coefficient without (a) and with (b) magnetic field. The curves from right to left correspond to increasingly large lattices as denoted in the figure. The dashed line corresponds to the exact result.

3.2 Coherence in the DKLM

The physics of the single magnetic impurity embedded in a metallic host is well understood [17]. At high temperatures the impurity spin is essentially free thus yielding a Curie law for the impurity spin susceptibility. Below the Kondo temperature $T_K \propto \varepsilon_f e^{-1/JN(\varepsilon_f)}$ the impurity spin is screened by the conduction electrons. Here, ε_f is the Fermi energy and $N(\varepsilon_f)$ the density of states taken at the Fermi energy. Above T_K, singular spin flip scattering off the impurity spin produces a logarithmic increase in the resistivity as a function of decreasing temperature. Below T_K the impurity spin is screened and thus acts as a potential scatterer leading to a saturation of the resistivity.

In the case of a lattice of magnetic impurities, the situation differs. After an initial increase there is a characteristic energy scale T^* below which the resistivity drops rapidly to ultimately follow the Fermi liquid T^2 law. It is the aim of the simulations to investigate the energy scale T^*. An important issue is to understand whether T^* is a new lattice energy scale or if is related to the single impurity Kondo temperature. The DKLM we consider is shown in Fig. 5 which defines the pattern of impurity spins we have chosen to consider.

Figure 6a plots the resistivity versus temperature curve. It allows us to define three energy scales. The highest corresponds to the resistivity minimum, T_C. For the considered coupling $T_C \sim 0.5t$. We will define $T^* \sim 0.1t$ as the temperature scale at which a maximum in the resistivity is observed. The magnetic scale $T_S \sim 0.025t$ is defined by the energy scale below which the $\rho(T)$ follows an *activated* behavior. Next, we consider thermodynamic properties and concentrate on their behavior at the above defined energy scales.

Figure 7 plots the specific heat C_v, charge and spin uniform susceptibilities (χ_c, χ_s) as well as the staggered spin susceptibility $(\chi_s(\boldsymbol{Q}), \boldsymbol{Q} = \boldsymbol{b}_1/2)$ as a function of temperature at $J/t = 1.6$. From the technical point of view, the specific heat is computed using the recently proposed ME based method of

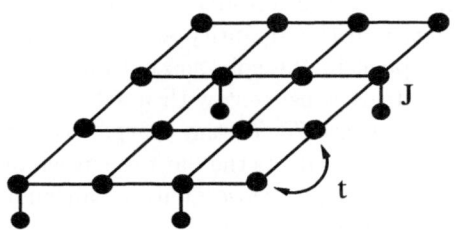

Fig. 5. The depleted KLM (DKLM). The corresponding Bravais lattice is spanned by $2\boldsymbol{a}_x$ and $\boldsymbol{a}_x + 2\boldsymbol{a}_y$. Each unit cell contains a single localized orbital for the impurity spin and four delocalized orbitals accommodating the conduction electrons. The top layer denotes the conduction electrons with hopping matrix element t. The bottom layer are the magnetic impurities which couple to the conduction electrons via the exchange J.

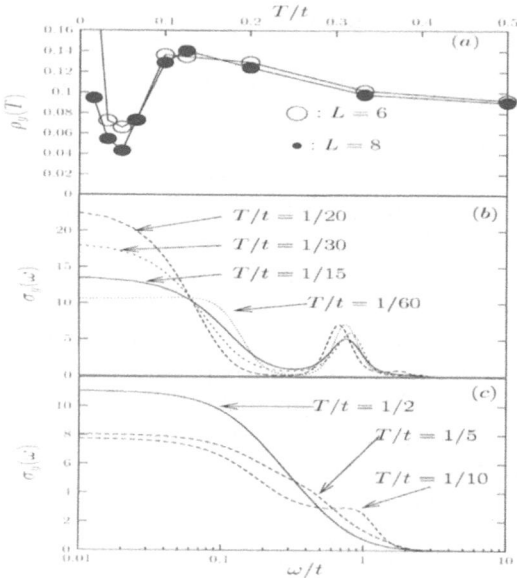

Fig. 6. Optical conductivity and related resistivity as a function of temperature at $J/t = 1.6$. The data in Figs. (b),(c) stem from simulations on lattices with 8×8 unit cells (L=8). In Fig. (a) both $L = 6$ and $L = 8$ results are included. As apparent and due to the inclusion of the magnetic field size effects are next to absent until the magnetic length scale exceeds the lattice size (see text).

[18]. As apparent in the temperature region $T_S < T < T_{\mathrm{coh}}$ one observes the following features. i) As indicated in Fig. 7c the specific heat is consistent with a γT law. ii) the uniform spin and charge susceptibilities after going through a broad maximum seem to saturate. Both points are the characteristics of a Fermi liquid. The staggered spin susceptibility essentially measures the magnetic length scale at the considered wave vector. The inset of Fig. 7a shows a marked increase in this quantity at the spin scale T_S. At the same energy scale, a *sharp* peak in the spin susceptibility is observed. Following the size effects, the data is consistent with a saturation of this quantity in the limit of zero temperature. At T_S the charge susceptibility 7b decreases rapidly. Upon analysis of size effects the data is consistent with the vanishing of this quantity at $T = 0$. Upon close analysis, an anomaly in the specific heat is apparent at T_S. Thus, T_S marks the onset of substantial spin fluctuations which triggers the formation of an insulating state which leads to the vanishing of the charge susceptibility and activated behavior of the resistivity below T_S.

Our QMC results are summarized in Fig. 8. The highest energy scale, T_C, corresponds to the resistivity minimum, triggered by enhanced spin flip scattering off the impurity spins. Below T_C the resistivity grows as a function of

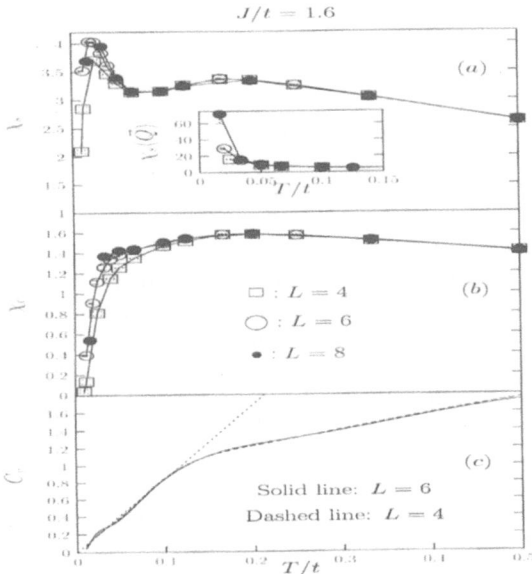

Fig. 7. (a) Uniform and staggered (inset) spin susceptibilities (b) charge susceptibilities and (c) specific heat for the DKLM at $J/t = 1.6$.

decreasing temperature. In this temperature range, the behavior of the resistivity is reminiscent of the single impurity case. For values of $J/t > 0.8$, the
onset of coherence is seen at T^*. In particular, at and below T^* the resistivity
drops sharply, the specific shows a linear in T behavior, and both the spin
and charge susceptibilities seem to converge to finite values. In a first approximation, this is what expects for a Fermi liquid. Given the limited data,
it is hard to pin down the functional form of T^* as a function of J/t. Nevertheless, an exponential fit $(e^{-1/\alpha J})$ accounts reasonably well for the data.
Due to the particle hole symmetry of the model which automatically leads to
nesting, we cannot follow the metallic state to $T = 0$. At T_S substantial spin-
spin fluctuations set in and drive the system to an insulator with activated
resistivity. For values of $J/t < 0.8$, the T^* drops below T_S. Hence after a
slow increase of resistivity below T_C, an activated behavior is expected below
T_S. Figure 8 equally plots the single impurity Kondo temperature T_K^I. The
important point, is that within our accuracy T^* compares very well with T_K^I.
This stands in rough agreement with experiments on $Ce_x La_{1-x} Cu_6$ [19].
Here the single impurity Kondo temperature $(x \ll 1)$ is given by $T_K^I \sim 3K$.
As x grows $(0.73 < x < 1)$, a maximum in the resistivity versus temperature
curve develops at energy scales in the range $T^* \sim 5 - 15K$. It is also worth
pointing out that below T^* the spin susceptibility in $CeCu_6$ seems to saturate
[19]. This stands in rough agreement with our results. Hence, T^* corresponds
essentially to the single impurity Kondo temperature T_K^I below which the

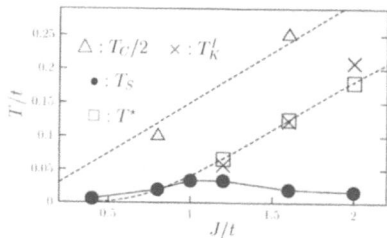

Fig. 8. Crossover scales for the DKLM as a function of J/t. T_C denotes the resistivity minimum. T^* corresponds to the energy scale at which a maximum in resistivity is observed and below which the specific heat is linear in T. Below T_S at marked increase in the magnetic length scale is observed. We use the form $e^{-1/\alpha J}$ to fit the T_{coh} and the form aT to fit the data for T_C. Finally, T_K^I corresponds to the single impurity Kondo temperature.

magnetic impurities are screened. It marks the onset on the heavy-electron state.

4   Conclusions

In conclusions, we have presented numerical simulations of the one-dimensional t-J model and of a depleted Kondo lattice model. In the one dimensional case the metallic state corresponds to a Luttinger liquid. We have presented some results on the single particle spectral function and shown how the finite doping modify this quantity with respect to the single hole case. For the Kondo Lattice model, we have been able for the first time to investigate numerically the onset on the heavy fermion state. We have seen that in the DKLM the formation of the heavy electron ground state starts at en energy scale T^* which compares remarkably well with the single impurity Kondo temperature.

Acknowledgements

This work was supported by the Deutsche Forschungsgemeinschaft and the Sonderforschungsbereich 382. The numerical calculations were performed at HLRS Stuttgart. We thank the above institutions for their support.

References

1. G. Khaliullin, JETP Lett. **52**, 389 (1990).
2. G. Sugiyama and S. E. Koonin, Anals of Phys. **168**, 1 (1986).
3. R. Blankenbecler, D. J. Scalapino, and R. L. Sugar, Phys. Rev. D **24**, 2278 (1981).

4. H. G. Evertz, in *Numerical Methods for Lattice Quantum Many-Body Problems*, *Frontiers in Physics*, edited by D. J. Scalapino (Addison-Wesley Publishing Company, Redwood City, 1997).
5. M. Brunner, F. F. Assaad, and A. Muramatsu, Eur. Phys. J. B **16**, 209 (2000).
6. M. Brunner, F. F. Assaad, and A. Muramatsu, Phys. Rev. B **62**, 12395 (2000).
7. M. Brunner, S. Capponi, F. F. Assaad, and A. Muramatsu, Phys. Rev. B **63**, R180511 (2001).
8. M. Jarrell and J. Gubernatis, Physics Reports **269**, 133 (1996).
9. R. Eder and Y. Ohta, Phys. Rev. B **56**, 2542 (1997).
10. J. Favand, S. Haas, K. Penc, F. Mila, and E. Dagotto, Phys. Rev. B **55**, R4859 (1997).
11. F. F. Assaad, Phys. Rev. Lett. **83**, 796 (1999).
12. S. Capponi and F. F. Assaad, Phs. Rev. B **63**, 155113 (2001).
13. M. Feldbacher and F. F. Assaad, Phys. Rev. B **63**, 73105 (2001).
14. F. F. Assaad, Phys. Rev. B **65**, 115104 (2002).
15. C. Lavalle, M. Brunner, F. F. Assaad, and A. Muramatsu, *High Performance Computing in Science and Engineering'00* (Springer-Verlag, Berlin, 2001), pp. 143–154.
16. E. Fradkin, *Field Theories of condensed matter systems*, *Frontiers in Physics* (Addison-Wesley Publishing Company, Redwood City, 1991).
17. A. C. Hewson, *The Kondo Problem to Heavy Fermions*, *Cambridge Studies in Magnetism* (Cambridge Universiy Press, Cambridge, 1997).
18. C. Huscroft, R. Gass, and M. Jarrell, Phys. Rev. B **61**, 9300 (2000).
19. A. Sumiyama, Y. Oda, H. Nagano, Y. Onuki, K. Shibutani, and T. Komatsubara, J. Phys. Soc. Jpn. **55**, 1294 (1986).

Self-Trapping of the Si(111)-(2×1) Surface Exciton

Michael Rohlfing

Institut für Festkörpertheorie, Universität Münster, Wilhelm-Klemm-Str. 10, 48149 Münster, Germany

Abstract. We discuss the localization of the surface exciton at the Si(111)-(2×1) surface due to self-trapping, which leads to a characteristic temperature-dependent linewidth of the optical response and to a significant Stokes shift of the luminescence. Self-trapping results in this case from a structural relaxation in the excited state, caused by the interplay between electronic and geometric degrees of freedom. The most significant contribution to this effect comes from one single geometric deformation mode which is driven by the internal electronic charge transfer in the self-trapped exciton. To study these mechanisms we employ computational ab-initio techniques designed for excited states (density-functional theory and many-body perturbation theory), combined with tight-binding representations that allows us to simulate enlarged supercells containing several thousand atoms.

1 Introduction

Optically excited electronic states in nanoscale low-dimensional condensed-matter systems, like surfaces, conducting polymers, and molecules, are of high interest in science and technology. The properties of such states depend sensitively on the structure of the underlying system, as resulting from the complex interrelation between atomic positions and electronic excitations. A careful theoretical analysis of this interdependence requires a comprehensive quantum-mechanical treatment of both the geometric and the electronic structure, including electronic many-body effects and coupling mechanisms. As a prototype example, we discuss the surface exciton at the Si(111)-(2×1) surface and its optical response (see also [1]).

The surface exciton at the Si(111)-(2×1) surface is one of the most prominent elementary excitations in low-dimensional semiconductor science [2–6]. The exciton produces a distinct signal in the optical reflectivity spectrum (see Fig. 1a). Very accurate and highly reproducible experimental data have been obtained from freshly cleaved Si(111) surfaces. A number of theoretical approaches have discussed the position and amplitude of the peak, which are well understood, by now (see Sec. 3). Its *line shape, width, and temperature dependence*, on the other hand, have not been addressed in detail by theory, so far. In [2], Ciccaci et al. measured the line width and its temperature-dependence (shown in Fig. 1), which were suggested to be related to electron-phonon interaction. Motivated by these measurements,

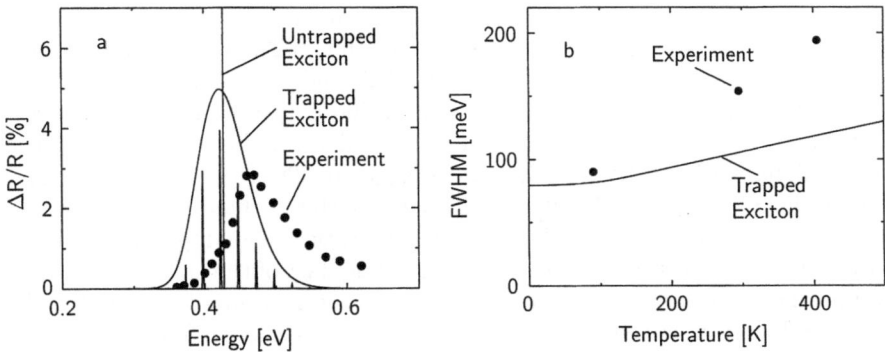

Fig. 1. (a) Calculated differential reflectivity spectrum (DRS) of the Si(111)-(2×1) surface exciton at T=90 K in comparison with the experimental data (•) of [2]. The vertical line at 0.43 eV indicates the energetic position of the calculated excitation of the untrapped state (see [6]). The peaks below the solid line indicate the Franck-Condon factors. For the resulting spectrum, additional broadening of 15 meV has been employed (see text). (b) Spectral linewidth (full width at half maximum, FWHM) of the DRS as a function of temperature, in comparison with the experimental data (•) of [2].

we present here a comprehensive theoretical study of these issues, employing computational ab-initio many-body approaches designed for the investigation of excited electronic states [7–12]. Our results indicate that the exciton observes spatial localization due to a self-trapping mechanism, which breaks the periodicity of the surface and leads to structural relaxation in the excited state. Apart from spectral broadening, this also causes a significant Stokes shift. We expect that such relaxation effects, that are well-known in molecules, also occur in other low-dimensional periodic systems and play a crucial role in light absorption and emission.

2   Computational Approach

The calculations presented here have been carried out by combining several computational approaches to condensed-matter physics. In particular, we employ ab-initio many-body perturbation theory (MBPT) specifically designed for the investigation of excited electronic states [8–12], combined with an efficient tight-binding approach to investigate excited states in enlarged supercells.

The properties of the unperturbed, periodic surface are addressed by electronic-structure techniques for periodic systems (density-functional theory (DFT) within the local-density approximation (LDA), the GW approximation (GWA), and MBPT). This yields the necessary information about the ground-state geometry, the quasiparticle band structure, and the unperturbed surface exciton. The calculations are carried out in a 10-layer slab configuration (containing 20 atoms in the unit cell), with 24 representative **k** points in the unit cell (both for LDA and for GWA), 30 Gaussian basis functions per atom, 40 occupied and 360 empty bands in the necessary band summations. The exciton is formed from the occupied and the empty dangling-bond bands (D_{up} and D_{down}) at 390 **k**-points. The calculation of the $390 \times 390 = 152100$ matrix elements of the electron-hole interaction, each requiring a 6-dimensional real-space integration involving four electronic wave functions, constitutes the most demanding part of the calculation. After computing the electron-hole interaction, the exciton results from solving the Bethe-Salpeter equation (BSE) of coupled electron-hole configurations.

One of the key concepts of the exciton-geometry interrelation is the investigation of the change in the surface geometry when electronic charge is locally re-distributed (this charge transfer is associated with the electronic transition from the D_{up} state to the D_{down} state induced by the exciton). We investigate this charge-geometry interrelation within DFT-LDA (in a similar way as discussed in [18]). Within the (2×1) surface unit cell, a certain amount of electronic charge (δq) is deliberately transferred from the D_{up} to the D_{down} band by fixing the corresponding occupation numbers, followed by structural relaxation [19]. This is done for several values of (δq), confirming that the relaxation is proportional to (δq). It allows to identify the deformation mode associated with the exciton, its vibrational frequency, and the relaxation energy gain depending on (δq). We assume that this mode (which is similar to an optical phonon mode at the Γ point of the Brillouin zone) is the relevant one for the self-trapping. Note, however, that this lattice deformation is not a phonon eigenmode of the electronic ground state [20].

Based on these findings, we finally investigate the self-trapping of the exciton, which requires a laterally enlarged supercell with a size of up to one thousand (2×1) unit cells. The direct use of ab-initio techniques is prohibitive for such large systems (which contain up to 20000 atoms). Instead, a tight-binding Hamiltonian for the dangling-bond orbitals is constructed from the ab-initio results of the periodic (2×1) surface (in the same way as suggested in [21]). The hole and electron states are thus expanded in a real-space basis set of two orbitals per (2×1) unit cell. The electron-hole interaction is given in the same basis. For orbital-orbital distances $d = |\mathbf{R}_\alpha - \mathbf{R}_{\alpha'}|$ large compared to the lattice constant, it is given by the Coulomb interaction $1/d$, modified by the effective dielectric constant of the surface [21]. Note that due to the large spatial extent of the exciton along the Pandey chain, this long-range behavior is the most relevant part of the interaction. For small distances $|R_\alpha - R_{\alpha'}|$, as well as for the onsite terms ($\alpha = \alpha'$), a finite limit is assumed. For a large supercell with unperturbed geometry, this Hamiltonian reproduces the previous ab-initio results for the surface exciton on the periodic surface [6]. When allowing for structural relaxations within the supercell, driven by the local charge rearrangement, the exciton localizes and forms a self-trapped state as discussed in the next paragraph. Since the wave function of the exciton and the underlying geometric structuredepend on one another, one has to carry out an iterative optimization procedure (of about 50 iterations) until self-consistency is reached.

The most time-demanding part of the work is the MBPT, which yields the reference data of the band structure and the exciton at the unperturbed, periodic surface and provides the necessary parameters to construct the tight-binding representation. Both the MBPT codes and the tight-binding approach are natural candidated for parallelization, running efficiently on up to 64 processors. The total computation time amounts to several thousand CPU hours.

3. Results and Discussion

From experimental studies [3,13–15] and computational ab-initio investigations [4–6] the basic physics of the surface and the exciton is known: (i) The surface is terminated by π-bonded Pandey chains, with two surface atoms per (2×1) surface unit cell (shown in Fig. 2a) [16]. In the electronic ground

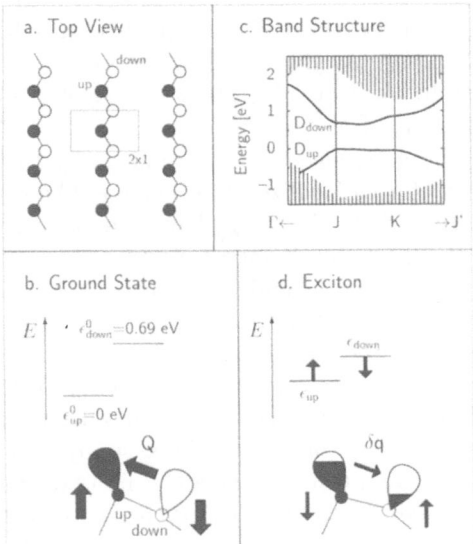

Fig. 2. Atomic and electronic structure of the Si(111)-(2×1) surface. (a) Schematic top view of the π-bonded zig-zag Pandey chains on the surface. (b) Schematic cross section perpendicular to the Pandey chain in the electronic ground state. (c) *GW* quasiparticle surface band structure of the D_{up} and D_{down} states. The shaded area denotes the projected bulk band structure. (d) Schematic cross section perpendicular to the Pandey chain in the *excited* electronic state, which is characterized by a reduced (down→up) charge transfer (δq), a reduced buckling, and a reduced gap between the onsite energies ϵ_{up} and ϵ_{down}.

state, the Pandey chain is buckled with a height difference of 0.51 Å between the up and down atom. (ii) Each surface atom exhibits a dangling-bond orbital, with onsite energies ϵ_{up}^0 and ϵ_{down}^0 (see Fig. 2b). In the band structure, the two orbitals produce two surface bands (shown in Fig. 2c), with a direct surface gap of 0.69 eV [4–6]. The orbitals of neighboring atoms are strongly coupled, leading to the significant band dispersion along the chain direction ($\Gamma \rightarrow J$ and $K \rightarrow J'$). (iii) The excitation of electrons from the D_{up} to the D_{down} band gives rise to a distinct peak at 0.45 eV in the optical spectrum (see Fig. 1a). This peak, however, does not simply correspond to vertical interband transitions (which would lead to a broad feature starting at an onset energy of 0.69 eV); instead, the interband transitions experience strong electron-hole interaction and form a coupled electron-hole pair state, i.e. an exciton with a binding energy of 0.26 eV [6].

In the electronic ground state, the D_{up} band (mainly formed from dangling-bond orbitals at the up atoms) is occupied while the D_{down} band is empty. This causes a significant electronic charge transfer Q from the down orbitals to the up orbitals (see Fig. 2b). This charge transfer can be considered to be a main driving force for the buckling. The buckling, in turn, is responsible for the difference between the onsite energies ϵ_{up}^0 and ϵ_{down}^0. One key observation of our work is that a reduction (or partial inversion) of the charge transfer (indicated as δq in Fig. 2d) leads to a reduction of the buckling by about $(0.27\text{Å}) \cdot (\delta q)$ (δq given in electrons per (2×1) surface unit cell). This is accompanied by a raising of ϵ_{up} by about $(0.35\text{eV}) \cdot (\delta q)$ in energy and a corresponding lowering of of ϵ_{down}.

Such a reverse charge transfer can be induced by the surface exciton – provided that the exciton is localized (see below). The exciton is composed of transitions from the D_{up} to the D_{down} band, i.e. it involves a hole which is distributed among the D_{up} orbitals and an electron among the D_{down} orbitals. The details of the distribution depend on the internal structure of the exciton, which is controlled by the single-particle properties of the electron and hole and by the electron-hole interaction [10–12]. In each unit cell, this hole and electron invoke a reverse charge transfer proportional to the weight of the exciton in the cell. If the exciton is freely mobile and delocalized over the entire surface (as for a two-dimensionally periodic system), the charge transfer in each unit cell would be infinitesimally small. A finite charge transfer requires that the exciton be localized and the periodicity disturbed. We propose that this localization arises from self-trapping of the exciton (see Fig. 3).

The localized exciton causes the reverse charge transfer δq which is now non-zero in about 30 unit cells (shown in Fig. 3a). This invokes a locally reduced buckling, as discussed in Fig. 3b. The reduced buckling causes local changes in ϵ_{up} and ϵ_{down}. This acts as an attractive potential for both the hole and the electron, which are trapped as localized single-particle states (see panel c). Since the surface exciton is now mainly formed between these localized states, the entire exciton becomes localized by its own presence, i.e. it becomes self-trapped. The exciton and its confinement potential must be

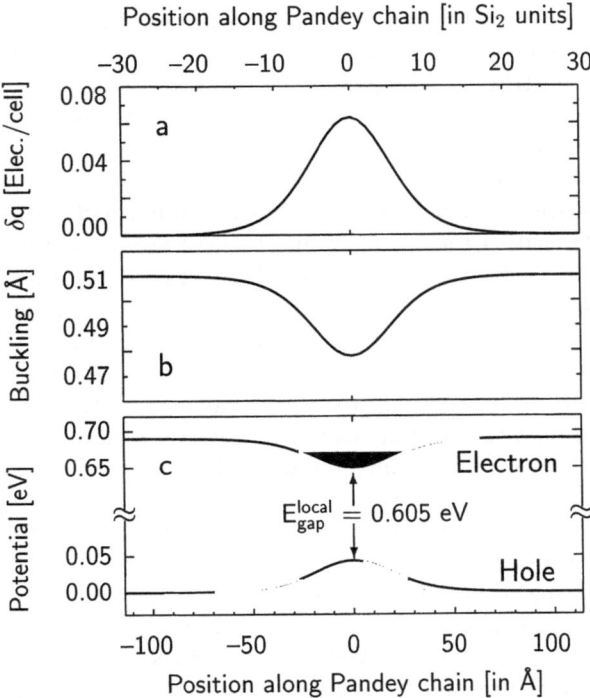

Fig. 3. Self-trapping mechanism of the surface exciton. (a) Local reduction δq of the charge transfer between D_{up} and D_{down} of the self-trapped exciton, displayed along one Pandey chain. δq is determined by the exciton wave function [22]. (b) Local change of the Pandey chain geometry (reduced buckling), induced by the charge transfer reduction of panel (a). (c) Attractive potential for the electron and the hole, caused by the locally reduced gap between ϵ_{up} and ϵ_{down}. The horizontal lines indicate the bound states of the electron and the hole (see text).

calculated self-consistently. In the final result, the reduction of the buckling amounts to 0.035 Å in the center of the exciton. Both the hole and the electron are trapped in corresponding single-particle potentials of 42 meV depth, which produce localized states of 17 meV binding energy for both particles. The exciton, which is now mainly formed between the trapped hole and the trapped electron, thus gains 2×17 meV by being trapped. It gains additional 30 meV because the electron-hole interaction between localized particles is enhanced (290 meV instead of 260 meV for the free surface exciton). The electronic transition energy for this modified geometry is thus about 65 meV smaller than the one of the original geometry. We therefore expect that, after structural relaxation, luminescence occurs at 365 meV instead of 430 meV, with a significant Stokes shift of 65 meV. To our knowledge, luminescence measurements have not been performed on the Si(111)-(2×1) surface, so far.

Apart from the Stokes shift between absorption (or reflectivity) and emission of light, the self-trapping effect has a strong influence on the line shape of the optical spectrum and leads, in particular, to spectral broadening. The present situation of two different equilibrium structures for the electronic ground and excited state (i.e. full buckling everywhere vs. locally reduced buckling) can be analysed in terms of the corresponding Franck-Condon factors. This procedure, which has been employed in molecular spectroscopy for a long time, considers the quantum-mechanical atomic motion around the two equilibrium positions. In our case, the atomic elongation mode of the self-trapping is associated with a vibrational frequency of about 25 meV [20]. The resulting exciton spectrum and temperature-dependent line width are displayed in Fig. 1. The calculated line shape and width are in reasonable agreement with the measured data. Note that at zero temperature the line width does not tend to zero but reaches a finite value of about 80 meV. Two issues, however, still remain unresolved: (i) Different from many molecular spectra, no individual vibrational lines are observed in the exciton spectrum, and (ii) the increase of the calculated line width with temperature is not as strong as found in experiment.

(i) The missing of individual vibrational lines may be due to additional inhomogeneous broadening. One possible source is given by surface defects (in particular, steps) on the cleaved Si(111) surface [17]. Such defects interrupt the Pandey chains and modify the electronic excitation energies due to quantum confinement, leading to a statistical ensemble of various excitation energies. It would be highly revealing if low-temperature optical experiments on a single chain (maybe by scanning microscopy techniques) could be performed to focus on a single excitation state. Further broadening originates from coupling to other phonon modes beyond the most relevant deformation mode discussed here. To account for both effects, additional artificial broadening of 15 meV has been included in the theoretical results presented in Fig. 1.

(ii) The increase of the calculated linewidth with temperature is not as strong as found in experiment. Above $T=100$ K we find a nearly linear increase with a gradient of about 1.5 k_B (with k_B the Boltzmann constant) while the measured FWHM increases with 3.8 k_B (see Fig. 1b). The difference in the gradient of about 2 k_B may again be related to the neglect of additional coupling of electrons and holes to other phonon modes.

In conclusion, we have addressed the excitation-induced influence of the geometrical structure on the surface exciton at Si(111)-(2×1). The coupling between the electronic and the geometric structure can lead to self-trapping of the exciton, accompanied by a significant spectral broadening and a Stokes shift between absorption and emission. The investigations have been carried out within a combination of highly reliable ab-initio approaches designed for excited electronic states, followed by a tight-binding formalism including the electron-hole interaction. Similar effects are expected to occur in other low-dimensional semiconductor materials, like conducting polymers and large conjugated molecules, as well. Localization due to self-trapping should be relevant whenever the system size is comparable to or larger than the intrinsic size of the exciton. The self-trapping mechanism discussed here should therefore be present in many optoelectronic materials and play a crucial role in light absorption and emission processes.

4 Acknowledgments

The author thanks F. Bechstedt, R. Del Sole, S.G. Louie, J. Pollmann, and L. Reining for fruitful discussions. This work was supported by the Deutsche Forschungsgemeinschaft under Grant No. Ro-1318/4-1 and Ro-1318/5-1 Computational resources have been provided by the Bundes-Höchstleistungsrechenzentrum Stuttgart (HLRS).

References

1. M. Rohlfing and J. Pollmann, Phys. Rev. Lett. **88**, 176801 (2002).
2. F. Ciccacci, S. Selci, G. Chiarotti, and P. Chiaradia, Phys. Rev. Lett. **56**, 2411 (1986).
3. P. Chiaradia, A. Cricenti, S. Selci, and G. Chiarotti, Phys. Rev. Lett. **52**, 1145 (1984).
4. L. Reining and R. Del Sole, Phys. Rev. Lett. **67**, 3816 (1991).
5. J.E. Northrup, M.S. Hybertsen, and S.G. Louie, Phys. Rev. Lett. **66**, 500 (1991).
6. M. Rohlfing and S.G. Louie, Phys. Rev. Lett. **83**, 856 (1999).
7. L. Hedin, Phys. Rev. **139**, A796 (1965).
8. M.S. Hybertsen and S.G. Louie, Phys. Rev. Lett. **55**, 1418 (1985).
9. M. Rohlfing, P. Krüger, and J. Pollmann, Phys. Rev. Lett. **75**, 3489 (1995).
10. S. Albrecht, L. Reining, R. Del Sole, and G. Onida, Phys. Rev. Lett. **80**, 4510 (1998).
11. L.X. Benedict, E.L. Shirley, and R.B. Bohn, Phys. Rev. Lett. **80**, 4514 (1998).
12. M. Rohlfing and S.G. Louie, Phys. Rev. Lett. **81**, 2312 (1998).
13. R.I.G. Uhrberg, G.V. Hansson, J.M. Nicholls, and S.A. Flodström, Phys. Rev. Lett. **48**, 1032 (1982).
14. P. Perfetti, J.M. Nicholls, and B. Reihl, Phys. Rev. B **36**, 6160 (1987).
15. R.M. Feenstra, W.A. Thompson, and A.P. Fein, Phys. Rev. Lett. **56**, 608 (1986);
16. K.C. Pandey, Phys. Rev. Lett. **49**, 223 (1982).
17. P. Chiaradia and S. Nannarone, Surface Science **54**, 547 (1976).
18. O. Pankratov and M. Scheffler, Phys. Rev. Lett. **75**, 701 (1995).
19. The self-trapped exciton is not a periodic object, i.e. the charge transfer (δq) varies from one unit cell to the next. The transfer of the *periodic* DFT results to this non-periodic situation relies on the assumption that the change of the geometric and electronic structure from one unit cell to the next, as well as the influence of such changes to the physics in one unit cell, are small enough to be neglected.

20. The deformation is mainly given by a raise of the down atom and a slight lowering of the up atom. From the total-energy curve of the lattice deformation we obtain a vibrational frequency of 25 meV. This deformation mode is lower in energy than the 31 meV surface phonon rocking mode discussed by Alerhand and Mele [Phys. Rev. B **37**, 2536 (1988)] and the 50 meV I'' mode discussed by Ancilotto *et al.* [Phys. Rev. Lett. **65**, 348 (1990)].

21. L. Reining and R. Del Sole, Phys. Rev. B **44**, 12 918 (1991).

22. The trapped exciton is nearly completely localized on one Pandey chain only, with less than 10 % weight on the neighboring chains.

Chemistry

Bernd A. Hess

Chair of Theoretical Chemistry
University of Erlangen–Nuremberg
Egerlandstr. 3, D-91058 Erlangen, Germany

The reliable calculation of compounds containing transition metals is notoriously difficult. Often, density functional theory provides a good compromise between accuracy and efficiency, in particular if ground states of not too complicated electronic structure are involved. In this case, even reactions involving transition metal compounds, often acting as catalysts, can be studied in detail. An example is given in this volume by Pelzer and van Wüllen, who studied the iron-III-catalyzed Michael reaction. Calculations of this type are usually carried out in close collaboration with experimental groups, and the results of the calculation are used as an input to the design of further experiments.

Excited states of transition-metal compounds require more sophisticated treatment. The reason is simple: Transition metal atoms (which are characterized by d orbitals which are not fully occupied) feature energetically close-lying excited states with different numbers of d electrons ("open shells"), typically derived from the atomic $s^2 d^n$, $s^1 d^{n+1}$ and $s^0 d^{n+2}$ configurations. These electronic configurations are markedly different as far as the so-called electron correlation is concerned. The latter concept adresses the deviation of the motion of the electrons from the model of mutually independent electrons, moving in an avaraged field created by the other electrons and the nuclei. The multi-particle effects attributable to the electron correlation are decisive for the energy separation of the excited states, up to the question about the nature of the ground state itself, which in many transition-metal compounds is just "the excited state with the least energy", and not, as in most ordinary chemical compounds, endowed with an electronic configuration in which all electrons are paired with opposite spin, forming "closed shells".

Capturing these complicated multiparticle effects requires the use of a quantum chemical methodology, which is based on a Hilbert space approach. I.e., the building blocks of the theory are not objects (particles, flows, etc.) in 3D real space, but rather multiparticle wave functions (configuration state functions), built from all possible tensor products of one-electron wave functions. For these multiparticle states, the Schrödinger equation has to be solved. The corresponding mathematical problem comprises a number of degrees of freedom which increases with the number of electrons like a factorial. In the best practically feasible quantum-chemical approximation methods taking electron correlation effects into account in a systematic manner, the computational expense scales like a power of n, the number of the electrons, typically proportional to n^6–n^7 for the most reliable of these methods. Needless to say that there is a lot of ongoing effort to develop efficient computer codes for handling systems of practical importance with these methods. In this volume, the development of an implementation of the Multi-Reference Configuration Interaction (MRCI) Method including applications to dinotrosoethylene and VF_2 is reported by Stampfuß, Vogel and Wenzel, and sophisticated applications of the MRCI method on vanadium oxide clusters V_nO_m are described by Pykavy and von Wüllen, making use of the MOLPRO program, a program suite featuring the most advanced quantum-chemical methods and based on work done by the Stuttgart Theroretical Chemistry group.

A completely different, but also N_p-hard problem is tackled in this volume by Hartke, namely the search for a *global* minimum of the energy as a function of the geometric configuration of the atoms in a cluster. In studies of this type, the interaction between the molecules is usually represented by model potentials of varying sophistication. Hartke's contribution reports on calculations of the structure of pure neutral water clusters making use of two different potentials, which were unexpectedly shown to lead to entirely different minimum structures for clusters with 17 water units and above.

Structural Trends and Transitions
in Water Clusters

Bernd Hartke

Institut für Physikalische Chemie, Christian-Albrechts-Universität,
Olshausenstraße 40, 24098 Kiel

Abstract. Global geometry optimization of pure neutral water clusters using the highly accurate but computationally expensive many-body TTM2-F potential have been performed within reasonable real time, using a parallelized and specialized implementation of evolutionary algorithms. In comparison to previous studies using highly approximate and cheap water potentials that work well for small clusters, qualitatively different structures result for larger clusters, exposing subtle failures of these cheap models. This unexpected result could not have been obtained within reasonable real time on a serial machine.

1 Clusters as Challenge for Theory

Single molecules as well as the "infinitely" extended periodic solid are well within the realm of routine calculations today. In spite of its overwhelming importance in many areas [1,2], the size region in between, however, still remains elusive. Disregarding outer and inner structure, cluster properties can be extrapolated from larger to smaller numbers n of particles, with some theoretical foundation [3]; but these smooth, extrapolable trends stop in the size region of $n = 10^3$–10^7 (depending on the physical observable) and are replaced by a seemingly irregular behavior. Obviously, for these smaller clusters, their individual, size-dependent structure is important for their properties. In fact, simple and long established theoretical arguments point to the general occurence of structural transitions in this size region [4], that is, sudden changes of cluster structure upon addition or removal of only one or a few particles.

Unfortunately, traditional experimental structure determination methods (NMR or X-ray diffraction) are not applicaple to the elusive small clusters envisaged here. Therefore, direct structural information is hard to obtain in cluster experiments; typically, rather elaborate theories are needed to provide structure and property data and thus to help in explaining the observations.

Ab initio electronic structure calculations are currently entering a new era, with their traditionally prohibitive scaling of computational expense with system size being replaced by the ultimately optimal linear scaling. Although these developments make applications of advanced electronic structure methods to larger, chemically relevant systems possible for the first time, such methods are still orders of magnitude more expensive than any empirical potential. But even with empirical potentials, theoretical treatments of clusters are inherently expensive: Their configuration space increases exponentially with cluster size [5,6], with estimates for the number of minima for a proto-type cluster of 100 Lennard-Jones atoms ranging from 10^{40} to 10^{80}. Molecular clusters constitute yet another level of difficulty: In addition to positional degrees of freedom, there are also three orientational degrees of freedom per particle; this leads to millions of chemically reasonable orientational isomers already for a small cluster of only 20 water molecules with fixed positions [7]. Clearly, this precludes applications of any method that needs coverage of representative parts of these vast configuration spaces.

2   Comparison of Theoretical Approaches

At first sight, molecular dynamics (MD) or Monte-Carlo (MC) methods may appear to be natural choices for a theoretical treatment of clusters. In fact, they promise direct access to thermodynamic data and to simulations of the system at experimentally relevant temperatures. However, in applications of these methods to more usual systems (even in difficult large-scale simulations of huge biochemical entities), a priori information on structures and a wide distribution of energy barriers conspire to create a massive but meaningful restriction of configuration space that needs to be covered. This is not the case for clusters: Usual chemical bonding rules do not apply, and MD/MC methods may have a hard time not getting trapped in subsections of configuration space (for example, already for 38 Lennard-Jones atoms, only one of the most recent, sophisticated MC sampling techniques ensures adequate coverage also of the global minimum [8]). Hence, according to the argumentation of the previous section, MD or MC methods are more difficult and more expensive to apply to clusters. To put this in perspective: If one does not resort to empirical potentials but attempts to do MD in conjunction with density functional (DFT) calculations, statistics sufficient for thermodynamics can only be generated for very small clusters on the order of just five atoms [9], even with the use of supercomputer centers.

In contrast, global optimization methods give direct access to the global minimum structure, and in practice also to lists of the most important local minima. Although it is often argued that experiments do not necessarily observe the global minimum structure, actually finding predominance of a local minimum structure is the exception rather than the rule. In many cases, these exceptions can be explained with a-posteriori corrections for zero-point energy and finite temperature, on top of global optimization results. In comparison to MD and MC, the obvious drawback of a global structure optimization approach is its static nature; without a-posteriori corrections it yields only the classical-mechanical minima on the potential hypersurface, without information on barriers and/or dynamics. This is argued to be fatal for fluxional clusters like the water clusters targeted here. But even there, knowledge of the actual potential energy minima is helpful to analyze the fluxional system and understand the interplay between effects from the potential and from the dynamics. Furthermore, fluxoniality is often less pronounced than one may assume, because of the low temperatures in typical cluster experiments. The mere occurence of "magic numbers" in such experiments (even with water clusters) is a strong hint at rather incomplete fluxoniality at the experimental temperature (which is often hard to determine even roughly).

In summary, MD/MC techniques and global structure optimization should be viewed as complementary approaches, with the latter yielding part of the information at a lower cost and hence constituting a natural choice as first step.

3 Water Clusters

Pure neutral water clusters have attracted considerable attention for a long time [1], from experimentalists and theoreticians alike. Obvious benefits of these studies include a better understanding not only of naturally occuring water clusters (water aerosols and "confined water", e.g. in zeolites and proteins) but also of bulk water structure and solvation. There are detailed comparisons between theory and experiment, with varying degrees of agreement, at least throughout the range $(H_2O)_n$, $n = 2 - 12$ [10–13]. With the exception of the notorious case of the water hexamer [12,14,15], there have been no surprises, though.

So far, larger clusters have been studied systematically only by global optimization methods on the very simple and computationally cheap TIP4P potential [7,16], mainly as a benchmark for the application of these methods to molecular clusters. The TIP4P model and similar, even more approximate models are often used in large-scale biochemical simulations. Therefore, it is of considerable interest in how far potentials of this sort are able to model local water structure accurately and reliably. At least in the size range mentioned above, $n = 2 - 12$, TIP4P has been found to predict all global minimum structures of water clusters qualitatively correctly, in comparison to ab-initio calculations and experiment. This is rather surprising, comparing its very simple functional form and physical background to more detailed approaches [17].

Therefore, it was not clear at all whether the counter-intuitive findings of the above-mentioned global optimization studies on the TIP4P potential [7,16] were real and reliable or indicate a breakdown of the model: Firstly, there is no obvious structural principle in the global minima of TIP4P water clusters; on the contrary, the whole cluster rearranges at each transition from n to $n + 1$. Secondly, there is a surprisingly strong preference for stacking of cubes and pentagonal prisms, resulting in marked deviations from near-spherical outer forms for several cluster sizes. Last but not least, TIP4P water seems to refuse to act as a solvent for itself, up to surprisingly large clusters: With the sole exception of $n = 19$, all global minimum structures from $n = 2$ to $n = 22$ have all water molecules sitting at the cluster surface, none in the interior. This is in stark contrast to protonated water clusters [18] or cation microsolvation clusters [19,20], where water cages around interior particles start to form at much smaller sizes and close up e.g. at $n = 16$ around K^+ or $n = 18$ around the much larger Cs^+. In particular, the famous dodecahedral "clathrate" structure predicted by Castleman [21] and found (with some modifications and restrictions) for these mixed systems [18–20], which features a central particle surrounded by a deformed dodecahedral water cage, is not an important minimum structure for pure neutral $(H_2O)_n$, $n = 21$, within the TIP4P model.

The aim of this study was to check these points by global geometry optimization of pure neutral water clusters on a more refined and computationally much more expensive water potential, TTM2-F by Burnham and Xantheas [22]. In contrast to TIP4P (featuring rigid monomers interacting only pairwise via point charges and a simple Lennard-Jones term), this model has flexible, polarizable monomers, uses "smeared-out" charges and polarizabilities (with the induction terms explicitly covering many-body interactions), and employs the highly accurate intramolecular ab initio potential and dipole surface of Partridge and Schwenke. We have obtained the original potential and gradient subroutines for TTM2-F from the authors.

TTM2-F performs extremely well compared to MP2 ab-initio calculations for small clusters, but also for a wide variety of other systems, up to bulk ice [22]. Therefore, it is considered one of the best universal water potentials currently available; in contrast to e.g. SAPT-5s, which is exceedingly good in simulating vibration-rotation tunneling spectra [23] that cover only restricted regions of the potential hypersurface but breaks down e.g. for radial distribution functions in the bulk [24]. Hence, global optimizations on the TTM2-F potential can reasonably be expected to provide results close to MP2 quality and reliability, at a fraction of the cost. However, its evaluation is about two orders of magnitude more expensive than that of the simple TIP4P model.

4   Global Geometry Optimization by Evolutionary Algorithms

Every global cluster geometry optimization method that can guarantee the eventual location of the global minimum has to cover at least a strictly representative part of configuration space, if not all of it. Hence, every such method faces the exponential increase of this space with cluster size and is applicable only to very small clusters. In fact, methods of this type have been successfully applied only to trivially small clusters, for example to Lennard-Jones clusters with up to only seven atoms [25].

Hence, in practice, one has to give up the guarantee for reaching the global minimum and to resort to methods combining stochastic and heuristic aspects. One promising approach in this category are evolutionary algorithms. Since their first application to the global cluster geometry optimization problem [26], various technical refinements have been applied. The resulting implementation of the present author is now applicable to pure and mixed atomic

[27] and molecular [7] clusters. Published benchmark data [27,7] show that the program is at least as good as the currently best global cluster geometry optimization programs available. Its particular strength is a very favorable scaling with cluster size for atomic van-der-Waals clusters [27] (only cubic, as compared to the theoretically expected exponential scaling).

Details of the algorithm have already been published [27,7]. For reference in the following section 5, some central steps should be mentioned: Global optimization proceeds by iterating a random starting "population" of clusters (typically 30, for the cluster sizes envisaged here) through many "generations". In each generation, all possible pairs of clusters are formed, subjected to a series of "genetic" operators ("crossover" exchanges cluster parts between the clusters of a pair, "mutation" randomly alters positions and orientations of particles in each cluster), and then locally optimized. This leads to an intermediate cluster pool, containing about 900 clusters. From this pool, the next generation of 30 clusters is selected, according to a combination of structural and energy criteria. These selected clusters are further refined by a series of postprocessing operations (e.g. directed improvement of obviously badly placed particles) and then again locally optimized. This generational scheme is iterated until the best energy does not improve further for a given number of generations. The genetic operators are simple manipulations of cartesian coordinates, with little or no cluster size dependence; hence, they typically take less than 1% of the total computation time. The 900 local optimizations per intermediate generation take up about 82% of the total computation time, and the postprocessing (including the local optimizations done there) the remaining 17%.

Therefore, total computation times are completely dominated by the time it takes to evaluate the potential (and its gradient) in the course of local optimizations of clusters. For the simple TIP4P model, a single global optimization run of this kind can be performed in serial fashion within a few days on a single Pentium-III processor [7]. Since TTM2-F is two orders of magnitude more expensive than TIP4P, it is obvious that this study could not be performed using such simple computational resources and methods.

5   Parallel Implementation

TTM2-F local geometry optimizations are dominating the computer time profile, but it is very hard to vectorize or parallelize these routines efficiently. They consist of various nested iterations, e.g. for the local geometry optimization itself and for the evaluation of the polarization contribution to the total energy, interspersed with direct calculation of terms of a totally different nature, e.g. Coulomb interaction between diffuse charge clouds. Therefore, we resorted to a different approach:

At a higher level of abstraction, with these local optimizations as given, indivisible units, the global optimization algorithm is embarrassingly parallel: Within each intermediate generation and within each postprocessing, all the local optimization tasks are completely independent of each other. At the same time, the amount of data to be communicated is small; per local optimization, it consists only of the final energy and the coordinates of the locally optimized cluster (60 real numbers for the positions and 60 real numbers for the orientations, for a cluster of 20 water molecules). Compared to this, the independent computation times are long: Already for a small random cluster of 6 water molecules, local optimization takes 13.2 seconds on average, on a single Cray T3E processor; this relation improves strongly with cluster size, since the scaling of local optimization is approximately cubic. Therefore, the program has been parallelized following a simple master-slaves model (using explicit MPI calls). This makes it possible to hand out a new local optimization task as soon as a slave finishes one; that is, we have an automatic load balancing built in.

Given the large number of computationally intensive but independent tasks and the low volume and frequency of communication, the ideal machine for such a program would be a cluster of several hundred personal computers. Unfortunately, such a machine is not (yet) available at computer centers in Germany. At the HLRS, the currently best approximation to such a machine is the Cray T3E-900/512. The rather small memory per processor on this machine is no impediment, since there is no need to store huge arrays at any time; actually, the whole executable program with its data takes only about 1.2 MBytes of memory.

Although the intermediate generation in our algorithm consists of about 900 clusters, corresponding to about 900 independent local optimization tasks, it would not be ideal to use all 512 processors on the Cray T3E, for various reasons: The length of the local optimization tasks varies rather strongly and unpredictably (the 13.2 CPU-seconds quoted above show a standard deviation of about 1.2 seconds, averaged over 100 runs with random starting geometries). Hence, the inherent automatic load balancing is only effective if each process gets several local optimizations tasks in each parallel phase, since this gives a chance for the time differences to average out.

Furthermore, there is a forced alteration between the intermediate generation stage with about 900 independent local optimization tasks and the postprocessing phase with only 60 such tasks (one could add on more postprocessing attempts, but this turns out not to be effective). Therefore, going for 400–500 processes would actually not result in a maximally efficient calculation (in terms of parallel speedup). From this point of view, 40–60 processes are a good compromise: One can expect good load balancing in the critical intermediate phase with about 10–15 local optimizations per process, at the cost of a non-optimal speedup in the considerably shorter and smaller postprocessing phase, but avoiding obviously idle processes there. Still smaller number of processes (10–20) would allow for some load balancing also in the postprocessing phase, but this would be outweighed by an increase in wall clock time of the dominating intermediate phase by a factor of 2–6. In fact, most of the production runs were done with 60 slave processes.

6   Results

In the first phase of this project (1.7.–31.12.2001), 40000 single-processor serial CPU hours were granted. Unfortunately, there was a delay in the processing of the continuation application, such that further calculation time was granted only days before the writing of this report. Therefore, this report can cover only this first phase. Due to the rather efficient parallel speedup of the program, the computer time of this phase was used up within only four months (August to November 2001). Initially, some of the time had to be used to hunt down an elusive floating-point exception, which was neither visible nor harmful on the platforms where the program was run before; on the Cray, immediate program termination upon encountering this exception could not be prevented, but the exact location and cause of this exception was not easy to determine. Also, simulations of this error situation on other machines turned out to be not possible. Finally, careful examination of the program revealed that the exceptions were actually due to not one but several places in the program where rare, accidental divisions by zero could occur.

Nevertheless, with the allotted computer time it was possible to perform global TTM2-F water cluster geometry optimizations for $n = 5, 6, \ldots, 25$, as projected in the application. The results contained quite a few surprises, some of which will be highlighted in the following. Full details and analyses as well as comparisons to other levels of calculation will be given in a regular journal publication [28].

Surprisingly, up to and including $n = 16$, all TTM2-F minimum structures are qualitatively identical to their TIP4P counterparts. Of course, the flexibility of the monomers in TTM2-F leads to slight distortions of the monomers, and also the intermolecular distances and angles are not exactly the same. In the sequence of low-energy local minimum structures, a few re-orderings can be observed. However, the qualitative nature of the global minimum structure is not affected in any of these cases.

Hence, it is astonishing that the qualitative nature of the global minimum structure does change drastically for $n = 17$: The best TIP4P geometry is a bent, distorted array of four cubes, with all water molecules at the surface of the cluster; local re-optimization of this structure shows that this is also a minimum on the TTM2-F potential, with an energy of -742.06 kJ/mol and an essentially unchanged geometry. This value is easily beaten by the -746.734 kJ/mol obtained from the global geometry optimization directly on the TTM2-F potential; it corresponds to a structure with one central water (cf. Fig. 1).

At $n = 19$, already TIP4P showed a water-centered structure, and this does not change by the use of TTM2-F. At the neighboring sizes $n = 18$ and $n = 20$, TTM2-F also does not change the nature of the global minimum, although in both cases the best structures have all water molecules at the surface and again show the preference for stackings of cubes and pentagonal

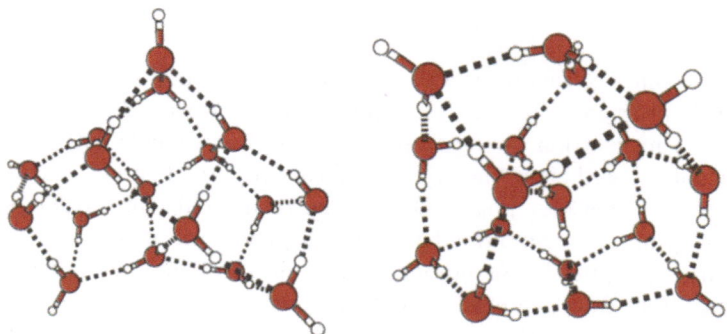

Fig. 1. Globally optimal $(H_2O)_n$ structures for $n = 17$: TIP4P structure with all molecules at the surface (left), and TTM2-F structure with one interior molecule (right).

prisms. However, for both cluster sizes, the lowest water-centered structure is now considerably closer in energy to the global minimum.

However, the biggest surprise comes at $n = 21$: TTM2-F local re-optimization of the best TIP4P geometry results in a total binding energy of -944.71 kJ/mol. The structure has all water molecules at the surface and is a distorted agglomeration of cubes, which is hard to locate for any global optimization algorithm. Unbiased global geometry optimization directly on the TTM2-F potential, however, yields a distorted centered dodecahedral structure with an energy of -949.264 kJ/mol in only 14 generations (cf. Fig. 2). This structure was not seen previously on the TIP4P potential, but apparently is very easy

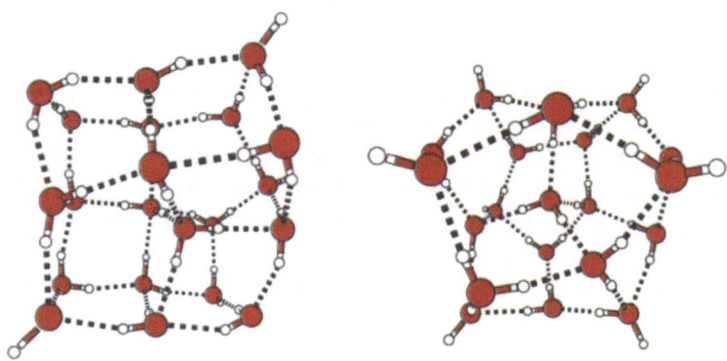

Fig. 2. Globally optimal $(H_2O)_n$ structures for $n = 21$: TIP4P structure with all molecules at the surface (left), and TTM2-F distorted dodecahedral structure with one interior molecule (right).

to locate on the TTM2-F potential, considering that global optimization runs for neighboring cluster sizes need about 50–100 generations to converge.

While larger TIP4P clusters seem to return to the structural patterns of stacked cubes and pentagonal prisms with all molecules at the surface [7], TTM2-F structures continue to be different: Some of the cube and prism stacking tendency is actually also present, but always in combination with one central water molecule surrounded by a water cage, similar to the $n = 21$ case. Preliminary results indicate that inclusion of a second water molecule into the cage may start at $n = 26$.

7 Summary and Conclusions

Our global geometry optimization algorithm for molecular clusters has successfully been adapted to a massively parallel environment, using a simple but efficient coarse-grained master-slave model with explicit MPI calls, exhibiting a fair amount of automatic load balancing for careful choices of the number of processes relative to the size of the population.

In a first real-world application of this code to pure neutral water clusters using the computationally expensive but highly accurate TTM2-F water model, it was shown that the popular TIP4P water model, which performs perfectly well for small clusters, leads to qualitatively and systematically wrong global minimum structures for larger water clusters, starting at $n = 17$. In particular, TIP4P has a marked tendency to favor structures with all water molecules at the surface of the cluster, at the expense of structures with one or more interior water molecules. In contrast, TTM2-F leads to an interior water molecule already at $n = 17$, and it also yields the expected (distorted) centered dodecahedral cage structure at $n = 21$. Since TTM2-F results are extremely close to ab-initio MP2 results for small clusters, there is reasonable hope that these results for larger clusters are also reliable.

Within a reasonable time frame, these calculations would not have been possible on a serial machine, even with a significantly faster processor. With 60 processors in parallel, however, such calculations can now be considered routine. Obvious applications include not only further homogeneous and heterogeneous atomic and molecular clusters, but also e.g. solvent effects on chemical reactions with many explicit solvent molecules or the structure and functional role of water molecules in active sites of proteins.

Acknowledgements

C. J. Burnham (then Pacific Northwest National Laboratory, now University of Utah) kindly provided several versions of his TTM2-F routines and patiently answered many questions arising during their inclusion in our global optimization program. A computer time grant on the Cray T3E of the HLRS Stuttgart is gratefully acknowledged.

References

1. H. Haberland (Ed.): "Clusters of atoms and molecules", Springer Series in Chemical Physics Vol. 52, Springer, Berlin, 1994.
2. T. P. Martin (Ed.): "Large clusters of atoms and molecules", Kluwer, Dordrecht, 1996.
3. J. Jortner, Z. Phys. D – Atoms, Molecules and Clusters 24 (1992) 247.
4. B. Hartke, Angew. Chem. 114 (2002) 1534.
5. M. R. Hoare and P. Pal, Adv. Phys. 20 (1971) 161; M. R. Hoare, Adv. Chem. Phys. 40 (1979) 49.
6. L. T. Wille, in: "Annual Reviews of Computational Physics VII", D. Stauffer (Ed.), World Scientific, Singapore, 2000; p. 25.
7. B. Hartke, Z. Phys. Chem. 214 (2000) 1251.
8. K. D. Jordan, personal communication, 2000.
9. D. W. Dean and J. R. Chelikowsky, Theor. Chem. Acc. 99 (1998) 18.
10. J. Sadlej, V. Buch, J. K. Kazimirski and U. Buck, J. Phys. Chem. A 103 (1999) 4933.
11. J. Sadlej, Chem. Phys. Lett. 333 (2001) 485.
12. J. Kim, D. Majumdar, H. M. Lee and K. S. Kim, J. Chem. Phys. 110 (1999) 9128.
13. B. Hartke, M. Schütz and H.-J. Werner, Chem. Phys. 239 (1998) 561.
14. K. Liu, M. G. Brown, C. Carter, R. J. Saykally, J. K. Gregory and D. C. Clary, Nature (London) 381 (1996) 501.
15. K. Nauta and R. E. Miller, Science 287 (2000) 293.
16. D. J. Wales and M. P. Hodges, Chem. Phys. Lett. 286 (1998) 65.
17. A. J. Stone: "The Theory of Intermolecular Forces", Clarendon Press, Oxford, 1996.
18. M. P. Hodges and D. J. Wales, Chem. Phys. Lett. 324 (2000) 279.
19. B. Hartke, A. Charvat, M. Reich and B. Abel, J. Chem. Phys. 116 (2002) 3588.
20. F. Schulz and B. Hartke, Chem. Phys. Chem. 3 (2002) 98.
21. P. M. Holland and A. W. Castleman, Jr., J. Chem. Phys. 72 (1980) 5984.
22. C. J. Burnham, J. Li, S. S. Xantheas and M. Leslie, J. Chem. Phys. 110 (1999) 4566; C. J. Burnham and S. S. Xantheas, J. Chem. Phys. 116 (2002) 1500, 5115.
23. E. M. Mas, R. Bukowski, K. Szalewicz, G. C. Groenenboom, P. E. S. Wormer and A. van der Avoird, J. Chem. Phys. 113 (2000) 6687, 6702.
24. K. Szalewicz, talk given at the workshop "Water in Confined Geometries", Telluride, July 31 – August 4, 2000.
25. C. D. Maranas and C. A. Floudas, J. Chem. Phys. 97 (1992) 7667.
26. B. Hartke, J. Phys. Chem. 97 (1993) 9973.
27. B. Hartke, J. Comput. Chem. 20 (1999) 1752.
28. C. J. Burnham, G. Rauhut and B. Hartke, manuscript in preparation.

Improved Implementation and Application of the Individually Selecting Configuration Interaction Method

P. Stampfuß, M. Vogel, and W. Wenzel

Forschungszentrum Karlsruhe, Institut für Nanotechnologie,
Postfach 3640, 76021 Karlsruhe

Abstract. We report on the progress of our implementation of the configuration-selecting multi-reference configuration interaction method on massively parallel architectures with distributed memory, which now permits the treatment of Hilbert spaces of dimension $O(10^{12})$ about 20,000,000 of which can be selected in the variational subspace. We provide scaling data for the CPU time of the code for the IBM/SP3 and the CRAY-T3E. We present benchmark results for two selected applications: the isomers of dinitrosoethylene and the electronic structure of two members of the transition metal dihalide family: VF_2 and VCl_2.

The multi-reference configuration interaction method (MRCI) is one of the established benchmark methods that offer a systematic approach for the calculation of the electronic structure of atoms and molecules [1–3]. An accurate quantum chemical treatment of complex molecules requires a balanced account of both dynamical and nondynamical correlation effects. The latter are particularly important when one wants to describe an entire potential energy surface, where bond-breaking or bond-rearrangements can occur. To adequately describe these effects, as well as for the treatment of electronically excited states, a wavefunction based method must accommodate the multi-reference nature of the electronic states. Dynamical correlation effects, i.e. the mutual influence electrons exercise on each other when they pass at close distance, are incorporated by considering excitations of the set of relevant reference configurations. The generic lack of extensivity of MRCI methods has at least been partially addressed with a number of a posteriori [4,5] corrections and through direct modification of the CI energy-functional [6–10].

Due to its high computational cost applications of the MRCI method remain constrained to relatively small systems. For this reason the configuration-selective MRCI-method (MRD-CI) [11–13], has become one of its most widely used versions. In this variant only the most important configurations of the interacting space of a given set of primary configurations are chosen for the variational wavefunction, while the energy contributions of the remaining configurations are estimated on the basis of second-order Rayleigh-Schrödinger perturbation theory [14,15]. Even within this approximation, the cost of MRCI calculations remains rather high. The development of efficient configuration-selecting CI codes [16,17,15,18–23] is inherently complicated by the sparseness and the lack of structure of the selected state-vector. In order to extend the applicability of the method, it is desirable to employ the most powerful computational architectures available for such calculations.

Here we report improvements of our massively parallel, residue-driven implementation of the MRD-CI method for distributed memory architectures [22]. In this implementation the difficulty of the construction of the subset of nonzero matrix elements is overcome by the use of a residue-based representation of the matrix elements that was originally developed for the distributed memory implementation of MR-SDCI [21]. This approach allows to efficiently evaluate the matrix elements both in the expansion loop as well as during the variational improvement of the coefficients of the selected vectors.

This manuscript is organized as follows: in section 1 we report recent improvements of the massively parallel MRD-CI implementation that permit the treatment of larger Hilbert spaces, i.e. the correlation of a larger number of electrons in larger basis sets. We demonstrate the scalability of the new integral driven version of the matrix element evaluation routine on two widely available massively parallel architectures, the CRAY-T3E and the IBM-SP3/SMP. We then elucidate applicability and limitations of MRD-CI in benchmark applications for three sensitive chemical problems: the ring-closure reaction of enedyienes and the electronic structure of benzofuroxan/dinitrosoethylene and the electronic structure of VF_2 and VCl_2.

1   Technical Improvements

In the transition residue driven approach each matrix element between two determinants (or configuration state functions $|\phi_1\rangle$ and $|\phi_2\rangle$) is associated with the subset of orbitals that occur in both the target and the source determinant [21,22]. This unique subset of orbitals called the *transition residue* is mediating the matrix element and serves as a sorting criterion to facilitate the matrix element evaluation on distributed memory architectures. For a given many-body state, we consider a tree of all possible transition residues as illustrated in Figure 1. For each such residue we build a list of *residue-entries*, composed of the orbital-pairs (or orbital for a single-particle residue) which combine with the residue to yield a selected configuration and a pointer to that configuration. While the number of transition residues is comparatively small, the overall number of residue-entries grows rapidly (as $N_{\text{selected}} \, n_e^2$) with the number of configurations N_{selected} and the number of electrons n_e.

Once the residue tree is available the evaluation of the matrix elements is very efficient. For each transition residue, each pair of entries of type (D) in Figure 1 generates a matrix element. The indices of the right- and left-hand side configurations are stored in the entries and the indices of the orbital pair generate the matrix element. If all entries of the residue tree are generated, its size limits the number of configurations in the variational subspace. In order to increase the size of the variational subspace we have removed all transition residues containing two external orbitals from the residue tree,

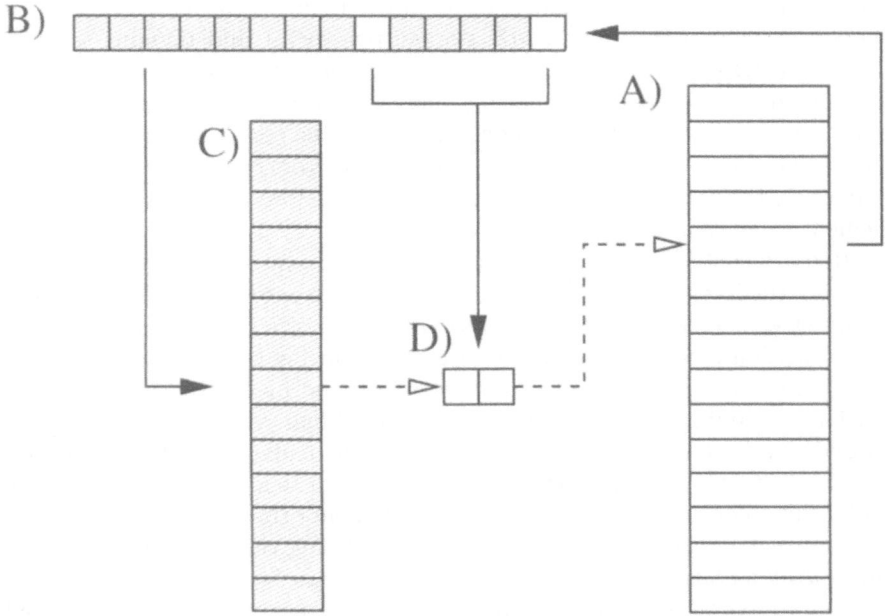

Fig. 1. Schematic representation of the residue tree: (A) list of globally indexed many body configuration, (B) a specific many body configuration with 14 electrons. Each possibility to remove two orbitals from the configuration (white boxes) generates an entry in the residue table (C) and an associated orbitals pair (D), which is stored together with the index of the original configurations. Solid arrows indicate the operations of building the residue tree dashed arrows the logical connection between its elements. The residue tree (C) is a list of transition residues. Associated with each residue is a list of entries, consisting of the orbital pairs and the index of the configuration that generated the entry.

reducing the size of the residue tree by more than an order of magnitude. Matrix elements mediated by such transition residues can be computed by gathering all elements of the coefficient vector that contain a particular external pair on a single node of the machine. We have implemented a scheme that efficiently evaluates the orbital difference map of the internal part of two such configurations to determine if a nonzero matrix element exists. The value of the associated integral element is easily obtained from the (small) table of all integrals with only internal indices which we replicate on each node. Implementing this change reduces the size of the residue tree significantly and allows to increase the size of the variational subspace by almost a factor of ten, to about 20 million configurations.

Improving the previous implementation [23], we now split the state vectors for the Davidson iteration across all nodes, such that only two copies of the total state vector are required on each node for the evaluation of the matrix elements. For 50 million configurations, a 64-bit representation of the state vector requires 400 MB storage per vector and remains the last large undistributed data element in our implementation. Even these can be eliminated, at the expense of locally increasing the residue tree. Instead of storing only the configuration index in the residue tree, one may also store the coefficient of the left- and right-hand side vectors of the matrix element, scattering the former and gather in the latter before and after the matrix element evaluation respectively.

Requirements for the integral storage could also be drastically reduced by splitting the integral file sorted in physics notation across all nodes. Once the residue tree has been constructed, its "heads", i.e. the information regarding the transition residue itself can be discarded. During the matrix evaluation, only the orbital lists are required. The space freed by eliminating the information pertaining to the content of the individual residues can be reused to address those orbital lists that contain a particular orbital pair. Once a set of integrals $\langle i_1 i_2 | i_3 i_4 \rangle$ with a given orbital pair, e.g. $i_1 i_2$, is served to the node, this information can be used to identify all orbital lists that generate nonzero matrix elements for this integral list. Using this information, an *integral driven* matrix element evaluation scheme can be implemented, during which all integral lists $\langle i_1 i_2 | ... \rangle$ are distributed across the nodes and rotated in a cyclic fashion until every integral list has visited every node. Since MPI permits a very fast cyclic data exchange and no search operations are required to identify nonzero matrix elements for a given set of integrals, this mechanism allows for efficient integral driven evaluation of all matrix elements not mediated by doubly external transition residues.

To demonstrate the performance of the improved code we have performed model calculations on two common massively parallel architectures with distributed memory, the CRAY T3E and the IBM-SP/SMP for a varying number of nodes. We report the total CPU (wallclock) time of each calculation and decompose it into several important, not necessarily contiguous portions, comprising (a) the generation of the integrals, SCF, four-index transforms, (b) the generation and distribution of the residue tree, (c) the iterative generation of approximate natural orbitals (NO) in BW-MRPT [24] (5 iterations of the NO loop, which is very similar to the MRD-CI selection step) and (d) the subsequent selection and iteration of a single state in the selected Hilbert space (5 iterations). In an ideal massively parallel implementation the total CPU time should be independent of the number of nodes (scalability) and competitive with a serial implementation (efficiency). The numerical cost of NO generation, expansion and iteration is usually directly proportional to the number of roots in one symmetry. The "logic" cost of generating the residue tree is proportional to the number of configurations multiplied with the number of electrons squared. The numerical cost of the iteration step depends on the average of the square of the the number of orbital pairs per transition residue, multiplied with the number of transition residues. In a non-selecting MR-SDCI calculation, this number would again scale as the

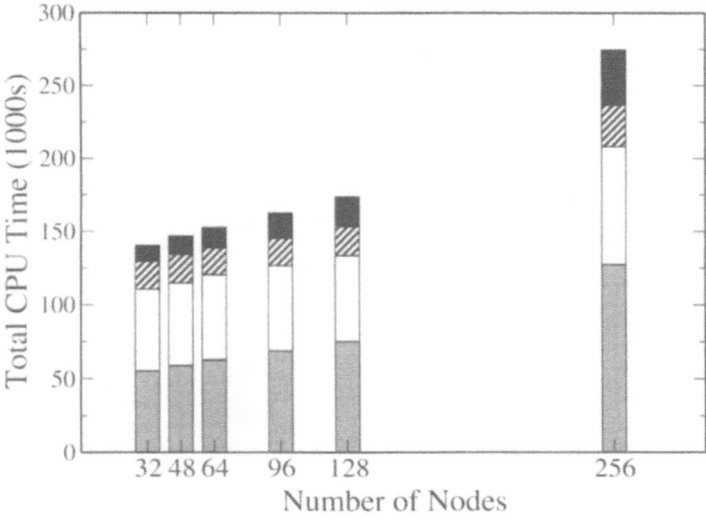

Fig. 2. Scaling plot of the total CPU time versus the number of nodes for a massively parallel MRD-CI calculation selecting 3.7×10^6 of 2×10^7 determinants running on 32-256 nodes on the CRAY-T3E. The individual contributions (from the top) are preparation (integral generation, SCF, four-index transform), logic (building and redistributing the residue tables) and configuration selection (including the perturbative calculation of the approximate natural orbitals in five iterations) and the iteration of the selected state-vector respectively.

number of orbitals squared. If comparatively few configurations of the overall Hilbert space are selected, the residue tree can be relatively sparse. In this case the numerical effort creating the residue tree becomes comparable to the effort evaluating the matrix elements.

Figure 2 shows a scaling plot for a calculation selecting 3.7×10^6 of 2×10^7 configurations running on 32 to 256 nodes of the CRAY TE3 of the NIC Jülich, which demonstrates that the calculation scales well from 32-128 nodes, with a significant loss of efficiency in the last doubling of the number of processors to 256 nodes. The fourfold increase of the number of nodes from 32 to 128 leads to total loss of efficiency of about 24% for the total calculation or 18% for the MRD-CI calculation alone. This reflects the fact that we have focused our efforts mostly on the performance of the MRD-CI code, rather than on the preceeding calculations. Among the components of the MRD-CI, the iteration loop scales worst, with a total loss of 35% in its efficiency. This results from the fact that the load-balancing of the matrix element evaluation in the integral driven mode becomes progressively difficult as the number of matrix elements is reduced. As the integrals are distributed cyclically in batches among the nodes, the entire calculation must wait for the slowest node to finish each integral batch. The size of the calculation was chosen to

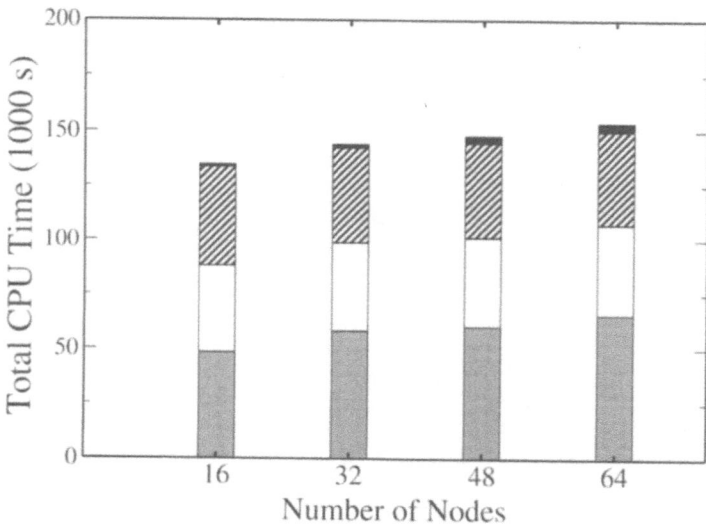

Fig. 3. Scaling plot of the total CPU time versus the number of nodes for a massively parallel MRD-CI calculation selecting 2×10^6 of 2.6×10^7 determinants running on 16-64 nodes on an IBM SP3/SMP. The individual contributions are labeled as in Fig. 2

make runs on relatively few nodes possible. It is therefore not surprising that this calculation does not perform well on 256 nodes, where about 80% of the efficiency is lost. if the size of the selected space is increased, the performance loss between 128 and 256 nodes is reduced, but the calculation no longer runs on 32 nodes.

Figure 3 shows a scaling plot for a similar calculation for dinitrosoethylene (DNE) on an IBM SMP/SP3 with 64 processors. Quadrupling the number of processors here, results only in a marginal loss of 12% of the overall efficiency of the calculation, the iteration loop again scales worst (34% loss). The better performance of the IBM for this calculation is largely attributable to its larger memory (1GB/processor as opposed of 513 kB on the T3E). Overall these data demonstrate a sufficient scalability of the code for a large range of processors for two important massively parallel architectures available today.

2   Applications

2.1   Dinitrosoethylene

Recent investigations into the ring-opening reaction of furoxan (oxadiazole-2-oxide) into dinitrosoethylene illustrated the theoretical difficulties encountered in the elucidation the reaction mechanism. There are six possible conformers of 1,2-dinitrosoethylene, all of which might in principle play a role

Table 1. Vertical and adiabatic $S_0 \rightarrow T_1$ excitation energies of the tct-DNE

Method	Basis	vertical	adiabatic
DFT	6-311+G(2df,2pd)	11.6	-0.9
CCSD(T)	6-311+G(2df,2pd)	25.1	18.2
icMRCI+P	6-311+G(2df,2pd)	24.4	NA
MRD-CI	aug-cc-pVDZ	28.5	31.2
	6-311+G(2df,2pd)	28.8	29.3
MRD-CI+D	aug-cc-pVDZ	22.1	21.0
	6-31++G**	21.3	18.8
MRD-CI+P	aug-cc-pVDZ	22.5	21.6
	6-31++G**	21.8	19.4

in the ring-opening reaction of furoxan, which may serve as a possible source of nitric oxide in biology. High level methods are required qualitatively account for the energetic ordering of the various conformers, in particular the relevance of cis-cis-trans (cct) dinitrosoethylene (DNE). We follow the notation of Ref [25] labeling the various conformers with their acronyms (c=cis, t=trans). In all there are 26 stationary points of the ground state potential energy surface of DNE.

Using configuration selecting CI we have investigated the relative energies and the singlet triplet splitting (both adiabatic and vertical) of the tct and ttt isomers in comparison with recent B3LYP, CASPT2, CASPT3, CCSD(T) and internally contracted MRCI (icMRCI) results. The geometries were optimized at the level of B3LYP DFT method with an augmented 6-311++G(d,p) basis set, for triplets the unrestricted version of this method was used. The MRCI calculations were performed in an aug-cc-pVDZ and 6-311+G(2df,2pd) basis set in approximate natural orbitals generated in Brillouin-Wigner multireference perturbation theory [26,24]. The last basis set was used to permit a direct comparison with earlier CCSD(T) and icMRCI calculations. We used a 9 orbital active space with reference selection for all calculations.

The results for the absolute energies of the conformers and electronic states are summarized in Tables 1 and 2. Due the presence of lone pairs and strong static correlation effects, excitation energies are difficult to determine for the various conformers of DNE. For tct-DNE we find a good agreement between the Davidson corrected vertical excitation energy computed with MRD-CI and the results of icMRCI. The results for the aug-pVDZ basis and the larger 6-311++G(d,p) are very similar, indicating convergence with respect to the basis set. There is a 3.2 kcal/mol difference to the CCSD(T) result. Similarly there is good agreement between CCSD(T) and MRD-CI+P for the adiabatic excitation energy, which has not been previously computed with icMRCI. For the vertical excitations energies of ttt we find again good agreement between the correlated methods, but signifcant deviations (4 kcal/mol) for the adiabatic excitation energy.

Table 2. Vertical and adiabatic $S_0 \rightarrow T_1$ excitation energies of the ttt-DNE

Method	Basis	vertical	adiabatic
DFT	6-311+G(2df,2pd)	15.1	-7.7
CCSD(T)	6-311+G(2df,2pd)	27.6	7.3
icMRCI+P	6-311+G(2df,2pd)	26.2	7.7
MRD-CI	aug-cc-pVDZ	30.6	14.6
	6-311+G(2df,2pd)	33.4	15.6
MRD-CI+D	aug-cc-pVDZ	26.0	13.2
	6-311+G(2df,2pd)	26.7	12.3
MRD-CI+P	aug-cc-pVDZ	24.9	13.1
	6-311+G(2df,2pd)	25.2	11.7

2.2   Transition Metal Dihalies

The transition metal dihalide family has received much experimental as well as theoretical attention over the last twenty years [27,28] (for an overview see [29] and references therein). Because these "hot" molecules form at over 1000 K, the resulting rotovibrational spectra are complex and a detailed experimental elucidation of their electronic structure is difficult. Information about their low-lying electronically excited states is nevertheless required to understand their behavior in chemical reactions and other applications, e.g. their use in laser transport [30]. Recent theoretical investigations have cast doubt over the accuracy and applicability of ligand field theory (LFT), which has been used for over a decade to assign the spectra of these molecules [27,31,32]. Because of their complexity, high-level theoretical studies of members of the MX_2 family remain relatively rare, in particular at the benchmark level [33–36].

In this report we aim to close part of this gap by reporting the first benchmark results for two members of the MX_2 family, VF_2 and VCl_2 using the configuration selective multi-reference configuration interaction method (MRD-CI) [12] in comparison with its internally contracted cousin (icM-RCI) [37]. We report vertical and adiabatic excitation energies as well as sections of the potential energy surfaces for the first 11 excited states of these molecules (which all lie within 2 eV of the ground state). We find a qualitative agreement with those states that could previously be computed in the framework of density functional theory (DFT) [29] but large quantitative deviations for the excitation energies. We discuss the influence of basis set effects, the choice of the active space, extensivity corrections, core-valence correlations and relativistic effects on the accuracy of these results.

Transition metal compounds are notorious for the importance of both static and dynamic correlation effects in their electronic structure [38]. For this reason, correlated multireference methods, such as MRCI, presently offer the best theoretical framework to study these molecules in the vicinity of their ground state geometry, where extensivity errors are not very relevant.

Table 3. Vertical excitation energies in eV of VF$_2$ in a DZP (with and without core-valence corrections) and a TZP basis sets relative to the ground state energy of (-1142.526720 H / -1142.606739) (r_{VF} = 1.8051976 Å). Adiabatic excitation energies in comparison to DFT [29] using the active space A described in the text.

Sym($D_{\infty h}$)	Sym(D_{2h})	E_v	E_v(CV)	E_v	E_{ad}	E_{ad}
		DZP	DZP	TZP	DZP	DFT
$^4\Sigma_g^-$	$^4B_{1g}$	0.00	0.00	0.00	0.00	0.00
$^4\Pi_g^{+,-}$	$^4B_{3g}$	0.42	0.33	0.47	0.36	1.12
$^4\Phi_g^{+,-}$	$^4B_{3g}$	0.99	0.88	1.06	0.96	—
$^4\Delta_g^+$	^4A_g	1.33	1.19	1.31	1.09	—
$^4\Delta_g^-$	$^4B_{1g}$	1.51	1.21	1.38	1.26	—
$^2\Sigma_g^+$	^2A_g	1.56	1.55	1.55	1.55	2.20
$^2\Sigma_g^-$	$^2B_{1g}$	1.63	1.62	1.62	1.63	—
$^2\Delta_g^+$	^2A_g	1.61	1.57	1.57	1.61	—
$^2\Delta_g^-$	$^2B_{1g}$	1.65	1.63	1.65	1.65	—
$^2\Pi_g^{+,-}$	$^2B_{3g}$	1.83	1.66	1.81	1.80	4.17

We have performed calculations using the configuration selective configuration interaction method (MRD-CI), in a recently developed a massively parallel implementation [22]. This implementation permits the treatment of Hilbert space of up to 10^{10} configurations and large active spaces. Both molecules are linear, therefore calculations were performed in D$_{2h}$ symmetry, the largest Abelian subset of the D$_{2\infty}$ symmetry group. The SCF configurations for VF$_2$ was ($7a_g^2, 3b_{3u}^2, 3b_{2u}^2, 5b_{1u}^2, 1b_{2g}^2, 1b_{3g}^2, 1a_g$), that of VCl$_2$ was ($9a_g^2, 5b_{3u}^2, 4b_{2u}^2, 6b_{1u}^2, 2b_{2g}^2, 1b_{3g}^2, 1a_g$). Unless otherwise noted we have frozen the (1s2s2p3s3p) orbitals of V, the 1s orbitals on F and the (1s2s2p) orbitals on Cl, resulting in core spaces (4,2,2,0,3,0,0,0) and (6,3,3,0,5,1,1,0) with symmetries ($a_g, b_{1u}, b_{2u}, b_{3g}, b_{3u}, b_{1g}, a_u$) for VF$_2$ and VCl$_2$ respectively (core valence corrections see below). We used a 17 orbital, 19 electron active space A=(6,1,1,2,3,2,2,0) comprising the (2s2p)/(3s3p) shells on F/Cl respectively and the (3d,4s) shell on V. In addition there were 8 inactive orbitals (2,1,1,0,2,1,1,0) that were doubly occupied in all references, but correlated in the CI calculation. The calculations were performed in state-averaged (within one symmetry representation) approximate natural orbitals computed in BW-MRPT [24]. We have used both augmented double (DZP) and augmented triple zeta (TZP) quality ANO basis sets [39,40] (for F:(14s,9p,4d,3f) / [5s,4p,3d,2f], for V:(21s,15p,10d,6f,4g) / [8s,7p,5d,3f,2g]). We computed the lowest three excited states for each representation for both the doublet and quadruplet spin-subspace. The calculations were performed using CAS-reference-spaces resulting in 280–330 references and Hilbert spaces of O(30 × 10^6) configurations, of which O(10^6) were selected at the smallest coefficient threshold of 3 × 10^{-5} that we used. At this threshold the remaining perturbative energy correction was less than 1mH. The calculations were

Table 4. Comparison of vertical excitation energies computed at $r_{VF} = 1.75$ Å) with MRD-CI and icMRCI using the active space A' described in the text. For comparison the excitation energies using the larger active space A in MRD-CI are also shown.

Sym($D_{\infty h}$)	Sym(D_{2h})	E_v MRD-CI A'	E_v icMRCI A'	E_v MRD-CI A
$^4\Sigma_g^-$	$^4B_{1g}$	0.00	0.00	0.00
$^4\Pi_g^{+,-}$	$^4B_{3g}$	0.61	0.58	0.42
$^4\Phi_g^{+,-}$	$^4B_{3g}$	1.00	1.02	0.99
$^2\Sigma_g^+$	^2A_g	1.58	1.56	1.56
$^2\Sigma_g^-$	$^2B_{1g}$	1.61	1.56	1.51
$^2\Delta_g^+$	^2A_g	1.60	1.64	1.61
$^2\Delta_g^-$	$^2B_{1g}$	1.69	1.63	1.64
$^4\Delta_g^+$	^4A_g	1.43	1.70	1.33
$^4\Delta_g^-$	$^4B_{1g}$	1.76	1.71	1.65
$^2\Pi_g^{+,-}$	$^2B_{3g}$	1.92	1.91	1.83

performed on using 64 nodes of the IBM/SP2 of the HLRZ Karlsruhe, where a calculation with four states converged typically in less than 2 hours. The accuracy of these results was independently verified by comparing with internally contracted (icMRCI) [37] using the MOLPRO [41] *ab-initio* quantum chemistry package. Because of the high-cost of these calculations, only a smaller active space A'=(4,1,1,2,1,2,2,0) could be used.

Using a augmented double zeta quality ANO basis set basis set and active space A we obtained $r_{VF} = 1.805$ Å for linear geometry ($D_{2\infty}$). Next we computed vertical excitation energies for both molecules using the augmented double zeta quality ANO basis set and active space A described above. For VF2 (see Table (3)), we found 14 excited states in D_{2h} within 2eV of the ground state, which reduce to 11 states when degeneracies of the $D_{2\infty}$ representations in D_{2h} are taken into account. As expected the spectrum of the transition metal dihalides is extraordinary dense. We have checked the accuracy of the calculation using a augmented triple zeta quality ANO basis set quality basis set and found discrepancies of maximally 0.12 eV between the two data sets, indicating that a augmented double zeta quality ANO basis set is sufficient to obtain a quantitative resolution of dynamical correlation effects for the excited state of this molecule (in vicinity of the ground state geometry). The order of the excitations was preserved. Next we considered the relevance of core-valence excitations, unfreezing the V (3s,3p) shell and including it in the inactive space of the MRD-CI calculation. This means that the reference set remained unchanged, but that additional dynamical correlations arising from the excitation of core electrons in the references were considered. Here we found a significant lowering of the first four excited states and for the $^2\Pi_g$ excitation by about 0.15 eV.

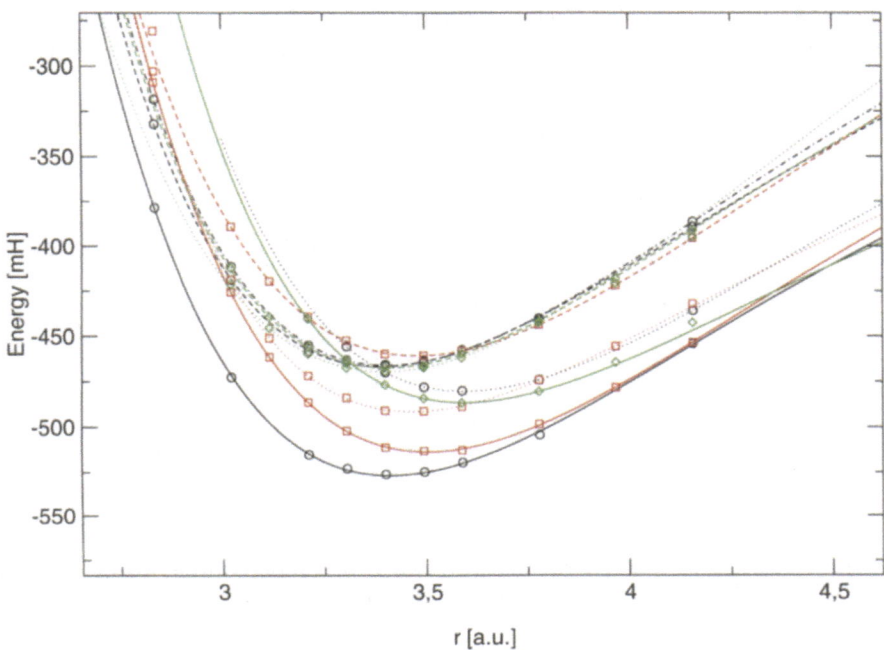

Fig. 4. Potential energy surfaces of low-lying excited states of VF_2 (see Table (3)) computed with MRD-CI using an augmented double zeta quality ANO basis set and active space A.

In order to determine the accuracy of these results we computed the vertical excitation spectrum at $r_{VF} = 1.75$ Å with icMRCI (see Table (4)) using the MOLPRO *ab-initio* program package. In order to make these comparisons we had to reduce the size of the active space somewhat. We found very good agreement between the two methods in the reduced active space A' – the largest energy difference was 0.1 eV. The orbitals for the icMRCI calculations were obtained from MCSCF calculations. Note however, that the MRD-CI calculations show significant deviations between the calculations of the two active spaces, so that the larger active space (which could not be used in our version of MOLPRO), was used for all subsequent calculations.

The potential energy surfaces for symmetrically stretched VF_2 as a function of bond distances are shown in Figure (4). As expected from the vertical excitation energies there is a clear separation of the lowest three states, but a very dense set of states between 1.5 and 2 eV. There is a large number of level crossings, even for the low-lying states, within 0.3 Å of the equilibrium bond distance. Table (3) also shows the adiabatic excitation energies of VF_2 in comparison with those obtainable as lowest states in their respective symmetry by DFT. The table shows that the order of states is preserved in the two methods, but large quantitative errors remain. These are attributable to

Table 5. Adiabatic and vertical excitation energies of VCl_2 computed in MRD-CI using a DZP-quality ANO basis with the active space A described in the text.

Sym($D_{\infty h}$)	Sym(D_{2h})	E_{ad} [eV]	E_v [eV]	$E_{ad}(DFT)$
$^4\Sigma_g^-$	$^4B_{1g}$	0.00	0.00	0.00
$^4\Pi_g^{+,-}$	$^4B_{3g}$	0.26	0.28	0.81
$^4\Delta_g^+$	^4A_g	0.60	0.67	—
$^4\Phi_g^{+,-}$	$^4B_{3g}$	0.61	0.62	—
$^4\Delta_g^{+,-}$	$^4B_{1g}$	0.77	0.84	—
$^2\Delta_g^+$	^2A_g	1.63	1.64	1.82
$^4\Pi_g^{+,-}$	$^4B_{3g}$	1.67	1.67	—
$^2\Delta_g^-$	$^2B_{1g}$	1.71	1.72	—
$^2\Delta_g^-$	$^2B_{1g}$	1.72	1.72	—
$^2\Delta_g^+$	^2A_g	1.74	1.75	—
$^2\Pi_g^{+,-}$	$^2B_{3g}$	1.80	1.80	3.35
$^2\Delta_g^+$	^2A_g	1.95	1.95	—

Table 6. Harmonic frequencies and dissociation energies for VF_2 and VCl_2. The frequencies for symmetric/asymmetric stretch were computed from fits of the PES in Fig. those for the bend from a 14 point fit of the bent PES.

	ω_{sym} [cm^{-1}]	ω_{asym} [cm^{-1}]	ω_{bend} [cm^{-1}]	D_e [eV]
VF_2 (MRD-CI)	582	770	189	11.50
VF_2 (DFT)	630	773	—	13.35
VCl_2 (MRD-CI)	322	499	91	9.62
VCl_2 (DFT)	345	492	48	9.53

large differential dynamic and non-dynamic correlation effects. The errors in the adiabatic excitation energy of the $^2\Pi$ doublet and $^4\Pi$ quadruplet state are more than 100% of the actual value, but the ground state symmetry and the order of the excitations is predicted correctly.

These calculations demonstrate that the excitation spectrum of VF_2 is resolved well using an augmented double zeta quality ANO basis set with active space A. We have therfore calculated the excitation sprectrum and PES of VCl_2 at this level of theory. We find the equilibrium bond distance of $r_{VCl} = 2.244$ Å slightly larger than the experimental value of $r_{VCl} = 2.172$ Å [42]. The vertical and adiabatic excitation energies are summarized in Table (5). The symmetry of the ground state and the lowest excitation in preserved, but the doublet $^4\Phi$ and $^2\Delta$ states are exchanged. With increasing energy, the correlation between the order of the states falls rapidly. Owing to the softer nature of the Cl (3s,3p) orbitals, the low-lying spectrum is compressed, the first excitation lies a mere 0.26 eV above the ground state.

We have also computed the vibrational frequencies of VF_2 and VCl_2 from fits to the PES. The results for both molecules are summarized in Table (6). We find rather good agreement between DFT and MCRI for the bond-stretching frequencies. Because the molecules are rather floppy, the bending frequencies are soft and difficult to calculate in DFT. However, it appears certain that both molecules are linear in their ground state. We have also computed the electronic dissociation energies of both molecules and find $D_e(VF_2) = 11.5eV$ and $D_e(Cl_2) = 9.6eV$ in comparison to an "estimated" experimental $D_e(VF_2) = 12eV$ [29] and a measured $D_e(VCl_2) = 10eV$.

3 Summary

Benchmark methods, such as CCSD(T) or MRCI, provide useful results to our understanding of the electronic structure of molecules and of chemical reactions despite their high computational cost. It is therefore worthwhile to devote significant effort to the implementation of these methods on the most powerful available computational architectures, i.e., presently massively parallel machines with distributed memory. Here we reported significant improvements of our massively parallel implementation of the configuration selective MRD-CI method, which presently permits the treatment of Hilbert spaces of up to 10^{10} configurations, about 2×10^7 can be selected into the variational subspace.

We have reported benchmark accuracy results for the singlet and triplet states of various isomers of dinitrosoethylene in good agreement with other theoretical methods. We have performed an exhaustive investigation of the low-lying electronically excited states of VF_2 and VCl_2 using the configuration selective and internally contracted MRCI methods. Little is presently known experimentally about these molecules. Because of their low coordination their spectrum is rather dense and strong dynamic and nondynamic correlation effects must be considered for its accurate quantitative description. We find that the spectrum is well resolved using a augmented double zeta quality ANO basis set and a sufficiently large active space. Neither core-valence correlations nor relativistic effects lead to a dramatic change in the non-relativistic results. The vibrational frequencies and dissociation energies are computed in good agreement with previous theoretical results. This study laid the ground work for a thorough comparative investigation of the entire transition metal dihalide family at the benchmark level, which is currently in progress.

Acknowledgments

We gratefully acknowledge helpful discussions with G. Rauhut and N. Dobrodey. Part of this work was supported by DFG Grants KEI-164/11-2 and by DFG grant WE 1863/10-1. We acknowledge the use of the supercomputer facilities at the HLRZ Karlsruhe and the NIC.

References

1. B. O. Roos. *Chem. Phys. Letters*, 15:153, 1972.
2. B. O. Roos and P. E. M. Siegbahn. The direct configuration interaction method. In H.F. Schaefer III, editor, *Methods of Electronic Structure Theory*, page 189. Plenum, New York, 1994.
3. I. Shavitt. In H. F. Schaefer III, editor, *Modern Theoretical Chemistry*. Plenum, New York, 1977.
4. S. R. Langhoff and E. R. Davidson. *Int. J. Quantum Chem.*, 8:61, 1974.
5. W. Butscher, S. Shih, R. J. Buenker, and S. D. Peyerimhoff. *Chem. Phys. Letters*, 52:457, 1977.
6. J. Cižek. *J. Chem. Phys.*, 45:4256, 1966.
7. R. J. Bartlett and I. Shavitt. *Chem. Phys. Letters*, 50:190, 1977.
8. R. Gdanitz and R. Ahlrichs. *Chem. Phys. Letters*, 143:413, 1988.
9. P. Szalay and R. J. Bartlett. *J. Chem. Phys.*, 103:3600, 1995.
10. J. P. Daudey, J.-L. Heully, and J. P. Malrieu. *J. Chem. Phys.*, 99:1240, 1993.
11. R. J. Bunker and S. Peyerimhoff. *TCA*, 12:183, 1968.
12. R. J. Buenker and S. D. Peyerimhoff. *Theor. Chim. Acta*, 35:33, 1974.
13. R. J. Buenker and S. D. Peyerimhoff. *Theor. Chim. Acta*, 39:217, 1975.
14. Z. Gershgorn and I. Shavitt. *Int. J. Quantum Chem.*, 2:751, 1968.
15. B. Huron, J.P Malrieu, and P. Rancurel. *J. Chem. Phys.*, 58:5745, 1973.
16. R. J. Buenker and S. D. Peyerimhoff. *New Horizons in Quantum Chemistry*. Reidel, Dordrecht, 1983.
17. J. L. Whitten and M. Hackmeyer. *J. Chem. Phys.*, 51:5548, 1969.
18. R. J. Harrison. *J. Chem. Phys.*, 94:5021, 1991.
19. S. Krebs and R. J. Buenker. *J. Chem. Phys.*, 103:5613, 1995.
20. M. Hanrath and B. Engels. *Chem. Phys.*, 225:197, 1997.
21. F. Stephan and W. Wenzel. *J. Chem. Phys.*, 108:1015, 1998.
22. P. Stampfuß, H. Keiter, and W. Wenzel. *J. Comput. Chem.*, 20:1559, 1999.
23. P. Stampfuß and W. Wenzel. *J. Mol. Structure*, 506:99, 2000.
24. W. Wenzel and M. M. Steiner. *J. Chem. Phys.*, 108:4714, 1998.
25. M. Schweizer J. Stevens and G. Rauhut. *J. Amer. Chem. Soc.*, 123:7326, 2001.
26. W. Wenzel and K. G. Wilson. *Phys. Rev. Letters*, 68:800, 1992.
27. C. W. DeKock and D. M. Gruen. *J. Chem. Phys.*, 44:4387, 1966.
28. S. H. Ashworth, F. J. Griemann, and J. M. Brown. *JACS*, 115:2978, 1993.
29. S. G. Wang and W. H. E. Schwartz. *J. Chem. Phys.*, 109:7252, 1998.
30. T. Tokuda, N. Fuji, S. Yoshida, K. Shimuzu, and I. Tanaka. *CPL*, 174:385, 1990.
31. R. J. Deeth. *J. Chem. Soc. Dalton Trans.*, page 1061, 1993.
32. A. Möller, M. A. Hitchman, E. Kraus, and R. Hoppe. *Inorg. Chem.*, 34:2684, 1995.
33. S. Larsson, B. O. Roos, and P. E. M. Siegbahn. *CPL*, 96:436, 1983.
34. S. Y. Shashkin and W. A. Goddard. *JPC*, 90:225, 1986.
35. C. W. Bauschlicher and B. O. Roos. *J. Chem. Phys.*, 91:4785, 1989.
36. B. O. Roos, K. Andersson, M. P. Fülscher, P.-A. Malmqvist, L. Serrano-Andres, K. Pierloot, and M. M'erchan. *Adv. Chem. Phys.*, 93:219, 1996.
37. H. J. Werner and P. J. Knowles. *J. Chem. Phys.*, 89:5803, 1988.
38. K. Andersson, B.O. Roos, P.-A. Malmqvist, and P.-O. Widmark. *Chem. Phys. Letters*, 230:391, 1994.

39. R. Pou-Amerigo, M. Merchan, I. Nebot-Gil, P.O. Widmark, and B. Roos. *Theor. Chim. Acta*, 92:149, 1995.
40. P.O. Widmark, P.A. Malmquist, and B. Roos. *TCA*, 77:291, 1990.
41. J. Almlöf, H.J. Werner, P.J. Knowles. *MOLPRO Manual*, 1998.
42. K. Kutchitsu. *Structure Data of Free Polyatomic Molecules*. Springer, 1992.

Accurate ab initio Calculations for Vanadium Oxide Clusters

Mikhail Pykavy and Christoph van Wüllen

Institut für Chemie, Fakultät II, Technische Universität Berlin,
Straße des 17. Juni 135, D-10623 Berlin, Germany

Abstract. Multi reference correlation calculations (MR-CI and MR-ACPF) have been performed for small V_nO_m clusters. VO_2 has two doublet states which are so close that it is difficult to predict the symmetry of the ground state. For $V_2O_4^+$ and V_2O_4 we find minimum energy structures of C_{2h} symmetry (trans bending of the vanadyl units) in contrast to what has been reported in the literature. The magnetic coupling of the electrons is such that low spin states are favoured (singlet for V_2O_4, doublet for $V_2O_4^-$).

1 Introduction

Many experimental and theoretical studies have been published recently targeting at the ground state properties of vanadium oxide clusters of different size and composition [1–15]. Most of the theoretical work is based on the density functional formalism and differs mainly by the chosen exchange-correlation functionals or (and) basis sets [11–15]. However, it is not always obvious whether the density functional methods based on the single reference formalism are suitable to describe the ground state properties of the different vanadium oxide species in the right way. Especially, the coupling of the unpaired 3d electrons of several vanadium atoms is a delicate question which in general is a subject for more sophisticated multi reference methods [16,17].

This report focuses on the ground state properties of VO_2 and of the neutral as well as singly charged V_2O_4 species. In the case of cationic $V_2O_4^+$ only one 3d(V) electron is present and the question about the possible spontaneous localisation of the latter at a single metal centre is of interest. In the neutral molecule V_2O_4 the coupling (high-spin or low-spin) of the two 3d(V) electrons has to be described properly. The anion $V_2O_4^-$ has again an odd number of 3d electrons, which may be localised at one of the metal centres or delocalised over the both of them. Furthermore, similar to the neutral molecule the spin-state of the the anion cannot be predicted by simple arguments, and careful calculations are required to make a theoretical prediction.

No uniform opinion exists in the literature about the symmetry of the ground state of the VO_2 molecule [2,12,14]. Two states of different symmetry, $^2A_1$ and $^2B_1$, seem to be energetically so close to each other that depending on the applied experimental or theoretical method one or another is found to be the ground state. Highly accurate calculations, which recover most of the correlation energy, are necessary to settle this difficult case.

This report is organised as follows: in section 2 a brief description of the computational methods that we use will be given, section 3 contains results of our calculations and discussion, followed by some technical notes on the installation of the program and the computational effort.

2   Method of calculation

In the introduction we have already mentioned that for the systems under study, we need multi reference methods to correctly describe the electronic structure. Since we have to describe both static and dynamic electron correlation effects, we must start from a multi configuration wave function and add a treatment of dynamic correlation on top. Thus, in a first step we construct molecular orbitals using Complete Active Space Self-Consistent Field (CASSCF) method. For VO_2, the active space includes the 3d and 4s orbitals at the vanadium centres, together with the 2p shells at oxygen. For the binuclear compound V_2O_4, it is no longer possible to include oxygenic orbitals in the active space, and one has to choose an active space upon careful analysis of the electronic structure. In the simplest case, we put the vanadium 3d and 4s orbitals in the active space, but sometimes it is also necessary to include charge-transfer configurations (with a hole in the oxygen 2p shells) in the reference function. Usually, a state averaged CASSCF calculation for many low lying states is used to generate start orbitals for the CASSCF calculation on the states of interest. These start orbitals form 3d(V), 4s(V) and 2p(O) shells as intended. Without such a two-step procedure, the nature of the active orbitals is sometimes quite different from what on had in mind.

The second step of our studies are the time consuming correlation calculations using multi configuration reference (MR) functions. As a reference the CASSCF wave function for the ground state of the molecule (from the orbital optimisation step) is used. Different methods are used to take into account electron correlation effects: The (singles and doubles) configuration interaction method (MR-CI), partly including the Davidson correction (MR-CI+Q), average quadratic coupled-cluster (MR-AQCC) and average coupled-pair-functional (MR-ACPF) methods. The latter methods are variants of the MR-CI approach that correct for the main defect of MR-CI, namely the missing size consistency. For all systems, all valence electrons (3d(V), 4s(V), 2s(O) and 2p(O)) were correlated in the CI-type calculations. All calculation have been performed using the MOLPRO (Version 2000.1) program system [18] which we installed at the HLRS (see sec. 3.3).

Most quantum chemical calculations published for V_nO_m species were based on the density functional method. This method, although being quite robust in the sense that spectacular failures are seldom, has the disadvantage that there is no internal consistency check and no route to systematic improvement, two properties that are inherent to wavefunction-based ab initio calculations. However, to exploit these features one must be able to estimate correlation energies at the basis set limit (BSL) by an extrapolation of results from limited-accuracy calculations with finite basis sets.

Two different extrapolation techniques have used in our studies. The first one is the Truhlar two-point-extrapolation [19]:

$$E_{corr}^{BSL} = \frac{E_{corr}^{X1} \times X_1^2 - E_{corr}^{X2} \times X_2^2}{X_1^2 - X_2^2}$$

where X_1 and X_2 are the so called cardinal numbers $(2, 3, 4, \ldots)$ of two (subsequent) basis sets, E_{corr}^{X1} and E_{corr}^{X2} are the values of the dynamic correlation energies obtained using the basis with cardinal number X_1 and X_2, respectively. The dynamic correlation energy is operationally defined as the difference between the CASSCF calculation and the MR-CI (or MR-ACPF, etc.) calculation. E_{corr}^{BSL} is then the estimation of the correlation energy value at the basis set limit (BSL). For this two-point extrapolation formula, we used results from the TZ and QZ basis sets throughout this work. The estimated basis-set limit for the dynamic correlation energy, E_{corr}^{BSL}, is then added to the CASSCF energy obtained with the largest basis set. This is a valid procedure since CASSCF energies – and therefore the static part of the correlation energy – are virtually converged already at the QZ basis set level.

Occasionally, we also tested a second extrapolation technique of Klopper et al. [20], based on the (analytically) known behaviour of the correlation energy of two-electron systems like the helium atom

$$E_{corr}^{X} = E_{corr}^{BSL} + c/X^n$$

The correlation energies E_{corr}^{X} were calculated for $X = 3, 4, 5$ (i.e., for the TZ, QZ and 5Z basis sets) and the parameters c and n and the basis set limit for the dynamic correlation energy, E_{corr}^{BSL}, were adjusted using standard non-linear fitting techniques.

It has been shown [20] that extrapolation techniques are only reliable if the larger and larger basis sets approach the basis set limit in a controlled way. The most important condition is the so-called correlation consistency which was extensively analysed by Dunning and coworkers [21, 22]. These authors have also developed series of the correlation consistent basis sets for the main group atoms up to the fourth period [22]. Unfortunately, no such basis sets exists for transition metal atoms [23] (apart from a few exceptions[1]). For this reason we have constructed a series of correlation consistent basis sets for vanadium atom. The composition of the constructed basis sets is shown in the Table 1.

The details of the basis set optimisation will be published elsewhere [25].

For oxygen we use cc-pvXZ basis sets of Dunning [21].

---

[1] T. Noro et al. [24] have developed a series of correlation consistent basis sets for vanadium which describe, however, only the correlation of the 4s-shell.

Table 1. Composition of the correlation consistent basis sets for vanadium atom

Basis set	un-contracted functions	contracted functions	extension
DZ	(14s 9p 4d)	⟨ 6s 4p 2d ⟩	1f
TZ	(17s 12p 6d)	⟨ 8s 6p 4d ⟩	2f 1g
QZ	(19s 14p 7d)	⟨ 9s 7p 5d ⟩	4f 2g 1h
5Z	(21s 16p 8d)	⟨ 10s 8p 6d ⟩	6f 4g 2h 1i

3 . Results and discussion

In this section we will present results of our calculations on the ground states of VO_2 and $V_2O_4^{+/0/-}$ molecules. As our work is still in progress and not yet finished, these results have still to be regarded preliminary, and the picture we present here is not fully complete. The results obtained so far nevertheless point out some salient features present in the compounds under study and have led to reconsideration [26] of density functional calculations published earlier.

3.1 The ground state of VO_2

The VO_2 molecule is a formal $3d^1$ system and at first glance, so one might guess that the treatment this system is not difficult. However, charge transfer effects from oxygen 2p orbitals to the vanadium 3d orbitals are important for the correct description of the two lowest states, $^2A_1$ and $^2B_1$, which are energetically very close to each other (ΔE is less then 0.1 eV). Furthermore, the mutual order of these two states[2] depends on the geometry of the molecule (see also [14]). A definite answer obviously requires geometry optimisation at a correlated level. Even for such a small molecule, MR-AQCC geometry optimisations are so expensive that we could not do it without the supercomputer power provided by HLRS.

In C_{2v} symmetry, there are only two geometrical degrees of freedom: the bond length R_{VO} and the bond angle $\angle OVO$. In order to get a closer insight into the geometry dependence of the energy separation and order of both $^2A_1$ and $^2B_1$ states, we performed partial geometry optimisations of the bond length for some fixed values of the angle. Besides being slightly faster than a two-dimensional full geometry optimization, this procedure yiels potential energy curves – the energy at a given $\angle OVO$ angle, with the VO distance optimized. Such potential energy curves for the $^2A_1$ and $^2B_1$ states have been obtained at different theoretical levels. The results of the MR-AQCC

---

[2] Note that these two states become a degenerate $^2\Delta$ pair if the molecule becomes linear.

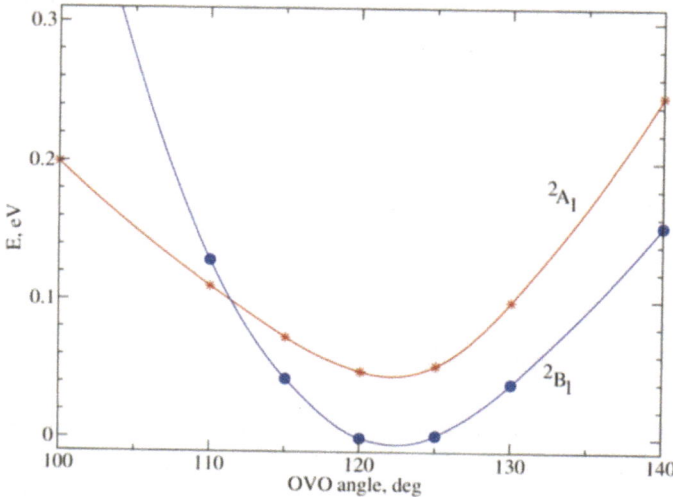

Fig. 1. Calculated MR-AQCC potential energy curves for the $^2A_1$ and $^2B_1$ states of VO_2 molecule

calculations are shown on the Fig. 1. It is clearly seen that the $^2B_1$ state is energetically lower than the $^2A_1$ state, but the energy distance between them is just 0.04 eV at the equilibrium bond angle 123° (which has almost the same value for both states!). The energetic separation between the two states is not constant, however. The both curves cross each other at 112°.

CI calculations with Davidson correction yield very similar curves and are not presented here. Pure CI calculations show slightly different equilibrium geometries, although the whole picture does not change, i.e. the lower state is still the $^2B_1$ state and the energetic distance to the $^2A_1$ state remains very small.

All calculations presented here were done using TZ basis sets, which is the smallest one that can reliably be used. Single-point extrapolations of the correlation energy to the basis set limit will be done at the minimum geometries obtained in the partial optimisations in the near future.

3.2   Ground state calculations for $V_2O_4^{+/0/-}$

A decent description of the electronic structure of these binuclear vanadium oxide clusters is quite difficult, if there is more than one 3d electron: on one side, both the low spin coupling cannot even qualitatively be described by single-reference methods. On the other hand, the dynamic correlation energy greatly affects the energy spacing and even the ordering of the states, mainly because dynamic correlation is different in states of different spin multiplicity[3]. The size of the V_2O_4 clusters is still so small that highly accurate correlation calculations can be performed. The small number of metal 3d electrons (1, 2, and 3 for the cationic, neutral and anionic cluster) also helps making the calculations tractable. A comparison of our high-level ab initio data and density functional results will give an insight into the ability of the density functional methods to describe properly the electronic properties of the vanadium oxide clusters. This is useful because there are competing recipes how to treat such situations at the density functional level, and insight gained for $V_2O_4^{+/0/-}$ can be transferred to larger clusters, where direct comparison between ab initio and DFT calculations is no longer possible.

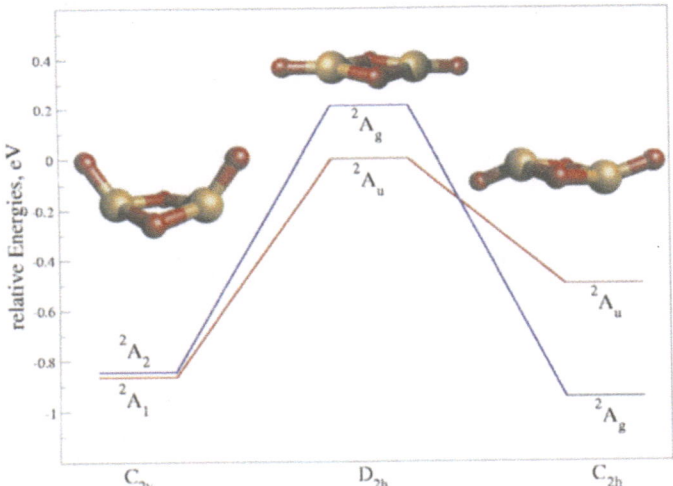

Fig. 2. Calculated MR-ACPF energies of two lowest states for different $V_2O_4^+$ conformers

Let us start with the ground state of the cation $V_2O_4^+$. This is formally a $3d^1$ system. This single unpaired electron occupies the lowest 3d(V)-like (delocalised) molecular orbital. A planar, D_{2h}-symmetric structure with two bridging oxygens and two exocyclic V=O bonds is not stable for the cation: the vanadyl (V=O) bonds bend away from the V_2O_2-plane (see Fig. 2). We have found that the trans-like bending (yielding a C_{2h}-symmetric conformer) is energetically favoured over cis-bending which leads to C_{2v} symmetry. The difference between these two conformers amounts to 0.25 eV (MR-AQCC with TZ basis). Our result differs from a recent DFT study in which only the cis-conformer was considered [12] and lead to a re-investigation of the

---

[3] This is a consequence of the Pauli principle, which keeps electrons of like spin apart. Therefore the dynamic correlation energy arising from the Coulomb repulsion is larger in the low-spin cases.

problem at DFT level [26] which improves the agreement between the two methods.

The question about the localisation of the single unpaired electron was also considered in our studies although only at the DFT level. So far we have no hint that this localisation occurs for $V_2O_4^+$. Multi-reference CI calculations on this problem would be quite expensive. As we mentioned in the section 2 we have to include charge transfer configurations into the reference wave function. For the positively charged species such as $V_2O_4^+$ their contribution cannot be neglected (about 15 % weight in the wave function). The importance of these configurations is even larger if the electron localises. The number of terms in the reference function grows exponentially with the number of active orbitals, this makes the subsequent MR-CI calculations quite expensive. We have to postpone this problem for the moment.

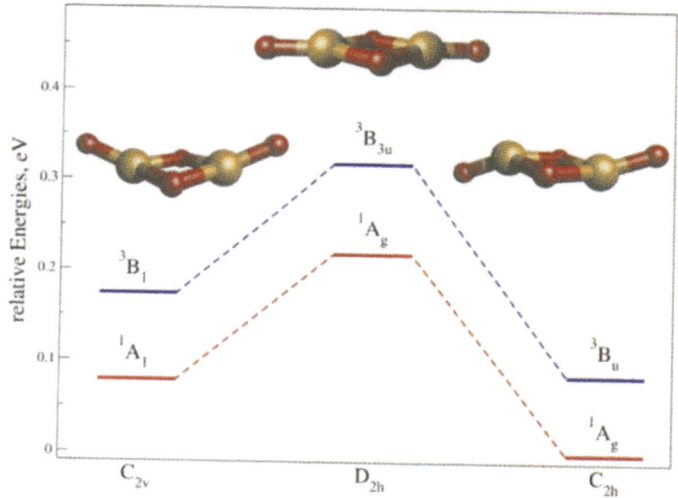

Fig. 3. Calculated MR-ACPF energies of two lowest states for different V_2O_4 conformers

For the neutral V_2O_4 molecule we have performed a full geometry optimisation at the MR-CI and MR-ACPF level of theory for the six lowest states. The results for two lowest states are shown on the Fig. 3. Similar to the cation the neutral molecule also prefers the C_{2h} conformer even though the energy difference to the C_{2v} conformer is rather small (slightly less than 0.1 eV). This result does not agree, however, with the published DFT calculations in which the C_{2v} conformer was postulated [12, 14].

Furthermore, we have found the low spin state of V_2O_4 to be energetically lower than the high-spin one, i.e. both unpaired 3d(V) electrons have antiparallel spins. Since DFT is inherently a single-reference method, it cannot describe such a situation unless one resorts to symmetry-broken descriptions, where one mixes states of different spatial and spin symmetry to get a single-configuration situation. Our data is now being used to test such approaches on small vanadium oxide clusters.

In contrary to the cation and the neutral molecule, the anion $V_2O_4^-$ prefers planar geometry. This result agree with the DFT studies of Vyboishchikov and Sauer [12]. As for the neutral, we find again the low-spin-coupling of unpaired 3d(V) electrons (i.e., a doublet state) to be energetically preferable compared to the quartet state, which was assumed in the DFT studies.

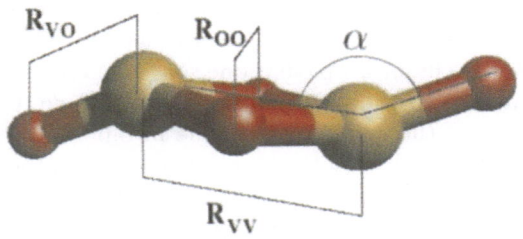

Fig. 4. Internal coordinates of V_2O_4

The optimised geometric parameters of all three species in their ground states are collected in the Table 2 (they are also shown of the Fig. 4). The bond distances in the inner V_2O_4 ring are quite similar for all three $V_2O_4^{+/0/-}$ species. On the other hand, the out-of-plane bending angle α and the exocyclic VO-bond distance depend strongly on the charge of the molecule. The change of the 3d-occupation from cation to the neutral molecule and anion is accompanied by the prolongation of the VO bond, which has a bond order larger than two in the cation, becomes two in the neutral molecule and less than two in the anion. Decrease of the bending angle in the $+/0/-$ series is due to the increasing occupation of the b_{3g} orbital which favours the planar geometry. An detailed analysis of this behaviour will be published elsewhere [26].

Table 2. Optimised (MR-ACPF) geometries of the $V_2O_4^{+/0/-}$ species

Species	ground state	R_{VV}, Å	R_{OO}, Å	R_{VO}, Å	α, °
cation	^2A_g (C$_{2h}$)	2.582	2.476	1.551	130.0
neutral	^1A_g (C$_{2h}$)	2.628	2.527	1.595	141.1
anion	^2A_u (D$_{2h}$)	2.620	2.592	1.651	180.0[a]

[a] This angle was not optimised due to the D$_{2h}$ symmetry.

3.3 Technical Part

Installation of MOLPRO (Version 2000.1) on NEC SX5 At the beginning of this project we became aware that the MOLPRO quantum chemical program system had only been installed on the SX4 at the HLRS, and since it used the obsolete *float2* floating point format, the program would not run on SX5 hardware. Even worse, only an outdated (1998) version of MOLPRO was available, and this installation has to be considered experimental since it does not pass the test jobs which come along with the program.

We therefore decided to install the current version (at that time, MOL-PRO2000) such that it would run both on SX4 and SX5. This requires small changes in the installation procedures. We still have to use the f77sx compiler – this is the main reason why we need the SX4 as well (f77sx is not installed on the HLRS SX5 machine). Note that one source module (cilsdm) has to be compiled without vectorisation, otherwise the test job `so.test` fails.

A large fraction of the computation done in our MOLPRO calculations are done within matrix multiplications, so we make use of the vendor-supplied support routines (`VDMXMA` and friends). Stimulated by the results of a profiling run, we then coded new routines for sparce matrix multiply (MOLPRO's `mxmas` and related routines) optimised for vector processors. After these optimisation, we profiled a typical MR-CI calculation and found that slightly more than 50% of the CPU time is spent in matrix multiplications, and 22% is spent in the vectorised routine `locate` which searches a target number (bit pattern) in a large integer vector. All "hot spots" of the program thus make efficient use of the SX5 hardware.

Calculations done in this project The memory requirements of our calculations were modest (400 MB for VO_2, 900 MB for V_2O_4), so all our calculations were done in single-processor jobs. For VO_2, altogether 14 partial geometry optimisations (of the bond length) were performed (for two states and seven fixed OVO angles), for each of the theoretical models MR-CI, MR-CI+Q and MR-AQCC. This required about 600h CPU time. For the binuclear V_2O_4 complexes, a typical single-point MR-ACPF calculation requires 20h CPU time, around 200h have to be spent if partial optimisation of a single coordinate is to be performed. Note that at least 22 orbitals have to be correlated in the MR-CI-type calculations. This is the reason for their high computational demand.

4   Conclusions

We have demonstrated the necessity of sophisticated multi reference correlation methods for the correct description of the ground states of small vanadium oxide clusters. The VO_2 molecule seems to be a challenge for the theoreticians due to the complexity of the problem which requires an enormous computational effort and care. In the case of $V_2O_4^{+/0/-}$ species we were able to identify possible problems in the density functional theory treatment of these systems. Multi reference methods give the "right answer for the right reason". Further studies are required, however, to make final conclusions about the ability of density functional methods to describe the ground states of vanadium oxide clusters with many metal centres. This work is in progress.

Acknowledgements

A grant of computer time for the SX5 at the HLRS is gratefully acknowledged. This research has been supported by the Deutsche Forschungsgemeinschaft through the Sonderforschungsbereich 546.

References

1. D. E. Clemmer, J. L. Elkind, N. Aristov, P. B. Armentrout, J. Chem. Phys. **95**, 3387 (1991)
2. L. B. Knight Jr., R. Babb, M. Ray, T. J. Banisaukas III, L. Russon, R. S. Dailey, E. R. Davidson, J. Chem. Phys. **105**, 10237 (1996).
3. G. V. Chertihin, W. D. Bare, L. Andrews, J. Phys. Chem. A **101**, 5090 (1997)
4. R. C. Bell, K. A. Zemski, A. W. Castleman, Jr., J. Phys. Chem. A **102**, 8293 (1998)
5. I. Kretzschmar, D. Schröder, H. Schwarz, C. Rue, P. B. Armentrout, J. Phys. Chem. A **102** 10060 (1998)
6. H. Wu, L.-S. Wang, J. Chem. Phys. **108**, 5310 (1998)
7. J.N. Harvey, M. Diefenbach, D. Schröder, H. Schwarz, Int. J. Mass. Spectr. **182/183**, 85 (1999)
8. R. C. Bell, K. A. Zemski, A. W. Castleman, Jr., J. Phys. Chem. A **103**, 1585 (1999)
9. R. C. Bell, K. A. Zemski, A. W. Castleman, Jr., J. Phys. Chem. A **103**, 2992 (1999)
10. S. E. Kooi, A. W. Castleman, Jr., J. Phys. Chem. A **103**, 5671 (1999)
11. G. L. Gutsev, B. K. Rao, P. Jena, J. Phys. Chem. A **104**, 11961 (2000)
12. S. F. Vyboishchikov, J. Sauer, J. Phys. Chem. A **104**, 10913 (2000)
13. E. Brocławik, T. Borowski, Chem. Phys. Lett. **339**, 433 (2001)
14. M. Calatayud, B. Silvi, J. Andrés, A. Beltrán, Chem. Phys. Lett. **333**, 493 (2001)
15. M. Calatayud, J. Andrés, A. Beltrán, J. Phys. Chem. A **105**, 9760 (2001)
16. J. Cabrero, N. Ben Amor, C. de Graaf, F. Illas, R. Caballol, J. Phys. Chem. A **104**, 9983 (2000)
17. F. Illas, I. de P. R. Moreira, C. de Graaf, V. Barone, Theor. Chem. Acc. **104**, 265 (2000)
18. MOLPRO is a package of ab initio programs designed by H.-J. Werner and P. J. Knowles. The authors are R. D. Amos, A. Bernhardsson, A. Berning, P. Celani, D. L. Cooper, M. J. O. Deegan, A. J. Dobbyn, F. Eckert, C. Hampel, G. Hetzer, P. J. Knowles, T. Korona, R. Lindh, A. W. Lloyd, S. J. McNicholas, F. R. Manby, W. Meyer, M. E. Mura, A. Nicklaß, P. Palmieri, R. Pitzer, G. Rauhut,

M. Schütz, U. Schumann, H. Stoll, A. J. Stone, R. Tarroni, T. Thorsteinsson, H.-J. Werner.

19. D. G. Truhlar, Chem. Phys. Lett. **294**, 45 (1998)
20. W. Klopper, "R12 Methods, Gaussian Geminals" in Modern Methods and Algorithms of Quantum Chemistry, NIC Series, Vol. 3 (2000), p. 181–230
21. T. H. Dunning, J. Chem. Phys. **90**, 1007 (1989)
22. T. H. Dunning, Jr., K. Peterson, A.K. Wilson, J. Chem. Phys. **114**, 9244 (2001) and references therein.
23. T. H. Dunning, Jr., private communication.
24. T. Noro, M. Sekiya, T. Koga, H. Matsuyama, Theor. Chem. Acc. **104**, 146 (2000)
25. C. van Wüllen, M. Pykavy, in preparation.
26. M. Pykavy, J. Sauer, and C. van Wüllen, in preparation

The Iron(III) Catalysed Michael Reaction

Silke Pelzer and Christoph van Wüllen

Institut für Chemie, Sekr. C3, Technische Universität Berlin,
Straße des 17. Juni 135, D-10623 Berlin, Germany

Abstract. Density Functional Calculations have been performed on the mechanisms of the Iron(III) catalysed Michael Reaction.

1 Introduction

The Michael reaction, whose general scheme is given in Fig. 1, is an important and versatile carbon-carbon coupling reaction with numerous applications in the synthesis of organic compounds.

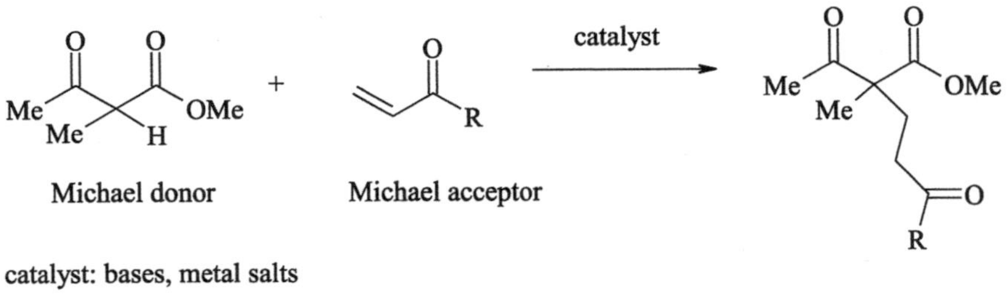

catalyst: bases, metal salts

Fig. 1. General scheme of a Michael reaction

Experimental work by Christoffers et al. [1] shows that ferric chloride hexahydrate is a very efficient catalyst for the Michael reaction with β-dicabonyl compounds (acting as Michael donors) and enones (as Michael acceptors). Under mild and non–basic conditions the reaction shows very high chemoselectivity making workup and purification easy to accomplish.

However, this success is tempered by several experimental difficulties encountered in the study of these reactions. Foremost, the reaction intermediates are unknown, and their structure and properties are difficult to elucidate due to the transient nature of the intermediates. Moreover, it has been found that there is no reaction between β-dicarbonyl compounds and acrylic acid alkyl esters such as methyl acrylate. This result is rather unexpected and could not be explained, as the reaction involving the related alkyl vinyl ketones occurs quickly and with high yields. Both of these quandaries are best addressed by theoretical calculations, which are more easily applied to unstable molecules and reaction mechanisms than experimental methods. The objective of this work is to investigate possible reaction channels and reasons why the acrylic acid alkyl esters are not reaction with the Michael donors. Understanding the inertness of methyl acrylate in this reaction may help in the rational design of experimental condition which let such species react.

Experimental findings suggest that both the enone (Michael acceptor) and the oxo-ester (Michael donor) are coordinated to the same metal center when the carbon-carbon coupling takes place. For example, enones such as cyclo-hexanone, where the two double bonds are fixed in a s-trans conformation, give no reaction. This suggests a catalytic cycle as depicted in Fig. 2.

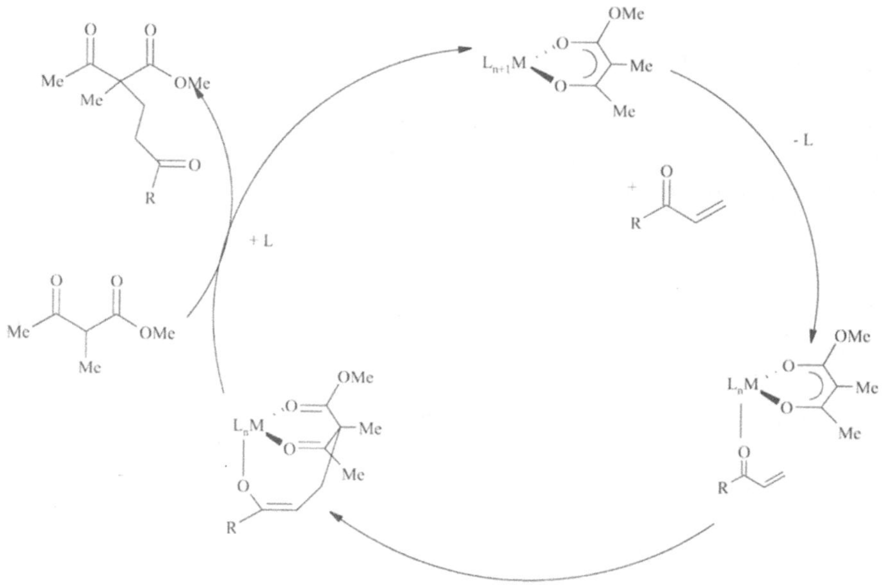

Fig. 2. Proposed reaction mechanism for the iron (III) catalysed Michael reaction

A quantum chemical modeling of such a catalytic cycle involves geometry optimizations of the various intermediates and the location of the transition states connecting the minima. These results can be cast into a reaction profile from which reactivities can be deduced semi-quantitatively.

2 Methods

With wave function based quantum chemical approaches it is notoriously difficult to describe transition metal compounds, escpecially those containing first-row transition metals such as iron. Hartree-Fock calculations and low-order correlation treatments such as MP2 are therefore not reliable enough for our purpose. Density functional calculations provide a robust and still computationally feasible alternative. We use the B3LYP hybrid exchange-correlation functional, whose usefulness in modeling transition metal chemistry has been demonstrated in numerous applications in the literature.

Calculations were carried out using the TURBOMOLE [2, 3] and Gaussian98 [4] program packages. In the TURBOMOLE calculations, we used our own densitiy functional code [5]. TURBOMOLE has been used to optimize geometries, while transition state searches, frequency calculations and IRC (reaction-path-following) calculations were done with Gaussian. Minimum geometries and transition states were located with the 6-311G* basis set, which is triple zeta in the valence shell and is augmented by a set of polarization functions for the non-hydrogen atoms. The nature of the minima and transition states has been established by a frequency analysis, and IRC calculations, starting at the transition states, were performed to verify that these transition states indeed connect the minima as shown in the reaction profile. Although this procedure implies a very high computational effort, it is the only way to be sure that the critical points (minima and transition states) belong to the same reaction path, and that there are no further unexpected minima in between.

Basis set effects on relative energies have been estimated by single-point calculations (at the minimum and transition state geometries) using the larger 6-311G**, 6-311+G** and 6-311++G** basis sets. These effects were rather modest and did not change the overall picture.

3 Results

Fe(III) is known to have a great tendency to bind β-diketones. The critical part of the reaction thus begins with the complex **1b** (see Fig. 3).

Note that there are three "spectator" ligands L at the Fe center which are not directly involved in the reaction. In a first series of calculations, these ligands were modeled just by water molecules. While a water molecule is probably a good model for *uncharged* oxygen donors, it is less likly a good model for negatively charged ligands. However, under the experimental conditions, a coordination of a second enolate occurs, and this may change the situation. We come back to this point at the end of this section. The deprotonation **1a** → **1b** is favoured by electrostatic effects (separation of two positive charges), and the (gas-phase) proton affinity of **2b** has been calculated as 314 kJ/mol compared to 1512 kJ/mol for the free diketo-enolate.

The ligand exchange **2b** → **3** most likely follows a dissociative pathway, that is, the barrier height for this step can be identified as the axial Fe-OH$_2$ bond dissociation energy which is about 60 kJ/mol. Thus coordination of the Michael acceptor is fast, and the barriers of the following steps are all higher.

For the carbon-carbon coupling step **3** → **4** a transition state (**TS3/4**) has been located. How the product is actually released from the metal center has not yet been completely determined, it seems clear that protonation at the oxygen coming from the acceptor is required before the Fe-O-bond can

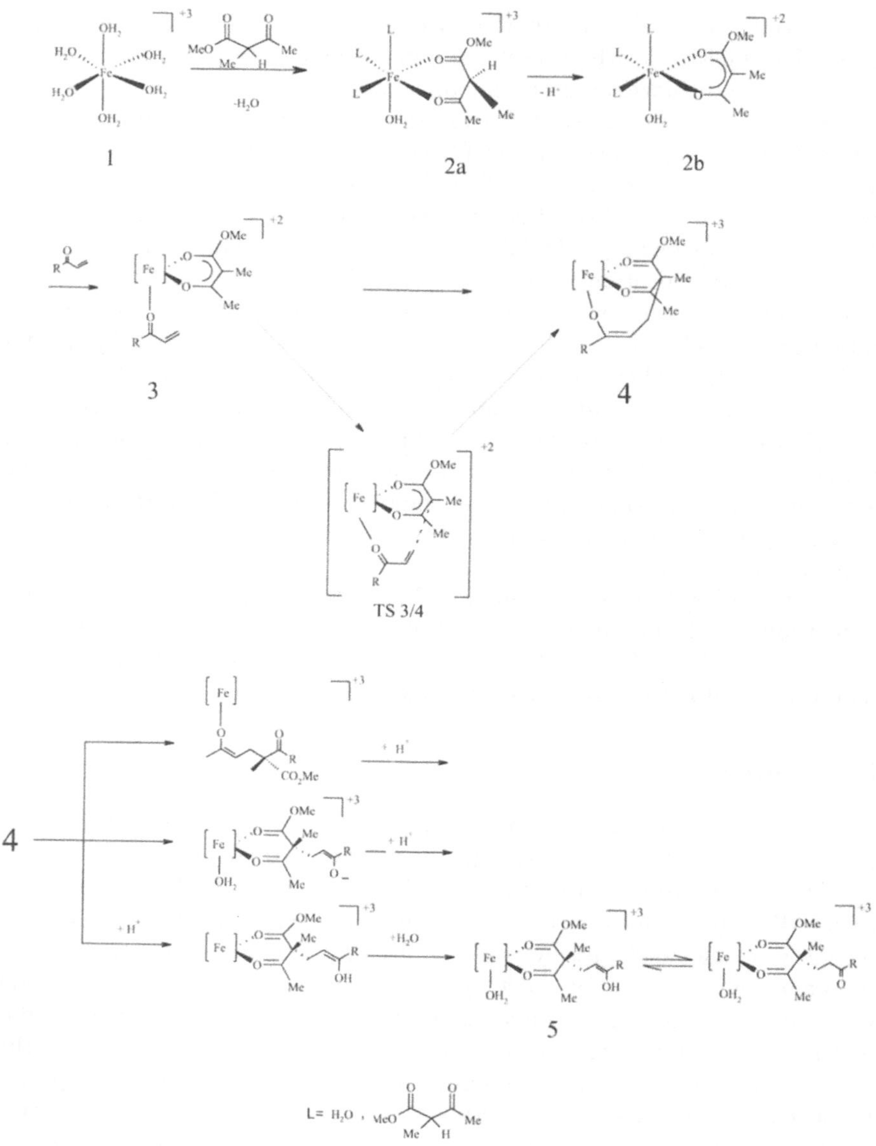

Fig. 3. Possible reaction channel which occurs for both Michael acceptors

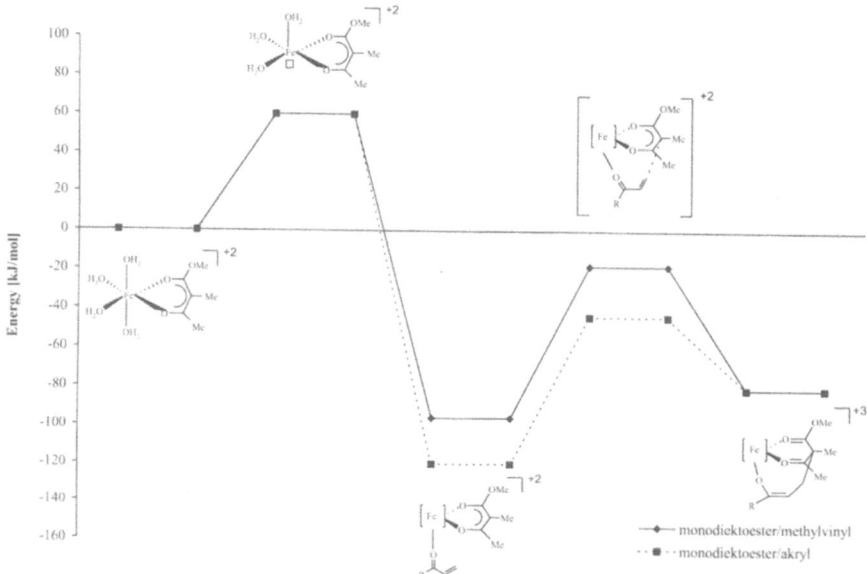

Fig. 4. Energy profile for the central part of the reaction

dissociate. Then a solvent molecule (water in our calculation) can coordinate to give complex **6**. Since **6** has lost π-conjugation compared to **2b**, release of the product from **6** should be facile, and the metal catalyst can enter the next cycle. Figure 4 shows the calculated reaction profile of these steps for two different Michael acceptors, namely methyl vinyl ketone (R=CH$_3$) and methyl acrylate (R=OCH$_3$). Experimentally, the second acceptor shows no reactivity at all, however, the reaction profiles which we calculated (see Fig. 4) show a very similar barrier in both cases. The only difference is that coordination of methyl acrylate is considerably more exothermic. Clearly, this computational result is not consistent with experimental observations.

Introducing a negatively charged oxygen donor as a "spectator ligand" is expected to alter the situation. Experimental observations are also consistent with the coordination of two enolate ligands. Therefore we also investigated the reaction starting from a variant of **2b**, where we substituted two water ligands by a diketo enolate anion. Two orientations of the two enolate ligands are possible, a coplanar geometry where the enolates are coordinated to four equatorial positions of the FeO$_6$ octahedron, and another geometry where three of these 4 sites form an octahedral face. The latter orientation was found more stable, reaction profiles starting from this species results are shown in Fig. 5. We note that these calculations were not only quite demanding in terms of CPU time, it was also hard to locate the transition state.

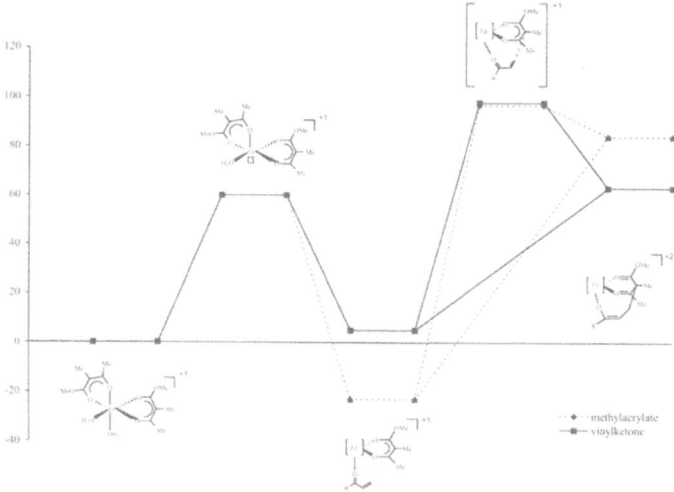

Fig. 5. Reaction profile for the bis-diketoenolate complex

Now the central step of the reaction is endothermic both for methyl vinyl ketone and methyl acrylate, and there is a large difference between these two substrates.

For methyl vinyl ketone, the largest barrier occurs for the C-C-coupling step and amounts to 90 kJ/mol, which can be overcome at ambient temparatures. Methyl acrylate on the other hand binds much stronger to the bis-diketoenolate complex and gives an unreactive intermediate – the barrier to C-C-coupling is 120 kJ/mol. This means that this step will be so slow at room temparature that product formation cannot be observed under typical experimental conditions (room temparature, 24 h reaction time). The picture is thus that methyl acrylate binds to the metal and does not want to react further, so with complex **4** a dead end has been reached. It can also be seen from Fig. 5 that methyl acrylate binds more strongly to the iron center than methyl vinyl ketone. As a consequence most of the catalyst should stay in the non-reactive state **4** with R=OCH$_3$, if both methyl acrylate and methyl vinyl ketone are in the reaction mixture.

4   Further experiments and outlook

While the results presented above are still somewhat preliminary, some of the implications were interesting enough to warrant new experimental work, which was untertaken in collaboration with J. Christoffers [6]. While it was known that methyl vinyl ketone is reactive and methyl acrylate is not, the new experimental results is that a mixture of both shows no reaction. This means that methyl acrylate blocks the iron centers, which is exactly what has been deduced from the computational results presented above.

Our results indicate that negatively charged spectator ligands can be used change the reactivity over a wide range. We also initiated new experiments to test this. For example, it has been found out that chloride-free iron salts – we tried $Fe(ClO_4)_3$ – are more efficient catalysts than ferric chloride. On the other hand, if excess chloride is given to the reaction mixture, the reaction rate substantially decreases. In our computational work, we have already begun to substitute one of the H_2O spectator ligands by chloride and to calculate the influence on the reaction profile. These examples show that our theoretical work has direct impact on future experimental studies, and vice versa.

5 · Technical section

The geometry optimizations were done with our own parallel version of the TURBOMOLE program package. In a density functional calculation, there are two parts in the calculation which account for the major part (98 %) of the CPU time in calculation on molecules of the size treated in this project. In energy calculations, the first task is the evaluation of the matrix elements of the Coulomb potential, which can be viewed as a contraction of a fourth-rank tensor $I_{ij,kl}$ (the two-electron integrals) with a second-rank tensor D_{kl} (the density matrix):

$$J_{ij} = \sum_{k,l=1}^{N} D_{kl} I_{ij,kl}$$

The dimension (basis set size) N typically ranges from 500 to 1000 in our calculations, and the two-electron integrals form a very sparse tensor, that is, most of them are negligibly small. Although the J matrix is formed in an iterative procedure many times (for varying density matrices D), the two-electron integrals are recalculated whenever they are needed, since there are still too many of them to be stored. Parallelization of this step is straightforward: the set of Integrals $I_{ij,kl}$ is divided into many small subsets, typically described by three integer parameters i_0, j_1, j_2. Such a triple specifies the subset of ij, kl for which $i = i_0$ and $j_0 <= j <= j_1$. As the calculation proceeds, a supervisor distributes these subsets among the working nodes which incrementally build private copies of J but share the input density matrix D. This also implements a primitive kind of load balancing. Note that there is no communication during this step except for the exchange of the index borders i_0, j_1, j_2. The distribution of the density matrices is done implicitly by the semantics of the `fork` system call, that is, all the working nodes share the same *physical* storage for the density matrix although they have different address spaces (of course, this only holds as long as the density matrix is not overwritte, which is the case). At the end these private copies must be summed up to give the full J matrix. This is the only place where large amounts of data go over interprocess communication lines. In gradient runs, a very similar strategy can be followed. In our B3LYP calculations, we also have to construct (scaled) exchange matrices, but this does not affect the parallelization strategy.

The other big task that we have parallelized is the numerical quadrature which is done in our density functional calculations. Each integral is actually calculated as a sum over the points n of the integration grid, for example the matrix element of the exchange-correlation potential for two basis functions χ_i and χ_j:

$$V_{ij} = \int \chi_i(\boldsymbol{r}) V_{xc}(\boldsymbol{r}) \chi_j(\boldsymbol{r}) \approx \sum_n \omega_n \chi_i(\boldsymbol{r}_n) V_{xc}(\boldsymbol{r}_n) \chi_j(\boldsymbol{r}_n)$$

Now each processor evaluates the private copies of the V matrix for subsets of grid points (allocated dynamically to implement load balancing), these partial V matrices have then to be summed up. V implicitly depends on the density matrix via the electron density

$$\rho(\boldsymbol{r}_n) = \sum_{k,l} D_{kl} \chi_k(\boldsymbol{r}_n) \chi_l(\boldsymbol{r}_n)$$

which in turn determines $V_{xc}(\boldsymbol{r}_n)$.

Since we are left with "serial" work which amounts to 1–2 percent of the total CPU time, we could only use 8 CPUs in parallel – otherwise the speedup factor would strongly decrease because of Amdahl's law. Using more than one node is possible with our program but is not efficient, also because interprocessor communication becomes considerably slower if one uses more than one node.

The code for the evaluation of the two-electron integrals is highly scalar and does not show optimum performance on the SR8000. Indeed, on a single processor this machine is slower than desktop hardware. On the other hand, practicall the whole CPU time in the numerical quadrature is spent on the evaluation of the densities $\rho(\boldsymbol{r}_n)$. This involves a (sparse) matrix-vector multiplication and is reasonably fast on the Hitachi hardware. In a typical calculation, we reach about 150 MFlops per processor which means that the performance is only slightly above 1 GFlops per (8-processor) node.

Acknowledgements

A large part of the calculations done in this project were done on the Hitachi SR8000 at the HLRS in Stuttgart. We are thankful to HLRS for a generous grant of computer time. Support from the Fonds der Chemischen Industrie is also gratefully acknowledged. We are thank Prof. J. Christoffers (University of Stuttgart) for many helpful discussions on this reaction, some of them made possible by a travel grant from the graduate college "Synthetic, Mechanistic and Reaction-Engineering Aspects of Metal Containing Catalysts" hosted at the Technical University in Berlin.

References

1. J. Christoffers, Synlett. **6** 723-732 (2001)
2. R. Ahlrichs, M. Bär, M. Häser, H. Horn, and C. Koelmel, Chem. Phys. Lett. **162**, 165 (1989)
3. M. Häser and R. Ahlrichs, J. Comput. Chem. **10**, 104 (1989)
4. M. J. Frisch, G. W. Trucks, H. B. Schlegel, G. E. Scuseria, M. A. Robb, J. R. Cheeseman, V. G. Zakrzewski, J. A. Montgomery, Jr., R. E. Stratmann, J. C. Burant, S. Dapprich, J. M. Millam, A. D. Daniels, K. N. Kudin, M. C. Strain, O. Farkas, J. Tomasi, V. Barone, M. Cossi, R. Cammi, B. Mennucci, C. Pomelli, C. Adamo, S. Clifford, J. Ochterski, G. A. Petersson, P. Y. Ayala, Q.Cui, K. Morokuma, D. K. Malick, A. D. Rabuck, K. Raghavachari, J. B. Foresman, J. Cioslowski, J. V. Ortiz, B. B. Stefanov, G. Liu,A. Liashenko, P. Piskorz, I. Komaromi, R. Gomperts, R. L. Martin, D. J. Fox, T. Keith, M. A. Al-Laham, C. Y. Peng, A. Nanayakkara, C. Gonzalez, M. Challacombe, P. M. W. Gill, B. Johnson, W. Chen, M. W. Wong, J. L. Andres, C. Gonzalez, M. Head-Gordon, E. S. Replogle, J. A. Pople, program Gaussian 98 (1999)
5. C. van Wüllen, Chem. Phys. Lett. **219**, 8 (1994)
6. S. Pelzer, J. Christoffers, and C. van Wüllen, unpublished work

Computational Fluid Dynamics (CFD)

Prof. Dr.-Ing. Siegfried Wagner

Institut für Aero- und Gasdynamik, Universität Stuttgart
Pfaffenwaldring 21, 70550 Stuttgart

Many basic problems in fluid dynamics can only be solved in a satisfactory manner by a close cooperation between theory, numerical simulation and experimental investigation. Computational Fluid Dynamics, CFD, can help to prepare complex wind tunnel and flight test, can give hints for the necessary resolution of the measuring equipment and can contribute in interpreting the measured data. Experimental data on the other hand serve to validate numerical results. Since most of the fluid mechanical problems are both very complex and unsteady, high performance computers (HPC), that have big storage and high computing performance, are necessary to get the required resolution of the problem, i.e. millions of grid points, and to get the answer of the computation in a reasonable turn-around time. With HPC and the big progress in the development of numerical methods and computational algorithms new insight into complex mechanisms of various fluid problems were possible.

Only a small portion of the submitted progress reports in CFD could be selected for publication. Although most of the reports revealed a very high scientific standard, those reports were preferably selected for publication, that demonstrated the unalterable usage of HPC for the solution of the problem. A big variety of flow problems is investigated. It starts with basic research problems, like laminar-turbulent transition, behaviour of turbulent and separated flows, heat transfer from droplets, reactive and non-reactive supersonic combustion and magneto-hydrodynamic flows. It continues with technological problems, i.e. investigation of wakes behind turbine blades, and flow in porous media. Finally flow simulations of technical applications are presented, e.g. two-phase flows in pipes, flows and heat transfer in high-temperature nuclear reactors, flow simulation around a shuttle-like configuration, fluid-structure interaction on helicopter blades and high-lift aerodynamics of transport aircraft.

Some specific findings with the help of HPC are highlighted in the following sections. Bonfigli et al. could clarify a previously unidentified shear-layer/vortex interaction mechanism that was discovered by Bippes at DLR in an experimental investigation of a yawed wing. The numerical investigation was only possible because of both the big experience of members of the transition group of the Institute for Aerodynamics and Gas Dynamics in using direct numerical simulations (DNS) of the Navier-Stokes equations and the big memory capacity of the NEC SX-5 (16 Processors, 32 GB RAM).

Eisenbach et al. demonstrate the capacity of large eddy simulations (LES). While the sub-grid- scale (SGS) models used show good results for attached flows, further research is necessary for separated flows. Up to 2.3 millions grid points were necessary to perform the computations.

Manhart investigated a turbulent separating boundary layer on a flat plate by DNS. By a local grid refinement he could reduce the original number of grid points by 1/3. But he still needed 36,7 Million grid prints to properly simulate the occurring separation bubble.

Hase et al. compute the heat transfer from droplets moving with transient velocity using the volume of fluid (VOF) method. They use up to 128 processors of the CRAY T3E/512-900 and a grid with up to 4.2 Million cells.

Gerlinger et al. simulate supersonic combustion as applied in scramjets. They investigate the optimum number of blocks in a domain decomposition procedure and use up to 256 nodes of the CRAY T3E/512-900 computer. The typical computation time using 117 blocks and 117 CPUs is 6 hours.

Dedner et al. investigate three-dimensional magneto-hydrodynamic (MHD) flow on an unstructured tetrahedral mesh with dynamic load balancing and demonstrate that control of the divergence of the magnetic field is absolutely crucial. They investigate the influence of the number of grids per processor on the performance and find out that the IBM SP-SMP at Karlsruhe will be favourable for their investigations both with respect to computing power and memory size.

Michelassi et al. use the HITACHI SR-8000 for their investigations of flow in a low pressure turbine with incoming wakes on the basis of LES. They applied domain decomposition with 32 sub-domains and a total number of 2 Million grid modes. The original performance could be increased by a factor of two with the help of members of the HLRS and ranges now between 14 and 15% of the computers peak performance. Due to the relatively good resolution both in time and space a set of interesting results was obtained, e.g. appearance of intermittent separation of the boundary layer and of large elongated flow structures.

Ölmann et al. used also the HITACHI SR 8000 to simulate the flow of two immiscible fluids in a spatial domain. They use the framework UG (Unstructured Grids) as a general software for configuring parallel adaptive multigrid methods.

Giese et al. performed three-dimensional simulation of two-phase flow in pipes including bubbly flow and stratified flow as well as cavitation. The computations on the NEC SX-4 of HLRS were necessary because of the big computational effort due to the two-fluid approach and the additional set of Navier-Stokes equations for the second phase. The segregated solver of the CFX-4 version is sensitive in cases of strong phase interactions and requires small iteration steps. A numerical mesh of more than 31400 nodes was necessary to properly resolve the flow features.

Becker et al. carried out three-dimensional numerical simulations of flow and heat transfer in high-temperature nuclear reactors on the NEC SX-4 and

NEC SX-5. The 360.000 control volumes with 40 blocks required a memory up to 1.5 GB.

Reinartz et al. investigated the supersonic flow around the shuttle-like technology demonstrator PHOENIX for a future European reusable space transportation system. 215 Million grid points are necessary for the inviscid solution. Preparations have begun to start Navier-Stokes computations.

Pomin et al. perform aeroelastic analysis of helicopter rotor blades on the basis of two comprehensive rotor analysis methods. They show that fluid-structure interactions are mandatory in helicopter aeromechanics and that viscous flow effects play in many respects an important role, that means the application of Navier-Stokes solvers and of proper coupling procedures. Up to 42.6 million grid cells are necessary to resolve the flow features. On the NEC SX-5 up to 16 CPUs are used which leads to a sustained performance of 27 GFLOPS with a memory of 23.5 GB. The authors of the paper consider HPC platforms with a reasonable (at the moment 16–32) number of high performance vector CPUs and an adequate memory per node to be an adequate facility. For future development of the present simulation tools and their application to increasingly complex configurations an increase of computer power both in storage and performance would be desirable.

Melber et al. of DLR investigate the viscous compressible flow around high lift configurations of transport aircraft. The TAU code uses a hybrid unstructured grid of 10 million grid points and reaches a vector-operation ratio of nearly 99%, 816 MFLOPS floating point performance and 37 MIPS on one processor of the NEC SX-5. To reach a fully converged solution at high angles of attack with partial flow separation 5000 iterations are required which corresponds to about 650 hours of CPU-time and 12 GByte of memory.

3-D-Boundary-Layer Transition Induced by Superposed Steady and Traveling Crossflow Vortices

G. Bonfigli, M. Kloker, and S. Wagner

Institut für Aerodynamik und Gasdynamik,
Universität Stuttgart, Pfaffenwaldring 21, D-70550, Germany

Abstract. Crossflow-induced laminar-turbulent transition and the initial stages of turbulence are investigated by means of spatial direct numerical simulations for the accelerating 3-D flat-plate boundary layer of the "Querströmungsprinzipexperiment" of the DLR-Göttingen [1]. The complete 3-D incompressible Navier-Stokes equations are solved marching in time on the basis of a 4th-order Runge-Kutta scheme. Fourier spectral expansions and 6th-order compact finite differences are used for discretization in space. In accordance with the experiment, both steady and traveling crossflow modes are excited. Attention is focused on the non-linear interaction between vortical structures generated by the primary instabilities. The growth of high-frequency secondary instabilities is also observed. Classical turbulent quantities are evaluated for the region downstream of the laminar breakdown.

1 Introduction

In aerodynamics, three-dimensional boundary layers characterize the flow around swept-back airplane wings. The favourable chordwise pressure gradient close to the leading edge on the wing upper side induces a velocity component orthogonal to the free-stream velocity, the crossflow component, which points everywhere towards the center of curvature of the boundary-layer-edge streamline (see Fig. 1). It is zero at the wall, reaches a maximum and attenuates then monotonically to zero at the boundary-layer edge.

The laminar-turbulent transitional region for 3-D boundary layers is characterized by vortical structures lying almost parallel to the free-stream velocity vector. The presence of many such vortices, one next to the other, at a distance of approximatively three boundary-layer thicknesses, was soon recognized by visualizing the wall shear stress, where they produce a characteristic stripe pattern. Experimental observations were confirmed by the theoretical analysis (linear stability theory, LST [12]) of the stability of the laminar solution with respect to infinitesimal wavelike perturbations (modes). According to the theory modal disturbances may undergo exponential growth or decay depending on their frequency and on their spatial wave vector. Both steady (CF vortices) and traveling modes (CF waves) may be amplified. Almost independently of the frequency, however, the wave vector of the most unstable modes is nearly orthogonal to the direction of the free-stream velocity. When perturbations reach relevant amplitudes, they become visible in form of the vortical structures observed in the experiment. The vortex axes then coincide with the wave fronts of the original mode.

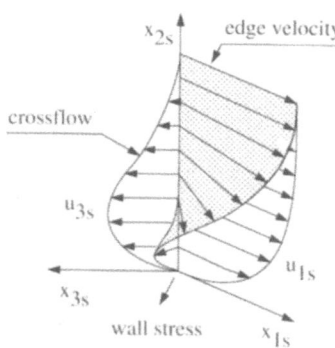

Fig. 1. 3-D boundary-layer velocity profile.

Recently, attention in crossflow-transition research has focused on the non-linear stages of transition, i.e. on the region of the flow where primarily unstable disturbances reach large amplitudes and a linearization of the governing equations is ruled-out. In general, saturation is observed for the primary modes, then the perturbation spectrum fills up progressively, and finally the flow breaks down into turbulence. Secondary instability of the flow field deformed by steady large-amplitude CF vortices has been shown to be one possible mechanism leading to the growth of high-frequency disturbances (secondary high-frequency instabilities, SHFI). Theoretical studies (secondary linear stability theory, SLST [7]) and direct numerical simulations (DNS, [4] [14] [16]) confirmed on this point previous experimental observations ([1], [8]). Up to present, however, theoretical, experimental and numerical results could not clarify in a satisfactory manner cases with superposed steady and traveling primary modes because the limiting cases (purely steady or traveling primary vortices) had not been scrutinized. This has been done in the course of this work and by DNS of Wassermann and Kloker [14] [15].

The present work considers the non-linear stages of crossflow induced transition for the case in which both a steady CF vortex and a CF wave (traveling CF vortex) reach relevant amplitudes in the saturation region. This reproduces the configuration of case "$(0,1)+(1,1)$" of the experimental investigations carried out at the DLR-Göttingen by Bippes and coworkers in the scope of the "Querströmungsprinzipexperiment" [1], which up to the present day represents the most complete experimental data-base for crossflow induced transition. However, also due to a lack of complete flow-field data, no complete picture could be given within the experiment. In the work presented here, the growth of secondary instabilities in the sense of the SLST is proven in spite of the fact that, since the primarily induced deformation of the velocity profiles varies both in time and along the downstream direction, the assumptions of the theoretical approach are not completely fulfilled. Moreover, an alternative mechanism leading to breakdown is found and highlighted. Thereby a strongly non-linear interaction between vortical structures evolving from the saturated primary modes generates complex structures, which in the end break down precipitating the onset of turbulence. The numerical simulation extends into the turbulent region. Results are shown for the turbulent 3-D boundary layer developing at the end of the integration domain. A comparison with numerical simulations by Spalart [13] and Huai, Joslin and Piomelli [3] for different turbulent boundary layers is provided.

2   Numerical method

The numerical simulation is performed in the rectangular domain shown in Fig. 2. The stationary laminar solution (base flow) is computed before starting the simulation of the unsteady disturbance flow. Only the procedure for the perturbance-flow computation will be presented in the following. The subscript "b" indicates quantities of the base flow, while a prime identifies perturbation quantities: $f = f_b + f'$. Figure 2 also shows the coordinate systems considered for the numerical procedure (x_i) and for post-processing ($x_{i\,s}$ and ξ_i). The orientation of $x_{1\,s}$ (angle φ_e) coincides with the local direction of the edge velocity. The ξ-system is used for flow visualizations. The position of its origin and its orientation will be specified when necessary.

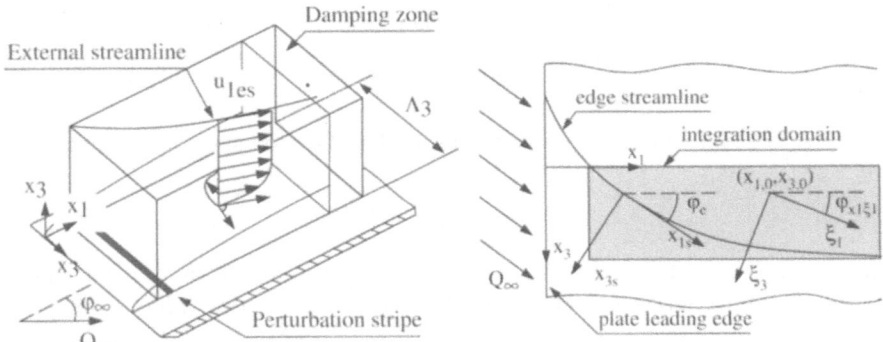

Fig. 2. Integration domain and coordinate systems.

Indicating dimensional quantities with overlined symbols, non-dimensional variables are introduced using the reference quantities $\overline{L} = 0.1\ m$, $\overline{U}_\infty = 14\ m/s$ and $\overline{\nu} = 1.5 \cdot 10^{-5} m^2/s$:

$$x_i = \frac{\overline{x_i}}{\overline{L}}, \quad t = \overline{t} \cdot \frac{\overline{U}_\infty}{\overline{L}}, \quad u_i = \frac{\overline{u_i}}{\overline{U}_\infty}, \quad Re = \frac{\overline{U}_\infty \cdot \overline{L}}{\overline{\nu}}. \tag{1}$$

The non-dimensional vorticity vector is negative for clockwise rotations when looking in the positive direction of the coordinate axes:

$$\omega_1 = \frac{\partial u_2}{\partial x_3} - \frac{\partial u_3}{\partial x_2}, \quad \omega_2 = \frac{\partial u_3}{\partial x_1} - \frac{\partial u_1}{\partial x_3}, \quad \omega_3 = \frac{\partial u_1}{\partial x_2} - \frac{\partial u_2}{\partial x_1}. \tag{2}$$

Fourier expansions are introduced for the spanwise direction. For a general quantity f' holds

$$f'(x_1, x_2, x_3, t) = \sum_{k=-K}^{+K} \hat{f}'_k(x_1, x_2, t)\, e^{ik\alpha_{03}x_3}, \qquad \alpha_{03} = \frac{2\pi}{\Lambda_3}, \quad (3)$$

where the fundamental wave length Λ_3 follows from the periodicity of the physical flow. Since all physical quantities are real, the relation $\hat{f}'_k = \hat{f}'^{*}_{-k}$ holds for all variables, and the complex Fourier coefficients $\hat{f}'_k$ have to be computed only for $0 \leq k \leq K$.

The resolving equations are obtained by projecting the vorticity formulation of the Navier-Stokes equations (see, e.g., [14]) onto the Fourier components of the spanwise discretization:

$$\frac{\partial \hat{\omega}'_{jk}}{\partial t} = \frac{1}{Re} \tilde{\Delta}\hat{\omega}'_{jk} + \hat{X}_{jk}, \qquad j = 1, 2, 3,\ 0 \leq k \leq K\ , (4a)$$

$$\frac{\partial^2 \hat{u}'_{1k}}{\partial x_1^2} - k^2\alpha_{03}^2 \cdot \hat{u}'_{1k} = -ik\alpha_{03} \cdot \hat{\omega}'_{2k} - \frac{\partial^2 \hat{u}'_{2k}}{\partial x_1 \partial x_2}, \qquad 1 \leq k \leq K, (4b)$$

$$\tilde{\Delta}\hat{u}'_{2k} = ik\alpha_{03} \cdot \hat{\omega}'_{1k} - \frac{\partial \hat{\omega}'_{3k}}{\partial x_1}, \qquad 0 \leq k \leq K., (4c)$$

$$\frac{\partial^2 \hat{u}'_{3k}}{\partial x_1^2} - k^2\alpha_{03}^2 \cdot \hat{u}'_{3k} = \frac{\partial \hat{\omega}'_{2k}}{\partial x_1} - ik\alpha_{03} \cdot \frac{\partial \hat{u}'_{2k}}{\partial x_2}, \qquad 1 \leq k \leq K. (4d)$$

Thereby $\tilde{\Delta}$ is the Laplace operator adapted for the used spectral formulation,

$$\tilde{\Delta} = \frac{\partial^2}{\partial x_1^2} + \frac{\partial^2}{\partial x_2^2} - k^2\alpha_{03}^2,$$

and $\hat{X}_{jk}$, $j = 1, 2, 3$, represent the spectral components of the non-linear terms of the Navier-Stokes equations (see [14]). The initial value problems

$$\frac{\partial \hat{u}'_{10}}{\partial x_1} = -\frac{\partial \hat{u}'_{20}}{\partial x_2}, \qquad \frac{\partial \hat{u}'_{30}}{\partial x_1} = \hat{\omega}_{20} \qquad (5)$$

are used for computing the harmonic components $k = 0$ of u'_1 and u'_3.

Dirichlet boundary conditions are set at the inflow requiring the perturbation to be zero for both velocity and vorticity. At the external boundary the vorticity vector is set to zero, and exponential decay is imposed for u'_2. No-slip conditions are given for u'_1 and u'_3 at the wall. Steady and unsteady perturbations may be introduced through suction and blowing ($u'_2 \neq 0$) within the perturbation stripe (see Fig. 2). The ordinary differential equations

$$\frac{\partial \hat{\omega}'_{1k}}{\partial x_1} = -\frac{\partial \hat{\omega}'_{2k}}{\partial x_2}, \qquad k = 0, \qquad (6a)$$

$$\frac{\partial^2 \hat{\omega}'_{1k}}{\partial x_1^2} - k^2\alpha_{03}^2 \hat{\omega}'_{1k} = -\frac{\partial^2 \hat{\omega}'_{2k}}{\partial x_1 \partial x_2} + ik\alpha_{03}\tilde{\Delta}\hat{u}'_{2k}, \quad k \neq 0, \qquad (6b)$$

$$\frac{\partial \hat{\omega}'_{3k}}{\partial x_1} = ik\alpha_{03}\hat{\omega}'_{1k} - \tilde{\Delta}\hat{u}'_{2k}, \qquad \forall k, \qquad (6c)$$

are used for computing the vorticity components ω_1' and ω_3' at the wall. A damping zone, where the disturbance vorticity is artificially suppressed by multiplication with a damping function, precedes the outflow boundary as shown in Fig. 2. The amplitudes of the perturbations convecting through the outflow boundary are negligibly small and thus no reflection is observed. The actual boundary conditions are of minor relevance as far as they do not contrast with the aimed decay of the perturbation amplitudes. The total chordwise extension of the region, where the solution is unphysically influenced by the damping, amounts typically to about $30\delta_1$.

Sixth-order compact differences are used for the chordwise (x_1) and wall-normal (x_2) directions [6]. While the grid step Δx_1 remains constant over the whole integration domain, Δx_2 is varied blockwise doubling it progressively, when getting away from the wall.

A 4-step 4th-order Runge-Kutta scheme is applied for the time integration. The computation of each new partial time step starts with the explicit computation of $\underline{\omega}'$ in all grid points except on the plate surface (equations (4a)). Thereafter the new vorticity distribution, with the exception of the harmonic component $k = 0$ of ω_2', is forced to zero within the outflow zone. The wall-normal velocity u_2' is then computed solving (4c) and the vorticity components ω_1' and ω_3' are evaluated at the wall by integrating equations (6a-c) and superimposing the damping function. As a final step, u_1' and u_3' are computed solving equations (4b), (4d) and (5).

3   Computational aspects

Owing to the Fourier expansion in spanwise direction the numerical procedure reduces the 3-D problem in physical space to a set of ($K+1$) complex 2-D problems for the single Fourier components. Coupling of the different modes is limited to the computation of the non-linear convective terms of the vorticity transport equations (4a-c), which are evaluated by means of a pseudospectral procedure applying an aliasing-free FFT. This allows an optimal moderate parallelization of the solving algorithm and fits perfectly to the characteristics of the supercomputers of the hww GmbH, Stuttgart: the NEC SX-4/32 (32 processors, 8 GB RAM) and the NEC SX-5 (16 processors, 32 GB RAM).

The wall-normal discretization, especially in the near-wall region, is crucial with respect to the resolution requirements. For the highest resolved simulation the grid spacing in the wall region was $\Delta x_1 = 0.130 \cdot 10^{-2}$ and $\Delta x_2 = 0.115 \cdot 10^{-3}$ (in wall coordinates for the turbulent regime $\Delta x_2^+ \leq 0.7$). For the spanwise Fourier-expansion 120 complex modes were used. The maximal time step was, in the end, enforced by the viscous time-step limit connected to Δx_2 and the number of time steps per fundamental disturbance period T ($\beta_0 = 2\pi/T = 6$) was 5000.

The computational costs and the memory requirements could be made affordable only by strongly coarsening the wall normal discretization in the outer region of the boundary layer (blockwise doubling of Δx_2) and by reducing the downstream extension of the integration domain (shifting the inflow boundary downstream and imposing unsteady inflow boundary conditions obtained from a previous run).

The serial code reaches about 1.8 GFLOPS (of 4 GFLOPS theoretical peak performance on the hwwsx5) at a vector operation ratio of 98% and an average vector length of 190. This corresponds to a computation time of 2.1 μs per grid point and time step and a memory requirement of about 156 Bytes per point. The Parallel version for the most expensive simulation of the turbulent region ran on 15 processors, reached 20 GFLOPS and required 30 GBytes working memory. The average speed up was about 0.75 per additional processor.

4 Numerical results

4.1 Baseflow

The baseflow considered in the scope of the present work reproduces the 3-D accelerating flat-plate boundary layer of the "Querströmungsprinzipexperiment" of the DLR-Göttingen [1] for a dimensional velocity at infinity equal to $Q_\infty^* = 19m/s$. The boundary layer edge velocity is:

$$u_{1be}(x_1) = 0.411(x_1 + 0.349)^{\frac{1}{2}}, \qquad u_{3be} = 0.916. = const. \qquad (7)$$

The most relevant parameters of the baseflow are plotted in Fig. 3. The chordwise velocity component increases by more than a factor of two within

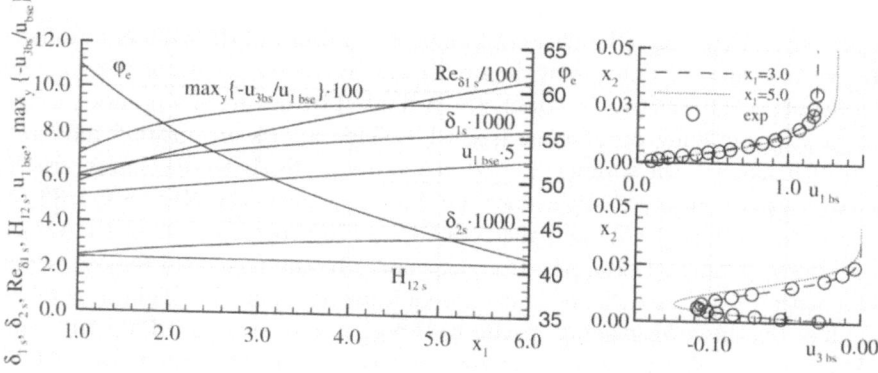

Fig. 3. Parameters of the baseflow in the streamline-oriented coordinate system (local sweep angle φ_e, displacement thickness δ_{1s}, momentum thickness δ_{2s}, local Reynolds number $Re_{\delta 1s}$, shape factor H_{12s}, maximum local crossflow velocity $\max_y\{-u_{3bes}\}$) and velocity profiles at different chord positions for DNS and experiment.

the integration domain. Correspondingly also the local Reynolds number $Re_{\delta 1\,s}$ grows, while the increase of the boundary layer thickness is limited because of the favourable pressure gradient. The local sweep angle decreases from over 60 deg. at the inflow to 40 deg. at the outflow. The maximal cross-flow velocity increases monotonously, initially faster than the downstream component (for $x_1 \leq 4.0$), further downstream in a direct proportional way. However, the ratio between both quantities is never larger than 10%. Velocity profiles at different x_1-positions are also shown in Fig. 3. Profiles for $u_{3\,b}$ show everywhere the classical crossflow shape. Experimental data for $x_1 = 3.0$ [11] are plotted for comparison.

4.2 Spatial development of the perturbation spectral components and general discussion of the transition mechanisms

Transition in the "DLR-Querströmungsprinzipexperiment" is characterized by the presence of primarily unstable stationary and unsteady modes with spanwise wavelengths approximately equal to $\Lambda_3^* = 12.0mm$. Indeed according to the LST wavelike perturbations with similar spanwise periodicity and with frequencies in the range $0.0Hz \leq f^* \leq 250.0Hz$ are strongly amplified over the whole extension of the integration domain.

The perturbation for the DNS was correspondingly defined. The fundamental wavenumber of the spanwise Fourier expansion was set equal to $\alpha_{0\,3} = 52.4 \approx 2\pi/\Lambda_3$ so that all relevant modes could be properly represented by the the the first harmonics of the expansion. The frequency $f^* = 133Hz$ ($\beta_0 = 6.0$) was chosen for the unsteady disturbance (hereafter T indicates the period of the perturbation: $T = 1/f$). Indicating by the index pair (h, k) the mode with frequency $\beta = h \cdot \beta_0$ and spanwise wavenumber $\alpha_3 = k \cdot \alpha_{0\,3}$, the steady mode $(0, 1)$ and the unsteady one $(1, 1)$ were perturbed at $x_1 = 1.56$ (midpoint of the perturbation stripe). The perturbation amplitudes were $A_{(0,1)} = 1.7 \cdot 10^{-3}$ and $A_{(1,1)} = 1.5 \cdot 10^{-4}$ respectively.

Figure 4 documents the downstream amplitude development for the perturbed modes. Bonfigli and Kloker [2] showed how the transition scenario may vary strongly depending on the relative amplitudes of the primary modes before and after the onset of saturation. In the present simulation (hereafter called case $(0, 1) + (1, 1)$, corresponding to case R5 in [2]) the CF vortex $(0, 1)$ is perturbed with distinctly larger amplitudes than the CF wave $(1, 1)$ and dominates the whole transitional region. At $x_1 = 2.6$ the amplitude of mode $(0, 1)$ is nearly 10% of the local free-stream velocity u_{bse}, non-linear effects become evident and the amplification of both primary modes is dampened. Simulations driven with slightly varied perturbation amplitudes showed that in vortex-dominated scenarios the amplification of the CF wave $(1, 1)$ always collapses when the mentioned threshold amplitude is reached by the CF vortex $(0, 1)$. The influence of mode $(1, 1)$ itself on the process, if any, is marginal.

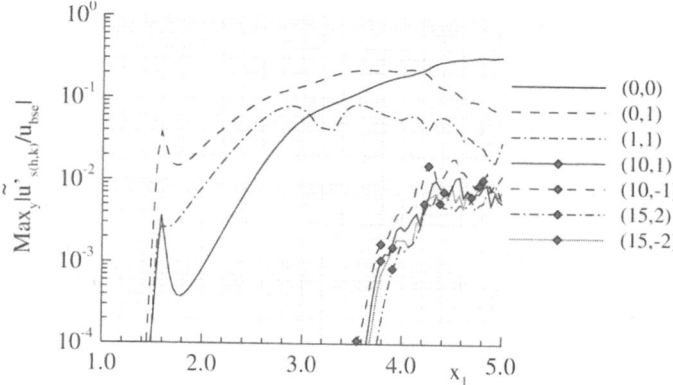

Fig. 4. Downstream amplitude development (max. over x_2) of the primarily perturbed modes and of some high-frequency modes.

More relevant is, on the contrary, the role of $(1, 1)$ in the saturation region ($x_1 \geq 2.6$ in Fig. 4), where non-linear interactions between the primary modes are responsible for the filling up of the perturbation spectrum and finally for breakdown. As a general rule it turned out that the larger the unsteady mode is, the faster transition may proceed. However qualitative differences are also possible and two different driving mechanisms can be outlined. In cases where the growth of mode $(1, 1)$ is stopped at low amplitudes, vortical structures associated with mode $(0, 1)$ dominate the flow development and the final breakdown is triggered by classical SHFIs of the flow deformed by mode $(0, 1)$. The role of mode $(1, 1)$ and of its superharmonics is limited to providing the necessary unsteady background perturbation. When, on the contrary, the amplitude of mode $(1, 1)$ is comparable to that of mode $(0, 1)$, strongly non-linear interactions between the primary modes lead the transition process. Complex vortical structures are initially generated by superposition of the flow fields associated with both primary instabilities (quasi-linear superposition) and develop then downstream by mutual induction (non-linear interaction). Its strength and complexity increases progressively until they break down and the turbulent regime is reached. Classical SHFIs are possible also in this case and in general always determine the growth of high-frequency spectral components in the most upstream part of the transitional region. The primarily deformed flow is then the result of the discussed interactions between the low-frequency primary disturbances and keeps strongly time-dependent in every inertial reference system.

The separation between the discussed transition mechanisms is clearly not sharp and the importance of SHFIs or primarily generated interaction vortical structures (IVS) varies continuously as a function of the ratio of the amplitudes of CF wave and CF vortex. In the numerical simulation, as well as in the experimental configuration of case "$(0, 1) + (1, 1)$" by Lerche [9], mode

$(1,1)$ reaches a considerable amplitude level ($7\% \cdot u_{bse}$) before its growth is suppressed by the dominating mode $(0,1)$. Both SHFIs and IVSs may develop within the saturation region as documented in the following sections.

4.3   Primarily generated interaction vortical structures

It has been said that the quasi-linear superposition of the disturbance fields associated with the primary modes $(0,1)$ and $(1,1)$ is the first step toward the generation of IVSs. Simulations have been driven where modes $(0,1)$ or $(1,1)$ were perturbed alone (respectively case $(0,1)$ and case $(1,1)$) and therefore could develop independently from any mutual influence. The resulting flow fields and the one obtained by superposing them linearly were then compared with case $(0,1) + (1,1)$.

Vortical structures in the region of incipient saturation ($2.13 \leq x_1 \leq 3.40$) are visualized in Fig. 5 by means of the λ_2-isosurfaces [5]. Here and in the following the reference axis ξ_1 (see Fig. 2) is nearly aligned with the axis of the stationary vortices and forms the angle $\varphi_{x_1 \xi_1} = 45$ deg. with the x_1-axis. The origin of the rotated system (ξ_1, ξ_3) lies in $x_{1\xi 0} = 2.13$, $x_{3\xi 0} = 0.07$. In both reference cases $(0,1)$ and $(1,1)$, elongated vortical structures lying parallel to the wave fronts of the perturbed modes extend over the whole domain (Figs. 5a and 5b). In case $(0,1) + (1,1)$ (Fig. 5c) the stationary perturbation dominates and the orientation of the vortex pattern is similar to the one observed in case $(0,1)$. However the λ_2 isosurfaces are discontinuous as a consequence of the modulation induced on the stationary vortex by the CF wave. The similarity in the region $\xi_1 \leq 0.8$ (i. e. $x_1 \leq 2.7$) with the visualization obtained for the linear superposition of the perturbance fields from cases $(0,1)$ and $(1,1)$ (Fig. 5d) confirms that the origin of the fragmentation of the main vortex is mainly due to linear effects.

The formation of local vorticity concentrations (shear layers or vortices) is also an effect of the linear superposition of CF wave and CF vortex and further downstream becomes determinant for the non-linear development of the flow. As Figs. 6a and 6b show for $\omega'_{\xi 1}$ (i. e. for the component of the perturbance vorticity parallel to the ξ_1-axis and nearly parallel to the axis of the most relevant vortical structures), both primary modes tend to generate stripe patterns in the vorticity distribution over every wall-parallel plane cutting the cores of the corresponding main vortices. Regions of negative $\omega'_{\xi 1}$ corresponding to the neighbouring CF main vortices are separated by regions of positive shear $\omega'_{\xi 1}$, which in case $(0,1)$ develop into couterrotating secondary vortices (Fig. 5a, $\xi_1 \geq 1.0$). When $(0,1)$ and $(1,1)$ are superimposed (Fig. 6c for $\xi_1 \leq 0.8$ and Fig. 6d) a checkered unsteady pattern emerges, where modulations are evident both in the direction normal and parallel to the axis of the dominating vortices $(0,1)$. Particularly relevant is the concentration of positive $\omega'_{\xi 1}$ marked as $M1$ in Fig. 6c, originating from the modulation induced by mode $(1,1)$ onto the counterrotating vortex generated by the main CF vortex $(0,1)$. The maxima $M2$ and $M3$ correspond

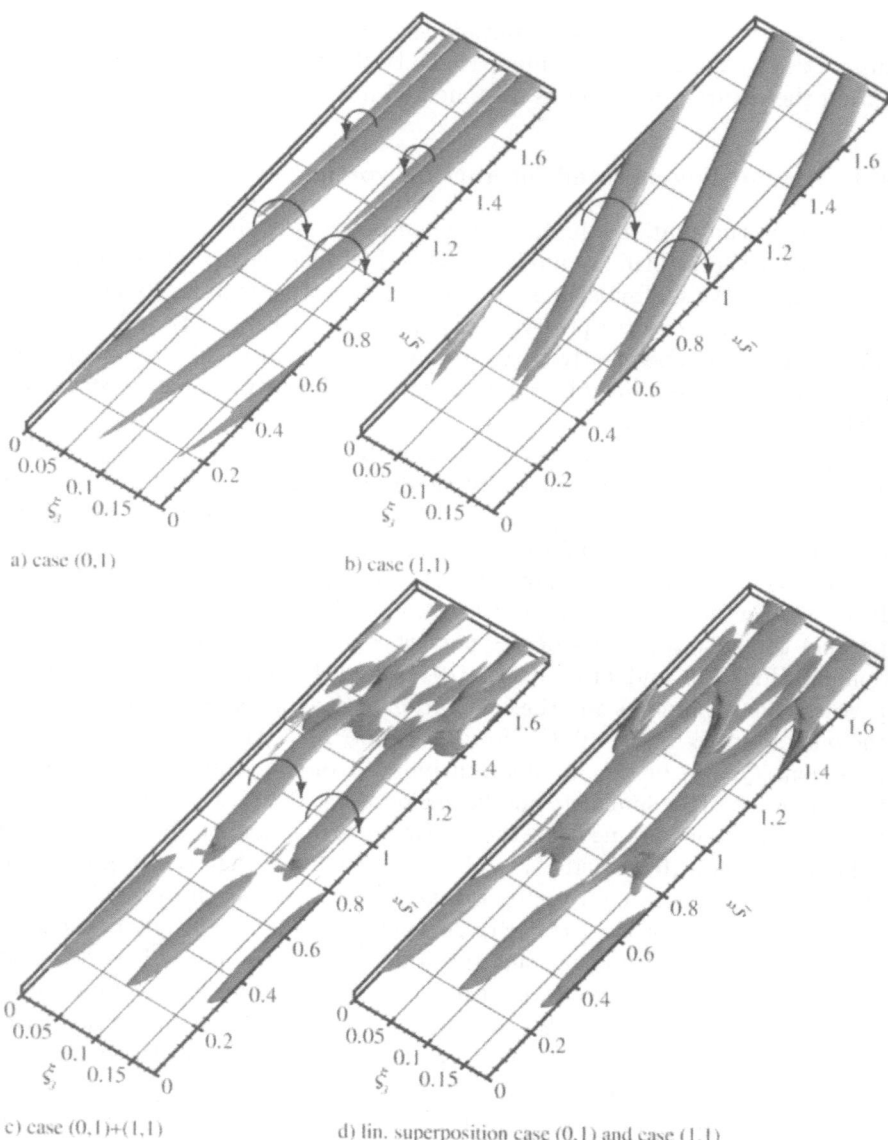

a) case (0,1)

b) case (1,1)

c) case (0,1)+(1,1)

d) lin. superposition case (0,1) and case (1,1)

Fig. 5. Visualization of vortical structures by means of λ_2-isosurfaces ($\lambda_2 = -4$); b)-d) represents snapshots.

to $M1$ after respectively one and two periods of the CF wave $(1,1)$ and are already strongly deformed by non-linear effects.

Concentrating on the simulation case $(0,1) + (1,1)$, Fig. 7 highlights the mechanisms leading to the development of a typical primarily generated in-

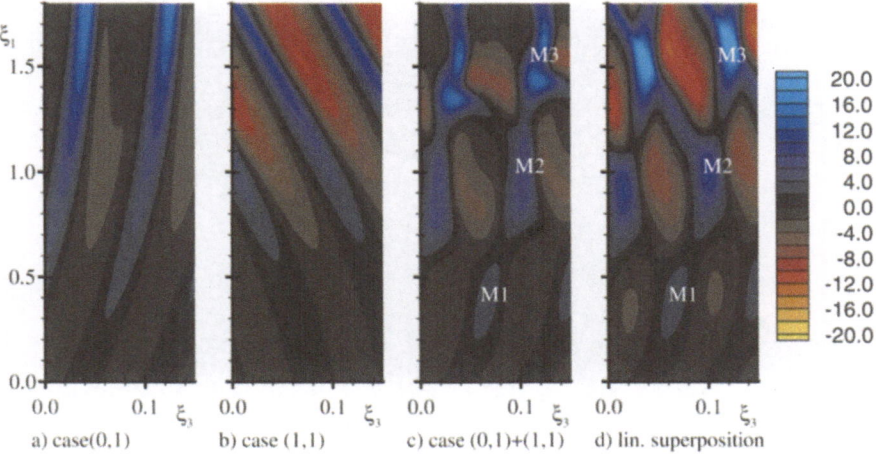

Fig. 6. Component $\omega_{\xi 1}$ of the perturbation vorticity in the plane $x_2 = 0.85 \cdot 10^{-2} \approx$ $0.25 \cdot \delta$ for the different simulation cases and for the linear superposition of the perturbation fields from cases $(0,1)$ and $(1,1)$ (Fig. d). The primary main vortices rotate clockwise when looking downstream and have negative ω_{ξ_1} values.

teraction vortical structure (IVS). To the maximum $M1$ of $\omega'_{\xi 1}$ in Fig. 6c there corresponds a local upheaval of the region of positive $\omega_{\xi 1}$ (Fig. 7a). This convects downstream and, as a consequence of the amplification of the primary modes, rises progressively into the boundary layer (Fig. 7b) forming an "overturning" wave shape. Up to a certain stage ($\xi_1 \leq 1.5$) the process is mainly governed by the linear superposition of modes $(0,1)$ and $(1,1)$, but when the region of positive vorticity reaches far enough into the boundary layer, the influence of the velocity field induced by the primary CF vortex becomes relevant. The vorticity concentration is then lifted first upwards and then in the direction opposite to the basic crossflow describing a nearly circular trajectory around the core of the CF vortex (Figs. 7c and 7d). At $\xi_1 \approx 2.1$ ($x_1 \approx 3, 6$) the overturning tip of the convected positive vorticity detaches from the originating concentration and forms a proper, spatially well-defined IVS (Fig. 7c), which clearly is a vortex. While undergoing deformation, also the intensity of the vorticity concentration increases (Figs. 7c and 7d).

The development of IVSs has been described on the basis of two-dimensional visualizations but is clearly a three-dimensional process. Figure 8 shows a close-up of the resulting IVS at two different stages of its development. The lifted head of the structures corresponds to the detached vorticity concentration, which can be seen in the plane of Fig. 7d. As a consequence of the induction connected with the velocity field of the primary vortex, the descending part of the structure steepens progressively producing strong gradients in the wall-parallel directions for all flow quantities. This increases the

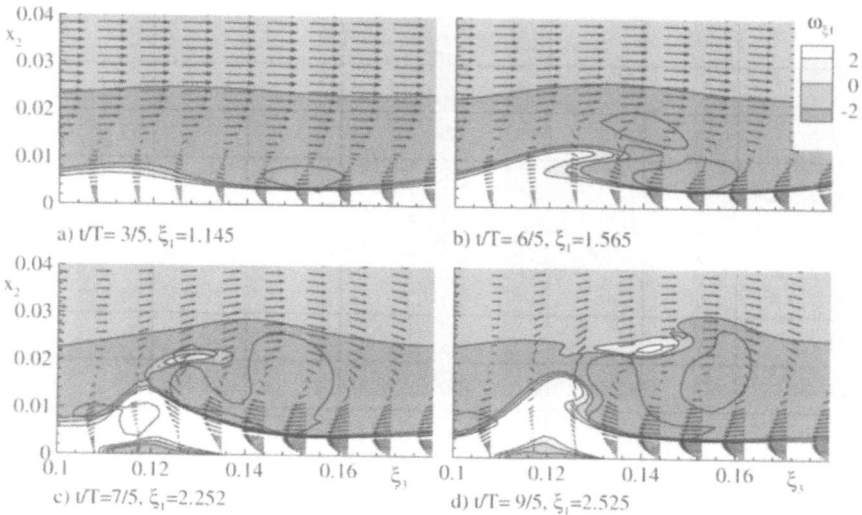

Fig. 7. Section planes across the region of positive vorticity at different times and downstream positions (the (ξ_2, ξ_3)-plane is nearly orthogonal to the axis of the primary CF vortex): vorticity component in the direction of the ξ_1-axis (shaded), projection of the velocity vector onto the section plane (arrows) and section of the λ_2-isosurfaces (thick lines, $\lambda_2 = -10$).

complexity of the flow and contributes to filling-up the perturbance spectrum by generating a spike-like velocity signal at the passage of the vortical structure (IVS-induced high-frequency disturbances, IVS-HFI). Eventually smaller structures develop close to the trunk of the IVS, which breaks down and initiates a turbulent spot.

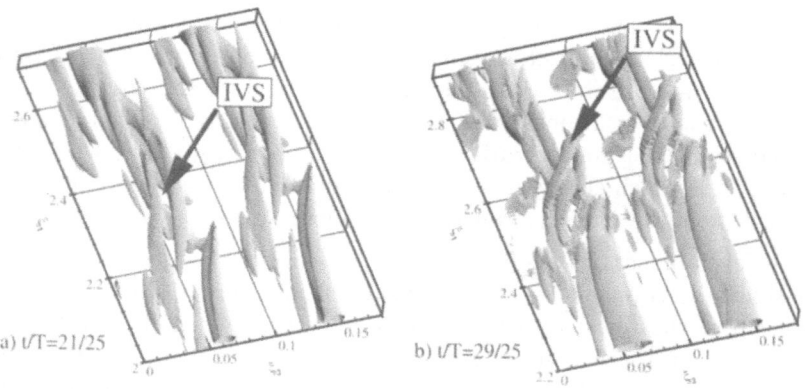

Fig. 8. λ_2–visualization of the IVS at two different time instances. The IVS rotates anti-clockwise, the main-vortex parts clockwise.

Upon employing SLST we find that the instantaneous flow field in the IVS region is secondarily unstable and that the late-stage IVS resembles a secondary finger vortex induced by low-frequency ($h \approx 5$) instability. Since higher frequency modes are generally more unstable and even low frequencies are more amplified in different regions, the observed individual structure is clearly not a "typical" finger vortex, in which case one would expect a streamwise row of similar structures, but rather an amplified IVS. We note that the IVS is not connected to the climbing structure visualized by Lerche, who extracted isosurfaces of the spanwise vorticity (Fig. 76 of [9]). Indeeed, the DNS showed that this kind of visualization smoothes out most of the delicacies of the actual flow field, including the IVS.

4.4 Classical secondary instability

The growth of SHFIs in the sense of the classical SLST [7] was verified in case $(0, 1) + (1, 1)$ by comparing, for a given frequency, the distribution over (x_2, x_3)-planes of the amplitudes of the temporal Fourier transform of the downstream velocity from the DNS with the eigenfunctions of the amplified modes resulting from the SLST.

Since the primarily deformed flow is not steady in any inertial reference system, the SHFIs develop in the form of short wave trains, well localized both in time and space. According to this, the Fourier transform of the data from the DNS were limited to short time intervals centered over the wave train to be investigated. At the boundaries of the considered interval the signal was smoothly brought to zero. The SLST-analysis was carried out on the basis of the instantaneous velocity profiles given from the DNS for the central point of the temporal window. The numerical code for the SLST was developed at IAG by Messing [10].

Figure 9 shows, for one particular temporal and spatial location ($x_1 = 3.47$ and $t/T = 14/25$), the primarily deformed profiles of the downstream velocity component, the distribution of the Fourier amplitudes from the DNS for the circular frequency $\beta = 120$ and the best fitting eigenfunction from the SLST. The amplification rates from DNS and SLST are respectively

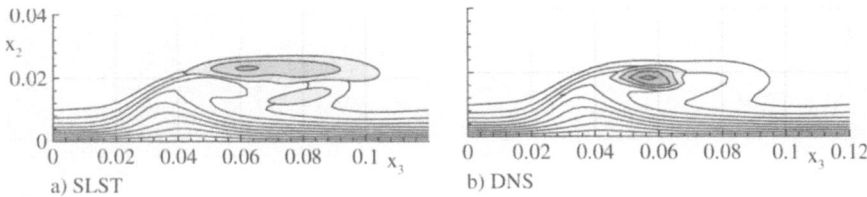

Fig. 9. Comparison of the SHFI shape from SLST and DNS: primarily deformed profiles ($u_{1\,s}/u_{1\,bse}$, $\Delta u_1 = 0.1$, lines) and normalized amplitudes of the SHFIs (shaded).

$\alpha_{i\ DNS} \approx 6.0$ and $\alpha_{i\ SLST} \approx 4.0$. The phase velocities are $c_{\xi_1\ DNS} = 0.82$ and $c_{\xi_1\ SLST} = 1.0$. Considering the differences between the temporal theoretical model and the spatially inhomogeneous and unsteady character of the primary perturbation in the DNS, the agreement is more than satisfactory.

The amplification of the observed secondary mode is clearly connected to the wall-normal gradient appearing in the instantaneous streamwise velocity profiles. This, on its turn, follows from the spanwise extended low velocity region ($x_2 \leq 0.25$, $0.03 \leq \xi_3 \leq 0.1$), which characterizes the profiles in the region between two consecutive sections of the discontinuous primary vortex (see Fig. 6c, $\xi_1 \approx 0.7$ and $\xi_1 \approx 1.4$). Such connection between the present SHFI and the wall-normal gradient in the velocity profile is given over the whole time interval during which the instability is amplified. Indeed the fact that the wavetrains of the SHFIs move downstream with the same velocity as the primary vortical structures and consequently always see similarly deformed velocity profiles is a precondition for the applicability of the SLST.

4.5   The turbulent 3-D boundary layer

At the end of the integration domain ($x_1 \approx 5.6$) the turbulent flow was well developed (maximal deviation of the time-averaged profiles from the spanwise mean lower than 10% of the edge velocity) and the local Reynolds number based on the momentum thickness reached the maximal value $Re_{\delta 2} \approx 950$.

Spanwise averaged mean profiles in wall coordinates for the velocity component parallel to the direction of the edge velocity are shown in Fig. 10 for different x_1-positions. DNS Results by Spalart [13] for a 2-D adverse pressure gradient boundary layer and LES results by Huai et al. [3] (case H-TF) for the boundary layer considered in the present work are also given for comparison. The good agreement between the DNS-profiles and the logarithmic law confirms that a well-developed turbulent state has been reached. The local

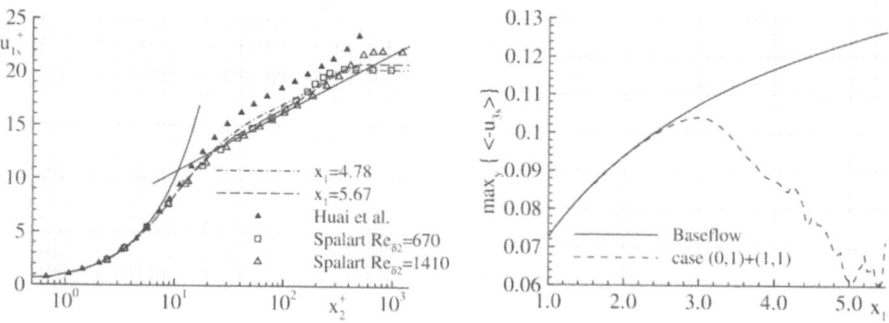

Fig. 10. Profiles of the downstream velocity component in the turbulent region of the integration domain (average in time and spanwise direction) and maximal crossflow component (maximum over x_2, average in time and spanwise direction) as a function of the downstream coordinate.

Reynolds number at $x_1 = 5.66$ ($Re_{\delta 2} \approx 950$) is intermediate between the values considered by Spalart [13], and intermediate is also the shape of the profile. The downstream component of the velocity does not seem to be significantly affected by the three-dimensionality of the boundary layer, which indeed is greatly reduced in the turbulent region as shown in Fig. 10. The poor agreement with Huai et al. is probably to be attributed to the coarse discretization used by the latter and to uncertainties in their LES-modeling.

The development of vortical structures in the turbulent region is shown in Fig. 11 by means of λ_2-isosurfaces. Some typical structures can be recognized, which are well known in the literature for characterizing turbulent boundary layers: elongated near-wall vortices (streaks) lying aligned with the flow direction and horse-shoe vortices moving away from the plate surface. Remindings of the spanwise periodicity connected with the primary mode $(0,1)$ are still present as shown by the alternation of regions with lower and higher streaks concentration. The distance between neighbouring streaks is nearly $\Delta_{3s}^+ \approx 100$, at least in the regions with higher structure concentration. Streaks rotating in both directions are observed. However, the ones presenting anticlockwise rotation tend to be better developed and to reach greater extensions in downstream direction. This might be a consequence of the asymmetry of the flow, which, even if reduced, is still present (see Fig. 10). Some climbing structures are visible which, on a smaller scale, vaguely

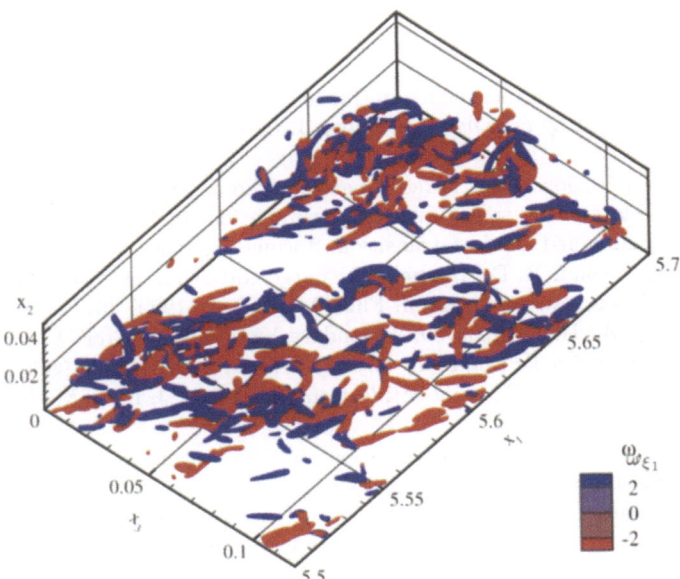

Fig. 11. Vortical structures in the turbulent region ($\lambda_2 = -3000$). The color gives the sign of the vorticity component ω_{ξ_1} (nearly parallel to the free-stream velocity) on the λ_2-isosurfaces. Notice the plate fixed coordinate system.

recall the IVS observed previously in the transitional region (see Fig. 8). In a similar way as discussed in that context, indeed, the mutual induction between neighbouring vortices (here streaks) is expected to play an important role also in self sustaining turbulence and might be connected to the generation of the well known ejection phenomena. Mutual interactions may be observed between streaks of any rotation.

5   Conclusions

The transitional and the turbulent regimes of the 3-D boundary layer of the "DLR-Prinzipexperiment" have been investigated by means of DNS. A shear-layer/vortex interaction mechanism, which dominates the final laminar breakdown in the presence of multiple saturated primary modes, has been highlighted. Unlike in the LES results of Huai et al. [3], the velocity profiles in the turbulent region have been shown to fit the law of the wall and the log law. The development of wall streaks however appears to be influenced by the asymmetry of the flow. All simulations have been run on NEC SX machines, whose vectorial architecture, with moderate parallelization and shared memory, perfectly fits the requirements of the numerical procedure. The greater memory capacity of the NEC SX5 was necessary for the most resolved computations.

References

1. Bippes, H.: Basic experiments on transition in three-dimensional boundary layers dominated by crossflow instability. Progress in aerospace sciences **35** 363–412.
2. Bonfigli, G.; Kloker, M.: Three-dimensional boundary-layer transition phenomena investigated by spatial numerical simulation. In Fasel, H. F.; Saric, W. S. (Eds.): Laminar-turbulent transition. Springer Verlag (2000) 619–624.
3. Huai, X.; Joslin, R. D.; Piomelli U.: Large-eddy simulation of boundary-layer transition on a swept wedge. J. Fluid Mech. **381** (1999) 357–380.
4. Högberg, M.; Henningson, D.: Secondary Instability of Cross-Flow Vortices in Falkner-Skan-Cooke Boundary Layers. J. Fluid Mech. **368** (1998) 339–357.
5. Jeong, J.; Hussain, F.: On identification of a vortex. J. Fluid Mech. **285** (1995) 69–94.
6. Kloker, M.: A robust high-resolution split-type compact FD-scheme for spatial DNS of boundary-layer transition. Applied scientific research **59** 353–377, special issue: LES and DNS of complex flows, Kluwer acad. publishers (1998).
7. Koch, W; Bertolotti, F. P.; Stolte, A.; Hein, S.: Non linear equilibrium solutions in a three-dimensional boundary layer and their secondary instability. J. Fluid Mech. **406** (2000) 131–174.
8. Kohama, Y.; Saric, W. S.; Hoos, J. A.: A high-frequency secondary instability of crossflow vortices that lead to transition. Proc. Roy. Aeron. Soc., Boundary layer transition and control, Cambridge, UK, 8-12 April 1991.

9. Lerche, T.: Experimentelle Untersuchung nichtlinearer Strukturbildung im Transitionsprozess einer instabilen dreidimensionalen Grenzschicht. Dissertation, Universität Göttingen (1997).

10. Messing, R.: private communication.

11. Müller, B.: Experimentelle Untersuchung der Querströmungsinstabilität im linearen und nichtlinearen Bereich des Transitionsgebietes. Dissertation, Universität Göttingen (1990).

12. Schlichting, H.: Grenzschicht-Theorie. Verlag G. Braun, 8. Auflage (1982).

13. Spalart, P. R.: Direct simulation of a turbulent boundary layer up to $Re_\theta = 1410$. J. Fluid Mech. **187** (1986) 61–98.

14. Wassermann, P.; Kloker, M.: Mechanisms and control of crossflow-vortex induced transition in a three-dimensional boundary layer. J. Fluid Mech. **456** (2002) 49–84.

15. Wassermann, P.; Kloker, M.: Transition mechanisms induced by traveling crossflow vortices in a 3-D boundary layer. Submitted to J. Fluid Mech., April 2002.

16. Wintergerste, T.; Kleiser, L.; Breakdown of a crossflow vortex in a three-dimensional boundary layer. In Chollet, J. P. et al. (Eds.): Direct and large-eddy simulation II. Kluwer academic press (1997) 179–190.

Large-Eddy-Simulations of Turbulent Wall Bounded Flow with and without Adverse Pressure Gradient

S. Eisenbach, M. Manhart, and R. Friedrich

TU-München
Fachgebiet Strömungsmechanik
Boltzmannstraße. 15
D-85748 Garching

Abstract. Studies on Large-Eddy-Simulations (LES) are presented of turbulent wall bounded flow at $Re_\tau = 590$ and $Re_\Theta = 670$ either with and without adverse pressure gradient (AGP). The Simulations of the Kalter and Fernholz Case [6] show the shortcomings of eddy viscosity subgrid scale models to predict pressure induced separation and reattachement.

1 Introduction

Separation of boundary layer flow is a common phenomenon in many fluid mechanical applications. It often inhibits the further improvement of devices like blades of turbo machines or aircraft wings in high lift configuration as it limits the exchange of momentum between the devices' surface and the flow. Additionally. it contributes to drag and noise. Unfortunately, it is exactly the appearance of flow separation which is one of the major challenges to numerical simulations, as its underlying physics are hard to model and direct numerical simulation (DNS) without a turbulence model is not affordable in high Reynolds number cases due to resolution requirements.

A promising approach for the numerical simulation of turbulent flow is large eddy simulation (LES). It is based on the idea that only large structures, containing most of the energy of the flow, need to be resolved by the chosen numerical grid, while the contribution of the smaller scales can be taken into account by a so called subgrid-scale (SGS) model.

This study aims at evaluating different SGS-models in the case of wall bounded turbulent flow with and without separation. Simulations are carried out for a channel flow and a zero pressure gradient boundary layer with different subgrid scale models. Then simulations of a boundary layer separating due to an adverse pressure gradient according to the DNS of Manhart and Friedrich [8] are done. These simulations are compared to an experiment of Kalter and Fernholz[6] for a higher Reynolds number.

Kalter and Fernholz studied the flow in an annular channel, where the separation of the boundary layer on the inner surface is caused by specially designed openings in the outer wall. Thus the separation develops slowly due to the induced adverse pressure gradient. This experiment differs from other cases of separating flow where the separation is caused by a local change of the geometry like the flow over a backward facing step or over an obstacle.

2   Numerical method

2.1   Basic computational code

The simulation is calculated using the code MGLET (Werner and Wengle [14], Manhart [7] and Manhart, Tremblay and Friedrich [9]). MGLET is based on a finite volume formulation of the Navier-Stokes equations for incompressible flows on a staggered cartesian, non-equidistant grid. For the spatial discretization a second order central scheme is implemented whereas for the time-discretization an explicit second-order scheme (leapfrog with time-lagged diffusion term) is used. In order to account for pressure gradients, a Poisson equation has to be solved. This is accomplished by a multigrid solver using a velocity-pressure iteration scheme (Hirt et al. [5]) as smoother.

2.2   Programming model

MGLET is designed to perform on a variety of different high performance computers (HPV) like vector, vector parallel or massively parallel computers. To ensure the portability of the code, it is written in Fortran and uses the Message Passing Interface (MPI) for parallelization as both are a common standard on HPC platforms. On the Hitachi SR–8000 of the HLRS the MPI-MPP programming model has been applied where one MPI process is started on every processor. The Compas shared memory paralellization, provided by Hitachi for the SR–8000 is not used. The parallelization is based on a domain decomposition method on a block structured grid. For the present simulation parallelization was applied in streamwise direction with sixteen blocks on the same number of processors distributed over two nodes. We use double precison data for MPI communication because this showed a performance gain on the SR–8000 of the LRZ in Munich. This resulted in a performance of 210 MFlops for the production runs.

The speedup, up to the number of processors used for the simulation, is shown in Fig. 1. A loss of performance is visible between eight and sixteen processors. This is due to the inter–node communication needed for runs with more than eight processors. The basic code without statistics has a memory requirement between 12 and 30 words per grid point, depending on the chosen domain decomposition and the SGS-model.

2.3   Boundary conditions

Channel flow For the computation of channel flow, periodic boundary conditions are used in spanwise and streamwise direction for both pressure p and

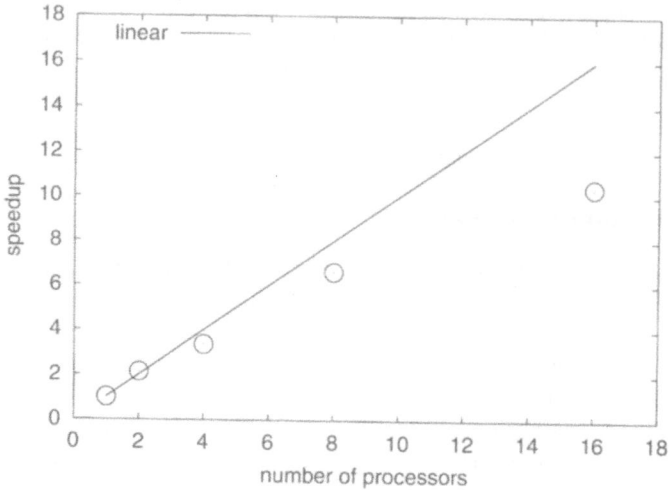

Fig. 1. Speedup of the MPI-MPP parallel code.

velocity components u_i in three directions. No slip and impermeability conditions apply for the velocities on the walls i.e. the upper and lower boundary of the channel while Neumann conditions are used for the pressure.

Boundary layer flow For boundary layer flow, periodic conditions can only be used in spanwise direction. At the inflow plane, time dependent, three dimensional boundary conditions are needed for the velocities. They are computed by superimposing fluctuations, taken from a position downstream of the inflow plane, onto a mean velocity profile that matches the desired inflow condition. For the presented simulations we use profiles from a direct numerical simulation of Spalart [13]. As the boundary layer thickness grows spatially, fluctuations outside the initial boundary layer thickness are damped exponentially. On the wall no-slip and impermeability conditions are applied whereas in the outflow plane zero-gradient conditions are used for the velocities. In the outer plane, parallel to the wall, the pressure is specified (Dirichlet condition) along with zero vertical gradients of the two tangential velocity components. In the spanwise direction, the flow is periodic. With these conditions the boundary conditions for the pressure Poisson equation are also fixed.

3   Subgrid scale models

An LES formulation can be derived by low-pass filtering the Navier-Stokes-Equations for incompressible fluids with top-hat filters of width proportional to the mesh size. In doing so, we obtain the tensor of the non resolved subgrid scale stresses which has to be modeled:

$$\tau_{ij}^{sgs} = -\rho(\overline{u_i u_j} - \overline{u_i}\,\overline{u_j}) \tag{1}$$

where ρ denotes the constant density of the fluid and the overbar the filtering procedure. The following SGS models are considered in this study:

- Smagorinsky model [12] with Van-Driest damping
- Dynamic formulation of the Smagorinsky model as proposed by Germano et al. [4]
- The recently developed Increment model of Brun and Friedrich [2]

The Smagorinsky model, whilst first published 1963, is still the most commonly used SGS model. It is based on the assumption, that the effects of small scale turbulence contribute to the filtered flow via an augmentation of viscosity and is thus called an eddy-viscosity model. It replaces the viscosity μ in the Navier-Stokes-Equations by an effective viscosity $\mu_{eff} = \mu + \mu_t$ where μ_t is called the turbulent viscosity. μ_t is calculated from

$$\mu_t = C_S^2 \rho \Delta^2 |\overline{S}| \tag{2}$$

where C_s is a model constant, Δ the filter length scale, i.e. the grid spacing, and S the strain rate. The value of the constant C_s must be chosen depending on the flow type. Additionally, since the contribution of the model does not tend against zero close to the wall as it should, Van-Driest damping of the model has been used.

A more general formulation of the Smagorinsky model can be found by determining the constant C_s dynamically from the flow. Germano et al. [4] derived an approach, where the constant is calculated by applying a test filter to the calculated flow field. Most implementations of this approach use clipping and/or averaging in homogeneous directions of the constant in order to augment the stability of the calculation. This formulation does automatically adjust to different kinds of flow and leads to reasonable results in wall bounded channel flow without the need of an explicit wall treatment [4].

The increment (INC) model is, in distinction to the Smagorinsky model, a scale similarity model. The name scale similarity model is derived from the basic assumption of such models, that the largest non–resolved scales are similar to the smallest resolved scales. The INC-model is based on the ansatz

$$\tau_{ij}^{sgs} = f\, \delta u_i \delta u_j \tag{3}$$

with the velocity increment δu_i defined as

$$\delta u_i = \sum_{k=1}^{3} [u_i(x_k + \Delta x_k) - u_i(x_k)] \tag{4}$$

with x_k the location and Δx_k the length of the gridcells in the kth direction.

The coefficient f is calculated dynamically at every time step from a test filter procedure with a top hat filter:

$$f(x,t) = \frac{1}{4} \frac{\overline{\overline{u_k u_k}} - \overline{\overline{u}}_k \overline{\overline{u}}_k}{\delta \overline{u}_l \delta \overline{u}_l} \tag{5}$$

Although this dynamic formulation does not show the need for clipping or averaging the constant, an exponential damping of the coefficient f for small values of $\delta \overline{u}_l \delta \overline{u}_l$, the trace of the velocity increments, has been implemented. The need for this treatment arises from the fact that equation (5) becomes undefined in the absence of turbulence as the numerator tends to very small values and the denominator tends against zero. For the present simulation an explicit filtering of the velocities has been applied to separate the contribution of the numerical scheme and the model.

One advantage of this model over eddy-viscosity models is the capability of reproducing energy transfer from smaller to larger scales, called back–scatter.

The model has proven to work well in turbulent pipe flow [3] and turbulent round jets [1]. As the implementation of the model into the code version used for the boundary layer flow is still in progress, results are shown only for channel flow.

4   Channel flow

For a first assessment of the properties of the different models, a turbulent channel flow at $Re_\tau = 590$ i.e. a Reynolds number comparable to the experiment of Kalter and Fernholz is conducted. The results of this configuration can be compared with a direct numerical simulation of Moser et al. [10].

4.1   Simulation Setup

The following parameters have been used to define the computational grid, with x, y, z denoting the streamwise, spanwise and the wall normal directions and δ_0 the boundary layer thickness at the inflow. The grid spacing is given in inner coordinates $\Delta^+ = \Delta u_\tau / \nu$ where u_τ is the shear velocity $u_\tau = \sqrt{\tau_w / \rho}$ and τ_w is the wall shear stress.

	x	y	z
domain size L/δ_0	6.4	3.2	2.0
number of gridpoints N	48	64	112
grid spacing Δ^+	78.5	29.5	1.77–30.1

4.2 Results

Mean streamwise velocity profiles in inner coordinates are shown in Fig. 2. All LES profiles are close to each other with the greatest differences to the DNS values in the buffer layer. The LES tends to predict higher velocities with the exception of the wake region were data are lower than the reference data. Results for the rms velocity fluctuations and the Reynolds shear stress $\langle u'w' \rangle$

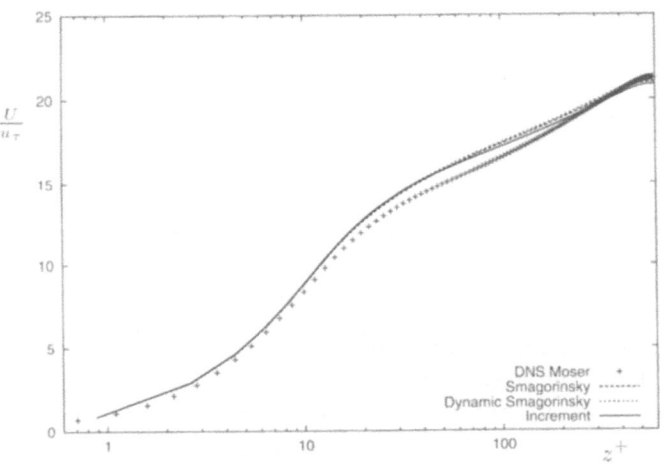

Fig. 2. Mean streamwise velocity profiles in inner coordinates.

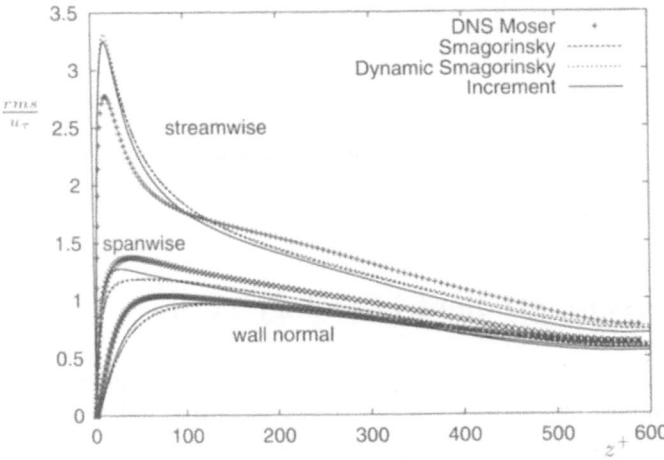

Fig. 3. Profiles of rms velocity fluctuations in inner coordinates.

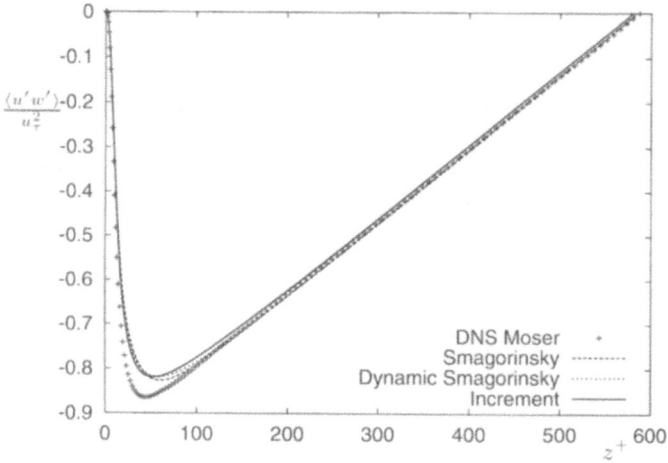

Fig. 4. Profiles of Reynolds shear stress $\langle u'w' \rangle$ in inner coordinates.

are presented in Figs. 3 and 4. Again all models give results in fair agreement with the DNS data. The streamwise rms velocity fluctuation is overpredicted in the viscous sublayer and the buffer layer while all other values remain lower than the DNS data. This is the usual behaviour seen in many computations in the literature. The INC model shows the best agreement with the DNS data up to the log–layer while both Smagorinsky models give slightly better values in the wake region and the outer flow. Results for $\langle u'w' \rangle$ are very close to each other and to the DNS. The INC model tends to show smaller values than the eddy viscosity models.

4.3 Conclusion

All models give reasonable results for the given grid resolution. Only small differences can be observed between the three models. None of the models is superior in all aspects.

5 Zero pressure gradient boundary layers

As the near wall resolution chosen for channel flow is not affordable for the separated boundary layer, tests are run to evaluate the different models on a coarser grid. Reference data for this test case are available from a DNS of Spalart et al. [13]. Greather differences among data obtained with the models and between LES and DNS data can be expected as the contribution of the model rises with decreasing grid resolution.

5.1 Simulation Setup

The parameters of this simulation are chosen to meet the planned resolution of the separated boundary layer. Thus the computational grid described by the following table is used.

	x	y	z
domain size L/δ_0	15	3.6	3.0
number of gridpoints N	64	40	78
grid spacing Δ^+	73	28	3.4–28

5.2 Results

Mean streamwise velocity profiles in wall normal direction at the position $x = 9.6\,\delta_0$ are shown in Fig. 5.

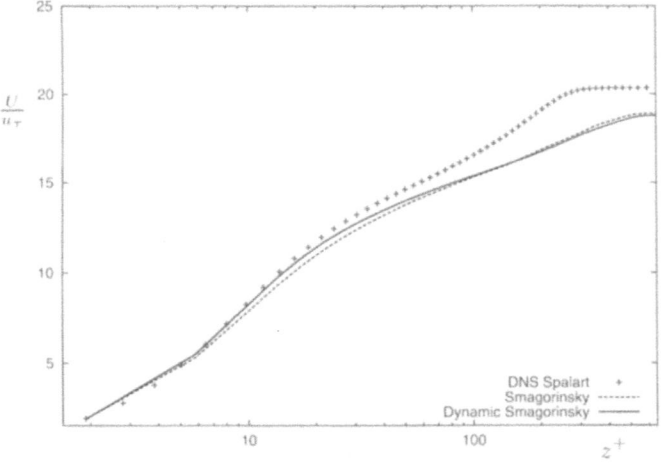

Fig. 5. Mean streamwise velocity profiles in inner coordinates.

The dynamic model shows slightly better results in the viscous sublayer than the Smagorinsky model with wall damping. The profiles start to deviate from the DNS data in the buffer layer leading to smaller than the DNS values in the log layer. These smaller values imply an overestimation in the contribution by the Reynolds stresses.

The streamwise and wall normal rms values are overpredicted by both models in the boundary layer while data points remain below the reference for the spanwise direction. $\langle u'w'\rangle$–profiles are overpredicted, too. For all values, the dynamic formulation gives better results than the basic model.

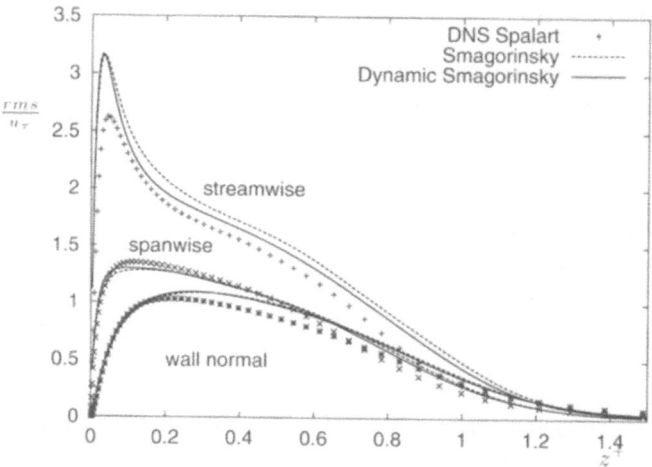

Fig. 6. Profiles of rms velocity fluctuations in inner coordinates.

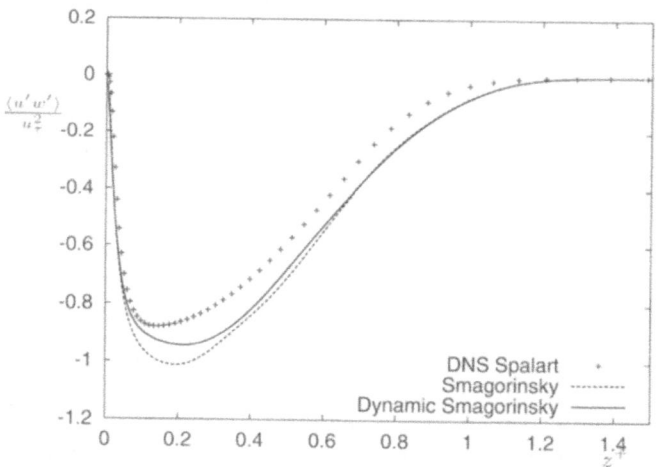

Fig. 7. Profiles of Reynolds shear stress $\langle u'w' \rangle$ in inner coordinates.

5.3   Conclusion

For the results shown here, a grid of only 70000 grid points has been used while the DNS of Spalart was calculated on a grid of 3.1 million grid points. The price to be payed for the savings in computational resources by the present LES is a remarkable degradation in accuracy which cannot be accepted from an academic point of view, but perhaps from a practical point of view. Irrespective of this point, the dynamic formulation of the Smagorinsky model tends to give better results than the basic model with Van-Driest damping without the need of an explicit correction near the wall. This is an important issue as Van-Driest damping is difficult to use in the case with pressure gradient, where u_τ and thus the distance to the wall in inner coordinates, changes in streamwise direction.

6   Flow with adverse pressure gradient

6.1   Parameters of the simulation

The major challenge of the chosen test case is the relatively high Reynolds number of $Re_\theta = 1500$ of the experiment in combination with the necessary size of the computational domain. On one hand, resolution requirements increase according to $Re^{1.8}$[11] as close to the wall the gradients of the velocities scale in inner coordinates z^+. On the other hand, the size of the domain is determined by outer lengths i.e. the boundary layer thickness and especially the length of the recirculation zone.

For tests of the SGS models, the Reynolds number has been chosen to $Re_\theta = 670$ at the inflow. Nevertheless more than ten million grid points are needed as can be seen in the following table describing the grid parameters.

	x	y	z
domain size L/δ_0	106	14.4	10.6
number of grid points N	512	160	128
grid spacing Δ^+	50–68	29	3.6–65

The pressure profile in streamwise direction remains constant for the first seven δ_0 of the domain, followed by a zone of $43\delta_0$ where the adverse pressure gradient is applied. On the last $55\,\delta_0$ the pressure remains constant again.

The timestep was set to $0.005\,U_\infty/\delta_0$. One timestep of the simulation took about 7 seconds on 16 PE, depending on the number of iterations needed to meet the required divergence criterion of $\text{div}(\boldsymbol{u}) < 0.001$.

6.2   Results

The skin friction coefficients for the two simulations with different sgs models are compared to experimental and DNS data on two different grids and shown in Fig. 8. All numerical simulations show higher values at the entry into the domain than the experiment which is partly explained by the lower Reynolds number of the simulations. The curves remain too high for both LES while both DNS show a steeper descent in the adverse pressure gradient zone, leading to detachment of the DNS but not of the LES.

Figures 9 and 10 show streamwise profiles of displacement thickness δ^* and momentum thickness θ.

All profiles remain in close agreement with the experimental data for the initial part of the adverse pressure gradient zone ($x \leq 20\,\delta_0$). In the remainder

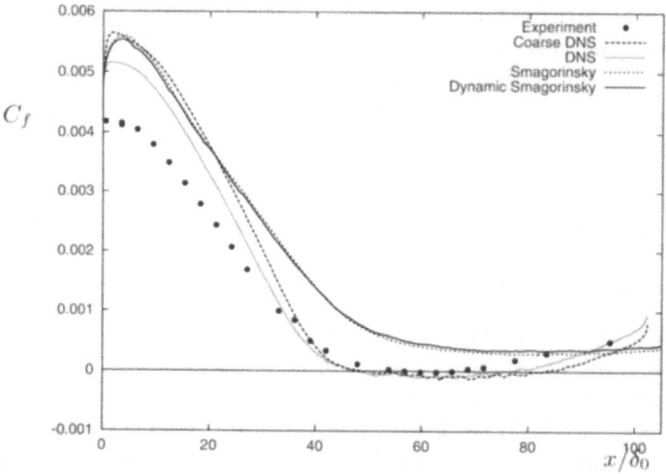

Fig. 8. Profile of the skin friction coefficient C_f in streamwise direction.

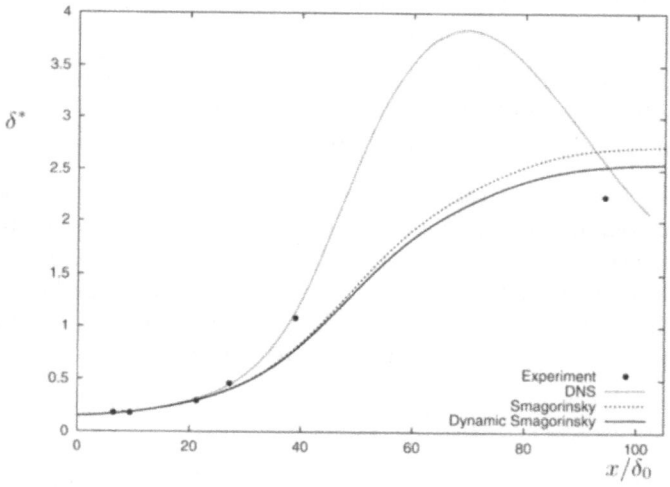

Fig. 9. Evolution of the displacement thickness δ^* in streamwise direction.

of the domain, DNS data show a peak for the displacement thickness at $x = 70\,\delta_0$ which is not reproduced by LES. Nevertheless, only small differences can be seen in the evolution of the momentum thickness until $x = 90\,\delta_0$ where the DNS data show a decrease while the boundary layer continues to grow in the LES.

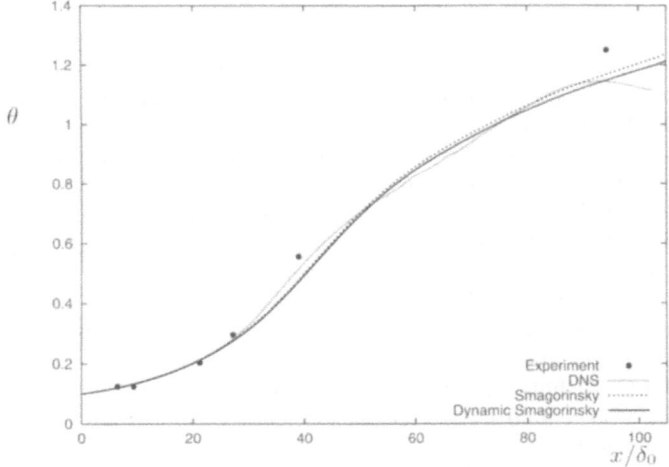

θ

Fig. 10. Evolution of the momentum thickness θ.

6.3   Discussion of the results

Neither of the two formulations of the eddy-viscosity model was capable of predicting the separation of the boundary layer. As it cannot be derived from cost efficient simulations without separation if a chosen model and numerical configuration is capable of reproducing the separation, a test was run with the same numerical setup, where the SGS model had been turned off. This resulted in a separation of the flow, confirming results of the coarse DNS of Manhart. The separation bubble continued to grow until it exceeded the size of the computational domain. This indicates, that the absence of separation was due to the influence of the SGS model but not due to the chosen grid resolution.

7   Conclusion and outlook

Large eddy simulations of turbulent boundary layer flow with and without adverse pressure gradient have been carried out. While the used SGS models show good results for attached flows further research is crucial for the prediction of pressure induced separation.

Despite not showing great differences to other subgrid scale models models in channel flow, the Increment model is one possibility for a step forward as its scale similarity ansatz might produce better results in separated flow than eddy-viscosity models.

References

1. C. Brun, C.B. Da Silva, and R. Friedrich. LES of a round jet with sgs models based on the velocity structure function. In *Euromech Colloquium 412*, Munich, Germany, 2000. Euromech.

2. C. Brun and R. Friedrich. The spatial velocity increment as a tool for SGS modeling. *Ercoftac Bulletin*, (48):16–22, March 2001.

3. C. Brun, T.J. Httl, and R. Friedrich. A-posteriori tests of a new subgrid-scale model: LES of fully developed pipe flow. In 8th *European Turbulence Conference*, Barcelona, Spain, 2000. Euromech.

4. M. Germano, U. Piomelli, P. Moin, and W.H. Cabot. A dynamic subgrid scale eddy viscosity model. In *Proceedings of the Summer Workshop*, Stanford CA, 1990. Center for Turbulence Research.

5. C.W. Hirt, B.D. Nichols, and N.C. Romero. Sola – a numerical solution algorithm for transient fluid flows. Technical Report LA-5852, Los Alamos Sci. Lab., Los Alamos, 1975.

6. M. Kalter and H.-H. Fernholz. The influence of free-stream turbulence on an axisymmetric turbulent boundary layer in, and relaxing from an adverse pressure gradient. In 5th *European Turbulence Conference, Siena 1994*, 1994.

7. M. Manhart. Direct numerical simulation of turbulent boundary layers on high performance computers. In E. Krause and W. Jaeger, editors, *High performance Computing in Science and Engineering 1998*. Springer Verlag, 1999.

8. M. Manhart and R. Friedrich. DNS of a turbulent separated boundary layer with separation. *To appear in: Heat and Fluid Flow*, 2002.

9. M. Manhart, F. Tremblay, and R. Friedrich. MGLET: a parallel code for efficient DNS and LES of complex geometries. In C.B. Jenssen, H.I. Andersson, A. Ecer, et al., editors, *Parallel computational fluid dynamics : trends and applications : Proceedings of the Parallel CFD 2000*, Trondheim, Norway, May 22-25 2001. NTNU, Elsevier.

10. R. D. Moser, J. Kim, and N. N. Mansour. Direct numerical simulations of turbulent channel flow up to $Re_\tau = 590$. *Physics of Fluids*, 11(4):943–945, April 1999.

11. U. Piomelli and E. Balaras. Wall-layer models for large-eddy simulations. *Ann. Rev. Fluid Mech.*, 34:349–374, 2002.

12. J. Smagorinsky. General circulation experiments with the primitive equations, I the basic experiment. In *Mon. Weath. Rev.*, number 91, page 99, 1963.

13. P.R. Spalart. Direct simulation of a turbulent boundary layer up to $Re_\theta = 1410$. *J. Fluid Mech.*, 197:61–98, 1988.

14. H. Werner and H. Wengle. Large-eddy simulation of turbulent flow over and around a cube in a plate channel. In F. Durst et al., editor, *Turbulent Shear Flows*, volume 8. Springer Verlag, Berlin, 1993.

Investigation of a Turbulent Separating Boundary Layer by Direct Numerical Simulation

M. Manhart

Fachgebiet Strömungsmechanik, Technische Universität München
Boltzmannstr. 15, D-85748 Garching, Germany

Abstract. A direct numerical simulation of a separating and reattaching turbulent boundary layer on a flat plate was conducted on a high performance computer. The separation has been introduced by a streamwise pressure gradient that corresponds to the one of an actually performed experiment (Kalter and Fernholz, [10]). The use of a locally refined grid near the wall is one of the key points that allowed the simulation on the presently available hardware. The comparison between experiment and simulation is fully satisfying considering that the Reynolds number of the experiment could not be achieved inthe simulation.

1 Introduction

Flow separation is one of the phenomena that cannot be predicted by engineering methods like Reynolds averaged Navier-Stokes equations (RANS). The flow physics in separation bubbles differ substantially from those in attached flows, a matter of fact that leads to the well-known breakdown of the law-of-the wall when flow reversal occurs [24]. The processes connected with flow reversal are not well understood, especially when the separation line is not fixed in space by geometrical constraints. It is known for a long time that the separation line on smoothly varying surfaces does not move forward and backward as a straight line but displays a highly irregular shape. Thus, separation can be defined only in a statistical sense, e.g. by the location where the time-averaged wall shear stress reaches zero.

The improvement of turbulence models in such flow regimes requires a deep understanding of the dynamics and mechanisms of separation and reattachment. Direct numerical simulation (DNS) can provide necessary data to improve our corresponding physical insight resolving accurately the three-dimensional and time-dependent flow fields. Up to now, only a limited number of direct numerical simulations of separated turbulent boundary layers are available, since length scales are growing rapidly during the separation process making a DNS extremely expensive. The separation of a turbulent boundary layer has first been numerically analyzed by [2]. Recently, [26] and [27] performed a DNS of a separated turbulent boundary layer. An extensive study has been performed by Na and Moin [22] at a low Reynolds number ($Re_\theta = 300$, based on inlet free-stream velocity and momentum thickness).

In their study, separation/reattachment have been enforced by a strong adverse/favourable pressure gradient in streamwise direction.

In contrast to previous studies, the study presented here is a DNS that predicts flow separation in a situation that corresponds to an actually performed experiment [10]. It has a significant higher Reynolds number and a weaker pressure gradient than the study of Na and Moin. This leads to a completely different flow balance such that the size and the shape of the separation bubble is mainly determined by Reynolds stresses instead of pressure gradient.

The present DNS is part of a series of simulations investigating the changes of turbulence structure in decelerated boundary layers. These DNSs have been performed on high performance computers and results have been used to evaluate the performance of RANS models [7,8,21] and wall boundary conditions for large eddy simulation (LES) [17]. A detailed verification of the results by comparison with experimental data can be found in [20].

2   Computational details

The following section describes the numerical method used in the simulations. It consists of a finite-volume method for incompressible flows formulated on a non-equidistant staggered mesh. A Poisson equation for the pressure correction is obtained by the projection-method according to [1]. A local grid refinement has been introduced in order to save computational resources while achieving the required resolution near the wall. This local grid refinement was crucial for performing the present DNS on the available hardware. For an efficient use of the hardware, the code has been fully vectorized and parallelised

2.1   Basic scheme

The flow variables, velocity components and pressure, are defined on a non-equidistant Cartesian mesh in a staggered arrangement. Principally, velocity components are stored in the centres of cell faces, while pressure is stored in the cell centers. The specific discrete formulations are derived by integrating the Navier-Stokes equations for an incompressible fluid over the corresponding control cells surrounding the definition points of the individual variables. We are using the mid-point rule for approximating the fluxes by the variables. The required interpolations and the approximation of the first derivatives are performed by linear interpolation and 2nd order central finite difference formulations, respectively. This altogether ensures 2nd order accuracy of the spatial discretisation [4].

The discrete solution is advanced in time by a leapfrog time step which is second order accurate with respect to the convection term:

$$u^{n+1} = u^{n-1} + 2\Delta t \left[C(u^n) + D(u^{n-1}) - G(p^{n+1}) \right] \tag{1}$$

C, D and G denote herein the discrete convective, diffusive and gradient operators. The pressure at the new time level $p^{n+1} = p^n + \Delta p^{n+1}$ is determined by the solution of the Poisson equation

$$Div\left[G(\Delta p^{n+1})\right] = \frac{1}{2\Delta t} Div(u^*), \tag{2}$$

where u^* is an intermediate velocity field obtained by solving equation (1) by using the pressure p^n at the known time level. A divergence-free field u^{n+1} is obtained after a velocity correction step.

$$u^{n+1} = u^* - 2\Delta t G(\Delta p^{n+1}) \tag{3}$$

The combination of central interpolation and a leapfrog time step is energy conserving for the one-dimensional pure convection equation. This is the reason why it is especially suited for LES and DNS. In combination with the diffusion operator, the leapfrog time step is slightly unstable (compare Fletcher, [5]). Therefore, the diffusive term is taken at the time level $n-1$ in equation (1). Every 41 time steps, an averaging step is performed in order to prevent $2\Delta t$ oscillations.

The Poisson equation (2) is solved by an iterative procedure accelerated by a multigrid cycle. The smoother is based on the velocity-pressure iteration presented by [6] with overrelaxation. This scheme gives the same convergence properties as a conventional Gauss-Seidel iteration with succesive overrelaxation (SOR). The advantage of the present algorithm is the easy treatment of boundaries, at which only velocity boundary conditions have to be specified. In order to improve the convergence of the iterative solver, a multigrid procedure has been implemented.

2.2   Local grid refinement

The introduction of a locally refined grid near the wall is one of the key points for an efficient simulation of the separated turbulent boundary layer. The refinement is achieved by dividing one coarse grid cell into 8 fine grid cells. The coarse and the fine grids are arranged in an overlapping way, so that the coarse grid is defined globally (global grid) and the fine grid is defined only locally (zonal grid). The coarse-grid and the fine-grid solutions are fully coupled. The coupling is achieved by transferring the fine-grid solution in the overlap region to the coarse grid each time before the algorithm changes from the fine grid level to the coarse grid level. We use averaging over four cell faces for the velocities and averaging over 8 grid cells for the pressure restriction. While solving the Poisson equation on both levels, we use the pressure correction on the coarse grid as a new pressure estimate for the fine grid in the multi-grid cycle.

The treatment of the interface between fine and coarse grid needs special attention. It appears at two different positions in the algorithm, (i) as pressure

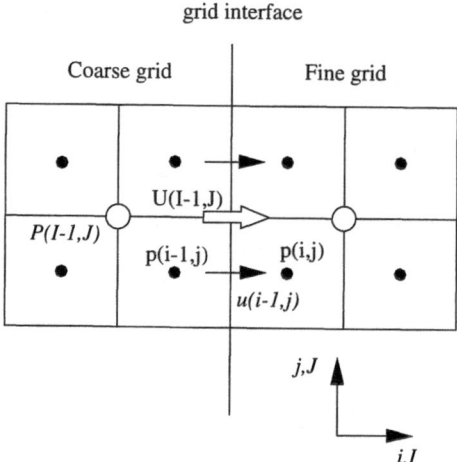

Fig. 1. Configuration of grid interface between local fine and global coarse grid.

boundary condition for the fine grid and (ii) as velocity boundary condition for the fine grid. We are using so-called ghost cells in order to manage the boundary conditions for the grid interface. The situation is displayed for a two-dimensional configuration in Fig. 1. Capital letters indicate coarse-grid variables and small letters indicate fine-grid variables. The fine-grid variables on the left hand side of Fig. 1 are called ghost cells. The coarse-grid variables on the right hand side of Fig. 1 are obtained from the fine-grid variables by a restriction. Due to the staggered arrangement of the variables and the special refinement strategy, both coarse-grid variables as well as fine-grid variables are located on the grid interface.

On the local grid, a Neumann boundary condition is used for the pressure at the fine grid/coarse grid interface. It has been found that this treatment is superior to a Dirichlet boundary condition for the fine grid pressure correction at the grid interface [12]. As a consequence, the fine-grid interface velocities $u(i-1,j)$ and $u(i-1,j+1)$ in Fig. 1 are determined by a suitable conservative prolongation from the coarse-grid interface velocity $U(I-1,J)$. Our tests showed that first order interpolation leads to smaller distortions at the grid interface than second order interpolations due to their inherent damping properties [18].

The local grid approach has been used successfully for turbulent channel flow [12], turbulent zero pressure gradient and adverse pressure gradient flow [14,15,13]. For a more detailed description of the algorithm see [18].

2.3   Boundary conditions

Since the flow is homogeneous in spanwise direction, periodic boundary conditions are prescribed in this direction. The wall is treated by a no-slip con-

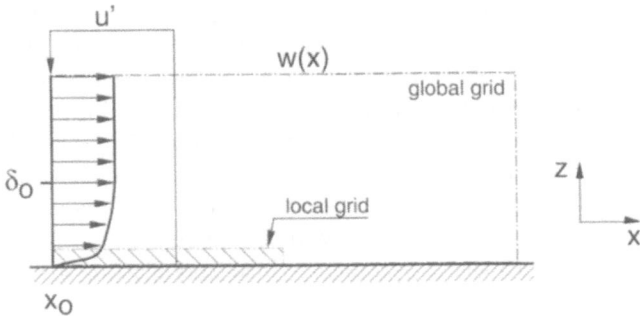

Fig. 2. Geometry of the boundary layer simulations (not to scale).

dition. At the outflow, a zero gradient condition is applied for the velocities, which compiles into a Dirichlet condition for the pressure ($p = 0$). Due to its non-physical behaviour, this condition produces a distortion of the flow traveling upstream. The region of influence visible in the skin friction coefficient extends over about two inflow boundary layer thicknesses, which is smaller than one outflow boundary layer thickness and corresponds to about 2 % of the computational domain (see e.g. Fig. 3).

At the upper boundary of the domain, the vertical velocity is prescribed. It has been obtained from a prior coarse grid simulation [19]. Due to different sizes of the separation bubbles in the prior and the present simulation, the vertical velocity had to be slightly adjusted in order to obtain the correct streamwise variation of the freestream velocity. In the prior simulation, the pressure gradient has been introduced by a Dirichlet condition for the pressure which has been derived from the variation of the freestream velocity using Bernoulli's equation.

Time-dependent inflow profiles are constructed by superposition of a time-mean profile and a fluctuation from a position $10\delta_0$ downstream (where δ_0 is the boundary layer thickness at the inlet plane, see Fig. 2). A detailed description of that method, which is a variant of the method proposed by [23], is given in Manhart and Friedrich ([19]). Herein, a comparison with the method proposed by [11] revealed no significant differences between both approaches.

2.4 Configuration

The present DNS follows an experiment of a turbulent boundary layer developing along a circular cylinder ($Re_\theta \approx 1500$) which is subjected to a streamwise adverse pressure gradient until separation occurs [10]. In the subsequent nearly zero-pressure gradient region the flow reattaches and slowly relaxes to a canonical boundary layer. The separation bubble is very thin and can be suppressed by a certain level of freestream turbulence. During the separation process the thickness of the boundary layer grows by a factor of about 9.

In the simulation, a rectangular domain has been selected covering the separation bubble. In addition to that, the following simplifications have been introduced. First, a flat plate boundary layer has been simulated instead of a boundary layer along a cylindrical body with spanwise curvature. The effect of this simplification is small because the ratio of boundary layer thickness/diameter is small. Second, the Reynolds number has been lowered in the simulation compared to the experiment, which may lead to some changes in the size of the separation bubble. So, we do not expect a one-to-one quantitative correspondence between simulation and experiment.

In terms of boundary layer thickness at the inlet δ_0, the reference position x_r in the simulation is located $10\delta_0$ downstream of the inlet (at $x/\delta_0 = 5.0$). Using the free-stream velocity U_0 at the inlet, the Reynolds numbers at this position are $Re_\theta = 870$ and $Re_\theta = 1560$ in the simulation and the experiment, respectively. The Reynolds number based on wall friction velocity is $Re_\tau = 319$ at the reference position. In terms of the displacement thickness δ_r^* at the reference position, the dimensions of the computational box are $L_x = 542\delta_r^*$ in streamwise, $L_y = 128\delta_r^*$ in spanwise and $L_z = 54\delta_r^*$ in vertical direction. In what follows, all quantities are normalized by U_0 and δ_0, respectively. ($\delta_r^*/\delta_0 = 0.188$).

We performed a preliminary simulation, denoted here as "coarse DNS" [19], in which a zonal grid covered the wall region from the inflow to the outflow plane. An analysis of the directional dissipation scale [16] in this simulation revealed that the local grid is necessary only in the first half of the computational domain. A second finer DNS was designed in which the zonal grid covers only the first half of the computational domain (Figure 2). The parameters of both simulations are shown in Table 1. With a wall normal grid spacing at the reference position x_r of $\Delta z_{min}^+ = 1.6$, the position of the first grid point is at $z^+ = 0.8$. The simulations have been carried out on a Fujitsu VPP/700 vector-parallel computer. By using 16 processors, about 12 Gflop/s were achieved. One time step of $\Delta t * U_0/\delta_0 = 0.005$ took about 28 CPU-seconds. Preliminary simulations have been carried out on a CRAY T3E-900 on which 70 Mflop/s have been achieved per processor. The size of the present simulation, however, required more memory than it was available

Table 1. Parameters of the numerical grids. Inner coordinates are based on the friction velocity at the reference position

		NX	NY	NZ	Δx^+	Δy^+	Δz^+
coarse	global	640	320	96	52.8	24.8	6.6
	local	1280	640	32	26.4	12.4	3.3
fine	global	1280	448	160	25.5	17.2	3.2
	local	1280	896	32	11.7	7.2	1.6

on the CRAY T3E-900. Even on the VPP/700, the present simulation would not have been possible without the use of the local grid refinement, since it saved a factor of more than three in computational resources.

3 Results

3.1 Statistical results

The development of the free-stream velocity is displayed in Fig. 3 (top), showing that after a region of deceleration, a constant free-stream velocity has been adjusted. The deceleration of the boundary layer leads to a continuosly decreasing skin friction coefficient c_f (Fig. 3, bottom) and to separation of the boundary layer ($c_f = 0.0$) at about 45 boundary layer thicknesses after the inlet. Since there is no noticeable variation of the free-stream velocity after separation ($x/\delta_0 \approx 40$), the boundary layer is slowly relaxing from the separated regime by diffusive and convective processes only. This leads to a long regime where c_f is nearly zero. In both simulations, the separation bubble as measured by a negative c_f-value is a little bit longer than in the experiment. The length of the recirculation zone doesn't seem to be appreciably influenced by the grid resolution. As seen in an analysis of the spanwise energy spectra in the prior coarse grid simulation [19], the grid resolution requirements become increasingly less demanding as one approaches the separation zone. The difference between fine grid DNS and experiment can be

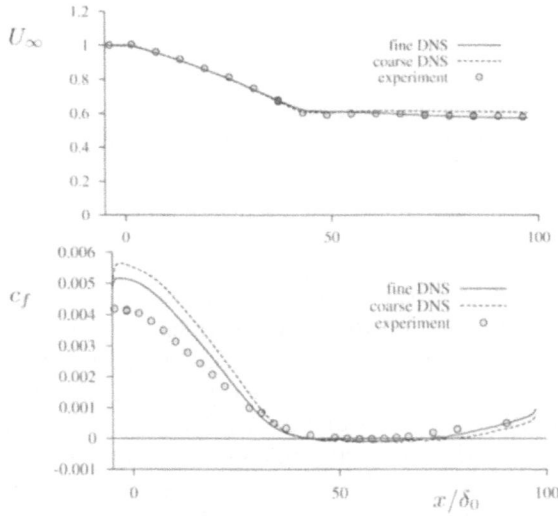

Fig. 3. Distribution of free-stream velocity (top) and skin friction coefficient (bottom) in the separated turbulent boundary layer

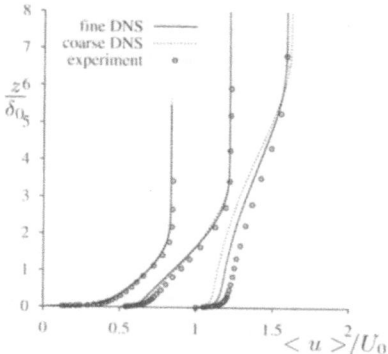

Fig. 4. Averaged streamwise velocity profiles in comparison with the experiment of [10]; from left to right: $x/\delta_0 = 22$, $x/\delta_0 = 34$ and $x/\delta_0 = 90$ (profiles are shifted by $< u > /U_0 = 0.5$)

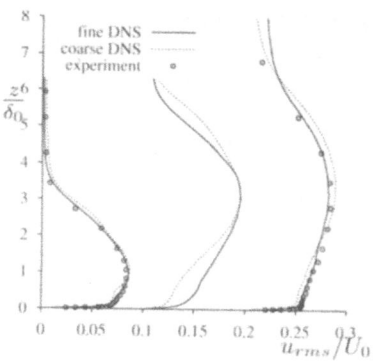

Fig. 5. RMS of the streamwise velocity component in comparison with the experiment of [10];from left to right: $x/\delta_0 = 34$, $x/\delta_0 = 50$ and $x/\delta_0 = 90$ (profiles are shifted by $u_{rms}/U_0 = 0.1$)

attributed to the different Reynolds numbers. Coles' empirical correlation [3] gives values of $c_f \approx 0.004$ for $Re_\theta = 1500$ which corresponds to the experiment and $c_f \approx 0.0047$ for $Re_\theta = 870$ which corresponds to our DNS. The coarse grid simulation produces a c_f-value too high in this region. It is slowly approaching the more realistic value of the fine grid DNS and, surprisingly, reaches separation at the same streamwise position. The relative unimportance of the grid resolution on the recirculation length leaves one main reason for the larger separation region in the simulation, namely the lower Reynolds number.

Profiles of streamwise velocity averaged in time and spanwise direction are compared in Fig. 4 with experimental data. Two positions are located in the adverse pressure gradient region ($x/\delta_0 = 22$ and $x/\delta_0 = 34$) and one after reattachment ($x/\delta_0 = 90.0$). In the separation zone, no experimental profiles are available. Before separation, good accordance between all results has been achieved. After reattachment, the profiles differ due to the different lengths of computed and measured separation bubbles. The profiles are still relaxing towards a canonical zero pressure gradient boundary layer. RMS-values of the streamwise velocity fluctuations are shown in Figures 5. Around the separation bubble, the grid dependence of the RMS-values is weaker than in the oncoming boundary layer and the agreement between fine DNS and experiment is obvious.

3.2   Instantaneous flow structures

The separation and reattachment lines are not fixed in space and time. In order to get an impression of the complicated nature of the instantaneous shape of the separation bubble, we show grey-scale plots of the instantaneous streamwise velocity component. In Fig. 6, perspective views of the streamwise velocity component in planes vertical to the coordinate axes are

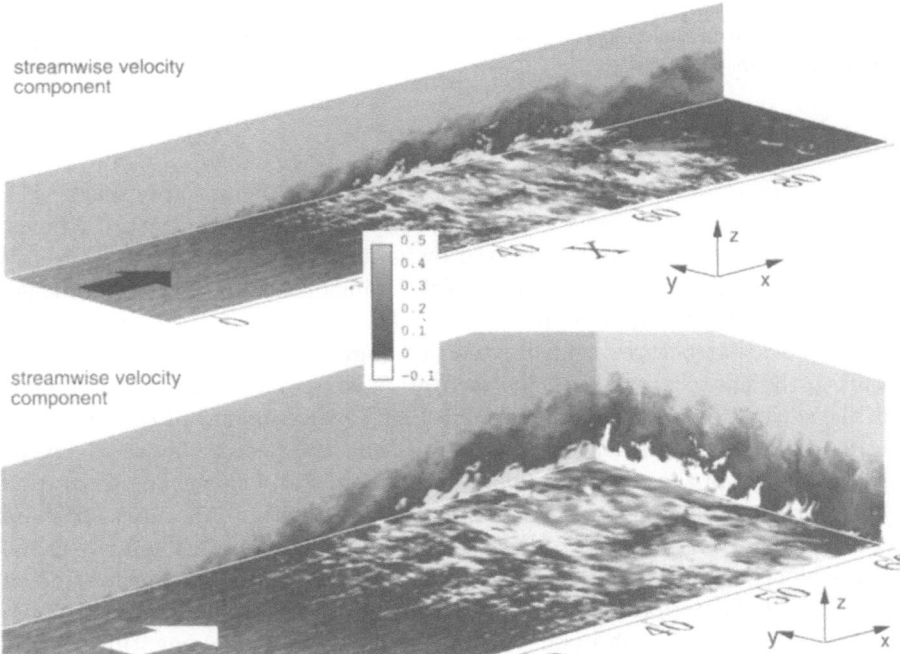

Fig. 6. Instantaneous velocity fields: perspective view of streamwise velocity component. Whole domain (top) and zoom (bottom).

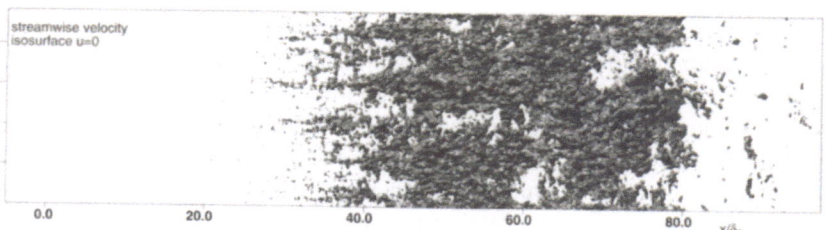

Fig. 7. Instantaneous velocity fields: Top view of isosurface $u = 0.0$ of streamwise velocity component

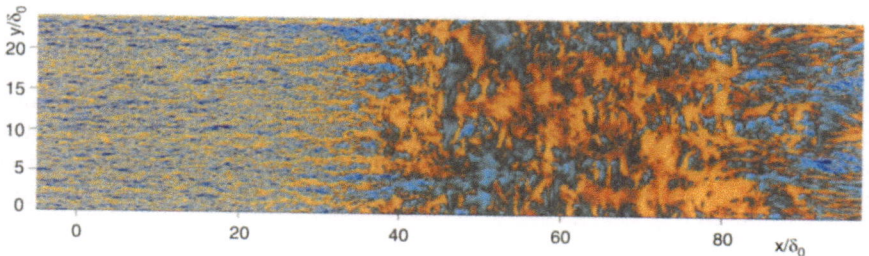

Fig. 8. Instantaneous velocity fields: Top view of streamwise velocity fluctuations in a plane $z/\delta_0 = 0.046$

shown. Different velocity magnitudes are coded by different grey scales. The freestream velocity is coded by a mild grey, slow velocities by dark tones. The backflow velocities are indicated by white spots within the dark, nearly black regions. In this view, the strong thickening of the boundary layer during the separation process becomes obvious. This is connected with the growth of large scale structures, which extend throughout the whole layer from the wall to the shear layer above the separation zone. The instantaneous shape of the separation bubble is displayed in Fig. 7 by a top view of the isosurface of $u = 0.0$. The main feature emanating from this picture is the highly irregular shape of the separation bubble. At the instant shown, the separation line meanders between $x/\delta_0 = 35$ and $x/\delta_0 = 45$ with a dominant spanwise wavelength of $5\delta^*$ (in terms of local displacement thickness). The holes in the zero-velocity surface are generated by blobs of forward flow reaching down to the wall. This supports the idea of a strong vertical momentum transport in the free shear layer through which positive streamwise momentum reaches the wall.

An interesting question concerns the evolution of the so-called streaks during the deceleration of the boundary layer. Streaks are elongated flow patterns observable in the wall region of turbulent boundary layers. The dynamics of these streaks are considered as most important for the recreation

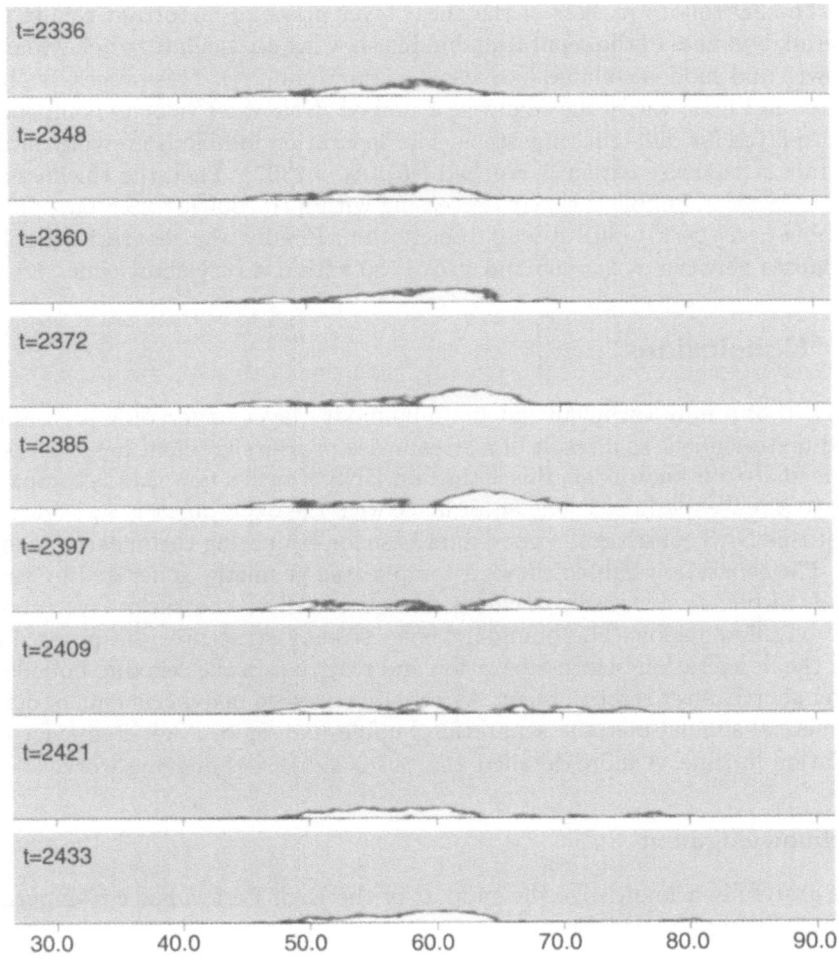

Fig. 9. Spanwise averaged streamwise velocity component; dark gray is from $-0.01 < u < 0.0$

cycle of near wall turbulence [9]. In Fig. 8 we made the streaks visible by plotting the streamwise velocity fluctuations in a plane parallel to the wall at $z/\delta_0 = 0.046$ (about 15 wall units in terms of friction velocity at the reference position). The spanwise spacing of the streaklike patterns is about 100 wall units in the initial phase of the boundary layer ($x/\delta_0 < 20$). Downstream, they are growing in spanwise direction and after $x/\delta_0 = 40$ they are replaced by completely different large scale structures. But, after reattachment, streaky structures are quickly reforming although at different spanwise wave length than before separation.

The 2D roll-up process of the shear layer plays an important role in the overall dynamics of the separation bubble showing an oscillatory behaviour of growth and sudden collapse (see also Na and Moin [22]). This process is illustrated in Fig. 9, where the spanwise averaged streamwise velocity component is displayed for different time steps. The separation bubble is growing until a certain streamwise extent is reached ($tU_0/\delta_0 = 2397$). Then the thickness of the separation bubble is slowly decreasing within the downstream part of the bubble ($x/\delta_0 > 65$) until it is extremely thin. Finally, the separation bubble collapses between $x/\delta_0 = 65$ and $x/\delta_0 = 80$ within a very short time.

4   Conclusions

A DNS of a fully turbulent flat plate boundary layer undergoing separation and reattachment as a result of a streamwise pressure gradient has been performed. To our knowledge, this is the first DNS of such a flow that is comparable to a real experiment. The accordance with the experiment is satisfying so that this DNS constitutes a good data base for evaluating turbulence models.

The separation bubble shows a complicated geometry generated by large scale turbulent structures reaching from the wall to the shear layer above the backflow region. The boundary layer streaky structures disappear during the deceleration before separation and reappear in the relaxing boundary layer shortly after reattachment. As already observed in experiments or other numerical simulations, the separation bubble undergoes a low-frequency oscillation in time. A more detailed analysis is subject of ongoing work.

Acknowledgment

We gratefully acknowledge the support of the High Performance Computing Centre Stuttgart (HLRS).

References

1. A. J. Chorin. Numerical solution of the Navier-Stokes equations. *Math. Computat.*, 22:745–762, 1968.
2. G.N. Coleman and P.R. Spalart. Direct numerical simulation of a small separation bubble. In C.G. Speziale and B.E. Launder, editors, *Near-Wall Turbulence Flows*, pages 277–286. Elsevier, 1993.
3. D. Coles. The turbulent boundary layer in a compressible fluid. In *Report R-403-PR*. The Rand Corporation, Santa Monica, CA, 1962.
4. J.H. Ferziger and M. Perić. *Computational Methods for Fluid Dynamics*. Springer, Berlin, 2nd edition, 1997.
5. C. A. J. Fletcher. *Computational Techniques for Fluid Dynamics*. Springer, 1988.
6. C.W. Hirt, B.D. Nichols, and N.C. Romero. Sola – a numerical solution algorithm for transient fluid flows. In *Los Alamos Sci. Lab.*, Los Alamos, 1975.

7. T. Hüttl, M. Manhart, R. Friedrich, G. Deng, and J. Piquet. Direct numerical and statistical simulation of turbulent boundary layer flows with pressure gradient. In E.H. Hirschel, editor, *Notes on numerical fluid mechanics*. Vieweg-Verlag, Braunschweig, 2000.

8. T.J. Hüttl, G. Deng, M. Manhart, and R. Friedrich. Testing turbulence models by comparison with DNS data of adverse-pressure-gradient boundary layer flow. In E. Krause and W. Jaeger, editors, *High Performance Computing in Science and Engineering 2000*. Springer Verlag, Berlin, Heidelberg, 2001.

9. J. Jiménez and A. Pinelli. The autonomous cycle of near-wall turbulence. *J. Fluid Mech.*, 389:335–359, 1999.

10. M. Kalter and H.H. Fernholz. The influence of free-stream turbulence on an axisymmetric turbulent boundary layer in, and relaxing from, an adverse pressure gradient. In *5th European Turbulence Conference, Siena 1994*, 1994.

11. T.S. Lund, X. Wu, and K.D. Squires. Generation of turbulent inflow data for spatially-developing boundary layer simulations. *J. Comp. Phys*, 140:233–258, 1998.

12. M. Manhart. Zonal direct numerical simulation of turbulent plane channel flow. In R. Friedrich and P. Bontoux, editors, *Computation and visualization of three-dimensional vortical and turbulent flows. Proceedings of the Fifth CNRS/DFG Workshop on Numerical Flow Simulation*, volume 64 of *Notes on Numerical Fluid Mechanics*. Vieweg Verlag, 1998.

13. M. Manhart. Direct numerical simulation of an adverse pressure gradient turbulent boundary layer on high performance computers. In E. Krause and W. Jäger, editors, *High Performance Computing in Science and Engineering '99*, pages 315–326, Berlin, Heidelberg, New York, 1999. Springer.

14. M. Manhart. Direct numerical simulation of turbulent boundary layers on high performance computers. In E. Krause and W. Jaeger, editors, *High performance Computing in Science and Engineering 1998*. Springer Verlag, 1999.

15. M. Manhart. Using zonal grids for direct numerical simulation of turbulent boundary layers with pressure gradients. In W. Nitsche, H.-J. Heinemann, and R. Hilbig, editors, *Vol. 72, Notes on numerical fluid mechanics*, pages 299–306. Vieweg-Verlag, Braunschweig, 1999.

16. M. Manhart. The directional dissipation scale: a criterion for grid resolution in direct numerical simulations. In C. Dopazo and et al., editors, *Advances in Turbulence VIII*, pages 667–670, Barcelona, 2000. Eighth European Turbulence Conference, CIMNE.

17. M. Manhart. Analysing near-wall behaviour in a separating turbulent boundary layer by DNS. In B. Geurts, R. Friedrich, and O. Metais, editors, *Direct and Large-Eddy Simulation IV*. Kluwer Academic Publishers, Dordrecht, 2001.

18. M. Manhart. A zonal grid algorithm for DNS of turbulent boundary layers. *Computers and Fluids*, (submitted), 2002.

19. M. Manhart and R. Friedrich. Towards DNS of separated turbulent boundary layers. In P.R. Voke, N.D. Sandham, and L. Kleiser, editors, *Direct and Large-Eddy Simulation III*, pages 429–440. Kluwer Academic Publishers, Dordrecht, 1999.

20. M. Manhart and R. Friedrich. DNS of a turbulent boundary layer with separation. *International Journal of Heat and Fluid Flow*, accepted for publication, 2002.

21. M. Manhart, R. Friedrich, G.B. Deng, and J. Piquet. Direct versus statistical simulation of accelerated/retarded and separating/reattaching turbulent

boundary layers. In E.H. Hirschel, editor, *Notes on numerical fluid mechanics, to appear.* Springer-Verlag, Berlin, 2002.

22. Y. Na and P. Moin. Direct numerical simulation of a separated turbulent boundary layer. *J. Fluid Mech.*, 370:175–201, 1998.

23. L. Schmitt, K. Richter, and R. Friedrich. Large-eddy simulation of turbulent boundary layer and channel flow at high Reynolds number. In U. Schumann and R. Friedrich, editors, *Direct and Large Eddy Simulation of Turbulence*, pages 161–176, Braunschweig, 1986. Vieweg.

24. R.L. Simpson. Turbulent boundary-layer separation. *Ann. Rev. Fluid Mech.*, 21:205–234, 1989.

25. R.L. Simpson. Aspects of turbulent boundary-layer separation. *Prog. Aerospace Sci.*, 32:457–521, 1996.

26. M. Skote, D. Henningson, N. Hirose, Y. Matsuo, and T. Nakamura. Parallel DNS of a separating turbulent boundary layer. In *Proceedings of the Parallel CFD 2000*, Trondheim, Norway, May 22-25 2000. NTNU.

27. M. Skote and D. S. Henningson. DNS of a separating turbulent boundary layer. In E. Lindborg and et al., editors, *Turbulence and shear flow phenomena. Second International Symposium*, Stockholm, 2001. KTH.

Predictions of the 3D Unsteady Heat Transfer at Moving Droplets

M. Hase and B. Weigand

Institute of Aerospace Thermodynamics, University of Stuttgart,
Pfaffenwaldring 31, 70569 Stuttgart, Germany

Abstract. A 3D numerical program for the transient simulation of the dynamic behavior of incompressible two-phase flows has been extended to the computation of heat transfer. In the program the VOF-method with interface reconstruction has been used for the calculation of the disperse phase. The governing equations and the implemented numerical model are described. Numerical results for a transient heat conduction problem of a rigid sphere show good agreement with analytical solutions. The predicted averaged Nusselt numbers for this problem from numerical simulations match well with experimental data from the literature. On the basis of two examples the difference between intermediate and high Reynolds number flow and heat transfer is pointed out. Finally, the influence of different initial droplet velocities on the time dependent temperature evolution is shown. The simulation has been performed on the Cray T3E/512-900 at the HLRS with up to 128 processors.

1 Introduction

The rate of heat transfer between the disperse and the continuous phase is a critical design parameter in many technical systems such as sprays in automotive engines or gas turbines. Due to high injection pressures in this systems the velocity of the disperse phase is far away from being steady state. The experimental investigation of this high speed process is very complicated. For high Reynolds numbers $Re > 270$ additionally the surrounding gas stream becomes asymmetric which makes it necessary to perform the computation also fully 3D [1]. Therefore, the heat transfer at droplets which move with unsteady velocity has been studied rarely experimentally and numerically in the past.

In the present paper the capability of the used in-house program FS3D (Free Surface 3D) for the computation of heat transfer from droplets to the surrounding gas has been investigated. For different initial droplet velocities with accelerated, decelerated and steady moving droplets the temperature evolutions have been examined. The program efficiency for the computation of droplet velocities has been presented already in [2,3]. The heat transfer implementation and validation is described in [4].

2   Analysis and numerical method

The simulation of two-phase flow problems at low Mach numbers requires the solution of the Navier-Stokes equations for incompressible flow including surface tension. At the Institute of Aerospace Thermodynamics an inhouse code FS3D has been developed in the last years, to handle this task. In the program the flow field is computed by direct numerical simulation. By using the direct numerical simulation, no turbulence model is required. The continuity and the momentum equation for incompressible flow are given by

$$\frac{\partial(\rho\mathbf{u})}{\partial t} + \nabla \cdot [(\rho\mathbf{u}) \otimes \mathbf{u}] = -\nabla p + \nabla \cdot \mu \left[\nabla\mathbf{u} + (\nabla\mathbf{u})^T\right] + \nabla \cdot \mathbf{T} \tag{1}$$

$$\nabla \cdot \mathbf{u} = 0 \quad , \tag{2}$$

where $\mathbf{T}$ is the capillary stress tensor and $\mathbf{u}$, ρ, μ and p are the velocity vector, the density, the dynamic viscosity and the pressure. For two-phase flows, additional information about the position of the interface is needed. In FS3D the Volume-of-Fluid method (VOF) [5] is used. For this method an additional equation has to be solved

$$\frac{\partial f}{\partial t} + \nabla \cdot (\mathbf{u}f) = 0 \tag{3}$$

for the volume fraction f (VOF-variable) of the dispersed phase. The VOF-variable is defined by

$$f = \begin{cases} 0 & \text{in the continuous phase} \\ 0 < f < 1 & \text{at the interface} \\ 1 & \text{in the dispersed phase} \end{cases} \tag{4}$$

With this variable, the change in density and viscosity over the surface can be computed by the equations

$$\rho(\mathbf{x}, t) = \rho_G + (\rho_L - \rho_G) f(\mathbf{x}, t) \tag{5}$$

$$\mu(\mathbf{x}, t) = \mu_G + (\mu_L - \mu_G) f(\mathbf{x}, t) \quad . \tag{6}$$

Other fluid properties like c_p and λ can be obtained in a similar manner.

To ensure a sharp interface and to suppress numerical dissipation of the dispersed phase during each time step, the interface was reconstructed by the PLIC-method (*Piecewise linear interface reconstruction computation*)[6] before the transport step of the dispersed phase was executed. The discretization takes place on a staggered grid (MAC-Grid) with the Finite Volume method. Additionally a coordinate transformation is applied which superposes the droplet velocity and solves the conservation equations in a moving coordinate system in order to track the droplet in the gas stream [7].

For an incompressible flow with constant fluid properties, apart from the variation over the interface which is described by the VOF-variable, the energy equation is decoupled of the momentum equations. To characterize the energy transport, the temperature form of the energy equation

$$\frac{\partial}{\partial t}(\rho c_p T) + \nabla \cdot (\rho c_p \mathbf{u} T) = \nabla \cdot (\lambda \nabla T) + \Phi \tag{7}$$

has been used. The dissipation term Φ can be neglected for low Eckert numbers, which is fulfilled for the here considered flow. To compute the transport of energy, the equation is divided into the convective and the heat conducting terms

$$\frac{(\rho c_p \acute{T})^{n+1} - (\rho c_p T)^n}{\delta t} = -\nabla \cdot (\rho c_p \mathbf{u} T)^{n+1/2} \tag{8}$$

$$\frac{(\rho c_p T)^{n+1} - (\rho c_p \acute{T})^{n+1}}{\delta t} = -\nabla \cdot (\lambda \nabla T)^n \tag{9}$$

where $\acute{T}$ is an intermediate result of the temperature. The 3D equations were replaced by a symmetrical sequence of three 1D equations. Simultaneously the equation

$$\frac{(\rho c_p)^{n+1} - (\rho c_p)^n}{\delta t} = -\nabla \cdot (\rho c_p \mathbf{u})^{n+1/2} \tag{10}$$

is solved, which is also replaced by three 1D equations. Furthermore, a divergence correction term was added to the eqs. (8) and (10) due to the fact that the divergence free condition of the velocity field is used to simplify the Navier-Stokes equations. The divergence free condition is not achieved after each one-dimensional transport step. After the computation of each one-dimensional transport step the temperature field is updated by

$$T^{n+1} = \frac{(\rho c_p T)^{n+1}}{(\rho c_p)^{n+1}} \qquad . \tag{11}$$

In this equation the nominator on the right hand side is obtained from eq. (9) and the denominator from eq. (10). To compute the value of T at the finite volume interfaces a Van-Leer-limiter method is used [8]. The discretization is second-order accurate in space and time both in the liquid and in the gas phase.

3   Results

The results shown here are for droplets with different initial velocities, different diameters and different viscosities. The investigated liquid is water with different dynamic viscosities $\mu_L = n \cdot \mu_{H_2O} = n \cdot 10^{-3}\,\text{kg}/(\text{ms})$. The surrounding gas is air. The initial temperature of the liquid is $T_L = 350\,\text{K}$ and $T_G = 293.15\,\text{K}$ for the gaseous phase.

The computational domain is displayed in Fig. 1. The 3D channel geometry for a droplet with the diameter $D = 1 \cdot 10^{-3}$ is $x = 1.2 \cdot 10^{-2}\,\text{m}$,

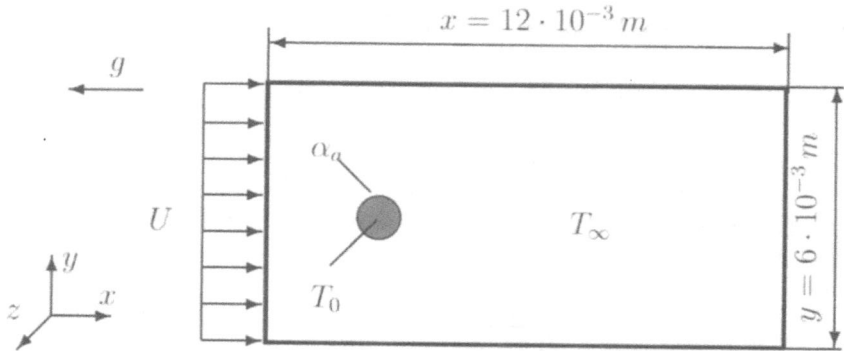

Fig. 1. Channel geometry and boundary conditions for a computation of a droplet diameter of $D = 1 \cdot 10^{-3}$ m.

$y = 0.6 \cdot 10^{-2}$ m and $z = 0.6 \cdot 10^{-2}$ m. For other droplet diameters the channel size has been chosen in a similar way to get the same geometric proportions. The gravitational force acts in the negative x-direction.

First a comparison between the numerical results and an analytical solution for the heat conduction problem in a sphere [9] has been performed. The problem includes the transient heat conduction in a solid together with a boundary condition of the third kind. To facilitate the use of Fourier's solution an asymptotic approximation has been used [10]. It contains a short-time and a long-time asymptote to describe the whole range for $0 < Fo < \infty$ with the Fourier number $Fo = at/R^2$ as a non-dimensional time. R is the droplet radius and a the thermal diffusivity of the liquid. The short-time asymptote results from the error function solution of the heat conduction equation which means that it is a more exact solution than a solution using the Fourier series for very short times. From the boundary conditions the Biot number $Bi = \alpha_a R/\lambda_L$ is a parameter in the asymptotic approximation, where α_a is the external heat transfer coefficient. To obtain the external heat transfer coefficient, empirical correlations for the Nusselt number of two different sources have been used [11,12].

The numerical simulation has been done on a $64 \times 32 \times 32$ grid for a water droplet in air with the properties given above. For obtaining an equivalent situation to the instantaneously heated sphere, which has been used as a comparison from the analytical solution, the liquid motion in the droplet has been suppressed. This is comparable with the situation of $\mu_L \rightarrow \infty$.

In Fig. 2 the non-dimensional mean droplet temperature

$$\bar{\Theta}(t) = \frac{\bar{T}(t) - T_\infty}{T_0 - T_\infty} \tag{12}$$

is displayed as a function of Fo, where $\bar{T}(t)$ is the mean volume averaged droplet temperature, T_∞ the temperature in the undisturbed gas stream and T_0 the initial temperature of the droplet.

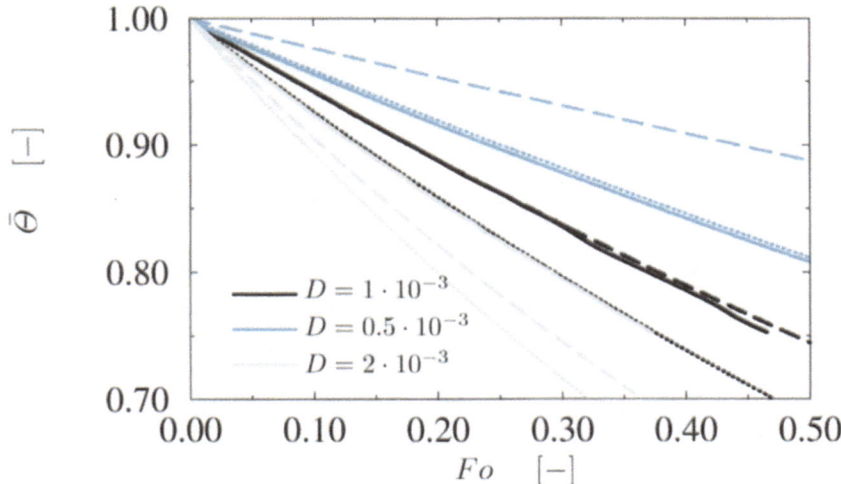

Fig. 2. Comparison of the temperature decrease of the non-dimensional mean droplet temperature $\bar{\Theta}$ in dependency on Fo .

The numerical results, represented by solid lines, are obtained from computations of droplets with different diameters moving with their steady velocity. For the analytical solution, with the given diameter and the velocity of the droplet, a heat transfer coefficient has been calculated from the empirical correlations for the Nusselt number from McAdam [11] (long dashed lines) and Baehr [12] (dotted lines) with the Prandtl number $Pr = \nu/a = 0.7$ for air. The correlations yield different results for the heat transfer coefficients and therefore also different temperature evolutions. For the droplet diameter $D = 1 \cdot 10^{-3}$ m a good agreement between the numerical result and the analytical solution with the correlation from McAdam is obtained. The correlation given by Baehr [12] varies about 25% from these values. For $D = 0.5 \cdot 10^{-3}$ m the simulation agrees well with the correlation given by Baehr [12]. However the correlation of McAdam differs by more then 50%. The reason for this large deviation is the wrong value of the heat transfer coefficient for $Re \rightarrow 0$ in the correlation of McAdam which is discussed later. For $D = 2 \cdot 10^{-3}$ m both correlations differ from the numerical results. The correlation of McAdam leads to a difference of about 20% and for the correlation given by Baehr of about 30%. The reason of this partially strong deviations in the temperature evolution are the different values of the Nusselt number from the different empirical correlations. To clarify this deviation, Fig. 3 shows the Nusselt numbers from these computations compared to data from literature and to the empirical Nusselt number correlations. The shown Nusselt numbers are time and space averaged (over the droplet surface). Additional to the correlations from Baehr [12] and McAdam [11] a third correlation from Whitaker [13] is added in Fig. 3. It can be seen that the

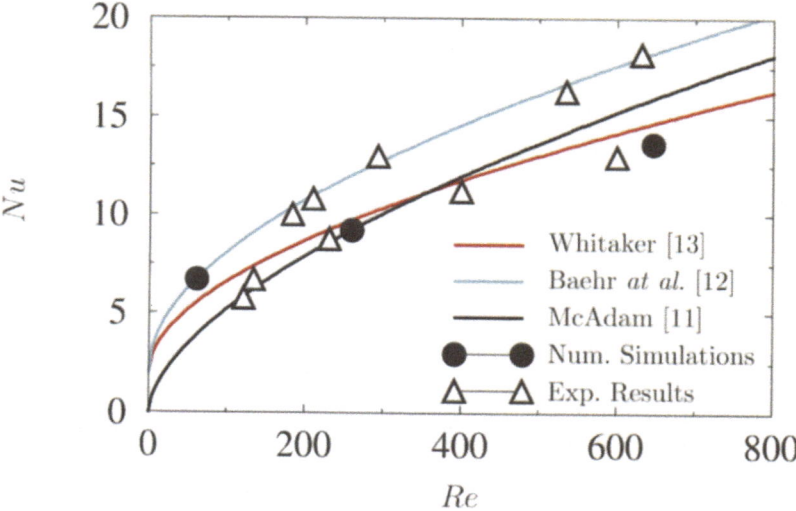

Fig. 3. Comparison of numerical and experimental results for the Nusselt number Nu in dependency of Re. The lines are correlations from the literature. The triangles are experimental data [13] and the filled spheres are results from own numerical simulations .

Nusselt number correlations differ substantially from each other. Especially the Nusselt number correlation of McAdam $Nu = 0.37Re^{0.6}Pr^{1/3}$ reaches a wrong value for heat conduction only ($Re \rightarrow 0$). The analytically obtained Nusselt number for $Re \rightarrow 0$ is $Nu = 2$. Due to this, the temperature evolution for low Reynolds numbers computed with the correlation of McAdam is not correct. This is the case for $D = 0.5 \cdot 10^{-3}$ m ($Re = 61$). The correlation of Whitaker [13] (red line) matches best with the Nusselt numbers from the numerical simulations (filled spheres). At higher Reynolds numbers $Re = 260$ and 660 the difference between the numerical results and this correlation is less then 10%. The different values from the experimental results (triangle) make clear why the three correlations differ so much from each other for the investigated Reynolds number range and that the numerical results are in the same range of accuracy than the experimental data.

In the first part of this section the accuracy of the new implemented energy equation has been demonstrated. A more detailed investigation including grid independency and a separated test of the heat conduction which indicates a good agreement with an analytical solution has been given in [4]. Now results of heat transfer computations and an examination of the temperature field around the droplet for two different Reynolds numbers will be presented.

In Figs. 4 and 5 the temperature and the velocity field of a droplet and the surrounding gas is depicted for two different Reynolds numbers. The Reynolds number of the droplet in Fig. 4 is $Re = 20$ and constant. The droplet in Fig. 5 is decelerated. Therefore this case is specified by the initial

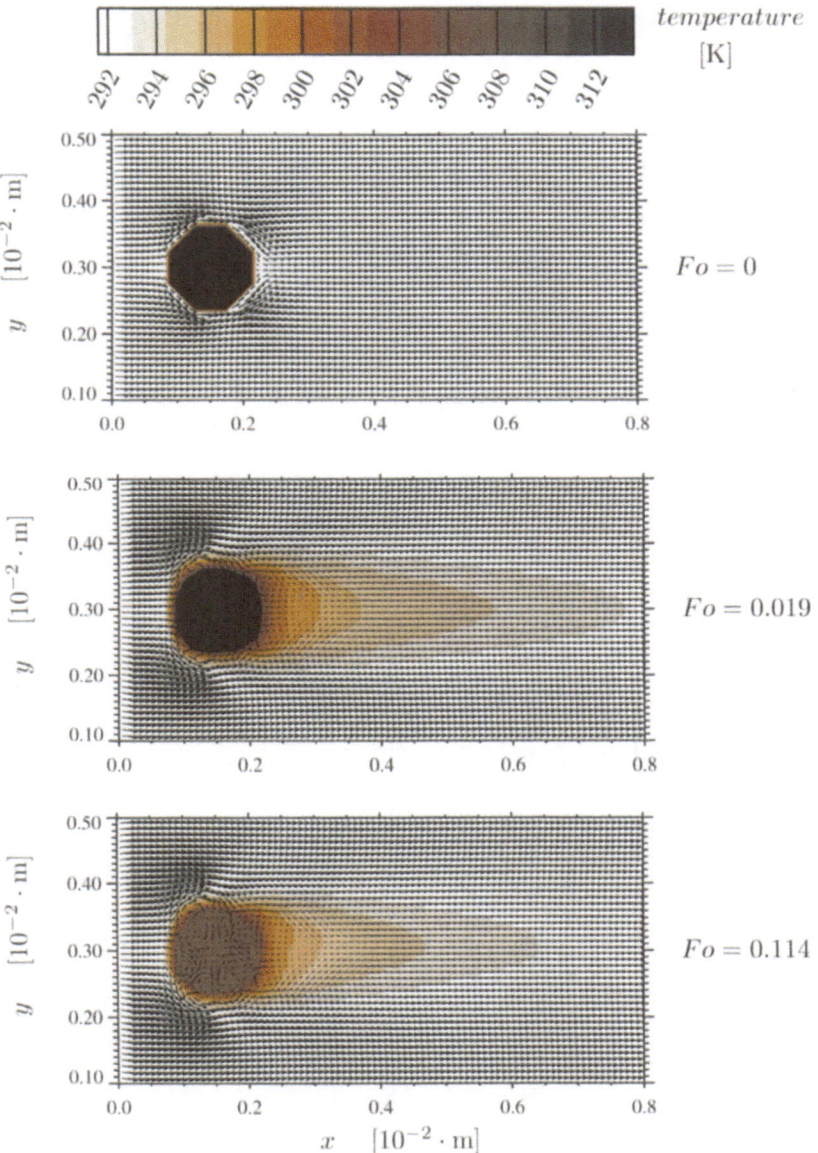

Fig. 4. Temperature and velocity field of a liquid droplet and the surrounding gas stream at different, non-dimensional times $Fo = 0$ (Fig. 4 a), $Fo = 0.019$ (Fig. 4 b) and $Fo = 0.114$ (Fig. 4 c) for $Re = 20$. Displayed is detail from a cut through the droplet center in the xy-plane.

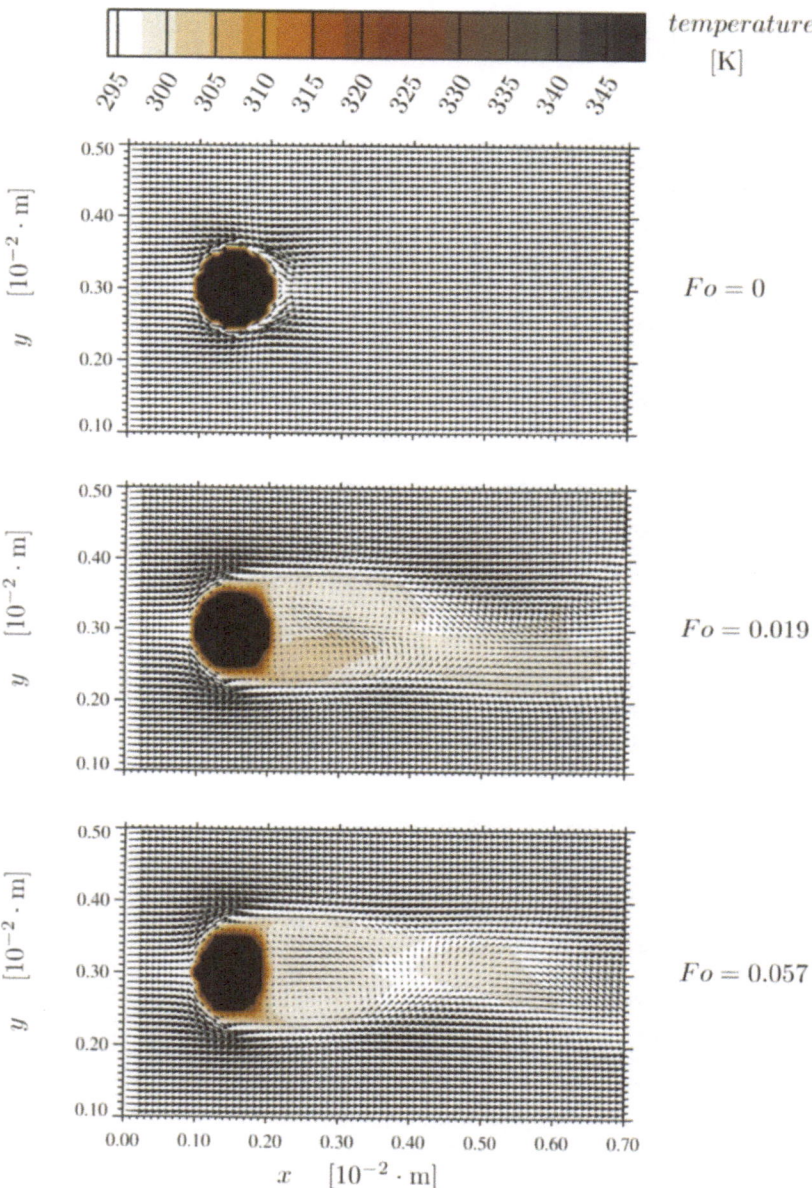

Fig. 5. Temperature and velocity field of a liquid droplet and the surrounding gas stream at different, non-dimensional times $Fo = 0$ (Fig. 5 a), $Fo = 0.019$ (Fig. 5 b) and $Fo = 0.057$ (Fig. 5 c) for $Re_0 = 660$. Displayed is detail from a cut through the droplet center in the xy-plane.

Reynolds number of $Re = 660$. In the figures the temperature and the velocity field are displayed at three different times as a detail from a cut through the droplet center in the xy-plane.

The series in Fig. 4 with $Re = 20$ is an example of a steady state, axissymmetrical flow around a droplet. The results presented here are obtained from a fully 3D simulation, to point out that the code is able to reproduce this flow type correctly. The flow in Fig. 5 is strongly different from steady state flow. The flow in this example is transient and non-symmetrical. The transition to this type of flow occurs between $Re = 210$ and 270 [1].

The initial conditions are displayed in Figs. 4a and 5a. In Figs. 4b and 4c the symmetrical flow field and the steady state velocity can be seen. Also the cooling of the droplet is clarified from the decrease of the droplet temperature and the shorter temperature wake. In Figs. 5b and 5c the 3D vortex street behind the droplet as well as the 3D evolution of the temperature field can been seen. The convective heat transfer in the flow field can also been observed. In particular in Fig. 5c a flow pattern with high temperature has been detached from the recirculation zone directly behind the droplet.

Now the temperature evolution for droplets with increasing or decreasing velocities will be studied. In order to achieve this, droplets with four different initial velocities $U_0 = 1, 4, 6$ and $10\,\mathrm{m/s}$ and with the initial temperatures $T_L = 350\,\mathrm{K}$, $T_G = 293.15\,\mathrm{K}$ have been computed. The liquid properties are the same as for water and the gas properties are the same as for air. The liquid viscosity is set to $\mu_L = \mu_{H_2O}$. The grid size is chosen as $64 \times 32 \times 32$ due to the computational costs for such a long run. As shown in [4], this grid size can

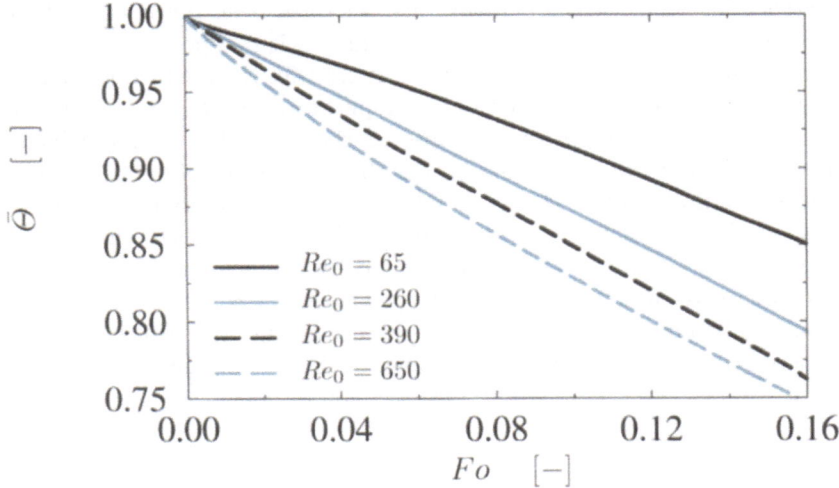

Fig. 6. Non-dimensional temperature difference of the droplet mean temperature of droplets with the diameter $D = 1 \cdot 10^{-3}$ m in dependency on the non-dimensional time Fo for different initial Reynolds numbers.

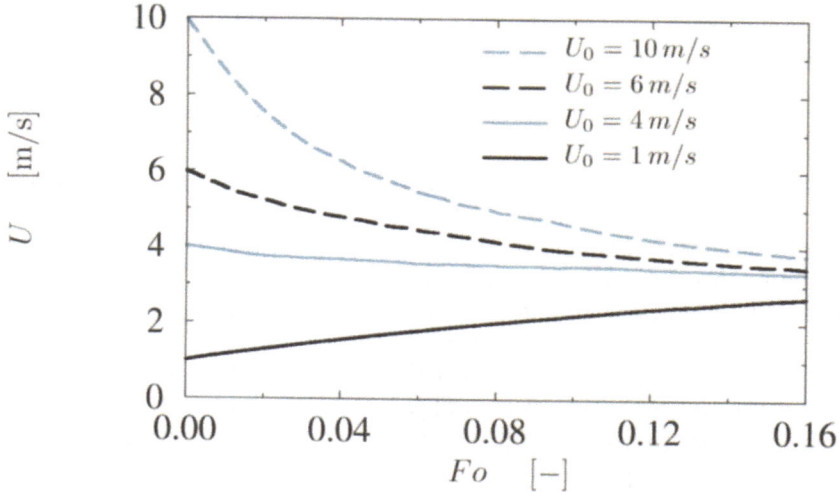

Fig. 7. Evolution of the droplet velocity U of four droplet with diameter $D = 1 \cdot 10^{-3}$ m and different initial velocities U_0 in dependency on the non-dimensional time Fo. The steady velocity of a droplet with $D = 1 \cdot 10^{-3}$ m is approximately $U = 4$ m/s.

guarantee qualitatively correct results. In Fig. 6 the temperature evolution for different initial droplet velocities represented by initial Reynolds numbers is shown in non-dimensional variables for the temperature $\bar{\Theta}$ and Fo.

Additionally the appropriate evolution of the droplet velocity is displayed in Fig. 7 as a function of Fo for the cases with the same initial droplet velocities. The smallest temperature decrease has been obtained at lowest initial Reynolds number $Re_0 = 65$ (solid line) with respect to the droplet velocity. With increasing Re_0, also the heat transfer rises, which is obviously because of the steeper decrease of the temperature for $Re_0 = 260$, 390, and 650 in Fig. 6. This is obvious because of the dependency of the Nusselt number on the Reynolds number and thus the dependency of the heat transfer coefficient on the droplet velocity. For $Fo = 0.12$ the velocity for the droplets with the initial velocity $U_0 = 4$, 6, 10 m/s is approximately the same. In the temperature evolution this is expressed by the comparable gradient in the temperature in contrast to the droplet with $U_0 = 1$ m/s which has an obviously different gradient in the temperature evolution at this time. At the considered time the velocity of this droplet is clearly different from the other cases.

4   Computational Resources

The simulation has been performed on the Cray T3E/512-900 at the HLRS. The CPU number differs between 16 CPU's and 128 CPU's depending on the job size and the required "real" time.

A typical simulation for this project takes place on a $128 \times 64 \times 64$ grid with 524288 cells on 32 parallel CPU's. An example for such a computation is the computation of the temperature and velocity field in Fig. 5. In this case the CFL-condition is the governing time restriction. The computational time is approximately 8h with 23469 time steps. To study the grid independency, same computations on a larger grid ($256 \times 128 \times 128$) with 4184304 cells on 128 parallel CPU's have been performed. Typical values for this case are a run of 12 h with 17139 time steps. Computations with this resolution haven't been shown in this paper but e.g. in [4].

Non typical simulations for this project are the simulations for the comparison with an analytical solution (Fig. 2). Here the mentioned, long "real" time is required to get comparable results and not only a very small temperature decrease. To get these results in an acceptable time small domains with $16 \times 16 \times 16$ have been used on each of the 16 CPU's. The required time for the three cases is between 10 and 20 h with 110.000 to 180.000 time steps. Additional in this computation the viscosity is very high to suppress the internal circulation so the momentum diffusion is the governing time restriction which leads to the small time steps and the long computation time.

In [14] the performance of the program FS3D has been analyzed in detail but without the energy equation.

5   Concluding Remarks

An existing 3D numerical program has been extended to enable computation of heat transfer. The comparison of the numerical results with an asymptotic solution for a heat conduction problem shows good agreement for droplets with different diameters. For different empirical correlations strong differences for the heat transfer coefficient have been observed. For unsteady moving droplets first results have been presented which show the expected behavior for different droplet velocities on the heat transfer.

Acknowledgments

The authors would like to thank the "Deutsche Forschungsgemeinschaft" (DFG) for the financial support of this project and the High-Performance Computing-Center Stuttgart (HLRS) for support and computation time on the high performance computers.

References

1. Bagchi, P., Ha, M.Y. and Balachandar, S.: Direct Numerical Simulation of Flow and Heat Transfer from a Sphere in a Uniform Cross-Flow. Journal of Fluids Engineering, 123:347–358, 2001.
2. Rieber, M., Graf, F., Hase, M., Roth, N. and Weigand, B.: Numerical simulation of moving spherical and strongly deformed droplets. In Proceedings ILASS-Europe 2000, 2000.
3. Rieber, M. and Frohn, A.: A numerical study on the mechanism of splashing. International Journal of Heat and Fluid Flow, pages 1–7, 1999.
4. Hase, M. and Weigand, B.: Numerical study of the temperature field of unsteady moving droplets and of the surrounding gas. In Proceedings ILASS-Europe 2001, Zuerich, 2001.
5. Hirt, C.W. and Nichols, B.D.: Volume of fluid (VOF) method for the dynamics of free boundaries. Journal of computational physics, 39:201–225, 1981.
6. Rider, W.J. and Kothe, D.B.: Reconstructing volume tracking. Journal of computational physics, 141:112–152, 1998.
7. Hase, M., Rieber, M., Graf, F., Roth, N. and Weigand, B.: Parallel computation of the time dependent velocity evolution for strongly deformed droplets. In High-Performance Computing in Science and Engineering 2001: Transactions of the High Performance Computing Center Stuttgart (HLRS), pages 342–351. Springer-Verlag, 2001.
8. Hirsch, C.: Numerical computation of internal and external flows, volume 2. John Wiley and Sons, 1990.
9. Fourier, J.-B.J.: Theorie Analytique de la Chaleur. Gauthier-Villars, Paris, 1822.
10. Martin, H. and Saberian, M.: Improved asymptotic approximations for transient conduction and diffusion processes. Chem. Engineering and Processing, 33:205–210, 1994.
11. McAdam, W.H.: Heat transmission. McGraw-Hill, New-York, 3rd Auflage, 1954.
12. Baehr, H.D. and Stephan, K.: Wärme- und Stoffübertragung. Springer-Verlag, Berlin, Heidelberg, New York, 1994.
13. Whitaker, S.: Forced convection heat transfer correlations for flow in pipes, past flat plates, single cylinders, single spheres and for flow in packed beds and tube bundles. AIChE Journal, 18(2):361–371, 1972.
14. Rieber, M. and Frohn, A.: Parallel Computation of Interface Dynamics in Incompressible Two-Phase Flows. In High-Performance Computing in Science and Engineering 99: Transactions of the High Performance Computing Center Stuttgart (HLRS), pages 241–252. Springer Verlag, 2000.

Implicit LU Time Integration Using Domain Decomposition and Overlapping Grids

Peter Gerlinger[1], Peter Stoll[2], Fernando Schneider[1], and Manfred Aigner[1]

[1] Institut für Verbrennungstechnik, DLR Stuttgart
   Pfaffenwaldring 38-40, 70569 Stuttgart, Germany
[2] Institut für Thermodynamik der Luft- und Raumfahrt
   Pfaffenwaldring 31, 70569 Stuttgart, Germany
   now Alstom Power

Abstract. An implicit LU-SGS (Lower-Upper Symmetric Gauss-Seidel) algorithm is used for the simulation of reactive and non-reactive three dimensional high speed flows (supersonic combustion). The numerical method is based on an all Mach number preconditioning to enable convergence of the compressible flow solver in the low Mach number limit. The code is fully vectorized and may be used on massively parallel computers using MPI. Parallelization is performed by domain decomposition which causes losses in efficiency of the implicit numerical solver. Therefore overlapping domains are introduced to reduce both losses in convergence rate as well as in robustness. A comparison with non-intersecting grids demonstrates the effectiveness of this method. To investigate both approaches a 3-D turbulent Mach 3.85 supersonic ramp flow with shock wave boundary-layer interaction is chosen. The simulations use up to 256 nodes on a Cray T3E and up to 16384 blocks for the discretization of the computational domain. In addition results for a 3-D supersonic ramp combustor are presented. The finite-rate chemistry reaction mechanism involves 20 reactions and 9 different species.

1 Introduction

For the simulation of complex three dimensional flows explicit numerical time integration schemes are often prefered because of their simplicity. They usually require less computer time per time step than implicit methods and are more easy to parallelize by domain decomposition. On the other hand explicit schemes suffer from the CFL criterion allowing only very small time steps in many cases. Moreover, an explicit time integration of chemical source terms in case of finite-rate chemistry is practically impossible. The set of governing equations usually becomes extremely stiff in case of combustion and therefore requires at least a point implicit numerical method. Stiffness problems also result from highly stretched grids or from low-Reynolds number turbulence closures in regions of flow separation. In such cases implicit solvers are superior to explicit ones at the cost of a higher numerical effort per time step.

A very efficient implicit method that is often used for high speed flows is the LU-SGS (Lower-Upper Symmetric Gauss-Seidel) algorithm proposed by Jameson and Yoon [1,2]. As for any implicit scheme there are data dependencies in case of parallelization by domain decomposition. There are several possibilities to deal with this problem. In case of a 3-D LU-SGS or a line Gauss-Seidel scheme the computational grid may be decomposed in two dimensional subregions without data dependencies. However, for a large number of domains this may result in a large number of small stripes making an implementation inefficient. Another possibility is the modification of the algorithm to make it data parallel [3,4]. The most simple and most general method for domain decomposition in the implicit part of a numerical scheme is an explicit coupling of the blocks at its boundaries. This technique is easy to implement and vectorization is possible within any domain. On the other hand robustness and efficiency decreases with an increasing number of domains [5,6]. For this reason overlapping grids are used in the present paper (for the implicit part of the numerical solver only) and a detaild comparison with non-overlapping grids is given. All simulations are performed an a Cray T3E at the HLRS Stuttgart.

2 Governing Equations

The simulation of supersonic combustion requires the solution of the full compressible Navier-Stokes, turbulence and species transport equations as given by

$$\frac{\partial \mathbf{Q}}{\partial t} + \frac{\partial (\mathbf{F} - \mathbf{F}_\nu)}{\partial x} + \frac{\partial (\mathbf{G} - \mathbf{G}_\nu)}{\partial y} + \frac{\partial (\mathbf{H} - \mathbf{H}_\nu)}{\partial z} = \mathbf{S} \qquad (1)$$

where the conservative variable vector is

$$\mathbf{Q} = \left[\bar{\rho}, \bar{\rho}\tilde{u}, \bar{\rho}\tilde{v}, \bar{\rho}\tilde{w}, \bar{\rho}\tilde{E}, \bar{\rho}q, \bar{\rho}\omega, \bar{\rho}\tilde{Y}_i \right]^T, \qquad i = 1, 2, ..., N_k - 1, \qquad (2)$$

$\mathbf{F}$, $\mathbf{G}$ and $\mathbf{H}$ are inviscid, and $\mathbf{F}_\nu$, $\mathbf{G}_\nu$ and $\mathbf{H}_\nu$ are viscous fluxes in x-, y- and z-direction, respectively. The source vector $\mathbf{S}$ is given by

$$\mathbf{S} = \left[0, 0, 0, 0, 0, S_q, S_\omega, \overline{S}_i \right]^T, \qquad i = 1, 2, ..., N_k - 1, \qquad (3)$$

where S_q and S_ω are source terms of the q-ω model [8] and S_i are source terms resulting from chemistry. The variables in eq. (2) are the density $\bar{\rho}$, the velocity components $\tilde{u}$, $\tilde{v}$, and $\tilde{w}$, the total specific energy $\tilde{E}$, the turbulence variables $q = \sqrt{k}$ (k = turbulent kinetic energy) and $\omega = \epsilon/k$ (ϵ = dissipation rate of k) and the species mass fractions $\tilde{Y}_i$. N_k is the number of different species. The simulation of hydrogen combustion involves a 9-species (N_2, O_2, H_2, H_2O, OH, O, H, HO_2, and H_2O_2), 20-step reaction mechanism [7]. For turbulence closure a two-equation low-Reynolds-number q-ω turbulence model [8,9] is employed.

3   Numerical Scheme

The computational domain is discretized by a cell centred finite-volume approach. To achieve a higher flexibility the structured grid may be decomposed into an arbitrary number of different blocks. For supersonic combustion the numerical method should be able to deal with a number of requirements:

- *numerical stiffness* of the set of governing equations due to chemistry,
- *high cell aspect ratio grids* that appear in case of high Reynolds number flows in all near wall regions,
- *large numbers of transport equations* requiring a fast and efficient method for time integration,
- *multigrid techniques* for convergence acceleration (the solver has to damp out efficiently high frequency error components),
- *vectorization* (the solver should be able to be vectorized both in the explicit and implicit part),
- *parallelization* (the solver should work efficiently on massively parallel systems.

While from the numerical point of view an implicit scheme for time integration should be used to fullfill most of these points, vectorization and parallelization becomes more difficult than for an explicit solver. The implicit LU-SGS algorithm seems to be a good compromize concerning all mentioned requirements and therefore is chosen for all investigations to follow. First we explain the numerical method for time integration and next modifications that improve the performance of the scheme in case of domain decomposition and a large number of different blocks.

3.1   The Standard LU-SGS Algorithm

After an implicit discretization of eq. (1), first order linearization in time for all non linear fluxes and all Mach number preconditioning [10] a linear system of governing equations has to be solved. This set of governing equations is expressed symbolically by

$$(D + L + U)\, \Delta\mathbf{Q}^{n+1} = -\Delta t\, \mathbf{R} , \qquad (4)$$

where the complete left hand side operator is decomposed into a diagonal operator D, a lower L, and an upper U triangular operator. These operators include Jacobians of inviscid and viscous fluxes as well as the source term Jacobian resulting from chemistry. Finally $\mathbf{R}$ is the discretized residual vector and $\Delta\mathbf{Q}$ the update during one time step Δt. In case of the LU-SGS algorithm eq. (4) is approximately factored and solved in two consecutive steps [1,11]:

Lower sweep: $\qquad\qquad (D + L)\; \Delta\bar{\mathbf{Q}} = -\Delta t\, \mathbf{R} \qquad\qquad (5)$

Upper sweep: $\qquad\qquad (D + U)\; \Delta\mathbf{Q}^{n+1} = D\; \Delta\bar{\mathbf{Q}} \qquad\qquad (6)$

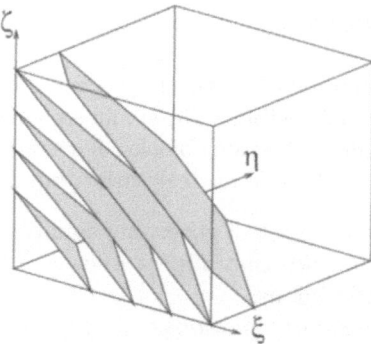

Fig. 1. Lower sweep along 2D computational diagonal planes through the 3D grid.

and the solution is updated by $\mathbf{Q}^{n+1} = \mathbf{Q}^n + \Delta\mathbf{Q}^{n+1}$. Chemistry is treated in an implicit fashion and fully coupled with the fluid motion. The source and viscous Jacobians add to the diagonal D forming a matrix which has to be inverted directly at every grid point. There are two major steps to be performed for any iteration: The purely explicit part of residual calculation (RHS – right hand side) and the implicit part (LHS – left hand side) where the L and U sweeps have to be performed. As to allow a good vector performance the implicit part (left sides of eq. (5) and (6)) has to be fully vectorized too. This is possible by sweeping along diagonal planes through the computational domain. The procedure is shown in Fig. 1 for the lower sweep (the upper sweep is performed in opposite direction). Because the volumes located on diagonal planes with $i+j+k = const.$ (on a structured grid) are computational independent it is possible to vectorize the implicit part for all volumes located within such a plane. With exception of the corner planes the vector length is high resulting in good vector performances. For 2D simulations the situation is worse due to shorter vector lengths [12]. Nevertheless, a complete vectorization is possible for the implicit LU-SGS scheme in both cases. Because the implicit part is data parallel for all volumes located in one plane, the algorithm could be parallelized by parts of these planes too. Unfortunately this is impracticable and inefficient to implement on computers with distributed memory.

3.2   LU-SGS Algorithm with Overlapping Grids

A basic problem of any implicit solver is that the line of information is interrupted at block boundaries in case of domain decomposition. This is especially severe if the number of domains is high. Additionally the vector performance decreases due to smaller vector lengths with decreasing block sizes. If the boundaries of a domain are treated explicitly the implicit solver will become more and more an explicit one if the number of blocks increases. For large three dimensional problems and moderate block numbers this is acceptable

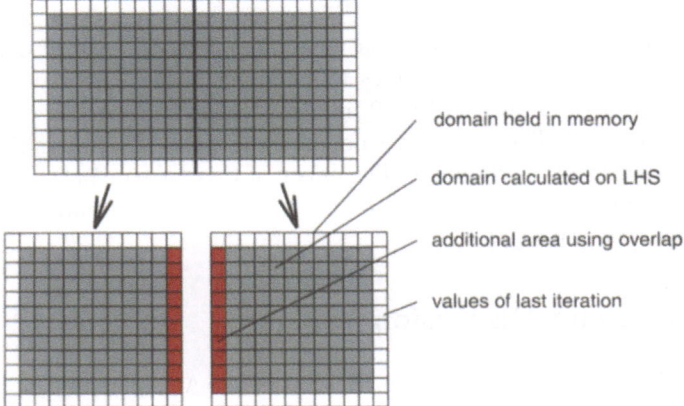

Fig. 2. Domain decomposition (2 domains) with and without overlapping grids used in the implicit part of the algorithm.

in most cases. However, for hundreds or thousands of blocks a significant reduction in convergence rate and stability can occur. In case of the LU-SGS algorithm a block size of $1 \times 1 \times 1$ cells would degrade the solver to a point Jacobi iteration. Moreover, due to the decreasing implicitness high CFL numbers may not be used any longer. On the other hand an explicit domain boundary treatment (in the implicit part) allows an easy parallelization even on massively parallel systems and therefore is prefered by most users.

One possibility to reduce losses caused by domain decomposition is the use of overlapping grids in the implicit part of the solver. Advantages and disadvantages of such a treatment are investigated for the LU-SGS algorithm. Both methods are identical for the RHS calculation but differ in the implicit part (LHS). Therefore all mentioned changes only refer ro the LHS. Instead of solving the linear set of governing equation for the exact size of a computational domain the modified version uses domains that are extended by one row of volumes in any coordinate direction. This is done as long as there are no physical boundaries. The advantage of this method is that better boundary values are available than in case of a purely explicit treatment. This will stabilize the algorithm and improve the convergence rate. The additional computational work is relatively low because the RHS and Jacobian calculations have to be performed anyway. Thus the additional work is limited to the extension of the L and U sweeps to the overlapping volumes. The procedure is demonstrated for a simple test case. Figure 2 shows a single grid (upper figure) where the grey part indicates the computational domain while the white volumes are surrounding boundary cells. The lower figures show the domain splitted into 2 blocks. The procedure with and without overlapping grids is quite similar. In both cases the calculation of the residuals **R** is performed first. This is done for the interior (grey) volumes of both blocks. Exactly the same volumes are updated by the LHS in case of non-overlapping grids. The LHS coupling between the blocks is purely explicit. In case of overlapping grids one additional line of volumes is included in the LHS calculation for both blocks. This is shown in Fig. 2 by the volumes indicated as "additional area using overlap". The LU sweeps now include the additional indicated volumes resulting from overlapping. The effect of this treatment is investigated for a three dimsional test case and is compared with the standard technique using non-overlapping grids.

4   Results and Discussion

Results will be presented for two 3D test cases: A supersonic nonreacting flow which serves to investigate the influence of the overlapping grids and a supersonic combustion chamber with ramp injector. The second simulation involves combustion based on finite-rate chemistry.

4.1   Non-Reacting Mach 3.85 Test Case

The experiment investigated is a Mach 3.85 fully turbulent flow over a double ramp system [13,5]. It includes shock wave boundary layer interactions, separation zones as well as subsonic zones with high turbulence intensites. Figure 3 shows a sketch of the geometry as well as inflow conditions and a strongly coarsened computational grid. This test case has been investigated experimentally by Garrison et al. [13,14] using planar laser scattering.

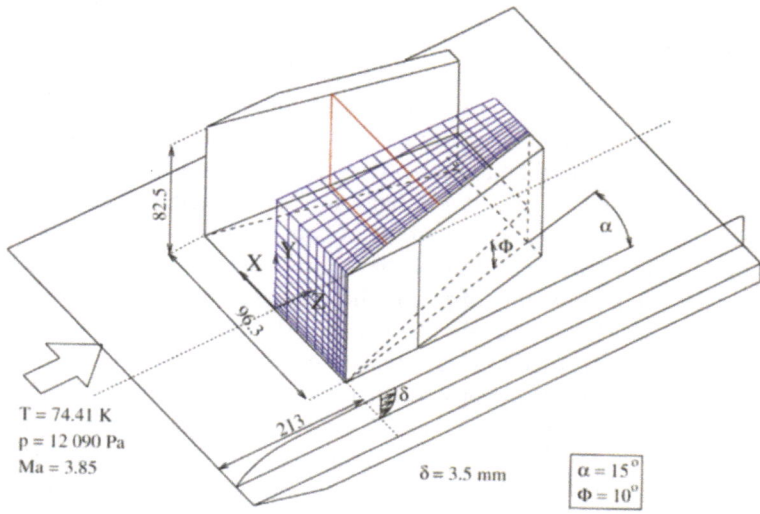

Fig. 3. Sketch of the experimental set up [13], inflow conditions, and computational grid (strongly coarsend) for the supersonic test case.

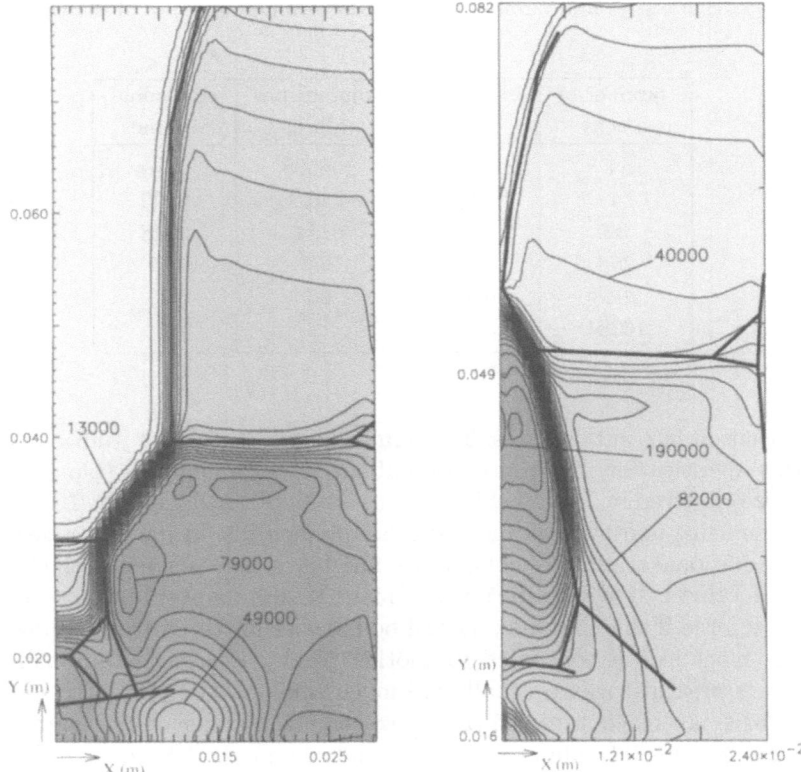

Fig. 4. Calculated pressure distribution (Pa) at $z = 70.35$ mm (left side) and $z = 89.95$ mm (right side). The thick lines correspond to measured flow structures [13,14].

Figure 4 shows calculated pressure distributions as well as experimentally obtained schock structures (thick lines). Simulation and experiment are in a good overall agreement. The grid size for the right part of the symmetric problem (see Fig. 3) is $48 \cdot 64 \cdot 64$ volumes in x-, y-, and z-direction, respectively. A constant grid spacing is used in z-direction. Due to the low-Reynolds number turbulence closure very fine grids are required at all near wall regions. For the present case the distance of near wall cell centres to solid wall is about $0.8 \cdot 10^{-6}$ m. The resulting cell aspect ratios are up to 1350. More datails concerning this simulation may be found by Stoll [5,10].

Domain decomposition

The effect of domain decomposition is investigated for the described test case. For this invesigation the computational grid is divided by domain de-

Table 1. Block sizes and number of blocks for 6 different types of domain decomposition.

number of blocks	block size	volumes per block	additional volumes
1	48·64·64	196608	0
4	48·32·32	49152	8192
32	24·16·16	6144	2048
256	12·8·8	768	512
2048	6·4·4	96	128
16384	3·2·2	12	32

composition into subdomains. The number of blocks ranges from 1 to 16384 in this investigation. All information about these simulations (block number and size) is given in Table 1. Simulations are performed with the LU-SGS algorithm using non-overlapping and overlapping grids. In the latter case there is an additional number of volumes for any block that is given in the last column of Table 1. However, the given number of additional volumes per block is only required if there are no physical boundaries for the corresponding block (every block side is connected to another block). Especially for small block numbers practical values usually are much smaller. The table also indicates that for small block sizes the ratio between the number of overlapping cells and the number of volumes increases and the method becomes inefficient. The chosen test case has about 200 000 volumes only and for moderate block numbers (256-512) the algorithm still works very efficient despite the relatively high number of additional volumes. The calculations are perforemd on a Cray T3E employing up to 256 nodes using MPI. Because the single block calculation required more memory than available on one node of this machine the one block simulation is performed on a work station. For the simulations with 2048 and 16384 blocks each of the 256 nodes worked with 8 and 64 blocks, respectively. As to achieve a similar interprocessor communication in case of more than 1 block per CPU, no neighbouring blocks are kept on one node. The CFL number was chosen to be 10 and the simulation was stopped when the density residual dropped below a prespecified limit. Figure 6 shows convergence histories versus the number of iterations without (left side) and with (right side) overlapping grids. In both cases convergence is possible using up to 2048 blocks. With 16384 blocks the overlapping grids still enables a residual drop of 3 orders of magnitude while the original method fails to converge. Moreover it may be seen that without overlapping grids there is a stronger slow down of the convergence rate if the number of blocks is increased. Figure 5 shows a comparison between both simulations. Plotted are the required numers of iterations for convergence versus the number of blocks employed. For the 2048 block case 12% more iterations are required (in comparison to the 1 grid simulation) if overlapping grids are used und

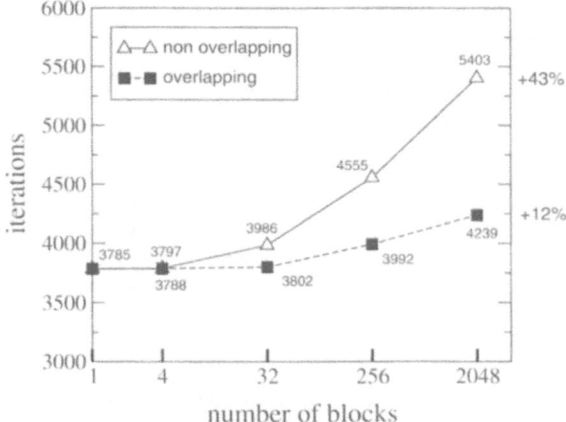

Fig. 5. Required numbers of iterations to achieve convergence in case of overlapping and non-overlapping grids versus the number of blocks.

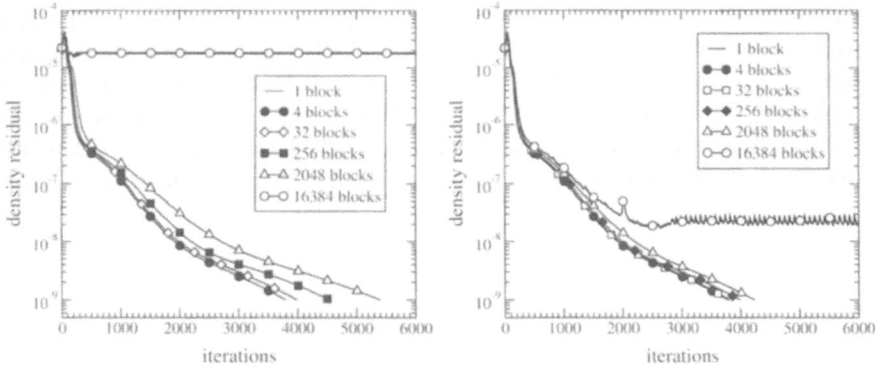

Fig. 6. Convergence histories using domain decomposition for different numbers of blocks. Left side without overlapping grids, right side with overlapping grids.

43% more iterations with non-overlapping grids. Thus the convergence rate as well as robustness is improved by the introduction of overlapping grids. This is paid for by an increase in required CPU time for one iteration. This may be seen from Table 2. The additional CPU time in case of overlapping volumes is relatively low because only about 20% of the total computational time is required for the L and U sweeps. Thus the use of overlapping grids is not only more robust but may also save CPU time (depending on the grid and block sizes). The differences in CPU time between both methods are small if up to 256 blocks are used. However, for the 2048 block case there is a reduction of 14% in CPU time by using overlapping grids.

Table 2. Number of iterations and computational times for convergence using non-overlapping and overlapping grids.

blocks	CPUs	non-overlapping		overlapping	
		iterations	time (h)	iterations	time (h)
4	4	3797	19.17	3788	19.62
32	32	3986	2.827	3802	2.791
256	256	4555	.4924	3992	.4889
2048	256	5403	.6425	4239	.5660

4.2   Ramp injector

The second test case considered is the simulation of a combustion experiment performed by McDaniel et al. [15] and Gauba et al. [16]. Figure 7 shows a sketch of the combustor with inflow conditions (upper figure) and the computational grid (lower figure, only every fourth grid line is plotted). Preheated hydrogen is injected with a Mach number of 1.7 into a supersonic air flow with a Mach number of 2.07. Downstream of the ramp a recirculation zone appears where ignition takes place. The simulation of hydrogen combustion is based on finite-rate chemistry [7]. To save CPU time the first channel part from the air nozzle throat to ramp injector is calculated in a separate 2D simulation. These results are used as inflow conditions for the subsequent 3D simulation that covers the downstream part of the combustor indicated as computational domain in Fig. 7. The grid ist strongly refined at all near

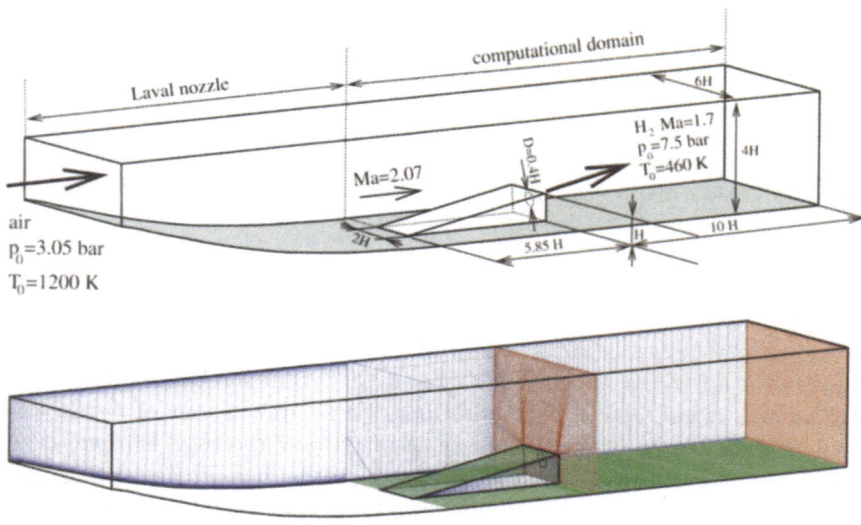

Fig. 7. Compustor geometry ($H = 6.35$ mm) and inflow conditions (upper figure). Computational grid (every fourth grid line is plotted, lower figure).

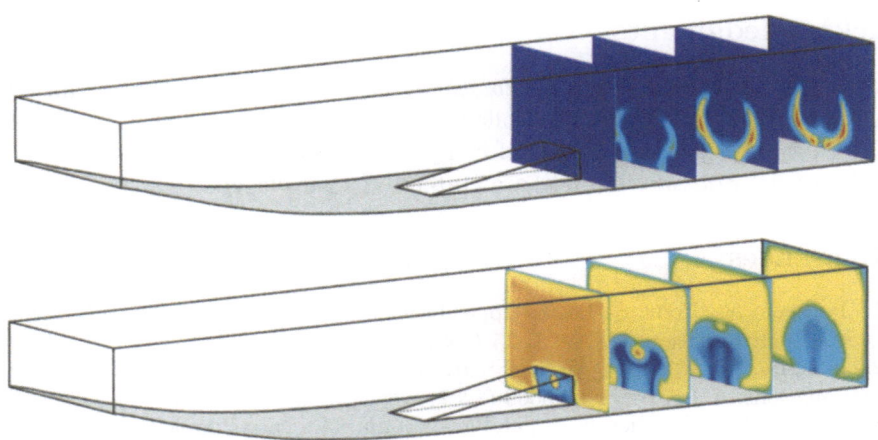

Fig. 8. Calculated OH distribution (upper figure) and Mach number distribution (lower figure) for a combustion chamber with ramp injector.

wall regions. For the 3D simulation about 1 million volumes are used. All y^+ values for the converged solution are below 2 what is a requirement for the low-Reynolds number turbulence closure. Due to physical boundary conditions a 7 block grid is created that is further divided into 117 blocks for the calculation on a Cray T3E [10]. The computational time for this simulation is about 6 h using 117 CPUs. Finally Fig. 8 shows the calculated OH molar fraction (upper figure) and Mach number (lower figure) distribution of this simulation. More details may be found by Stoll [10].

5   Conclusions

A domain decomposition method for parallel simulations of turbulent flows with an implicit method for time integration (an LU-SGS algorithm) has been presented. In the implicit part of the algorithm the use of non-intersecting and overlapping grids is investigated. It is shown that the convergence rate as well as robustness is improved by using overlapping grids. This is paid for by higher CPU times per iteration. While the difference between both methods is small for small block numbers the advantage of overlapping grids increases with increasing block numbers. For the test case investigated a 12% reduction in CPU time was obtained for a grid consisting out of 2049 blocks. The same kind of domain decomposition is used successfully for the simulation of high speed reactive flows with finite-rate chemistry.

Acknowledgments

We wish to thank the *Deutsche Forschungsgemeinschaft (DFG)* for financial support of this work within the Collaborative Research Center SFB 259 at the University of Stuttgart. Grants of computer time from the *Höchstleistungs-rechenntrum Stuttgart (HLRS)* are gratefully acknowledged.

References

1. Jameson, A. , Yoon, S.: Lower-Upper Implicit Schemes with Multiple Grids for the Euler Equations. AIAA J. **25** (1987) 929–937
2. Jameson, A. , Yoon, S.: An LU-SSOR Scheme for the Euler and Navier-Stokes Equations. AIAA paper 87-0600 (1987)
3. Wright, M. J., Candler, G. V., and Prampolini, M.: Data-Parallel Lower-Upper Relaxation Method for the Navier-Stokes Equations. AIAA J. **34** (1996) 1371–1377
4. Candler, G. V., Wright, M. J., and McDonald, J. D.: Data-Parallel Lower-Upper Relaxation Method for Reacting Flows. AIAA J. **32** (1994) 2380–2386
5. Stoll, P., Gerlinger, P., and Brüggemann, D.: Domain Decomposition for an Implicit LU-SGS Scheme using Overlapping Grids. AIAA paper 97-1896 (1997)
6. Jenssen, C. B.: Implicit Multiblock Euler and Navier-Stokes Calculations. AIAA J. **32** (1994) 1808–1814
7. Jachimowski, C. J.: An Analytical Study of the Hydrogen-Air Reaction Mechanism with Application to Scramjet Combustion. NASA TP 2791 (1988)
8. Coakley, T. J., Huang, P. G.: Turbulence Modeling for High Speed Flows. AIAA paper 92-0436 (1992)
9. Gerlinger, P., Brüggemann, D.: An Implicit Multigrid Scheme for the Compressible Navier-Stokes Equations with Low-Reynolds-Number Turbulence Closure. J. Fluids Eng. **120** (1998) 257–262
10. Stoll, P.: Entwicklung eines parallelen Mehrgitterverfahrens zur Simulation der Verbrennung in kompressiblen und inkompressiblen Strömungen. VDI Fortschritt-Berichte, Nr. 411, Reihe 7, Strömungstechnik (2001)
11. Shuen, J. S.: Upwind Differencing and LU Factorization for Chemical Non-Equilibrium Navier-Stokes Equations. J. Comp. Phys. **99** (1992) 233–250
12. Gerlinger, P., Aigner, M.: Assumed PDF Modeling for Finite-Rate Chemistry. High Performance Computing in Science and Engineering '01, Edts. E. Krause and W. Jäger (2001) 315–328
13. Garrison, T. J., Settles, G. S., and Horstman, C. C.: Measurements of the Triple Shock Wave/Turbulent Boundary Layer Interaction. AIAA J. **34** (1996) 57–64
14. Garrison, T. J., Settles, G. S., and Horstman, C. C.: Measurements of the Triple Shock Wave/Turbulent Boundary Layer Interaction. AIAA paper 94-2274 (1994)
15. McDaniel, J. C., Gauba, G., Quagliaroli, T. M., Grinstead, J. H., Laufer, G., Kreauss, R. H., Whitehurst, R. B., and Victor, K. G.: Combustion of Hydrogen in Mach 2 Air Using an Unswept Ramp Fuel Injector: A Test Case for CFD Validation. AIAA paper 94-2521 (1994)
16. Gauba, G., Haj-Hariri, H., and McDaniel, J. C.,: Numerical and Experimental Investigation of Hydrogen Combustion in a Mach 2 Airflow with an Unswept Fuel Ramp Injector. AIAA paper 95-2562 (1995)

Efficient Divergence Cleaning in Three-Dimensional MHD Simulations

A. Dedner, D. Kröner, C. Rohde, and M. Wesenberg

Institut für Angewandte Mathematik, Universität Freiburg,
Hermann-Herder-Str. 10, 79104 Freiburg, Germany

Abstract. We present the results of first realistic simulations using our state-of-the-art MHD code on unstructured tetrahedral meshes in 3d. The code incorporates local grid adaption with dynamic load balancing and relies on a recently proposed approximate Riemann solver. We demonstrate that it is absolutely crucial to control the divergence of the magnetic field and that our new hyperbolic divergence cleaning approach works well also in 3d.

1 Introduction

Our project is part of a cooperation with the group of M. Schüssler at the Max-Planck-Institut für Aeronomie in Katlenburg-Lindau under the auspices of the DFG priority research program "Analysis and Numerics for Conservation Laws". The aim of this project is to develop and implement numerical methods which allow simulations within two different parts of the solar atmosphere: In the *lower convection zone*, which starts $2 \cdot 10^5$ km below the visible surface and has a height of 10^5 km, we have to resolve localized structures in a huge (physically unbounded) domain. This requires local grid adaption and a suitable treatment of the artificial boundaries. The second domain is the solar *photosphere* ranging from 1000 km below to 1000 km above the visible surface. Here we have to deal with shocks and a complex equation of state due to the partial ionization of the plasma. Moreover, for the balance of energy we have to consider the non-local transport of radiation. In both domains we have to ensure that the discrete divergence of the approximation to the magnetic field remains small.

In recent years we developed new approaches to all of the challenges mentioned above. In [3] we proposed transparent boundary conditions suitable for stratified atmospheres. A very efficient approximate MHD Riemann solver for use in one- and multidimensional codes was introduced in [12]. Extensions of MHD solvers to the case of non-perfect gases were studied in [4, 7]. A class of new solvers for the radiation transport equation was introduced in [6]. Finally, we suggested the hyperbolic divergence cleaning [2, 5]. We implemented and tested all these schemes in our 2d code. Our 3d finite volume code is based on the locally adaptive solver for gas dynamics with dynamic load balancing developed in [10] using MPI. We are now ready to tackle 3d-simulations in the lower convection zone; the necessary extensions for the photosphere (radiation transport and non-perfect gases) are imminent.

The physical mechanism behind the development of sunspots is one of the questions one seeks to answer by numerical simulations. The mechanism currently conjectured works as follows: In the lower convection zone localized magnetic field concentrations, so-called *magnetic fluxtubes*, develop, which rise through the solar atmosphere and finally form the tiny filaments at the boundary of sunspots. However, this only works if the initially weak magnetic field is amplified by several orders of magnitude, which could be caused by conversion of potential energy [9]. The 3d-simulation we present also fits into this context. However, our main concern is not to discuss and interpret the physical relevance of our results, but to demonstrate the effectiveness of our new parallel 3d MHD solver with divergence cleaning within this setting.

The outline of this paper is as follows: In Sect. 2 the MHD equations are introduced; the numerical scheme is presented in Sect. 3. A brief description of the hyperbolic divergence cleaning is contained in Sect. 4. The setting and the results for our example, an exploding magnetic fluxtube, are discussed in Sect. 5. Finally, our conclusions are summarized in Sect. 6.

2   Governing Equations

The ideal MHD equations are a model for a compressible, inviscid, and electrically conducting gas in the presence of a magnetic field. They are derived from the Euler equations of gas dynamics and the Maxwell equations. Written in divergence form they read

$$\partial_t \rho + \nabla \cdot (\rho \mathbf{u}) = 0 \qquad \text{(conservation of mass),} \qquad (1a)$$

$$\partial_t (\rho \mathbf{u}) + \nabla \cdot (\rho \mathbf{u} \mathbf{u}^T + \mathcal{P}) = -\rho \mathbf{g} \qquad \text{(balance of momentum),} \qquad (1b)$$

$$\partial_t \mathbf{B} + \nabla \cdot (\mathbf{u} \mathbf{B}^T - \mathbf{B} \mathbf{u}^T) = 0 \qquad \text{(induction equations),} \qquad (1c)$$

$$\partial_t (\rho e) + \nabla \cdot (\rho e \mathbf{u} + \mathcal{P} \mathbf{u}) = -\rho \mathbf{g} \cdot \mathbf{u} \qquad \text{(balance of energy),} \qquad (1d)$$

$$\nabla \cdot \mathbf{B} = 0 \qquad \text{(divergence constraint).} \qquad (1e)$$

For the density ρ, the velocity $\mathbf{u} = (u_x, u_y, u_z)^T$, the magnetic field $\mathbf{B} = (B_x, B_y, B_z)^T$, the total energy e, and the unit tensor $\mathcal{I}$, the pressure tensor $\mathcal{P}$ is defined as

$$\mathcal{P} := \left(p + \frac{1}{8\pi} |\mathbf{B}|^2 \right) \mathcal{I} - \frac{1}{4\pi} \mathbf{B} \mathbf{B}^T. \qquad (2)$$

The total energy is the sum of the internal, kinetic, and magnetic energies

$$e = \varepsilon + \frac{1}{2} |\mathbf{u}|^2 + \frac{1}{8\pi\rho} |\mathbf{B}|^2. \qquad (3)$$

Since we consider the case of a perfect gas the gas pressure is given by the equation of state

$$p = (\gamma - 1)\rho\varepsilon \qquad (4)$$

for a constant adiabatic exponent $\gamma > 1$. $\mathbf{g} = \mathbf{g}(z) = \left(0, 0, g(z)\right)^T$ with $g(z) \geq 0$ is a prescribed function describing the gravitational force which acts in negative z-direction. All quantities depend on the space variables $\mathbf{x} = (x, y, z)^T \in \mathbb{R}^3$ and the time $t \geq 0$. The system is closed by suitable initial and boundary conditions.

The induction equations (1c) can also be written in the form $\partial_t \mathbf{B} + \nabla \times (\mathbf{B} \times \mathbf{u}) = 0$, which yields directly $\partial_t(\nabla \cdot \mathbf{B}) = 0$. Therefore the divergence constraint (1e) can be considered to be a condition for the initial data on the analytical level. Nevertheless, this condition requires special attention in numerical simulations, see Sect. 3. For the purpose of this paper it is convenient to rewrite the system (1a)–(1d) in the equivalent form

$$\partial_t \mathbf{U} + \partial_x \mathbf{F}_x(\mathbf{U}) + \partial_y \mathbf{F}_y(\mathbf{U}) + \partial_z \mathbf{F}_z(\mathbf{U}) = \mathbf{S}(\mathbf{x}, \mathbf{U}), \tag{5}$$

where $\mathbf{U} := (\rho, \rho u_x, \rho u_y, \rho u_z, B_x, B_y, B_z, \rho e)^T$ is the vector of conservative variables. $\mathbf{S}(\mathbf{x}, \mathbf{U})$ consists of zeros and the gravitational source terms. $\mathbf{F}_x$, $\mathbf{F}_y$, and $\mathbf{F}_z$ are the fluxes in x-, y-, and z-direction, respectively.

Within the construction of the numerical scheme we exploit the fact that the system (1a)–(1d) without the gravitational source terms is a hyperbolic system of conservation laws and is invariant with respect to rotations in the following sense: For $\mathbf{n} = (n_x, n_y, n_z)^T \in \mathbb{R}^3$ with $|\mathbf{n}| = 1$ the matrix

$$\mathcal{B}(\mathbf{n}) := \begin{cases} \begin{pmatrix} n_x & n_y & n_z \\ -\dfrac{n_y}{\sqrt{1-n_z^2}} & \dfrac{n_x}{\sqrt{1-n_z^2}} & 0 \\ -\dfrac{n_x n_z}{\sqrt{1-n_z^2}} & -\dfrac{n_y n_z}{\sqrt{1-n_z^2}} & \sqrt{1-n_z^2} \end{pmatrix} & (|n_z| < 1) \\[4ex] \begin{pmatrix} 0 & 0 & \operatorname{sgn}(n_z) \\ 0 & 1 & 0 \\ -\operatorname{sgn}(n_z) & 0 & 0 \end{pmatrix} & (|n_z| = 1) \end{cases}$$

is an orthogonal matrix, i.e., $\mathcal{B}^{-1} = \mathcal{B}^T$, with $\det \mathcal{B} = 1$. Thus the linear mapping $\mathbf{y} \mapsto \mathcal{B}(\mathbf{n})\mathbf{y}$ for $\mathbf{y} \in \mathbb{R}^3$ is a rotation. Using

$$\mathcal{R}(\mathbf{n}) := \begin{pmatrix} 1 & \mathbf{0}^T & \mathbf{0}^T & 0 \\ \mathbf{0} & \mathcal{B}(\mathbf{n}) & \mathcal{O} & \mathbf{0} \\ \mathbf{0} & \mathcal{O} & \mathcal{B}(\mathbf{n}) & \mathbf{0} \\ 0 & \mathbf{0}^T & \mathbf{0}^T & 1 \end{pmatrix} \tag{6}$$

for the zero matrix $\mathcal{O} \in \mathbb{R}^{3 \times 3}$ we find

$$\mathbf{F}_x(\mathbf{U})n_x + \mathbf{F}_y(\mathbf{U})n_y + \mathbf{F}_z(\mathbf{U})n_z = \mathcal{R}^{-1}(\mathbf{n})\mathbf{F}_x\left(\mathcal{R}(\mathbf{n})\mathbf{U}\right) \tag{7}$$

by straightforward calculations.

3   Numerical Scheme

We discretize the computational domain by a finite set of tetrahedrons $\mathbb{T} = \{T_j | 1 \leq j \leq N\}$ and use an explicit finite volume scheme with constant data on each tetrahedron: We set

$$\mathbf{U}_j^0 := \frac{1}{|T_j|} \int_{T_j} \mathbf{U}_0 \tag{8}$$

for the given initial data $\mathbf{U}_0$ where $|T_j|$ denotes the volume of T_j. The boundary of each tetrahedron T_j consists of four triangles $S_{j1}, \ldots, S_{j4}$ with area $|S_{j1}|, \ldots, |S_{j4}|$. In order to simplify the notation let us assume that the tetrahedral mesh $\mathbb{T}$ is uniform, i.e., it contains no hanging nodes and that for $1 \leq l \leq 4$ there is a uniquely defined neighbour $T_{jl} \in \mathbb{T}$ of T_j with $S_{jl} = T_j \cap T_{jl}$. For $1 \leq l \leq 4$ the unit vector $\mathbf{n}_{jl}$ is the normal on S_{jl} pointing from T_j to T_{jl}. For the time partition $0 = t^0 < t^1 < \cdots < t^n < t^{n+1}$ the size of the timestep $\Delta t_n := t^{n+1} - t^n$ is limited by a suitable CFL condition. A formal integration of (5) over $T_j \times [t^n, t^{n+1}]$ yields

$$\int_{T_j} \mathbf{U}(\cdot, t^{n+1}) - \mathbf{U}(\cdot, t^n) + \int_{t^n}^{t^{n+1}} \sum_{l=1}^{4} \int_{S_{jl}} \mathcal{R}_{jl}^{-1} \mathbf{F}_x (\mathcal{R}_{jl} \mathbf{U}) = \int_{t^n}^{t^{n+1}} \int_{T_j} \mathbf{S}(\cdot, \mathbf{U}).$$

Here we have used Gauss' theorem, (7), and $\mathcal{R}_{jl} := \mathcal{R}(\mathbf{n}_{jl})$. If we now introduce the piecewise constant volume values $\mathbf{U}_j^n$, the values on S_{jl} are no longer well-defined. Thus we replace the exact flux $\mathbf{F}_x$ by a suitable numerical flux $\mathbf{G}_x$ which is based on the approximate solutions to the local Riemann problems with left-hand states $\mathcal{R}_{jl} \mathbf{U}_j^n$ and right-hand states $\mathcal{R}_{jl} \mathbf{U}_{jl}^n$. We obtain the explicit formula

$$\mathbf{U}_j^{n+1} = \mathbf{U}_j^n - \frac{\Delta t_n}{|T_j|} \sum_{l=1}^{4} |S_{jl}| \mathcal{R}_{jl}^{-1} \mathbf{G}_x (\mathcal{R}_{jl} \mathbf{U}_j^n, \mathcal{R}_{jl} \mathbf{U}_{jl}^n) + \frac{\Delta t_n}{|T_j|} \int_{T_j} \mathbf{S}(\cdot, \mathbf{U}_j^n). \tag{9}$$

Since for $\tilde{\mathcal{R}}_{jl} := \mathcal{R}(-\mathbf{n}_{jl})$ all numerical fluxes $\mathbf{G}_x$ which we consider satisfy

$$\mathcal{R}_{jl}^{-1} \mathbf{G}_x (\mathcal{R}_{jl} \mathbf{U}_j^n, \mathcal{R}_{jl} \mathbf{U}_{jl}^n) = \tilde{\mathcal{R}}_{jl}^{-1} \mathbf{G}_x (\tilde{\mathcal{R}}_{jl} \mathbf{U}_{jl}^n, \tilde{\mathcal{R}}_{jl} \mathbf{U}_j^n)$$

the scheme (9) is conservative and it suffices to calculate the numerical fluxes only *once per interface* S_{jl}.

While any numerical quadrature rule which is at least exact for linear functions suffices for replacing the integral in (8), any suitable discretization of the integral in (9) has to ensure that stationary solutions to (5) are exactly preserved by the scheme (8), (9). (Otherwise, simulations of structures which are embedded in a static background atmosphere like the fluxtube considered in Sect. 5 are not possible.) The most simple solution to this problem is to calculate the error which would be introduced for the stationary solution in each tetrahedron and to subtract these errors within each timestep [3]. However, this method is computationally expensive if local grid adaption is used. Therefore we now use the following approach: Within each cell we store the *deviation* from the background solution instead of the conservative variables. For the calculation of the fluxes and the source terms these values are then expanded by the *exact background values at the required positions* in such a manner that the preservation of the background atmosphere is again guaranteed. This new approach yields very promising results and will be discussed in more detail in a forthcoming paper. A further advantage of this technique is that now the indicators used for the local grid adaption can also be based on the deviations. Therefore strong variations in the background atmosphere do no longer cause unnecessary grid refinements.

Our 3d tetrahedral finite volume code with local grid adaption and dynamic load balancing is an extension of the code presented in [10]. It is written in C++ using MPI and was originally developed on the IBM SP-256 at Karlsruhe. Since neither the grid structure nor the communication itself has been changed significantly, the results concerning the scalability of the code obtained in [10] are still valid. Some of these results are contained in Tables 1 and 2. Note that for minimizing the communication required for the local grid adaption hanging nodes are allowed within the code. If hanging nodes occur we compute the fluxes from the smaller into the larger elements.

For $\mathbf{G}_x$ we use the MHD-HLLEM flux introduced in [12]. This scheme clearly outperforms all the other approximate Riemann solvers considered in [12] in terms of resolution *and* computational efficiency.

4 Hyperbolic Divergence Cleaning

It is a well-known fact that, without taking special measures, in MHD simulations errors in $\nabla \cdot \mathbf{B}$ will usually occur and increase with time. This can happen even if the projection of the initial data on the computational grid is still divergence-free. Divergence errors may lead to strongly unphysical solutions as discussed in [1] or even to a crash of the simulation. Therefore, a wide variety of approaches for curing this problem is discussed in the literature, see [11] and the references therein. In [2,5] we proposed a new approach which only requires the additional solution of a linear 2×2-system. It can be used as simple add-on with any existing solver. Moreover, it seems to yield superior results. In the following we try to give a brief but self-contained summary of this new method. For more details please refer to [2,5].

The main idea of our new *Generalized Lagrange Multiplier (GLM)* divergence cleaning approach is to couple the divergence constraint (1e) for the magnetic field $\mathbf{B}$ with the system (1a)–(1d) by means of introducing an additional unknown function ψ: Equation (1e) and the equations for the magnetic

Table 1. Speedup by parallelization with respect to runs with the serial code and the parallel code on a single processor, respectively. Simulated problem: Euler equations of gas dynamics, forward facing step in 3d. Used hardware: IBM RS/6000 SP-256 at Karlsruhe, WideNodes@77MHz. This table is taken from [10, Table 6.2].

Measurement of speedup with approx. 760,000 elements				
number of grids / processors	time [s]	speedup with respect to		theoretical maximum
		1st line	2nd line	
serial code	16663	1.00		–
1	19479	0.86	1.00	–
2	9913	1.68	1.96	1.8 - 2
3	6747	2.46	2.89	2.7 - 3
4	5323	3.13	3.66	3.6 - 4
5	4278	3.90	4.55	4.5 - 5
6	3646	4.57	5.34	5.5 - 6
7	3180	5.24	6.13	6.4 - 7
8	2784	5.98	7.00	7.3 - 8
10	2357	7.07	8.26	9.1 - 10
14	1771	9.41	11.00	12.8 - 14
20	1314	12.68	14.82	18.2 - 20
32	910	18.31	21.41	29.1 - 32
comparison: globally refined grid with approx. 4.4 million elements, no grid adaption, no load balancing				
32	2016		–	

Table 2. Speedup by parallelization with respect to the runs with the parallel code with 20 processors. Note that for the run on 120 processors we have approximately 10,000 elements per processor. Therefore, the costs for the communication dominate and we cannot expect any further speedup. Simulated problem: Euler equations of gas dynamics, forward facing step in 3d. Used hardware: IBM RS/6000 SP-256 at Karlsruhe, ThinP2SCNodes@120MHz. This table is taken from [10, Table 6.4].

Measurement of speedup with approx. 1,220,000 elements			
number of grids / processors	time [s]	speedup with respect to 1st line	
		measured	theoretical
20	3161	1.00	
40	2007	1.58	1.8 - 2
80	1623	1.95	2.7 - 4
120	1502	2.10	4.0 - 6

induction (1c) are replaced by

$$\partial_t \mathbf{B} + \nabla \cdot \left(\mathbf{u} \mathbf{B}^T - \mathbf{B} \mathbf{u}^T \right) + \nabla \psi = 0, \tag{10}$$

$$\mathcal{D}\left(\psi \right) + \nabla \cdot \mathbf{B} = 0, \tag{11}$$

where $\mathcal{D}$ is a linear differential operator. We try to choose $\mathcal{D}$ and the initial and boundary conditions for ψ in such a way that a numerical approximation to (10), (11) is a good approximation to the original Eqns. (1c), (1e). We checked different possibilities for $\mathcal{D}$ and found the choice of

$$\mathcal{D}\left(\psi \right) := \frac{1}{c_h^2} \partial_t \psi + \frac{1}{c_p^2} \psi \tag{12}$$

for some positive constants c_h and c_p most satisfactory, since it leads to both *transport* and *damping* of divergence errors.

If we consider the one-dimensional system, i.e., (1a), (1b), (10), (1d), (11) with vanishing derivatives in y- and z-direction, we see that the equations for B_x and ψ are decoupled from the remaining system. Thus the hyperbolic part of these equations

$$\partial_t \begin{pmatrix} B_x \\ \psi \end{pmatrix} + \begin{pmatrix} 0 & 1 \\ c_h^2 & 0 \end{pmatrix} \partial_x \begin{pmatrix} B_x \\ \psi \end{pmatrix} = \begin{pmatrix} 0 \\ 0 \end{pmatrix} \tag{13}$$

can be solved *independent* of the remaining 7×7-system. The eigenvalues and corresponding right eigenvectors of (13) are $\pm c_h$ and $(1, \pm c_h)^T$, respectively. Thus it is easy to verify that the solution to the local Riemann problem for (13) with left-hand state $(B_{x,l}, \psi_l)^T$ and right-hand state $(B_{x,r}, \psi_r)^T$ has the value

$$\begin{pmatrix} B_{x,m} \\ \psi_m \end{pmatrix} = \begin{pmatrix} B_{x,l} \\ \psi_l \end{pmatrix} + \begin{pmatrix} \frac{1}{2}(B_{x,r} - B_{x,l}) - \frac{1}{2c_h}(\psi_r - \psi_l) \\ \frac{1}{2}(\psi_r - \psi_l) - \frac{c_h}{2}(B_{x,r} - B_{x,l}) \end{pmatrix} \tag{14}$$

on the cell interface. Therefore, we obtain $(\psi_m, c_h^2 B_{x,m})^T$ as numerical flux. If we add a ninth entry corresponding to ψ to the vector of conservative variables $\mathbf{U}$, the resulting numerical flux $\mathbf{G}_x^{\mathrm{GLM}}$ for (9) reads

$$\mathbf{G}_x^{\mathrm{GLM}}(\mathbf{U}_l, \mathbf{U}_r)$$
$$:= \mathbf{G}_x^{\mathrm{MHD}}(\mathbf{U}_l, \mathbf{U}_r; B_x^{1d}) + (0, 0, 0, 0, \psi_m, 0, 0, 0, c_h^2 B_{x,m})^T . \tag{15}$$

For $\mathbf{G}_x^{\mathrm{MHD}}$ we use the MHD-HLLEM flux for the one-dimensional MHD equations from [12] with an additional zero for the flux in ψ. Due to (1c) and (1e) we find that B_x has to be constant in 1d. Thus it enters the corresponding flux function as constant parameter B_x^{1d} which is used for the evaluation of $\mathbf{G}_x^{\mathrm{MHD}}$, whereas the values $B_x(\mathbf{U}_l)$ and $B_x(\mathbf{U}_r)$ are disregarded. In the GLM case, the choice $B_x^{1d} := B_{x,m}$ is the natural one. If no divergence correction is used, the last term in (15) is not present and we proceed as in [12].

To complete the implementation of the GLM-MHD system we have to handle the source term in the equation for ψ. A simple and unconditionally stable discretization is obtained by using an operator splitting approach. We first solve the homogeneous system by means of (9) and (15), which leads to the value $\psi_j^{n^*}$ for ψ on T_j. In a second step we solve the initial value problem

$$\partial_t \psi_j = -\frac{c_h^2}{c_p^2} \psi_j \tag{16}$$

in $[0, \Delta t_n]$ with $\psi_j(0) = \psi_j^{n^*}$. We use the exact solution to (16) and set

$$\psi_j^{n+1} := e^{-\Delta t_n \, c_h^2/c_p^2} \, \psi_j^{n^*} . \tag{17}$$

We choose c_h to be the maximal speed which ensures that none of the local "divergence waves" travels faster than the remaining MHD waves. Instead of prescribing c_p directly we fix $c_r := c_p^2/c_h \in (0, \infty)$ which maintains a fixed ratio between transport and damping. Motivated by numerical tests we use $c_r := 0.18$ within this paper. Since we found our scheme rather insensitive to the boundary conditions prescribed for ψ, we simply treat ψ analogous to ρ at the boundaries.

5   Simulation of an Exploding Magnetic Fluxtube

The setup for the simulation of an exploding magnetic fluxtube has been supplied by M. Rempel [8]. At the beginning we compute a stratified, hydrostatic background atmosphere which is a stationary solution to the MHD system (1a)–(1d). Then the initial values for the fluxtube are computed assuming hydrostatic equilibrium along the central field line. Caused by the deflection of the tube, the entropy within the tube is lower than in the surrounding atmosphere and the strength of the magnetic field increases with height. The radius of the fluxtube is determined by the conservation of magnetic flux. Initially, the *total pressure* within the tube (consisting of gas pressure p plus magnetic pressure $|\mathbf{B}|^2/(8\pi)$) is in equilibrium with the gas pressure of the background, which is equal to the total pressure since the atmosphere itself contains no magnetic field. However, at a certain height (the "explosion height") the *gas pressures* become equal and the equilibrium of the total pressures cannot be maintained any longer. This leads to an "explosion" of the central part of the tube, whose entropy above the explosion height is larger than in the atmosphere. For more details concerning the physical background and the motivation behind this setting please refer to [8]. We just use this problem as a testcase for our 3d code; the physical interpretation of the results is beyond the scope of this paper.

We used the following definitions for prescribing the initial conditions of an exploding magnetic fluxtube (see also Fig. 2 for t = 0):

- **background atmosphere:**
Pressure and density are given as solutions to

$$\bar{p}'(z) = g(z)\bar{\rho}(z), \quad \bar{\rho}(z) := \bar{p}(z)^{\frac{1}{\gamma}}$$

for $\gamma := \frac{5}{3}$ and the boundary condition $\bar{p}(0) = 1$. The height-dependent gravitation is given by

$$g(z) := 0.25\left[1 + \tanh(20z)\right]\left[1 - \tanh\left(20(z - 2)\right)\right].$$

- **magnetic fluxtube:**
  - *central field line:* $f(y) := -0.25 + 0.4 \exp\left(-\left(2(y - 3)\right)^2\right)$
  - *magnetic pressure:*

  $$p_m^z{}'(z) = -\frac{\bar{\rho}(z)g(z)}{\gamma}\left(\frac{p_m^z(z)}{\bar{p}(z)} - 1\right)$$

  with boundary condition $p_m^z(z_0) = 0.1\,\bar{p}(z)$ for $z_0 := f(0)$. We use p_m as a function of y by dint of $p_m(y) := p_m^z\left(f(y)\right)$.
  - *radius of tube:* $R(y) := 0.15\left(\frac{p_m(0)}{p_m(y)}\right)^{0.25}$
- **initial conditions:**
For

$$T(x, y, z) := \exp\left(-\left(\frac{(z - f(y))^2 + x^2}{R(y)^2}\right)^4\right)$$

we define

$$B_x(x, y, z) := \frac{f(y) - z}{R(y)}\sqrt{8\pi p_m(y)}\, T(x, y, z),$$

$$B_y(x, y, z) := \cos\left(\arctan(f'(y))\right)\sqrt{8\pi p_m(y)}\, T(x, y, z),$$

$$B_z(x, y, z) := \left[\sin\left(\arctan(f'(y))\right) + \frac{x}{R(y)}\right]\sqrt{8\pi p_m(y)}\, T(x, y, z),$$

$$\rho(x, y, z) := \bar{\rho}(z)\left(1 - \frac{|\mathbf{B}(x, y, z)|^2}{2\bar{p}(z)}\right)^{\frac{1}{\gamma}}\exp\left(-\frac{T(x, y, z)}{\gamma}\right),$$

$$u_x(x, y, z) := 0, \quad u_y(x, y, z) := 0, \quad u_z(x, y, z) := 0,$$

$$\left(\rho e\right)(x, y, z) := \left(\bar{p}(z) + 0.5\,(\gamma - 2)\,|\mathbf{B}(x, y, z)|^2\right)\frac{1}{\gamma - 1}.$$

- **computational domain:** $[-1.5, 1.5] \times [0.0, 6.0] \times [-0.6, 2.4]$.
- **boundary conditions:**
Periodic boundary in y-direction. Across the remaining boundaries the *normal* components of the moments and *all* components of the magnetic field are assumed to maintain their absolute value but to change their sign; the remaining components are simply copied to ghost cells.

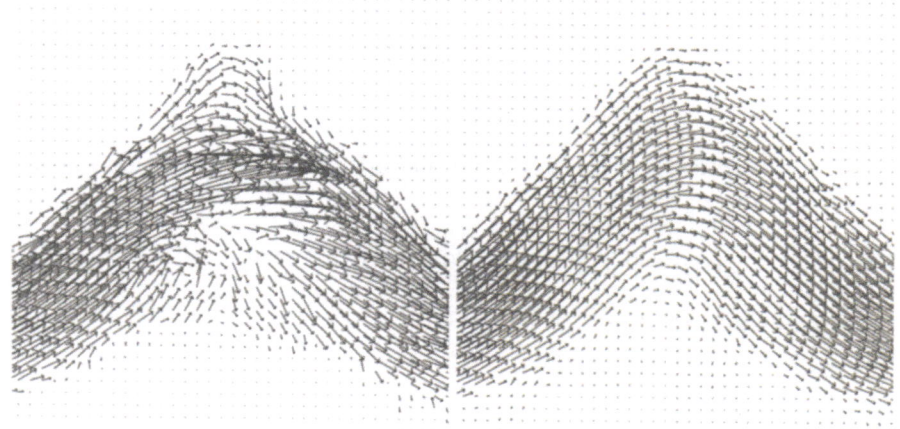

Fig. 1. Magnetic field of the exploding magnetic fluxtube at time $t = 1.5$. Left: Without divergence correction the magnetic field is strongly disturbed and the simulation crashes shortly after $t = 1.5$. Right: With the new hyperbolic divergence cleaning the magnetic field is undisturbed and the simulation succeeds, see also Fig. 2.

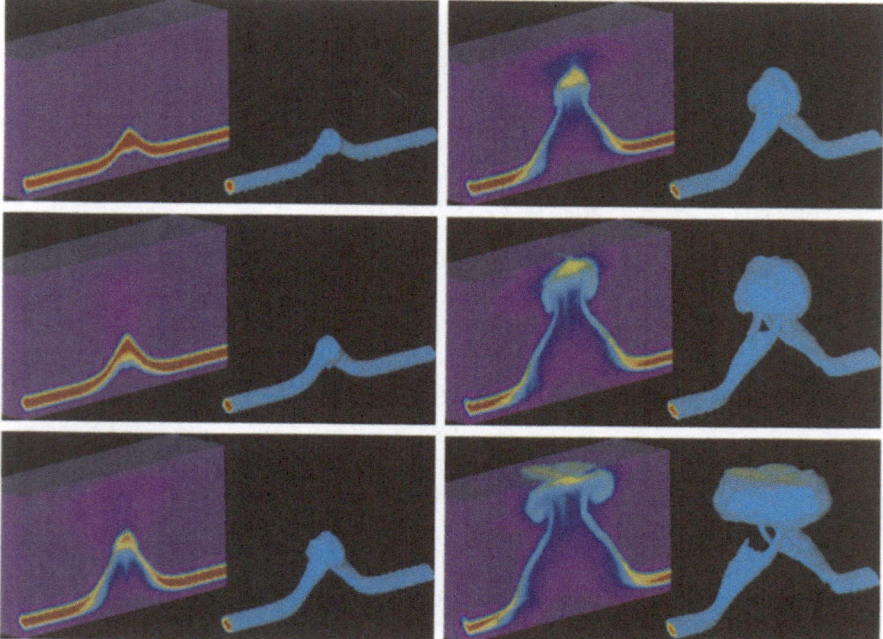

Fig. 2. Time sequence of an exploding magnetic fluxtube in 3d with hyperbolic divergence cleaning (isolevels and isosurface of entropy; $t = 0.0$, 1.5, 3.0, 4.5, 6.0, and 7.5). Used hardware: SGI Origin 2000 at Freiburg, R10000-CPUs@250MHz. Real time required for simulation: approx. 1 day using 10 CPUs.

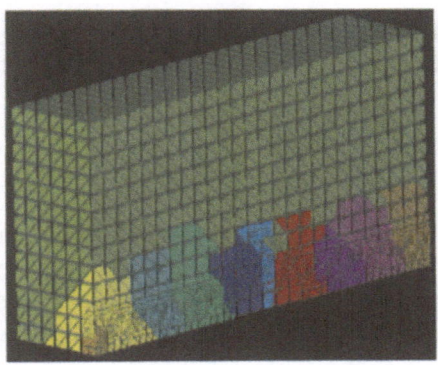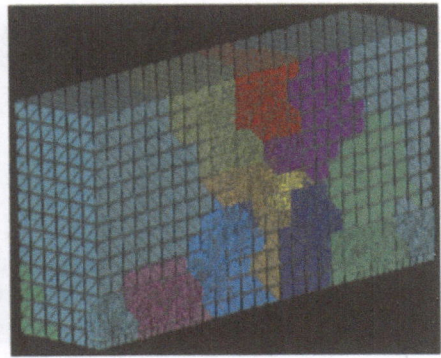

Fig. 3. Locally adapted grids and grid partitioning for the fluxtube simulation shown in Fig. 2 at time $t = 0.0$ (left) and $t = 5.0$ (right), respectively.

In Fig. 1 we see the influence of divergence errors on the magnetic field: without correction, the magnetic field is strongly disturbed and the simulation crashes shortly after $t = 1.5$. On the other hand, we see no disturbances if our new hyperbolic divergence cleaning (cf. Sect. 4) is applied. In this case, the simulation runs up to the final time without any problems, see Fig. 2. At the beginning we observe a moderate rise of the initially Ω-shaped magnetic fluxtube. As soon as the explosion height is reached, the upper parts of the tube "explode" and finally hit the top of the computational domain. The locally adapted tetrahedral grid (used indicator: variation of entropy) together with the partitioning created by the automatic load balancing can be seen in Fig. 3.

During the development of our 3d-code it was convenient to use the SGI Origin 2000 which was locally available at Freiburg. (For example, the simulation shown in Fig. 2 required approximately 1 day in real time on 10 250MHz-R10000 processors.) However, up to now the future availability of this platform is uncertain, and with its only 46 CPUs and 12 GB of memory it is too small for realistic long-time simulations. Therefore we intend to use the IBM SP-SMP at Karlsruhe for our future production runs. For further development, simple tests, and data evaluation we will profit from our IBM P270-workstation with four 375MHz Power3-II CPUs, since it is compatible with the SP-SMP on binary level.

6   Conclusions

We have seen that divergence cleaning is absolutely essential in multidimensional MHD simulations and that our new approach works well even in 3d. The successful calculation of an exploding magnetic fluxtube shows that we are now ready to tackle simulations of settings which are relevant for solar physics with high resolutions. The resulting requirements of both computing power and memory size favour the future use of the IBM SP-SMP at Karlsruhe for production runs.

Acknowledgements

The authors were partially supported by the DFG priority research program "Analysis and Numerics for Conservation Laws." They thank B. Schupp for supporting the extension of his code to MHD and M. Rempel for supplying the setting for the exploding magnetic fluxtube.

References

1. J.U. Brackbill and D.C. Barnes, *Note: The effect of nonzero* $\nabla \cdot \mathbf{B}$ *on the numerical solution of the magnetohydrodynamic equations*, J. Comput. Phys. **35** (1980), 426–430.
2. A. Dedner, F. Kemm, D. Kröner, C.-D. Munz, T. Schnitzer, and M. Wesenberg, *Hyperbolic divergence cleaning for the MHD equations*, J. Comput. Phys. **175** (2002), no. 2, 645–673, doi:10.1006/jcph.2001.6961.
3. A. Dedner, D. Kröner, I.L. Sofronov, and M. Wesenberg, *Transparent boundary conditions for MHD simulations in stratified atmospheres*, J. Comput. Phys. **171** (2001), no. 2, 448–478, doi:10.1006/jcph.2001.6779.
4. A. Dedner, C. Rohde, and M. Wesenberg, *Efficient higher-order finite volume schemes for (real gas) magnetohydrodynamics*, submitted to proceedings of HYP'2002, June 2002.
5. A. Dedner, C. Rohde, and M. Wesenberg, *A new approach to divergence cleaning in magnetohydrodynamic simulations*, submitted to proceedings of HYP'2002, July 2002.
6. A. Dedner and P. Vollmöller, *An adaptive higher order method for solving the radiation transport equation on unstructured grids*, J. Comput. Phys. **178** (2002), 263–289, doi:10.1006/jcph.2002.7001.
7. A. Dedner and M. Wesenberg, *Numerical methods for the real gas MHD equations*, Hyperbolic Problems: Theory, Numerics, Applications (H. Freistühler and G. Warnecke, eds.), ISNM, vol. 140, Birkhäuser, 2001, pp. 287–296.
8. M. Rempel, *Struktur und Ursprung starker Magnetfelder am Boden der solaren Konvektionszone*, PhD thesis, Georg-August-Universität, Göttingen, June 2001, http://webdoc.sub.gwdg.de//diss/2001/rempel/index.html.
9. M. Rempel and M. Schüssler, *Intensification of magnetic fields by conversion of potential energy*, Astrophys. J. **552** (2001), no. 2, L171–L174.
10. B. Schupp, *Entwicklung eines effizienten Verfahrens zur Simulation kompressibler Strömungen in 3D auf Parallelrechnern*, Phd thesis, Albert-Ludwigs-Universität, Mathematische Fakultät, Freiburg, Dezember 1999, http://www.freidok.uni-freiburg.de/volltexte/68.
11. Gábor Tóth, *The* $\nabla \cdot B = 0$ *constraint in shock-capturing magnetohydrodynamics codes*, J. Comput. Phys. **161** (2000), 605–652, doi:10.1006/jcph.2000.6519.
12. M. Wesenberg, *Efficient MHD Riemann solvers for simulations on unstructured triangular grids*, J. Numer. Math. **10** (2002), 37–71.

LES of Flow in a Low Pressure Turbine with Incoming Wakes

Vittorio Michelassi[1,2], Jan Wissink[1], and Wolfgang Rodi[1]

[1] Institut für Hydromechanik, Universität Karlsruhe, Kaiserstrasse 12,
   76128 Karlsruhe, Germany
[2] On leave from the University of Roma Tre, Italy

Abstract. The flow in a low-pressure turbine rotor blade with incoming periodic wakes is computed by means of LES based on a dynamic SGS model. The computations are performed by using 32 processors of a parallel super-computer. The optimisation and tuning of the algorithm for the HITACHI-SR8000 environment allowed reducing the CPU time of a factor 2. The simulations identify several relevant phenomena thanks to the good resolution of the problem in both space in time. The computed results are compared with experiments in terms of phase- averaged and mean quantities.

1 Introduction

The growing demand of high performance power generation and propulsion systems increased the research effort in the computer simulation of flow in gas turbines. Particular attention has been devoted to the wake-blade interaction phenomenon, which can have a considerable impact on the performances of both stator and rotor rows. In this view, experimental and computational analyses evolve from isolated row towards full-stage (stator plus rotor) investigations to capture the mutual interaction between the rows [3]. Intensive numerical investigations have been carried out in the past, especially in the field of LP turbines [5,8,2], which have been devoted much more attention than HP turbines [10]. Nevertheless, the stator-rotor interaction, and the wake-rotor blade interaction in particular, still requires some additional effort to shed further light on this intriguing complex phenomenon. As a consequence of the increased availability of experimental data, advanced Computational Fluid Dynamics (CFD) tools with refined turbulence and transition models have been developed for the simulation of stator-rotor interaction [11]. The modelling assumptions and the intrinsic limitations of the Reynolds Averaged Navier-Stokes (RANS) simulations often did not ensure the necessary degree of accuracy in both steady and unsteady simulations. Although the Direct Numerical Simulation (DNS) of a full stage is beyond the capabilities of modern super-computers, it is possible to compute the wake-blade interaction in a linear cascade, as done by Wu and Durbin [16] at a Reynolds number (Re) of 1.48×10^5. In their simulation the linear cascade T106, experimentally tested by Stadtmüller [12], was computed with incoming wakes generated by previous Large Eddy Simulation (LES) of a model problem that mimics the far-field behaviour of a turbulent wake. The data produced by such simulations can be conveniently used to develop models to be cast into RANS (or LES) computer codes, which should dramatically reduce the computational effort with respect to DNS. The flow around the T106 blade was measured at $Re = 6 \times 10^4$ [12,13] and 2×10^5 [13] without and with incoming wakes. Unfortunately, in the larger Re case the wake-to-blade pitch ratio is an odd number, which requires the simulation of more than one vane, and it was therefore abandoned to reduce the computational effort. Therefore, we selected the T106 geometry operating at $Re = 6 \times 10^4$ since it allowed LES to be performed with reasonable computational effort, and to analyse the flow structures and the unsteady boundary layer of the rotor blade in detail. The test case with incoming wakes was selected because it is representative of real operating conditions and it does provide important information on the wake-blade interaction. In this view, the aim of the simulations is twofold: first analyse the simulation data against the experimental data set, and second assess the CPU time required for such a challenging simulation in view of a massive use of LES for design and testing purposes.

2   The test case

The measurements refer to a low-pressure linear turbine test rig composed of seven aft-loaded blades [12]. The blade aspect ratio (h/c) is 1.76, according to which the assumption of a nearly two-dimensional flow at mid-span appears reasonable. The pitch-to-chord ratio (t/c) is 0.799, which clearly places this blade in the range of mildly loaded low-pressure turbine blades. Since the test-rig allows controlling the fluid density independently of the fluid velocity, it has been possible to perform the measurements at the relatively low Reynolds number of 6×10^4 (based on the blade chord C) and with an isentropic exit Mach number of 0.4. The stagger angle, γ, inlet, β_1, and outlet, β_2, blade angles are 30.72-deg, 37.7-deg and 63.2-deg respectively, defined with respect to the axial direction (see Fig. 1). The test rig allows simulating the effect of upstream blade rows by a moving bar wake generator with a bar diameter to blade chord ratio of $db/c = 0.02$. This system allows producing wakes very similar to those produced by an airfoil, although their strength is probably weaker than that of real IGV wakes. The test rig does not allow to reproduce any sort of static pressure interaction between stator and rotor rows, which might play some role in triggering transition on the front part of the rotor blade. Still, the low-pressure nature of the test indicates that this effect should be mild. Measurements include the time averaged distribution of the static pressure around the blade, the time averaged distribution of the losses in a cross section downstream of the blade, and the time resolved quasi-wall-shear stress on the suction side of the blade [12]. The measurements have some difficulties in identifying the actual inlet conditions in both terms of inlet flow angle and total pressure. Apparently, the actual inlet flow angle is not 37.7-deg as designed, but 45.5-deg, as suggested by a number of RANS simulations performed by Stadtmüller [12]. The increased inlet flow angle is probably a consequence of the low Reynolds number that causes a migration of the flow towards the bottom of the test rig (see [12]).

3 Computational aspects

3.1 Computational grid

Following the simulation of Wu and Durbin [16] the grid was carefully selected by using the elliptic grid generation algorithm proposed by Hsu and Lee [6]. Although a multi-block mesh would be the best option (with an O-grid wrapped around the blade and an H-type mesh for the flow core), the elliptic mesh generation ensured a nearly orthogonal (and Cartesian) grid close to the blade walls and it was therefore adopted for the simulations. The grid was carefully selected by analysing the results of a previous DNS [14] so as to ensure the correct resolution of flow structures. This preliminary check indicated that a grid with $448 \times 144 \times 32$ points in the stream-wise, pitch-wise, and span- wise directions respectively, would provide a sufficient resolution for LES. The resolution of the LES grid, which extends in the span-wise direction by $0.15C_{ax}$ and is shown in Fig. 1, was carefully tested in terms of non-dimensional grid size. For the present set of calculations the non-dimensional grid cell sizes are $x^+ \leq 40$, $y^+ \leq 1$, $z^+ \leq 15$ (these values are computed by taking u_τ on the suction side close to the trailing edge, which is the most demanding in terms of grid resolution). These values were deemed suited for providing the necessary resolution of fluid structures. The grid in the span-wise direction was selected upon the assumption of a homogeneous flow.

3.2 Implementation of boundary conditions

In all the computer simulations a no-slip boundary condition is enforced on the blade. The assumption of a periodic flow in the pitch-wise direction is not critical for LES since the size of the expected fluid structures is a small percentage of the blade pitch. Different problems arise for the span-wise direction where the periodic flow assumption needs to be carefully tested. In fact the desired reduction of the number of grid nodes and domain extension in the span-wise direction in order to reduce the computational time, should not affect the development of stream- and span-wise flow structures. The incoming wakes at the inflow boundary are enforced by using the data base provided by Wu and Durbin [15] who generated the incoming wakes with a preliminary LES of a flow which exhibits the typical far field behaviour of a turbulent wake. In the current simulations the wake half-width and maximum wake deficit were $0.03C_{ax}$ and 25% respectively. The wake pitch is $\frac{1}{2}$ of the blade pitch (see Fig. 1), and the non-dimensional tangential velocity of the

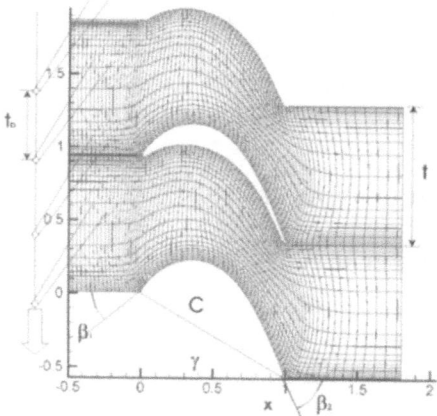

Fig. 1. Elliptic grid used for LES (every 6 nodes are shown in both stream-wise and pitch-wise directions

wake was $u_{bar} = 0.41$. These values have been selected in order to fit the experimental incoming wakes described by Stadtmüller [12] for measurements of the T106 blade.

3.3   The calculation method and computational effort

The LES of the flow around the T106 turbine blade has been performed by using the code LESOCC [1]. The code discretizes the incompressible Navier-Stokes equations using a cell-centred finite volume approach. The implicit solution of the Poisson equation for the pressure correction (SIMPLE) originally adopted in [1] is complemented with a Fourier solver in the span-wise direction [9] which substantially reduces the computational effort to enforce mass conservation. The equations are solved by marching in time with a three-stage Runge-Kutta algorithm. Mass conservation is enforced only after the final Runge-Kutta step. The Sub-Grid-Scale (SGS) model used in the LES is the dynamic model by Germano et al. [4]. Further details about the code may be found in [1,9]. In order to properly resolve the wake in both space and time, one period, i.e. a half-blade-pitch sweep of the wake, was resolved with 4800 time steps. This choice implies a maximum CFL number of approximately 0.30–0.50. The flow was allowed to develop for five periods, i.e. 24.000 time steps. Once the time-periodic flow was developed, phase-averaging was carried out for 10 periods, i.e. 48.000 time steps for a total of 15 periods. Instantaneous flow fields were also stored: for each of the last two periods 100 equally spaced-in-time fields were stored on the hard-disk to be able to produce animations which largely help in the interpretation of the results. Each wake passage was subdivided into 240 phases, i.e. one phase every 20 time steps. To save memory, the velocity field was averaged in the homogeneous spanwise direction before updating the phase-averaged statistics. The first implementation of the code was tested using the grid and operating conditions described above. The total number of grid nodes is 2.064×10^6, with 8.26×10^6 unknowns. The problem was split into 32 sub-domains, using 8 processors in the x-direction, 4 processors in the y-direction and 1 in the z-direction. The latter choise optimizes the performance of Fourier solver for the Poison equation applied in the spanwise direction. The inter-processor communication was performed with MPI. Each sub-domain has approximately 6.4×10^4 nodes, which easily fit into the memory of one processor. This domain decomposition does not necessarily minimise the number of inter-processor data exchange, but it has the less impact on the convergence rate of the mass conservation step. The simulations required a total number 110.400 time steps. It was possible to get 32 processors for 24 hours a day for an elapsed time of approximately 330 hours ($\approx$ 14 days). The total CPU time (for the 32 processors runs) was therefore approximately 11.000 hours. Before the optimisation the performance of the code was approximately 7% of the computer peak performances. This was considered not entirely satisfactory. Therefore, with the help of the technical support of the Rechenzentrum Universitt Stuttgart, the code was optimised by an appropriate choice of libraries, by switching on the pseudo- vectorisation of the do-loops and by removing integer-to-real multiplication in the routines which enforce the mass conservation. This last step was particularly effective since the mass conservation step alone consumes $60 - 70\%$ of the overall CPU time. After the modifications the extrapolated new performances reached $14 - 15\%$ of the computer peak performance with a reduction of more than a factor two in the computational time.

4   Results

4.1   Flow visualisation and flow structures

Jeong and Hussain [7] suggested that the local vortical structures of a three-dimensional flow can be conveniently identified by computing the second largest eigenvalue, λ_2, of $S^2 + \Omega^2$, in which S and Ω are the symmetric and anti-symmetric parts of the velocity gradient tensor, respectively. Since Wu and Durbin [16] found no differences when using either the instantaneous field or by subtracting the span-wise mean velocity at $Re = 1.48 \times 10^5$, λ_2 was computed using the instantaneous velocity field. The iso-surfaces of $\lambda_2 < 0$, shown in Fig. 2, allow a very clear detection of the vortices and their dynamics while being swallowed into the blade vane. The four phases show how the wake gradually enters into the blade vane and rotates counter-clockwise. As observed in the larger Reynolds-number case, the approaching wake does not show any preferential vortex direction as long as it does not feel the pressure gradient. As soon as the blade curvature provokes strong flow acceleration, the portion of the wake close to the pressure side is aligned with the principal axis along which the flow experiences a strong stretching. This phenomenon

Fig. 2. Instantaneous λ_2 iso-surfaces inside the blade vane. Arrow indicates incoming flow direction. ($t/T = 12.00$, 12.25, 12.50, 12.75 from top-left clockwise)

strongly amplifies vorticity in the stream-wise direction both in the present case at $Re = 5.18 \times 10^4$ and at $Re = 1.48 \times 10^5$ [16]. The shape of these vortical structures develops into tube-like filaments which are clearly visible also in Figs. 3 and 4, which show the axial vorticity iso-surfaces. The further development of the wake can be followed by relating it to the static pressure on the pressure side (see Fig. 6). The vortical structures are mildly elongated as long as the static pressure is substantially constant, while, for $x/C_{ax} > 0.7$ the strong flow acceleration on the pressure side provokes straining. On the pressure side, elongated structures clearly stem from the incoming wake, which triggers the overall phenomena. To facilitate the understanding of the overall flow development, Fig. 3 also shows the span-wise vorticity on a plane orthogonal to the blade, which allows a clear detection of the wake position in the vane. In addition to what was observed by Wu and Durbin [16], the lower Reynolds number simulations with the incoming wakes frequency twice as high, show that nearly the entire pressure side is continuously affected by the vortices. The figures also suggest that the vortices appearing on the pressure side are generated by the interaction of the wake with the strong flow acceleration initiated by the blade leading edge. While being transported

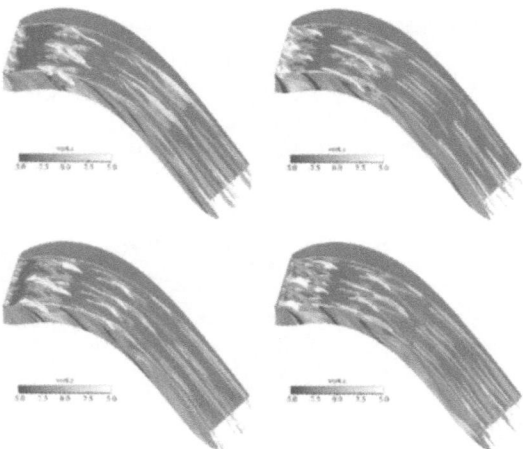

Fig. 3. Instantaneous iso-surface of axial vorticity on pressure side. The blade-normal plane shows the spanwise vorticity. (t/T = 12.00, 12.25, 12.50, 12.75 from top-left clockwise)

downstream, the wake vortices gradually approach the blade pressure side until they manage to enter the boundary layer. The strong acceleration of the flow core on this side of the blade is obviously responsible for the virtual absence of any significant turbulence activity. This explains why the elongated vortices do not manage to entrain turbulence into the pressure side boundary layer, which remains basically laminar until the trailing edge. Figure 4 shows the considerable differences between the pressure and suction side boundary layers. Apparently, when the wake wraps around the leading edge it is subject to a strong acceleration on both the suction and pressure sides. On the suction side the wake is elongated while being strongly accelerated, whereas the portion close to the pressure side is lagging behind. The wake distortion is quite strong up to $x/C_{ax} \approx 0.6 - 0.65$ and then, due to the reduction in the flow curvature, it largely reduces. In the portion of the suction side between the leading edge and $x/C_{ax} \approx 0.7$, the boundary layer does not experience strong disturbances. The instantaneous flow visualisation allowed identifying some random turbulence spots in the LES, which did not survive long and were quickly convected downstream. These spots appear only when the wake reaches the blade crown. Still, their effect is marginal and no transition of the boundary layer is observed down to $x/C_{ax} \approx 0.8$. Figures 3 and 4 show also that on the pressure side small-scale turbulence is absent, while it clearly appears toward the trailing edge on the suction side. This interesting series of snapshots of the instantaneous flow field on both pressure and suction sides allowed a clear representation of the large differences between the two.

Observe that the appearance of axial vorticity on the suction side boundary layer is clearly triggered by the incoming wakes. But this triggering is

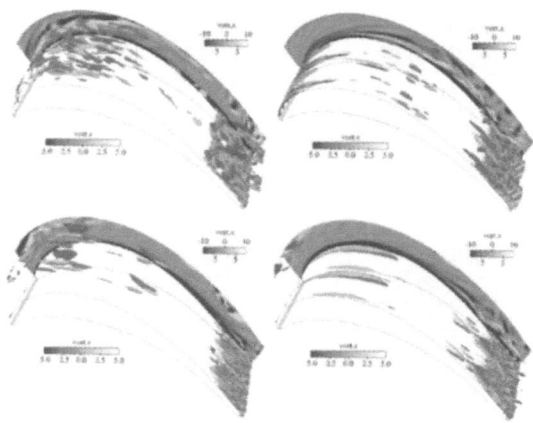

Fig. 4. Instantaneous iso-surface of axial vorticity on suction side. The blade-normal plane shows the span-wise vorticity. (t/T = 12.00, 12.25, 12.50, 12.75 from top-left clockwise)

not only linked to the position of the current wake inside the vane and above the boundary layer, but also to the position of the following incoming wakes with respect to the leading edge. In a further attempt to show how the wakes affect the boundary layer on both the suction and pressure sides, Fig. 5 shows the axial vorticity computed from the instantaneous flow field at the same phases of Figs. 3 and 4. The plots include four cross-flow sections located at approximately 5%, 30%, 60% and 95% of the axial chord. The figures clearly show that on the suction side the incoming wakes carry large structures, which, although present all along the blade, are smaller in magnitude with respect to pressure side. Moreover, the disturbances do not seem to reach the boundary layer, or, if they do, the strong flow acceleration prevents any significant small-scale turbulence from growing. Only in the proximity of the trailing edge does the visualisation show a significant interaction between the incoming wake and the vorticity created in the suction side boundary layer. Conversely, the pressure side reveals significantly stronger structures (see the levels of the axial vorticity), in agreement with what Fig. 3 suggested. Still, these are large-scale structures, which, although able to penetrate the boundary layer, clearly do not promote any fine-scale turbulent motions. The plots relative to the pressure side also show that these structures apparently survive for long time.

With respect to what was found by Wu and Durbin [16], the doubled incoming wakes frequency does not provide enough time to allow the pressure side boundary layer to dissipate the convected wake tube-like structures. Despite the large disturbances, the boundary layer remains substantially laminar since the wake structures do not seem to convect any small-scale turbulence.

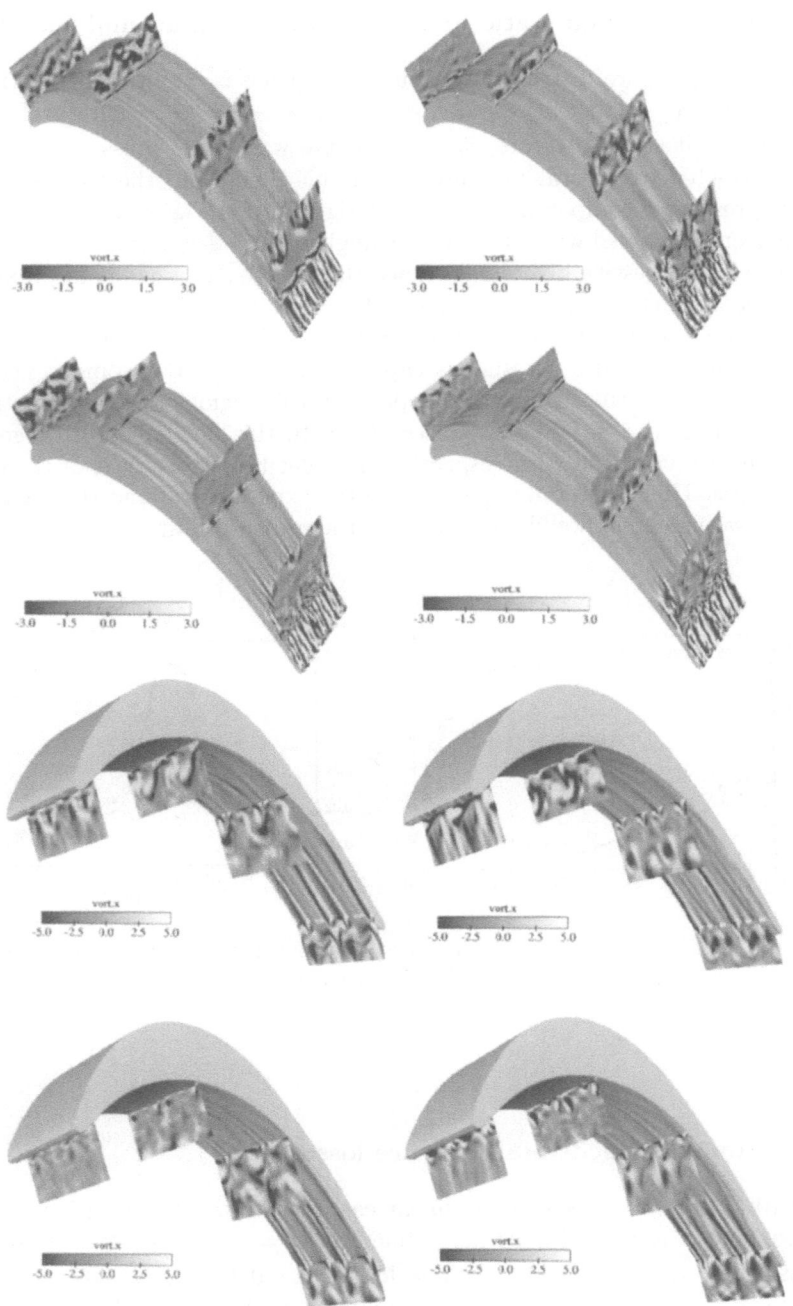

Fig. 5. Axial vorticity along suction and pressure sides and in four cross-flow planes. ($t/T = 12.2 - 12.4 - 12.7 - 12.9$ from top-left clockwise)

4.2   Time averaged static pressure distribution around the blade

The static pressure distribution around the blade is monitored in terms of static pressure coefficient, C_p defined as $C_p = \frac{p_w - p_{ref}}{p_{0-b} - p_{ref}}$, where pref is the reference exit static pressure, p_w is the blade wall static pressure, and p_{0-b} is a corrected inlet total pressure. With this definition, the time-averaged static pressure is compared with the test-rig values in Fig. 6. The simulations have been performed with an inlet flow angle of 45.5-deg. However, this angle produces an excessive blade load, especially on the suction side in proximity of the leading edge (see Fig. 6 (left)). The shift between the measurements and the LES clearly stems from the incompressible nature of the simulations. In fact the exit static pressure is captured very well, but some deviations from the data on the suction side appear. In this regions the local isentropic Mach number of the flow may exceed 0.45. In this flow velocity range, the local density variation with respect to the inlet total density is of the order of $5 - 10\%$. This variation, together with the experimental uncertainty in the inlet flow angle, may well be the cause of the pressure shift observed in Fig. 6 (left).

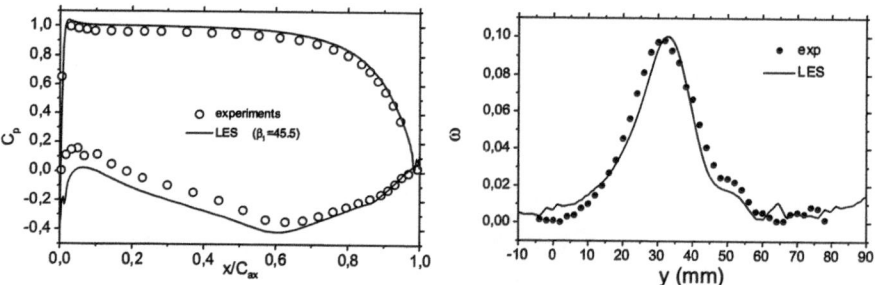

Fig. 6. Left: time averaged static pressure coefficient distribution around the blade, Right:Kinetic loss coefficient $40\%C_{ax}$ downstream of the trailing edge; (pitch is 80mm)

4.3   Time averaged total pressure losses

The total pressure losses in the linear cascade are monitored using the so-called kinetic loss coefficient, ω, defined as $\omega = \frac{p_{0-inlet} - p_{0-exit}}{p_{0-inlet} - p_{ref}}$. The total pressure p_{0-exit} is taken at a plane located at $0.4C_{ax}$ downstream of the trailing edge. Although this is often addressed as the 'profile' loss coefficient, in this case it includes also the mixing losses occurring from the trailing edge to the measurement plane. Figure 6 (right) compares the measured kinetic loss coefficient distribution in the pitch-wise direction against the LES. All the reported values are time averaged. The measured wake is shown to be quite thick, and the losses quite large, thereby illustrating the heavy off- design conditions at which the blade is operated. The predictions are in very good agreement with the measured data. This result constitutes a considerable improvement with respect to existing RANS simulations, which are often unable to give a fair representation of the flow downstream of the blade trailing edge.

5   Conclusions

The flow in a low-speed turbine blade linear cascade with incoming wakes revealed a number of interesting phenomena. The detailed analysis of the flow field indicated several features, part of which were already observed in [16] at a higher Reynolds number, which are itemised as follows:

- The dynamic SGS model is able to predict the suction side boundary layer state in fair agreement with experiments. The unavailability of detailed experimental data on the blade suction side prevents any further comparison with experiments.
- The boundary layer undergoes an intermittent separation. The flow then reattaches while being only marginally turbulent.
- The flow visualisation showed the presence of elongated flow structures stemming from the interaction of the incoming wake vorticity and the stream-wise pressure gradient in the blade vane. These structures, already observed in [16] at $Re = 1.48 \times 10^5$, are quite strong and persist long in time along the pressure side.

In conclusion, the results suggest that LES provides a fairly accurate insight into the details of the flow. Although this test case cannot provide data for real turbine operating conditions, it can be conveniently used to test existing transition models in unsteady wake-affected accelerating boundary layers. Further developments are expected from the simulation of higher Reynolds number cases. The increased value of the Reynolds number will hopefully meet the needs of real industrial applications. With such a computational efficiency, the experience gained so far allows to forecast a more intensive use of LES for industrially relevant flows.

Acknowledgements

This work was supported by the German Research Foundation (DFG). The steering committee of the Computer Centre of the University of Stuttgart is gratefully acknowledged for having provided the computing time on the HITACHI SR-8000 parallel computer. The authors would also like to thank Dr. Jochen Fröhlich for the fruitful discussions on SGS closures.

References

1. Breuer, M., Rodi, W.: Large eddy simulation for complex turbulent flows of practical interest. Flow Simulation with High Performance Computers II, Notes on Num. Fluid Mechanics, Vieweg Verlag, 1996.
2. Cho, N.-H., Liu, X., Rodi, W., Shönung: Calculation of Wake-Induced Unsteady Flow in a Turbine Cascade. ASME Paper 92-GT-306
3. Emunds, R., Jennions, I.K., Bohn, D., Gier, J.: The Computation of Adjacent Blade-Row Effects in a 1.5-Stage Axial Flow Turbine. ASME Journal of Turbomachinery, **121** (1999) 1–10
4. Germano, M., Piomelli, U., Moin, P., Cabot, W.H.: A dynamic subgrid-scale eddy viscosity model. Physics of Fluids A **3**, number 7 (1991) 1760–1765
5. Harvey, N.W., Cox, J.C., Schulte, V., Howell, R., Hodson, H.P.: The role of research in the aerodynamic design of advanced low-pressure turbine. Proceedings of the 3rd European Conference on Turbomachinery (1999) 123–132
6. Hsu, K., Lee, L.: A numerical technique for two-dimensional grid generation with grid control at all of the boundaries. J. Comput. Phys., **96** (1991) 451-469
7. Jeong, J., Hussain, F.: On the identification of a vortex. J. Fluid Mech., **285** (1995) 69–94
8. Liu, X., Rodi, W.: Velocity Measurements of Wake-Induced Unsteady Flow in a Linear Turbine Cascade. Experiments in Fluids, **17** (1994) 45–48
9. Mellen, C.P., Fröhlich, J., Rodi, W.: Computations for the European LESFOIL project. E.Krause and W. Jger (eds.), Scientific Computation in 2000, Springer, (2001).
10. Michelassi, V., Martelli, F., Dnos, T. Arts, C.H. Sieverding: Unsteady Heat Transfer in Stator-Rotor Interaction by Two Equation Turbulence Model. Transaction of the ASME Journal of Turbomachinery, **121**, (1999)
11. Rhie, C.M., Gleixner, A.J., Spear, D.A., Fishberg, C.J., Zacharias, R.M.: Development and application of a multistage Navier-Stokes solver: Part I Multistage modelling using body-forces and deterministic stresses. ASME Journal of Turbomachinery, **120** (1998) 205–214
12. Stadtmüller, P.: Investigation of Wake-Induced Transition on the LP turbine Cascade T106A-EIZ. DFG-Verbundproject Fo 136/11, Version 1.0.
13. Stadtmüller, P.: Investigation of Wake-Induced Transition on the LP turbine Cascade T106A-EIZ. DFG-Verbundproject Fo 136/11, Version 1.1.
14. Wissink, J.G.: DNS of a separating low Reynolds number flow in a turbine cascade with incoming wakes. To be published in: Proceedings of V International Symposium on Engineering Turbulence Modelling and Measurements, Mallorca (Spain), 16–18 September (2002)
15. Wu, X., Jacobs R.G., Hunt J.C.R., Durbin P.A.: Simulation of boundary layer transition induced by periodically passing wakes. Journal of Fluid Mechanics, **398** (2001) 109–153
16. Wu, X., Durbin P.A.: Evidence of longitudinal vortices evolved from distorted wakes in a turbine passage. Journal of Fluid Mechanics, **446** (2001) 199–228

Parallel Two-Phase Flow Simulations in Porous Media

U. Ölmann[1], R. Hinkelmann[1], and R. Helmig[1]

Institute of Hydraulic Engineering, Hydromechanics and Modeling of
Hydrosystems, University of Stuttgart, Germany

Abstract. MUFTE-UG, a parallel numerical simulator for multiphase flow and
transport processes in porous media, is introduced. The basic PDEs for two-phase
flow together with a discretization and solution scheme are presented. Aspects of the
implementation of the advanced numerical techniques of UG on parallel hardware
are shown and the simulator's parallel performance is demonstrated using a 2D
example.

1   Introduction

In recent decades, the relevance of simulating processes in all areas of engineering has increased enormously. This increase has been accompanied on the
one hand by a continuously improving performance of hardware architectures
as well as by substantial progress in available numerical techniques on the
other. Both support the application of mathematical models as prediction
tools within engineering practice, which in turn encourages the sophistication of the underlying mathematical models. We particularly observe this
general trend in the context of subsurface flow and transport modeling. It is
a reflection of the strong demand for computer programs which carry out simulations concerning the management of groundwater systems, the evaluation
of remediation techniques etc.

It seems to be natural that the engineer's desire for a level of detail within
these simulations exceeds the capacity of the available workstations. This
holds especially in two- and multiphase problems which are focal points in
practice. The classical answers to this dilemma are the utilisation of up-to-
date numerical concepts like multi-level methods and adaptivity as well as the
parallelisation of the algorithms these techniques are based upon. The latter
is becoming more and more realistic as the number of parallel computers
increases, multiprocessor PCs are available off-the-shelf and computers in
existing pools can be clustered to efficient units using freely available tools.
The integrated implementation of the above-named strategies can hardly be
coped with by individual engineers and up to now cannot be called state of
the art. As a result, the demand for reliable universal and therefore reusable
software with this purpose grows.

The parallel numerical simulator MUFTE-UG [4,7], which is the foundation of this report, is a combination of the MUFTE and the UG tools. The

intention of the framework UG (Unstructured Grids) is to fulfil the above-mentioned requirements in a flexible way. The features of MUFTE (Multiphase Flow, Transport and Energy) consist of different model concepts, discretization techniques and numerical schemes together with constitutive relationships and refinement criteria.

2   Two-Phase Flow Model

MUFTE-UG covers many aspects of subsurface flow. Its application areas range from isothermal 1-phase 2-component processes to 2-phase processes with and without phase transitions and non-isothermal 3-phase 3-component processes with phase transitions. As the present report is primarily concerned with some of the parallel aspects the underlying framework UG provides, comparatively simple two-phase flows are sufficient as an object of study.

2.1   Basic Model Equations

The flow of two immiscible fluids in a spatial domain Ω during a time interval I can be described with the equations of conservation of mass of the *wetting phase (w)* and the *nonwetting phase (n)*

$$\frac{\partial}{\partial t}(\rho_\alpha \phi S_\alpha) + \nabla \cdot (\rho_\alpha \boldsymbol{v}_\alpha) = \rho_\alpha q_\alpha \qquad \text{in } \Omega \times I,\, \alpha = \text{n}, \text{w} \tag{1}$$

in combination with the extended version of *Darcy's* law

$$\boldsymbol{v}_\alpha = -\frac{k_{r\alpha}}{\mu_\alpha}\boldsymbol{K}\left(\nabla p_\alpha - \rho_\alpha \boldsymbol{g}\right) \qquad \alpha = \text{n}, \text{w} \ . \tag{2}$$

In these equations, S denotes the unknown saturation, ϕ the porosity, ρ the density, t the time, $\boldsymbol{v}$ the velocity, q a source or sink term, k_r the relative permeability, $\boldsymbol{K}$ the intrinsic permeability tensor, μ the dynamic viscosity, p the unknown pressure and $\boldsymbol{g}$ the vector of gravitational acceleration.

This system of PDEs is completed by appropriate initial and boundary conditions as well as the following two algebraic relationships

$$S_\text{n} + S_\text{w} = 1 \tag{3}$$

$$p_\text{c} = p_\text{n} - p_\text{w} \tag{4}$$

which state that the void space in a porous medium is completely filled by the two phases and that a functional correlation exists between the capillary pressure and the difference of the phase pressures. With these algebraic relations, two of the four unknowns in the PDEs can be eliminated. For the case of p_w and S_n as the primary variables, the equations are as follows

$$(-1)^{\delta_{\alpha\text{w}}}\phi\frac{\partial\left(\rho_\alpha S_\text{n}\right)}{\partial t} - \nabla \cdot \left(\rho_\alpha \lambda_\alpha \boldsymbol{K}\left(\nabla p_\text{w} + \delta_{\alpha\text{n}}\nabla p_\text{c} - \rho_\alpha \boldsymbol{g}\right)\right) - \rho_\alpha q_\alpha = 0$$

$$\text{in } \Omega \times I,\, \alpha = \text{n}, \text{w} \ . \tag{5}$$

In this equation, δ stands for the Kronecker-δ and the mobility is given by $\lambda_\alpha = k_{r\alpha}/\mu_\alpha$.

Constitutive relationships for two quantities supplement the above-mentioned set of equations: the relative permeability $k_{r\alpha}$ ($\alpha = n, w$) accounts for the reduction of the porous medium's permeability for each of the two phases because of the mutual hindrance of the phases. The capillary pressure on the continuum scale describes the difference in phase pressure of both phases as it does on the pore scale. In the simulations in this report, the approach of *Brooks-Corey* is applied; this models both quantities as functions depending only on the saturation. Specific soil values, which enter these functions as parameters, control the character of these constitutive relations. For more details on them, see, for example, [7].

2.2   Discretization Scheme

Time and space directions are treated differently in the discretization of the resulting PDEs within MUFTE-UG: a fully implicit *Euler* scheme is used for the former whereas, for the latter, either the finite volume (alias box) method or the control volume finite element method is applied (see [6]). Within the conducted simulations, the node-centered fully upwind box scheme based on the p_w-S_n-formulation (5) was chosen. This leads to the following nonlinear system of algebraic equations for each time step:

$$(-1)^{\delta_{\alpha w}} \phi \left\{ \left[\rho_\alpha \hat{S}_n \right]_i^{n+1} - \left[\rho_\alpha \hat{S}_n \right]_i^n \right\} \frac{B_i}{\Delta t} -$$

$$- \sum_{j \in \eta_i} (\rho_{\alpha ij} \lambda_{\alpha ij})_{\mathrm{ups}}^{n+1} \oint_\Gamma \boldsymbol{K} \nabla N_j \boldsymbol{n} \, d\Gamma \, (\psi_{\alpha j} - \psi_{\alpha i})^{n+1} - B_i \left[\rho_\alpha \hat{q}_\alpha \right]_i^{n+1} = 0$$

$$\alpha = n, w \, . \quad (6)$$

In these equations, the symbol ^ indicates that the discrete value of a variable at node i is used. n represents the current time step of the iteration. B_i stands for the volume of a box that is assigned to node i. η_i is the set of nodes neighboring i whose boxes share an interface with B_i. ups indicates that an upwind technique is applied for the flux terms. The basic function is referred to as N_j, and $\boldsymbol{n}$ denotes the unit vector perpendicular to boundary Γ of box B_i. $(\psi_{\alpha j} - \psi_{\alpha i})$ is the direction of the discrete flow of phase α with $\psi_{\alpha i} = p_{wi} + \delta_{\alpha n} p_{ci} - \rho_{\alpha i} \boldsymbol{g}_i$.

2.3   Strategy for Solution of Discretized Equations

The approximate solution of the nonlinear system (6) with a dimension twice the number of nodes of the finite volume grid is reduced to the solution of a series of linear problems applying a damped Newton-Raphson method. In this report, these linear systems are treated with the BiCGSTAB method (see [9]) which in turn is preconditioned using a multigrid solver (MGS), [5]. Here, the MGS follows a V-cycle with two steps of pre- and postsmoothing by a modified ilu procedure. On the coarsest grid of the chosen V-cycle, an exact band-lu decomposition is utilized. For further reading, see [2].

3 Aspects of Parallelisation on a MIMD Architecture

UG provides a transparent interface for configuring parallel adaptive multigrid methods. Parallel preprocessing (generation of grids) as well as postprocessing (visualization) are also covered (see [1]). Concurrently, UG provides different static and dynamic load-balancing techniques, a detailed overview can be found in [8].

The parallelisation is carried out in the algebraic operations, such as scalar or matrix vector products. Moreover, the parallel programming model DDD [3] which is specialized on graph based data structures was developed. DDD enables portability to various HPC architectures which use different communication interfaces, such as MPI, PVM, NX or SHMEM. DDD has increased the measure of abstraction for the parallelisation and is a step in the direction of automatic parallelisation for message-passing programming models.

From a user's point of view the script-file controlling the simulation run only has to be extended by a few lines triggering the load-distribution among the available processors to change the sequential simulation into a parallel simulation run.

4 Application to Multi-Step-Outflow Experiment (MSO)

4.1 Experiment and Result

A laboratory-scale method for determining soil parameters is the so-called multi-step-outflow experiment: the water-phase pressure at the bottom of an initially water-saturated cylindrical porous medium located in a tube is lowered in discrete steps. As a consequence, water flows through the bottom surface while air enters the medium at the top. Using the measured water outflow as a response of the system, one can deduce the soil parameters by inverse modeling.

Within the scope of a research project dealing with upscaling techniques, MSOs are performed with artificial media and are reproduced in simulation runs. A 2D model serves as an object of study in this work. Figure 1 shows the geometry of the medium, which consists of a lens of coarse and fine sand embedded in a medium sand as background material, and the associated initial and boundary conditions. The soil parameters are given in Table 1. In Fig. 2, the pressure lowerings together with the resulting cumulative outflow are presented: one can clearly see that after the first step an equilibrium

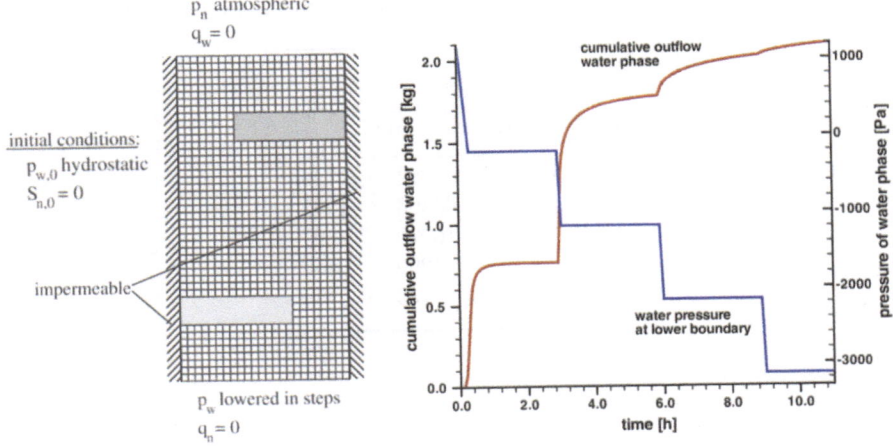

Fig. 1. Geometry, Initial and Boundary Conditions

Fig. 2. Pressure Steps and Cumulative Outflow

Table 1. Soil Parameters

parameter	fine sand	medium sand	coarse sand
porosity ϕ	0.5	0.4	0.35
absolute permeability K $[m^2]$	$1.427 \cdot 10^{-12}$	$1.426 \cdot 10^{-11}$	$1.66 \cdot 10^{-10}$
Brooks-Corey p_d [Pa]	1668.3	606.0	166.8
Brooks-Corey λ	1.29	1.29	1.29
residual saturation of (n)	0	0	0
residual saturation of (w)	0.5	0.1	0.05

has nearly been established, whereas the system needs more time to reach this state in the next steps when the overall water saturation has already decreased.

4.2 Performance Analysis

The simulation runs were executed on the Hitachi SR8000 at HLRS Stuttgart, which is a parallel computer consisting of 16 SMP-nodes connected via a high-speed (1 GB/s) inter-node network. Each node, equipped with 8 GB of RAM, offers the user 8 microprocessors with a single address space (64-bit PowerPC based instruction set, extended with mechanisms for prefetch and preload to supply pseudo-vectorization). An implementation of the MPI standard is available and is used by MUFTE-UG. Other parallel-programming models are provided as well.

Table 2. CPU Time For a Multigrid Method with Different Grid Levels

multi-grid levels	number of unknowns on the coarse grid	CPU time [s]
1	263682	1749
2	66306	1251
3	16770	657
4	4290	**365**
5	1122	574

Firstly, the level of the coarse grid within the MGS was varied in parallel simulation runs on four nodes (32 processors). In each case, the execution time for the first five time steps of the MSO was measured. A multigrid hierarchy of five grid levels served as a basis. Here, enormous differences arose; the optimum is attained by a four-grid solver, see Table 2. This is an effect of load-balancing. The theory of sequential MGS proposes running times that decrease with the number of grid levels. For a direct coarse-grid solver, this number should be chosen in such a way that the number of unknowns on the coarse grid is in the range of a few hundreds. If the number of unknowns on the coarse grid is much larger, the direct solver becomes too expensive w.r.t. CPU time.

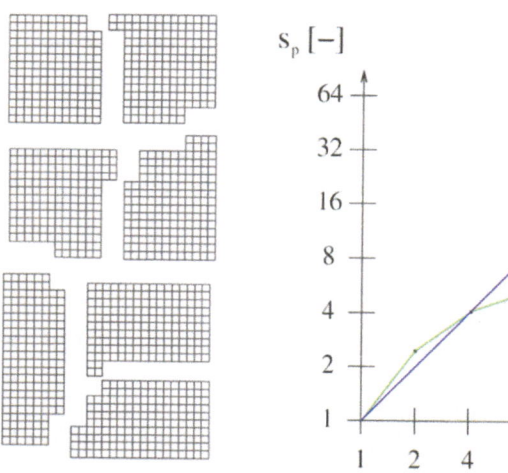

Fig. 3. Load Distribution for Seven Processors

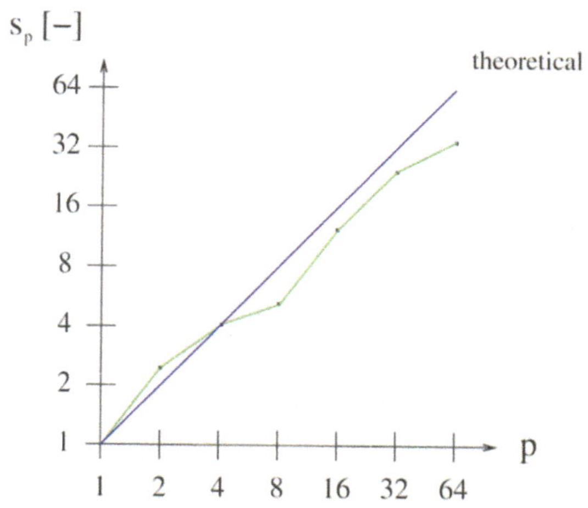

Fig. 4. Parallel Speedup of the Four-Grid Solver

The parallel speedup is investigated for the four-grid solver. A static load distribution was determined with the recursive inertial bisection method; this is shown in Fig. 3 for seven processors. The coarse-grid solution is computed on one processor. In Fig. 4, the parallel speedup is shown. It has an overlinear course for 2 and 4 processors which is probably caused by cache effects. For increasing numbers of processors, the efficiency decreases continuously. It must be taken into account that the convergence of the solver depends on the number of processors, i.e. the number of iterations increases with an increasing number of processors and the speedup is thus worsened.

5 Conclusions

The implementation of the parallel algorithm has proven to scale well. It could be shown that in contrast to theoretical expectations coming from the sequential MGS the optimal multigrid method w.r.t. CPU time is not the one that includes the whole grid hierarchy. Here, much potential for fine tuning of voluminous simulations is assumed.

References

1. Bastian, P., Birken, K., Johannsen, K., Neuss, N., Rentz-Reichert, H. and Wieners, C.: UG – A Flexible Software Toolbox for Solving Partial Differential Equations. Computation and Visualization in Science 1, 1997.
2. Bastian, P.: Numerical Computation of Multiphase Flows in Porous Media. Habilitation thesis, Christian-Albrechts-Universität Kiel, Germany, 1999.
3. Birken, K.: Ein Modell zur effizienten Parallelisierung von Algorithmen auf komplexen, dynamischen Datenstrukturen. Ph.D. thesis, Rechenzentrum, Univ. of Stuttgart, Germany, 1998.
4. Breiting, T., Hinkelmann, R. and Helmig, R.: Modeling of Hydrosystems with MUFTE-UG: Multiphase Flow and Transport Processes in the Subsurface. Fourth International Conference on Hydroinformatics, Iowa, USA, 2000.
5. Hackbusch, W.: Multi-Grid Methods and Applications. Springer-Verlag, Heidelberg, 1985.
6. Helmig, R.: Multiphase Flow and Transport Processes in the Subsurface. Springer-Verlag, Heidelberg, 1997.
7. Helmig, R., Class, H., Huber, R., Sheta, H., Ewing, J., Hinkelmann, R., Jakobs, H. and Bastian, P.: Architecture of the Modular Program System MUFTE-UG for Simulating Multiphase Flow and Transport Processes in Heterogeneous Porous Media. Mathematische Geologie 2, 1998.
8. Lang, S.: Parallele numerische Simulation instationärer Probleme mit adaptiven Methoden auf unstrukturierten Gittern. Ph.D. thesis,Institute of Hydraulic Engineering, Univ. of Stuttgart, Germany, 2001.
9. Van der Vorst, H.: BiCGSTAB: A Fast and Smoothly Converging Variant of Bi-CG for the Solution of Non-symmetric Linear Systems. SIAM J. Sci. Stat. Comput. 13 (1992), 631-644.

Three-Dimensional Simulation of Two-Phase Flow in Pipes

Tobias Giese and Eckart Laurien

Institute for Nuclear Technology and Energy Systems (IKE), University of Stuttgart, Pfaffenwaldring 31, D-70550 Stuttgart, Germany;
E-Mail: giese@ike.uni-stuttgart.de, laurien@ike.uni-stuttgart.de

Abstract. A two-phase flow of water in a pipe is investigated experimentally and numerically. A three dimensional simulation is performed using an enhanced two-fluid model. To take phenomena like phase change and a possible change of the flow regime into account, adequate inter-phase exchange terms were implemented. The results show that three dimensional phenomena in pipe elements can have a significant influence on the integral parameters of the flow situation.

1 Introduction

Two-phase flows in pipes are a matter of particular interest for the design of technical applications like power plants and chemical process plants. In contrast to single phase flow cases, the analysis of multiphase flow is difficult due to the modeling of the interaction of the phases in momentum, energy and turbulence. Phase change phenomena like boiling, cavitation and condensation may occur. Additionally, possible changes in flow regime of pipe flow i.e. from bubbly flow to stratified flow raise the complexity and prevent simple approaches for flow calculation.

In the past, one dimensional approaches of pipe flow were used and one dimensional simulation codes have been developed. However, their calculations are still based on empirical correlations gained from experiments. To use these correlations, it is necessary to neglect the influence of flow regime or to define the character of the flow a priori, e.g. using a flow pattern [1]. In a complex pipe system with phase interaction phenomena, it is not possible to get the necessary information about the flow regime in advance. To overcome the semi-empirical procedures, an approach with CFD is promising. The two-fluid model is deemed to be a general approach for multiphase problems and is described in Chapter 3.

The usage of CFD for problems with phase change phenomena makes it necessary to implement adequate models in the two-fluid description. For the flow case that was analysed experimentally, two models were developed: A model for the momentum interaction in different flow regimes and a model for the description of cavitation phenomena, see section 4 and 5.

To avoid that effects like secondary flows and gravity have to be neglected and to use the code for applied problems, a three dimensional approach for a complex pipe geometry is necessary. As a consequence, the computational effort is enormous and the usage of the HLRS supercomputers is necessary.

2   Analysed flow case and experimental results

The experiment was carried out by the authors at a test facility at Siemens, Erlangen in December 1999. Within the facility, a PVC pipe was installed which links a tank to a lower sited second tank. The height difference betweeen the tanks was 13 m, the pipe diameter 0.1 m and the overall length about 30 m.

Two flow cases were compared: A cold water (18 °C) single phase flow and a warm water (99 °C) two phase flow, both cases near atmospheric pressure (1 bar). The mass flux in the warm water flow case is reduced significantly in comparison to the cold water flow case. The difference can be explained by phase change phenomena such as boiling and cavitation. If the local pressure

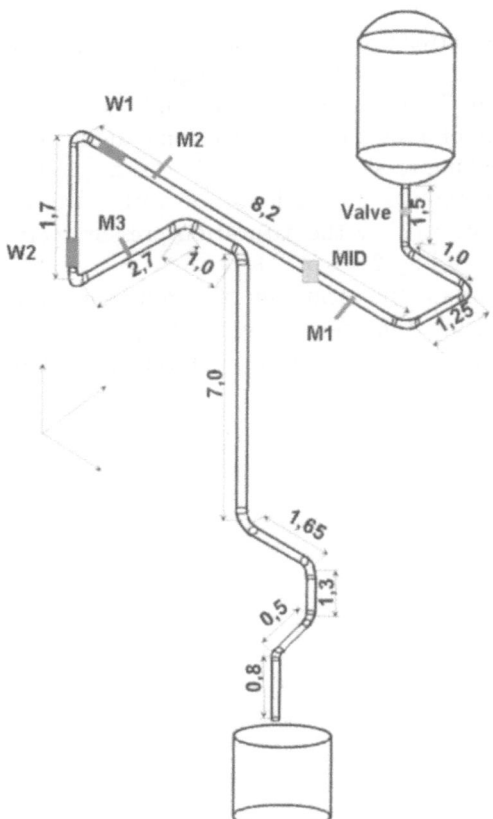

Fig. 1. Experimental Setup

in the pipe drops below the saturation pressure, cavitation phenomena occur. In this two phase flow case, the prediction of mass flux is difficult due to the coupling between the phase change phenomena depending on the local static pressure in a pipe segment and the pressure loss in the entire pipe.

The first experiment was performed with cold water (18 °C) which can be assumed to be a single phase flow case due to the absence of phase change phenomena. In the second case, the water in the upper tank is heated close to its saturation temperature ($\approx$ 99 °C). This high temperature increases the saturation pressure of water and causes cavitation phenomena in low pressure regions of the pipe. Due to the fact that a two phase character of the flow increases the pressure loss, a decreased mass flux is expected.

Figure 1 is a schematic picture of the experimental setup. The instrumentation consists of a magnetic inductive flow sensor (MID) which was sited in a pipe segment with single phase flow character. Three local temperature and pressure measuring devices where placed at M1, M2 and M3. Additionally, at two locations of the pipe the volume fraction of steam is measured with a simple wire-pair probe. To get an impression of the character of the flow, two windows (W1 and W2) allow the observation of the flow.

The differences between the two flow cases are significant. In the single phase flow case (cold water) low pressure regions appear (0.3 bar) in certain sections of the pipe. These pressure values and the gained mass flux agree well with the values of a simple one dimensional analysis with the Bernoulli equation. The visual impression of the flow meets the picture of a single phase flow with a few air bubbles. In the flow case with water close to 99 °C, the pressure does not drop below the saturation pressure corresponding to the local temperature, see Fig. 2.

The reason for this is the formation of steam which is visible in the windows of the pipe. A view on the jet leaving the pipe confirms that in a flow case with warm water, a gaseous phase is present. The main consequence of

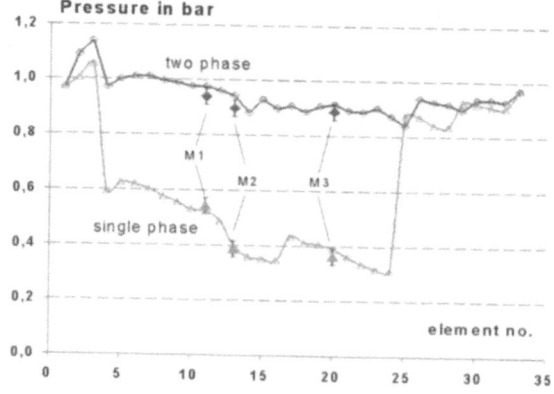

Fig. 2. Static pressure over pipe elements

the different character of the flow is the reduced mass flux in the case with high temperature. In comparison with the low temperature case, the mass flux is reduced from $36\,\mathrm{kg/s}$ to $23\,\mathrm{kg/s}$. This is a reduction of nearly 40% and can not be explained by measurement errors (error margin for the mass flux: less that 5%). The reason for this reduction is the influence of the multiphase character of the flow e.g. the production of steam by cavitation and, as a result, the increase of pressure loss in the pipe.

3 Two-Fluid Model

To model the two phase character of the flow, different approaches are possible. The Euler-Lagrange approach is adequate for the simulation of drops or bubbles in a continuous fluid, but it fails if complex flow regimes occur. To overcome this, the Euler-Euler approach is used. In this approach, both phases, liquid (index L) and steam (index G, gas), are regarded as two continua interpenetrating each other. With the help of two phase functions ε_L defined unity in pure liquid and zero in pure gas, and ε_G, defined unity in pure gas and zero in pure liquid, the basic equations for the average (denoted by an overbar) two-phase state quantities ($\alpha_L = \overline{\varepsilon_L}$, $\alpha_G = \overline{\varepsilon_G}$) can be formulated ([2],[3]). The three dimensional mass, momentum and energy conservation equations are integrated for each phase.

The mass conservation equation for the liquid phase

$$\rho_{\mathrm{L}}\frac{\partial \alpha_{\mathrm{L}}}{\partial t} + \rho_{\mathrm{L}}\nabla \cdot (\alpha_{\mathrm{L}}\overline{\boldsymbol{u}}^{\mathrm{L}}) = \Gamma_{\mathrm{L}} \tag{1}$$

and for the gas phase

$$\rho_{\mathrm{G}}\frac{\partial \alpha_{\mathrm{L}}}{\partial t} + \rho_{\mathrm{G}}\nabla \cdot (\alpha_{\mathrm{G}}\overline{\boldsymbol{u}}^{\mathrm{G}}) = \Gamma_{\mathrm{G}} \tag{2}$$

the momentum conservation equations for the liquid phase

$$\rho_{\mathrm{L}}\frac{\partial}{\partial t}(\alpha_{\mathrm{L}}\overline{u}_m^{\mathrm{L}}) + \rho_{\mathrm{L}}\nabla \cdot (\alpha_{\mathrm{L}}\overline{\boldsymbol{u}}^{\mathrm{L}}\overline{u}_m^{\mathrm{L}}) = -\frac{\partial (\alpha_{\mathrm{L}}\overline{p})}{\partial x_m} + \nabla \cdot \left(\alpha_{\mathrm{L}}\left(\underline{\overline{\tau}}^{\mathrm{L}} + \underline{\tau}^{\mathrm{Re\,L}}\right)\right) \tag{3}$$
$$+ M_m^{\mathrm{L}} + f_m^{\mathrm{L}}$$

and for the gas phase

$$\rho_{\mathrm{G}}\frac{\partial}{\partial t}(\alpha_{\mathrm{G}}\overline{u}_m^{\mathrm{G}}) + \rho_{\mathrm{G}}\nabla \cdot (\alpha_{\mathrm{G}}\overline{\boldsymbol{u}}^{\mathrm{G}}\overline{u}_m^{\mathrm{G}}) = -\frac{\partial (\alpha_{\mathrm{G}}\overline{p})}{\partial x_m} + \nabla \cdot \left(\alpha_{\mathrm{G}}\underline{\overline{\tau}}^{\mathrm{G}}\right) \tag{4}$$
$$+ M_m^{\mathrm{G}} + f_m^{\mathrm{G}}$$

the energy conservation equation for the liquid phase

$$\rho_{\mathrm{L}}\frac{\partial}{\partial t}\left(\alpha_{\mathrm{L}}\overline{h}^{\mathrm{L}}\right) + \rho_{\mathrm{L}}\nabla \cdot \left(\alpha_{\mathrm{L}}\overline{u}_m^{\mathrm{L}}\overline{h}^{\mathrm{L}}\right) = \nabla \cdot \left(\alpha_{\mathrm{L}}\left(\overline{\boldsymbol{q}}^{\mathrm{L}} + \overline{\boldsymbol{q}}^{\mathrm{Re\,L}}\right)\right) + E_{\mathrm{L}} \tag{5}$$

and for the gas phase

$$\rho_G \frac{\partial}{\partial t}\left(\alpha_G \overline{h}^G\right) + \rho_G \nabla \cdot \left(\alpha_G \overline{u}_m^G \overline{h}^G\right) = \nabla \cdot \left(\alpha_G \overline{q}^G\right) + E_G \qquad (6)$$

form the partial differential equation system that has to be solved.

In the equation for the liquid phase, the additional terms for turbulent fluxes $\tau^{Re\,L}$ and $q^{Re\,L}$, describing the Reynolds-stresses and the turbulent heat fluxes, appear. The flow in the liquid phase is modelled as a turbulent flow with the k-ε-model.

The two-phase character of the flow can be found in the inter-phase exchange terms. These terms model the momentum exchange and the heat transfer between the phases (k=L,G). The terms that model the inter-phase exchange can be summarized in the following vector:

$$\boldsymbol{A}^k = \begin{bmatrix} \Gamma_k \\ M_1^k \\ M_2^k \\ M_3^k \\ E_k \end{bmatrix} . \qquad (7)$$

The inter-phase exchange terms in this approach are well known for some of the flow cases, for example for dilute bubbly flow. But for many phenomena like boiling, cavitation and condensation and for different flow regime the terms are unknown.

4   Momentum Exchange

In the following, a momentum phase exchange model is discussed. The inter-phase exchange term in Eq. (7) can be replaced by volume forces corresponding to the assumed physical effect. The momentum exchange depends on the flow regime. Figure 3 shows a flow situation with the liquid phase mainly in the lower part of the pipe and the gas phase mainly in the upper part of the pipe.

For bubbly flow, the main interaction consists of the drag force of the bubbles moving relative to the liquid. With the assumption that the bubble

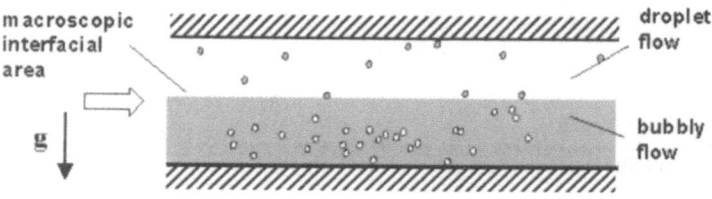

Fig. 3. Complex flow situation with bubbly and droplet flow

is spherical, the flow of water around a sphere can be used in analogy. The integration over all bubbles with the diameter d_B in certain control volume leads to a phase interaction force for bubbly flow

$$F_{\mathrm{B}} = \frac{3}{4} \frac{C_{\mathrm{D}}}{d_{\mathrm{B}}} \alpha_{\mathrm{G}} \rho_{\mathrm{L}} \left| \overline{u}_{\mathrm{G}} - \overline{u}_{\mathrm{L}} \right| \left(\overline{u}_{\mathrm{G}} - \overline{u}_{\mathrm{L}} \right) \ . \tag{8}$$

For droplet flow, the phase interaction force can be derived in almost the same manner. A single drop is moving in a gaseous surrounding. Summarizing all drops in a control volume this results in

$$F_{\mathrm{D}} = \frac{3}{4} \frac{C_{\mathrm{D}}}{d_{\mathrm{D}}} \alpha_{\mathrm{L}} \rho_{\mathrm{G}} \left| \overline{u}_{\mathrm{G}} - \overline{u}_{\mathrm{L}} \right| \left(\overline{u}_{\mathrm{G}} - \overline{u}_{\mathrm{L}} \right) \ . \tag{9}$$

The decision whether the term for the bubbly flow or the term for the droplet flow is used in the equations (3) and (4) can be made in dependency of the local volume fraction. Large values for α_L and small values for α_G indicate a bubbly flow. For values $\alpha_G < 0.25$ the bubbly flow correlation is used. For $\alpha_G > 0.75$, the droplet flow correlation is used. In between, an approach according to

$$F_{\mathrm{P}} = \left(\frac{\alpha_{\mathrm{G}} - 0.25}{0.5} \right) \cdot F_{\mathrm{D}} + \left(1 - \left(\frac{\alpha_{\mathrm{G}} - 0.25}{0.5} \right) \right) \cdot F_{\mathrm{B}} \ . \tag{10}$$

is used. Equation (10) is valid for all regions in Fig. 3 with the exception of the macroscopic interfacial area of stratified flow. In this area, a correlation for stratified flows according to Wang [4] is used

$$F_{\mathrm{Str}} = \frac{1}{2} A_{\mathrm{i}} f_{\mathrm{i}} \rho_{\mathrm{G}} \left| \overline{u}_{\mathrm{G}} - \overline{u}_{\mathrm{L}} \right| \left(\overline{u}_{\mathrm{G}} - \overline{u}_{\mathrm{L}} \right) \ . \tag{11}$$

A_{i} is the interfacial area and f_{i} an interfacial friction coefficient. The decision between (10) and (11) is based on an indicator for stratified layers. At the interfacial area of stratified flows, the gradient of the volume fraction has a maximum value. As a first step,

$$M^{\mathrm{K}} = F_{\mathrm{P}} \cdot \left(1 - \max \left(0; \min \left(\frac{|\nabla \alpha_{\mathrm{G}}|}{C_{\mathrm{Str}}}; 1 \right) \right) \right)$$
$$+ F_{\mathrm{Str}} \cdot \left(\left(0; \max \left(\frac{|\nabla \alpha_{\mathrm{G}}|}{C_{\mathrm{Str}}}; 1 \right) \right) \right) \tag{12}$$

with a parameter C_{Str} adapted to the grid resolution is used.

5   Cavitation Model

At IKE, attempts to analyse a two phase flow case in pipes led to the development of a cavitation model for water with high temperature and, as a consequence, high saturation pressure. The basic idea of the developed model is that the phase change in a cavitating flow is based on thermodynamic processes and not on bubble dynamics [5]. As described in [6], heat transfer between a bubble and surrounding liquid can be used to quantify cavitation phenomena.

Recent work extended this approach on phase change phenomena of drops in a gaseous surrounding to simulate flow situations shown in Fig. 3. To derive the inter phase mass transfer, the fact that the interfacial area of bubbles and drops is at saturation temperature T_{sat} is used. The saturation temperature depends on the local static pressure. The variation of the pressure, (assuming equal pressure for both phases) or the temperature of either steam or water leads to a heat transfer, which can be balanced with an inter-phase mass transfer. Assuming spherical bubbles and drops with diameters d_B and d_D, the heat transfer between the dispersed phase and its surrounding can be derived for both flow regimes. The inter-phase heat transfer coefficient α_B of a spherical bubble in motion with a relative velocity to the surrounding fluid can be approximated by

$$\alpha_B = \frac{Nu_B \cdot \lambda_L}{d_B} = \frac{\lambda_L}{d_B} \cdot \left(2 + 0.6 \cdot Re_B^{1/2} Pr_L^{1/3}\right) \ . \tag{13}$$

In this model, the Reynolds number Re_B and the Prandtl number Pr_L are used to predict the convective heat transfer according to the Ranz-Marshall correlation [7]. In equilibrium the heat flux $\Delta \dot{h}_B$ towards the interphase of one bubble can be used to predict the inter-phase mass transfer $\dot{m}_B$ at the bubble which is the basic value for the different source terms in the equation sets

$$\Delta \dot{h}_B = \Delta h_{LG} \cdot \dot{m}_B = \alpha_B \cdot (T_L - T_{sat}) \cdot \pi d_B^2 \ . \tag{14}$$

If λ_L is the thermal conductivity of the water and Δh_{LG} the evaporation enthalpy, this leads to the equation

$$\Gamma_B = (T_L - T_{sat}) \cdot \frac{6 \cdot \alpha_G \cdot \lambda_L}{d_B^2 \cdot \Delta h_{LG}} \cdot \left(2 + 0.6 \cdot Re_B^{1/2} Pr_L^{1/3}\right) \tag{15}$$

for the mass transfer at the interfacial area of all bubbles per volume. If the liquid is superheated ($T_L > T_{sat}$), it is assumed that the transported heat (from the liquid to the interfacial area of the bubble) is used for further vaporization of the liquid phase. The Ranz-Marshall correlation used for the droplet region leads to the equation

$$\Gamma_D = (T_G - T_{sat}) \cdot \frac{6 \cdot (1 - \alpha_G) \cdot \lambda_G}{d_D^2 \cdot \Delta h_{LG}} \cdot \left(2 + 0.6 \cdot Re_D^{1/2} Pr_G^{1/3}\right) \ . \tag{16}$$

The decision which flow regime correlation, bubbly or droplet flow, is used for the prediction of inter-phase mass transfer is performed by the volume fraction of steam according to

$$\Gamma = \left(\frac{\alpha_G - 0.25}{0.5}\right) \cdot \Gamma_D + \left(1 - \left(\frac{\alpha_G - 0.25}{0.5}\right)\right) \cdot \Gamma_B \ . \tag{17}$$

The derivation of the mass transfer between the phases based on bubbles and drops is misleading in the region of the macroscopic interfacial area of stratified flows. In this region, which can be identified by a large gradient of the phase fractions, a simple correlation for free surface flow can be used.

Due to the exchanged mass, momentum and energy is transported between the phases. However, the thermal energy of the system has to be modified in the case of mass transport. The reason is that for the vaporization of liquid, the latent heat Δh_{LG} has to be removed from the system. To enable the calculations of different flow regimes, an approach

$$E_{\mathrm{L}} = \alpha_{\mathrm{L}} \Delta h_{\mathrm{LG}} \Gamma_{\mathrm{L}} \tag{18}$$

$$E_{\mathrm{G}} = -\alpha_{\mathrm{G}} \Delta h_{\mathrm{LG}} \Gamma_{\mathrm{G}} \tag{19}$$

was implemented.

6 Simulation

To make a simulation in a complex geometry possible, the commercial code CFX-4.2 is used [8]. The code provides the basic two fluid equations and simple models for bubbly flow. However, for the simulations mentioned below, the terms for the momentum, mass and energy exchange were implemented by additional FORTRAN subroutines and the models of Chapter 4 and 5 were integrated. With these additional implementations, the simulation of the two phase flow in the experimental pipe is possible.

To test the capabilities of the cavitation model, a parameter study of the two phase flow in a pipe bend was performed ([6], [9]). The influence of the bubble diameter and nucleation density on the development of steam in water near saturation temperature was studied.

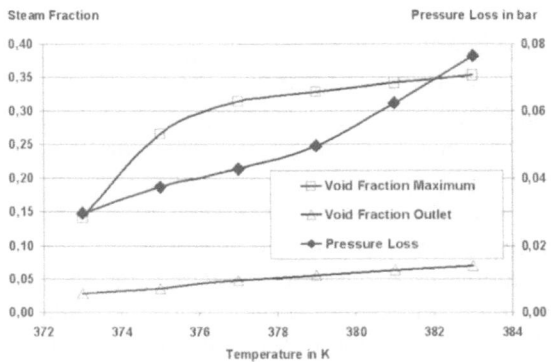

Fig. 4. Variation of water temperature

A simulation of the flow of hot water in a 90° pipe bend was simulated with the models of section 4 and 5. Due to the fact that the local static pressure falls below the saturation pressure, steam bubbles occur. In Fig. 4 the maximum steam fraction, the average steam fraction at the outlet and the pressure loss in the bend is shown when the temperature of the fluid is varied. The intensity of cavitation phenomena and the formation of steam increases in dependency of the temperature. The pressure loss is doubled due to a temperature rise of 10 K.

The outlet pressure determines the pressure level in the in the bend. At lower pressure level, the cavitation phenomena is intensified and the pressure loss is increased, see Fig. 5. Even with water at 373 K and an outlet pressure of 1bar, steam develops in the low pressure region near the inner radius of the bend. Due to the three dimensionality of the problem and secondary flows in the pipe bend, an increase of pressure loss according to a single phase flow case can be detected.

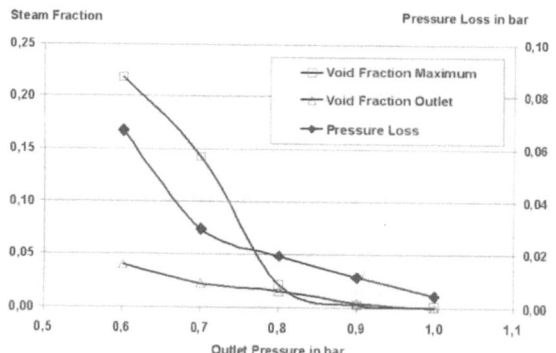

Fig. 5. Variation of outlet pressure

The application of the cavitation model and stratified flow model on the experimental pipe flow was performed [10]. A numerical mesh of more than 314000 nodes was necessary to take effects like secondary flows in bends into account. These effects seem to have an important influence on the production and the location of the steam in the pipe. A plot of the volume fraction of the steam at the pipe wall of a certain section of the pipe is provided in Fig. 6.

In the lower horizontal pipe section shown in Fig. 6, the flow regime converses from a bubbly flow to a stratified flow. The effect of gravity in this process is negligible in comparison to the effect of secondary flows in bends that transports the steam fraction to the inner radius of the bend. The mass flux of the recent calculations is 25 kg/s in the two phase flow case and differs from the experimental result (23 kg/s).

The simulation of the experimental pipe was performed on a NEC SX-4 at the High Performance Computing Center Stuttgart. The computational effort is large due to the two- fluid approach and the additional set of Navier-Stokes equation for the second phase. 300 CPU-hours were necessary for the Simulation of the experiment. The segregated solver of the version CFX-4 is sensitive in cases of strong phase interaction. This requires conservative underrelaxation to keep the simulation stable and a large number of iterations. The new version CFX-5 with a coupled solver may be able to improve the stability of the simulation [11]. The grid for the large geometry of the experiment requires a memory of 1GB.

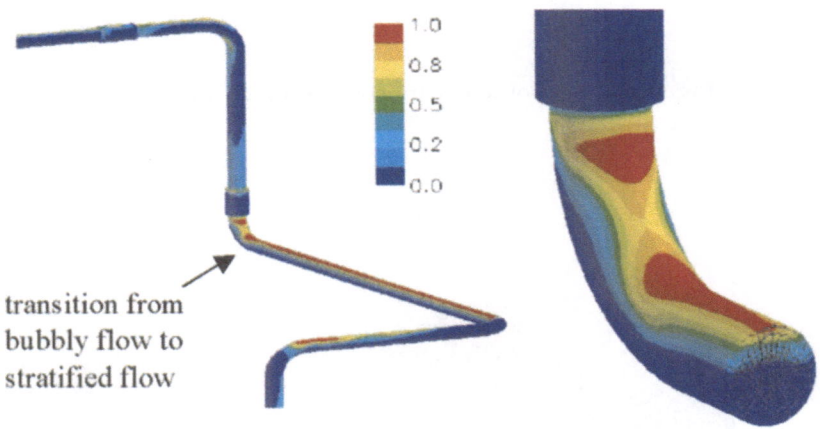

transition from bubbly flow to stratified flow

Fig. 6. Volume fraction of steam at pipe wall

As an example, the region after the second window element is shown in Fig. 6. After the second window causing an increased pressure loss, the static pressure decreases and the steam production is enforced. Due to the secondary flow vortexes, the steam accumulates at the inner radius of the bend and causes a stratified flow in the following horizontal pipe section.

Another important result of the analysis of secondary flow is the interaction of pipe elements. If the secondary flow of a pipe element enters the following pipe element, the pressure drop in this next element is influenced. E.g. if flow containing a single vortex enters a bend, one of the two new secondary flow vortexes is strengthened and the other is degraded due to superposition. As a result, the steam distribution in the bend is influenced, see Fig. 7. The great improvement of three dimensional CFD towards one dimensional calculations can be found in the possibility to get into details if necessary. In the following picture a two phase flow of water and steam flows through an inclined pipe segment with bends in near neighbourhood is visible (Fig.8). Only a three dimensional simulation can answer the question

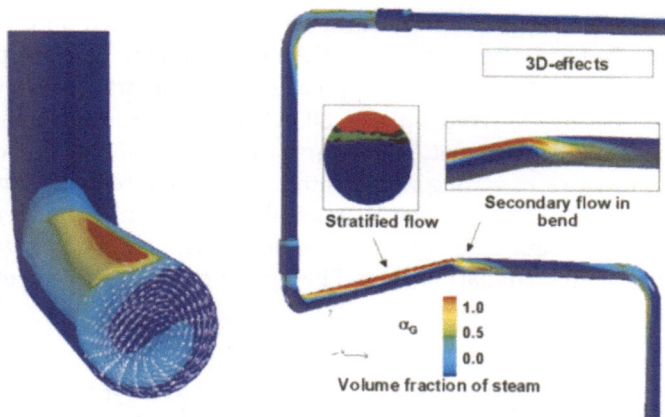

Fig. 7. Influence of secondary flows on void distribution

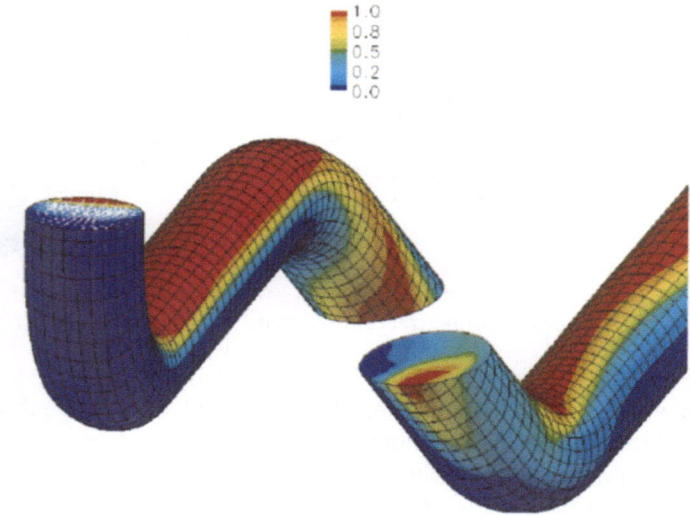

Fig. 8. Volume fraction of steam at pipe wall

on which pipe walls the liquid or gas phase is present. Gravity, secondary flows and phase transition affect the distribution of steam [12].

Of course, most technical applications can be designed without such a detailed and expensive analysis. But on the other hand, some applications require this knowledge desperately. It is worth to provide this analysis opportunity.

7   Computational Resources

The simulation was performed on a NEC SX-4 at the High Performance Computing Center Stuttgart. CFX-4.2 is partially vectorised and the authors vectorised their FORTRAN subroutines as far as possible. The computational effort is large due to the two-fluid approach and the additional set of Navier-Stokes equation for the second phase. Additionally, the segregated solver of the version CFX-4 is sensitive in cases of strong phase interaction. This requires small iteration steps. The new version CFX-5 with a coupled solver will overcome this problem [11].

8.   Conclusion

The experimental results of a gravity driven pipe flow including cavitation phenomena are presented and analyzed. For an appropriate simulation of the experiment, the two-fluid approach was used. The momentum interaction was modeled including different flow regimes, e.g. bubbly flow and stratified flow. Additionally, a thermal based model for cavitation in hot water was developed and tested. The models for different flow regimes and cavitation have been implemented using the FORTRAN interface and were verified through the experimental data. The deviation of the mass flux of the three dimensional simulation including theses models from the experimental results are small. An analysis of the simulation results shows the influence of secondary flows on steam distribution and, as a consequence, on cavitation and recondensation phenomena, on pressure loss and integral mass flux. Furthermore, the interaction of pipe elements caused by secondary flows and its effects on steam distribution was demonstrated.

In order to continue our model development and to apply the code to new geometric situations, the computational effort must be reduced in near future. The most promising approach to reach this aim is the usage of a coupled flow solver instead of an segregated solver.

Acknowledgement

This study has been funded by the German Electricity Association (VDEW) and the Gemeinschaftskraftwerk Neckar GmbH (GKN)

References

1. Friedel, L.: Druckabfall bei der Strömung von Gas/Dampf-Flüssigkeits-gemischen in Rohren. Chem.-Ing.-Tech. **50** (1978) 167–180
2. Lahey, R.T., Drew, D.A.: The analysis of two-phase flow and heat transfer using a multidimensional, four field, two-fluid model. Nuclear Engineering and Design **155** (1995)

366   T. Giese and E. Laurien

3. Laurien, E., Saptoadi, D.: On the fundamental two-fluid equations to model three-dimensional bubbly flows. Fourth International Conference on Multiphase Flow (ICMF 2001), New Orleans (USA)

4. Wang, M., Mayinger, F.: Simulation and Analysis of Thermal-Hydraulic Phenomena in a PWR Hot Leg related to SBLOCA. Nuclear Engineering and Design **155** (1995) 643–652

5. Plesset, M.S., Prosperetti, A.: Bubble Dynamics and Cavitation. Ann. Rev. Fluid Mech. **155** (1995) 145–185

6. Giese, T., Laurien, E.: A Thermal Based Model for Cavitation in Saturated Liquids. Jahrestagung der Gesellschaft für angewandte Mathematik und Mechanik, Göttingen (2000)

7. Ranz, W. and Marshall, W.: Evaporation from Drops. Chemical Engineering Progress **48** (1952) 142–180

8. AEA Technology: User's Manual CFX Version 4.2. AEA Technology, Harwell, UK (1999)

9. Hirsch, T.: Einfluss des Modellblasendurchmessers bei der konstruktiven Auslegung zweiphasig durchströmter Anlagenkomponenten. Studienarbeit. IKE, Universität Stuttgart (2001)

10. Giese, T., Laurien, E.: A Gravity Driven Pipe Flow with Phase Change Phenomena. Proc. Annual Meeting of Nuclear Technology, Dresden (2001)

11. Yin, D., Burns, A. D., Splawski, B. A., Lo, S. M. and Guetari, C.: Modeling of Complex Multiphase Flows: A Coupled Solver Approach, Fourth International Conference on Multiphase Flow, New Orleans, 2001

12. Giese, T., Laurien, E.: Experimental and Numerical Investigation of Gravity-Driven Pipe Flow with Cavitation. Tenth International Conference on Nuclear Engineering (ICONE 10), Alington, VA, April 14–18, 2002

Three-Dimensional Numerical Simulation of Flow and Heat Transport in High-Temperature Nuclear Reactors

Sascha Becker and Eckart Laurien

Institute for Nuclear Technology and Energy Systems (IKE),
University of Stuttgart, Pfaffenwaldring 31, D-70550 Stuttgart, Germany;
E-Mail: becker@ike.uni-stuttgart.de, laurien@ike.uni-stuttgart.de

Abstract. In next generation nuclear High-Temperature Reactors an annular nuclear core consisting of a central column of graphite spheres and a surrounding ring of fuel pebbles is employed. Due to the complex feeding and shutdown mechanisms three-dimensional effects of heat production, gas flow and heat transport may become important for safety analysis. To simulate flow and heat transport in the core and the surrounding graphite reflector a new code system based on CFX-4 has been developed and run on the NEC-SX4 and NEC-SX5 supercomputers. The simulations are performed with the Heterogeneous Model of porous media. The program has been verified by comparison with two-dimensional simulations of the HTR-MODUL using the well established thermal analysis code THERMIX. A sensitivity study of several models for pressure drop and heat transfer on a simplified model of an HTR-MODUL is performed. Additionally the influence of the variation of the volume porosity near walls on flow and heat transport is analysed. In order to demonstrate the simulation of three-dimensional effects the influence of a package of fuel pebbles located asymmetrically in the central column is investigated. A significant influence on the temperature distribution and the maximum temperature core is found.

1 Introduction

The correct simulation of flow and heat transport is of great importance for the next generation high temperature gas-cooled nuclear reactors (HTR) such as the "Pebble Bed Modular Reactor (PBMR)" [1]. This reactor is currently under development in South Africa on the basis of the HTR-Modul of Siemens/Interatom [2] but with increased thermal and electrical power. In the European High Temperature Reactor Network (HTR-TN) reactors with relatively high power will also be investigated. Higher thermal power than of the HTR-MODUL under the condition of limited maximum core temperature can be obtained by using an annular core consisting of a central column of 'blind' graphite pebbles with a surrounding ring of fuel pebbles, see Fig. 1. However, three-dimensional effects of flow and heat transport may arise, e.g. by a package of eccentrically misplaced fuel pebbles in the central graphite column or by insertion of just one absorber rod shutdown system.

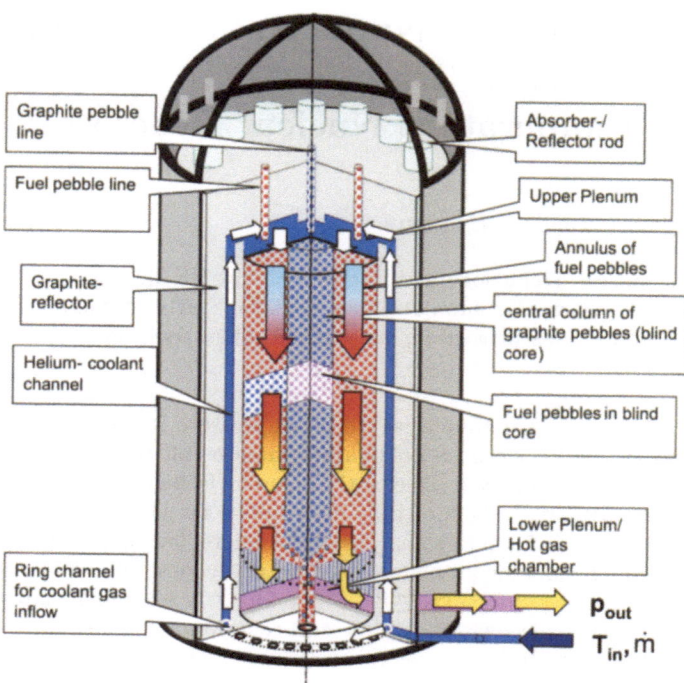

Fig. 1. Sketch of a HTR with annular core

Currently only two-dimensional special purpose codes like THERMIX/-KONVEK of Forschungszentrum Jülich are available for thermal analysis. However, three-dimensional effects of flow and heat transport may arise, e.g. by non-axisymmetric feeding of graphite and fuel pebbles through the nine feeding lines or by insertion of just one shutdown system. Unlike as during the development of the HTR-Modul in the 1980's it is now possible to develop three-dimensional (3D) numerical tools and to investigate three-dimensional effects on flow and heat transport in nuclear High-Temperature Reactors.

In this work the already existing code CFX-4 for three-dimensional flow analysis from AEA Technology [3] has been employed. The necessary models for pebble beds and the corresponding flow and heat transport phenomena have been implemented. The code is run on the NEC-SX-4 and the NEC-SX-5.

The new code has been validated for several steady state and transient high-temperature experiments for Helium and Nitrogen flow in a pebble bed conducted at the SANA (**S**elbsttätige **A**bfuhr von **Na**chwärme) test rig of Forschungszentrum Jülich, Germany [4]. In the experiments the heat transport has been investigated for a simplified model of a HTR-MODUL 3-4 hours after reactor shutdown. The simulations have shown good agreement to the experimental results for both gases [5].

In this paper the influence of various models for heat transport and pressure drop on the simulation result is investigated for a simplified model of a HTR during normal operation. The variation of the volume porosity near walls is also investigated. Furthermore results of the new code for a simplified model of the South-African PBMR as of 2000 with 268 MW thermal power are discussed. Steady simulation results show the influence of a package of fuel pebbles arranged non-axisymmetric in the central core on the maximum temperature for (i) a short time after injection and for (ii) an instant shortly before release [6].

2 Mathematical Model

The flow and heat transport in a High-Temperature Reactor is simulated unsteadily by the Heterogeneous Model, consisting of two sets of equations for both media, the gas, eqs. (1)-(3), and the pebble bed together with the solid parts, eq. (5). The volume porosity φ is used to determine, whether a single phase gas flow ($\varphi = 1$), flow and heat transport through the pebble bed ($0 < \varphi < 1$) or heat transport in the solid parts ($\varphi = 0$) are simulated. The heat transport in the pebble bed and the solid parts of the reactor like the side reflector are described by a temperature equation. The interaction of both media, gas and pebble bed, is modelled by interaction terms for pressure drop B_j, eq. (4), and heat exchange $\dot{q}'''$.

The Heterogeneous Model is similar to the Two-Fluid-Approach [7] but with a "frozen" second phase. The second phase, the solid part, is set to be fixed with no movement, because the pebbles move much slower than the fluid, so it is not necessary to solve the momentum equations for this phase.

The gas flow in the upper plenum, the coolant channels and the pebble bed are described by the three-dimensional spatially averaged conservation equations for flow through a porous medium as given below

$$\frac{\partial \varphi \rho}{\partial t} + \frac{\partial \varphi \rho u_i}{\partial x_i} = 0 \tag{1}$$

$$\left(\frac{\partial \varphi \rho u_j}{\partial t} + u_i \frac{\partial \varphi \rho u_j}{\partial x_i}\right) = B_j - \varphi \frac{\partial p}{\partial x_j} + \varphi \frac{\partial}{\partial x_i}\left(\mu\left(\frac{\partial u_i}{\partial x_j} + \frac{\partial u_j}{\partial x_i}\right)\right) \tag{2}$$
$$+ \varphi \rho g_j$$

$$\left(\frac{\partial \varphi \rho h}{\partial t} + \varphi \frac{\partial \rho u_j h}{\partial x_i}\right) = \varphi \frac{\partial}{\partial x_i}\left(\frac{\lambda}{c_p}\frac{\partial h}{\partial x_i}\right) + \dot{q}''' \tag{3}$$

with fluid density ρ, velocity u, static pressure p, viscosity μ, enthalpy $h = c_p\left(T - T_{\mathrm{ref}}\right)$, heat conductivity λ, heat capacity c_p, volumetric exchange heat flux $\dot{q}''' = \alpha\left(T_{\mathrm{por}} - T\right)$ between pebbles and gas, the volume porosity φ and the additional body force B_j in j-direction due to additional flow resistance caused by the pebble bed, modelled by Ergun's law

$$B_j = -\frac{150\mu(1-\varphi)^2}{d_p{}^2\varphi^3}u_j - \frac{1.75\rho(1-\varphi)}{d_p\varphi^3}|u|\cdot u_j \tag{4}$$

with the diameter $d_p = 0.06\,\text{m}$ of the pebbles. Several models for the additional flow resistance like Darcys law or the KTA-rule 3102.3 are implemented.

The pebble bed and the solid parts (graphite reflector) are described by the continuous porous medium approach and the heat transport is modelled by

$$(1-\varphi)(\rho c)_{\text{por}}\frac{\partial T_{\text{por}}}{\partial t} - (1-\varphi)\lambda_{eff}\frac{\partial^2 T_{\text{por}}}{\partial x_i^2} = \dot{q}_{\text{N}}''' - \dot{q}''' \tag{5}$$

with T_{por} as the temperature of the pebble bed, the exchanged volumetric heat flux $\dot{q}'''$ between pebble bed and gas, the volumetric nuclear heat production $\dot{q}_{\text{N}}'''$ and the effective heat conductivity λ_{eff} of the porous medium including a model for radiation described by the correlation of Zehner and Schlünder [8]. The variation of porosity near the walls is modelled by the correlation of Cheng and Hsu [9] resulting in a 50% higher maximum porosity than in the interior ($\varphi_\infty = 0.4$). The heat transfer from the porous medium to the fluid is described by a volumetric heat transfer coefficient α in accordance to KTA-rule 3102.2 or is taken from a correlation of Gnielinski [10]. Thermal dispersion effects due to additional mixing of gas in the core is taken into account by a model of Bauer [11] by increasing λ.

3   Numerical Method

The problem has been modelled with the set of equations as given above by the CFD-code CFX-4 for steady state and for transient cases [3]. The program CFX-4 is a vectorized and parallelized block structured Finite-Volume-code.

The equations have been solved using the SIMPLEC-algorithm. The solid parts in the reactor model are described as conducting solids and are solved comprehensively. All necessary models have been implemented by the USER-FORTRAN interface and vectorized.

The computational grid for the full-scale model of an HTR has been composed of approximately 1700 blocks. After optimization by the program Meshimport we achieved a reduction to 40 blocks which has supported a vectorized simulation with long vector lengths. Parallelized runs have failed due to our complex structure of additional FORTRAN subroutines and their required command files.

4   Investigation of Models

The influence of different flow phenomena and models on flow and heat transfer has been investigated using a simplified model of an HTR-MODUL with

220 MW thermal power (+10% excess power) and a Helium mass flux of 81 kg/s with an inflow temperatur of 250°C as shown in Fig. 2. This models consists only of the pebble bed with a diameter of 3 m and a height of 9.5 m. The flow enters the system from the top and exits at the bottom. The surrounding graphite parts with their conductivity and heat capacity are neglected but their heat transfer is modeled by a conducting boundary condition. A power distribution similar to results from neutronic simulation runs by codes like ORIGEN have been applied and is shown in Fig. 2 right.

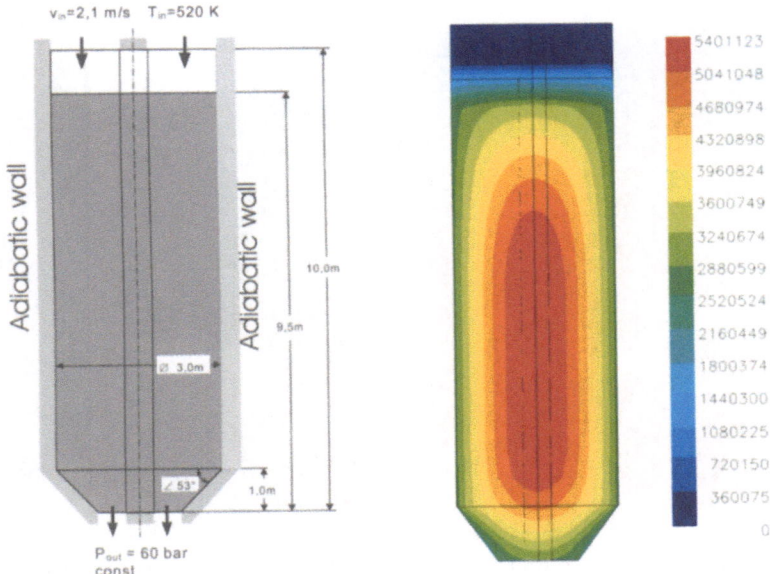

Fig. 2. Sketch of the simplified model of an HTR (left) and the power distribution (right) in W [13]

In numerous literature it is outlined that the variation of volume porosity near walls ("Channeling") can have a significant influence on flow and heat transport. Near the wall the package of the pebbles can not as dense and three-dimensional than in an infinite layer. This has been simulated [13] for (i) constant porosity of 0.4, see Fig. 3, left and (ii) variable porosity in accordance to the model of Hsu and Cheng, see Fig. 3, right with the temperature distribution in the pebble bed. The difference in the top part, the upper plenum, between Fig. 3 left and right comes from the initial guess and this part is not calculated, since there is no porous medium.

It can be seen that the temperature gradients are steeper near the walls if a variable porosity is assumed. Additionally a higher maximum temperature of 1250 K for variable porosity in comparison to 1235 K for constant porosity

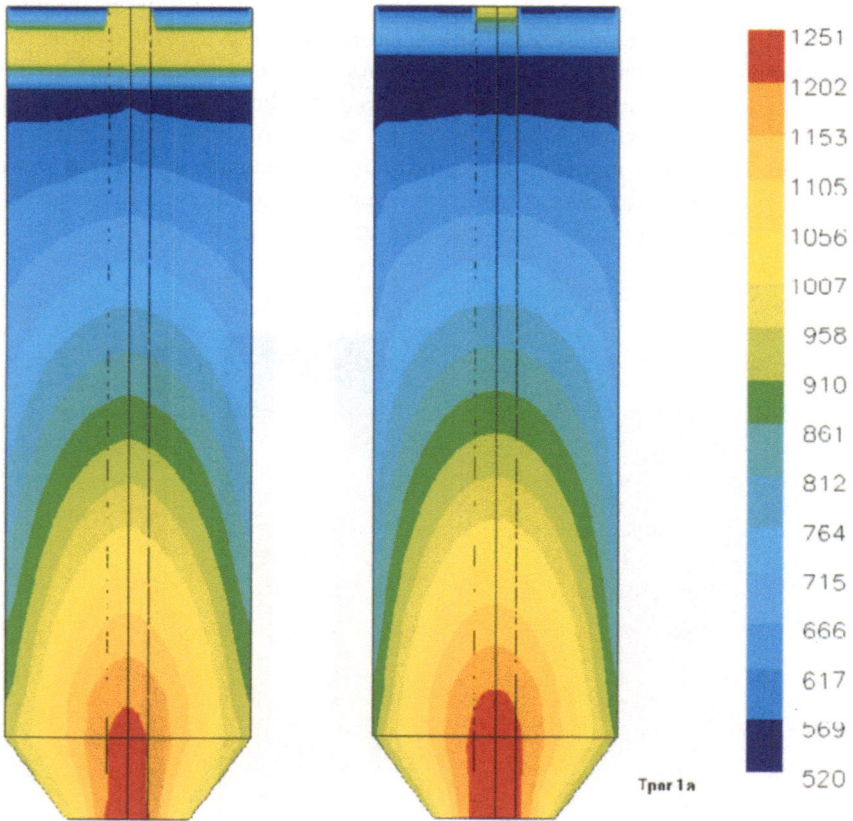

Fig. 3. Temperature distributions of the pebble bed (left: constant porosity, right: variable porosity)

is found. The pressure drop is significantly lower than in case of constant porosity. These effects are caused by the higher porosity near the walls which means that the fluid has more free space to flow than in the interior and thereby the mass flux near the walls is much higher than in the centre. So the heat transport by the fluid flow in the core is strongly affected by the distribution of the mass flux in case of the variable porosity as it is shown in Fig. 4.

If the correlation for the additional pressure drop of the flow due to the presence of the pebble bed is changed from the ERGUN-equation to the KTA-rule 3102.3 no effect on temperature and velocity field of gas and solid is found.

The opposite is found when the correlation for the effective heat conductivity in the pebble bed is varied. An own simplified equation has been compared to results achieved with the widely used correlation of Zehner and

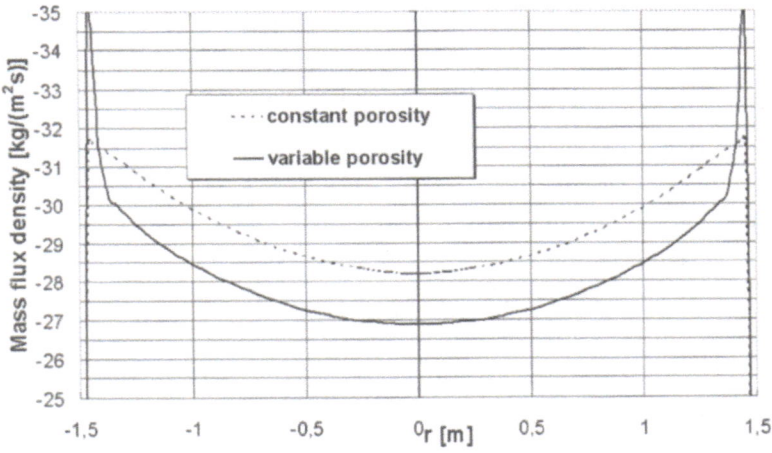

Fig. 4. Mass flux density in 3m height above the outlet

Schlünder. In Fig. 5 the difference in the temperatures in the pebble bed between both results are shown for the simplified equation compared to the results achieved by the correlation of Zehner and Schlünder.

Thermal dispersion by additional mixing of the gas in the pebble bed leads to smoother temperature profiles with less steep gradients as shown in

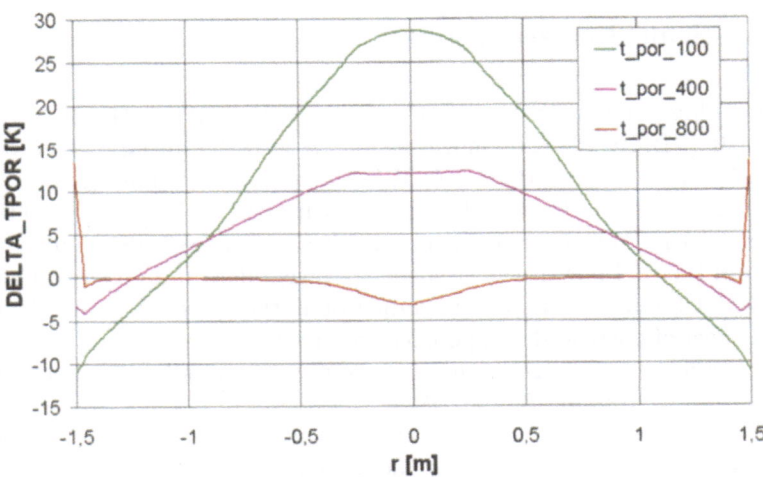

Fig. 5. Difference of the temperature in the pebble bed between result achieved with simplified correlation to result achieved with the correlation of Zehner and Schlünder in different heights above exit (100 cm, 400 cm and 800 cm above exit)

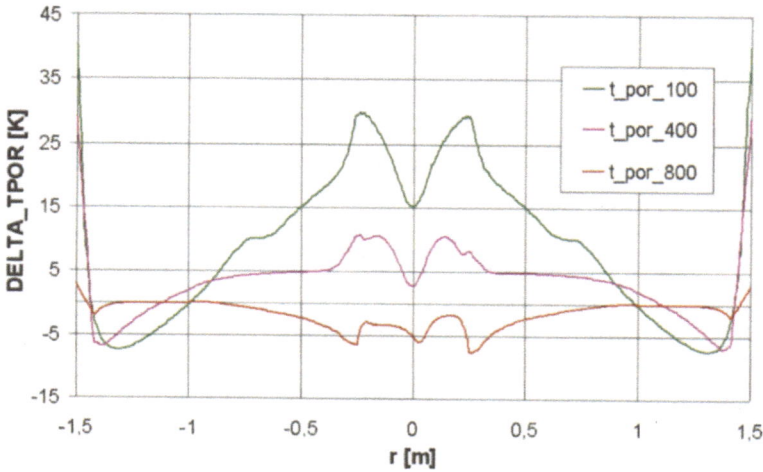

Fig. 6. Difference of the temperature in the pebble bed between result achieved with thermal dispersion to result achieved with no thermal dispersion in different heights above exit (100 cm, 400 cm and 800 cm above exit)

Fig. 6 with the difference in the temperatures of the pebble bed for the result of one simulation with thermal dispersion compared to the result with no dispersion. So thermal dispersion can not be neglected otherwise incorrect temperature distributions as shown in Fig. 6 can be obtained.

5 Full-Scale Investigations

The three-dimensional reactor models of the HTR-MODUL (HTR-200) and the PBMR (PBMR-268) are shown in Fig. 7 with the initially given power distribution [6]. For the HTR-MODUL the power distribution was taken from neutronic Monte-Carlo-calculations with ORIGEN as used in THER-MIX/KONVEK. For the PBMR an axisymmetric power distribution with given total thermal power was assumed to have a distribution similar to the HTR-MODUL but with no power production in the central column. To simulate the effect of misplaced fuel pebbles an eccentric power generating section in form of a quarter cylinder with 1 m height as shown in Fig. 7 right is used, resulting in a total power of 280 MW.

First the operational state of the reactor has been simulated as a basis for later transient accident simulations. The simulation results for the operational state of the HTR-MODUL with 200MW thermal power and a mass flux of 80 kg/s at 250°C is given in Fig. 8 with the distribution of the power and of the temperature in the pebble bed and the velocity distribution. The pressure drop due to the core of 0.6 bar was in good agreement to documented

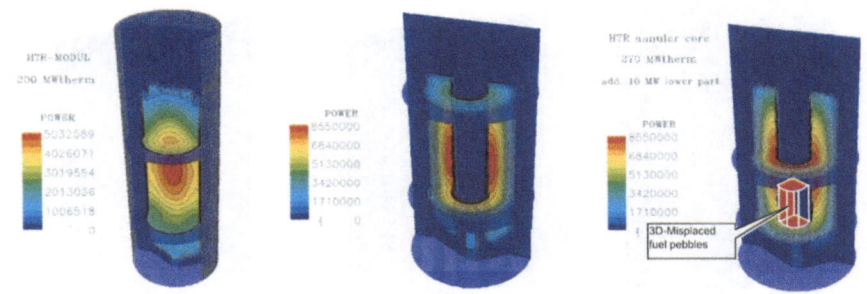

Fig. 7. Model of HTR-200 (left) and of the PBMR-268 (middle: nominal power, right: with additional fuel pebble package in the lower part of the core)

data. The maximum temperature of 1130 K agrees well with 1139 K from THERMIX/KONVEK.

As an initial condition for the unsteady cases of the PBMR-268 steady simulation with 126kg/s inlet mass flux of Helium at 500°C was performed. The symmetric case at nominal power of 268 MW_{therm} is shown in Fig. 9 left. A maximum temperature of around 1230 K has been calculated. In case the misplaced fuel pebble package is located in the upper part of the core which corresponds to an instant shortly after injection we have calculated a increase in the maximum temperature of 20 K compared to the standard operational case. The maximum temperature is almost not affected if the additional fuel pebbles are in the lower part of the reactor core which corresponds to a later time shortly before the pebbles will be taken out of the reactor core.

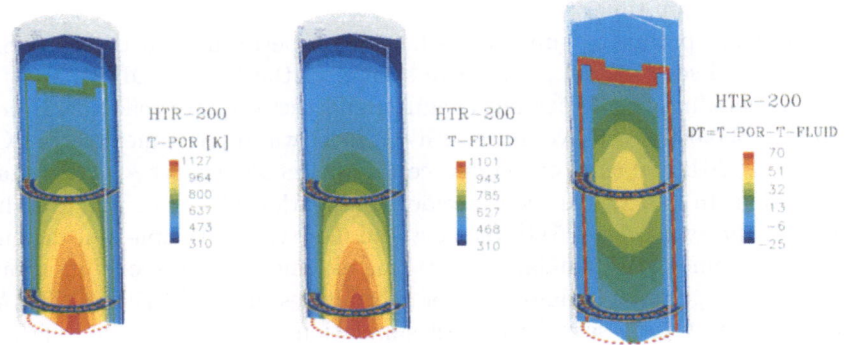

Fig. 8. HTR-200 — Distribution of temperature in the solid part and pebble bed (left), fluid (middle)and the temperature difference between pebble bed and fluid (right)

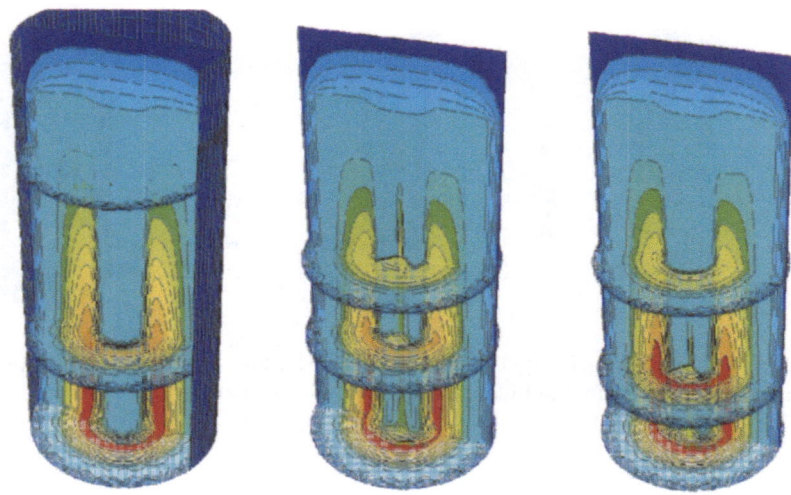

Fig. 9. Temperature distribution in the solid part and pebble bed of the PBMR-268 (left: nominal configuration, middle : with misplaced fuel pebble package in the upper part, right: misplaced fuel pebble package in the lower part)

Unsteady simulations of accident cases has been performed for the HTR-200 [6] and they have shown good agreement to previous results from [12].

6   Computational Requirements

The simulations of the simplified model of an HTR with 110.000 control volumes have been performed on the NEC-SX-4. The computational effort of this case was for steady state simulations approximately 500 MB with about 10 cpu-hours. For transient simulations the computational effort has increased to double the cpu-time.

The more promising simulations have been performed on the NEC-SX-5 for the full-scale model of the core system of the HTR-MODUL and the PBMR including the 36 coolant channels and the graphite reflector. Here a grid with 360.000 control volumes with 40 blocks was used. A memory of 1 GB and about 20 h cpu-time on one processor was necessary for a steady state calculation. In case of transient accident simulation [6] each simulation has required approximately 1.5 GB memory and more than 40 h cpu-time depending on the time to be simulated. It can be assumed that the computational effort was large due to many calls of subroutines of the USER-FORTRAN which has been necessary for correct simulation of the pebble bed with all phenomena and interactions. Currently we have used the code only with one processor. Until now it was impossible to use the program in parallel mode due to our complex structure of subroutines and their necessary command files.

7   Conclusion and Outlook

It has been shown that the CFD-code CFX-4 with our extensions predicts well the heat transfer in modular HTRs. These simulations are only possible on supercomputers like NEC-SX-4 or NEC-SX-5. The computational requirements are high but still limited on the use of a single processor with lots of memory. The influence of the variation of the volume porosity near walls and of various models for heat transfer and pressure drop in porous media and pebble beds has been investigated and a significant influence on the temperature distribution in the pebble bed is found.

The steady simulations of three-dimensional effects for the PBMR have shown the influence of non-symmetric power distributions on the temperature distribution and the maximum temperature in the reactor.

Unsteady simulations are performed to show the applicability of the new code to simulate accidents of a High-Temperature reactor [6].

References

1. Reactor safety analysis report of the South-African Pebble-Bed Modular Reactor (PBMR), Rev. E, PBMR Ltd., Centurion, South-Africa, 2000
2. Reutler, H.; Lohnert, G.H.: *Der modulare HTR — Ein neues Konzept für den Kugelhaufenreaktor*, Atomwirtschaft, Januar 1982, pp. 18-21
3. AEA Technology: User's Manual CFX Version 4.4. AEA Technology, Harwell, UK (2001)
4. Stöcker, B.: *Untersuchungen zur selbsttätigen Nachwärmeabfuhr bei Hochtemperaturreaktoren unter besonderer Berücksichtigung der Naturkonvektion*, Dissertation RWTH Aachen, Forschungszentrum Jülich, Jül-3504, 1998
5. Becker, S.; Laurien, E.: *Numerical simulation of the convective heat transport in pebble beds at high temperatures*; Annual Meeting Nuclear Technology 2001, May 14-17, 2001, Dresden, Germany
6. Becker, S. ; Laurien, E.: *Three-dimensional numerical simulation of flow and heat transport in a high-temperature reactor*, 1st Int. Topical Meeting High Temperature Reactor Technology (HTR) HTR-2002, Petten, Netherlands, April 22–24, 2002
7. Laurien, E.; Saptoadi, D.: *On the fundamental two-fluid equations to model three-dimensional bubbly flows*, Fourth Interantional Conference on Multiphase Flow (ICMF 2001), May 27 – June 1, 2001, New Orleans, USA
8. Zehner, P.; Schlünder, E.U.: *Wärmeleitfähigkeit von Schüttungen bei mäßigen Temperaturen*, Chemie-Ing.-Techn., pp. 933–941, 1970
9. Cheng, P.; Hsu, C.T.: *Fully-developed, forced convective flow through an annular packed- sphere bed with wall effects*, Int. J. Heat Mass Transfer, Vol. 29, pp. 1843–1853, 1986
10. Gnielinski, V.: *Wärme- und Stoffübertragung in Festbetten*; Chem.-Ing.-Tech., Vol. 52, No. 3, pp. 228–236, 1980
11. Bauer, R.: *Effektive radiale Wärmeleitfähigkeit gasdurchströmter Schüttungen mit Partikeln unterschiedlicher Form und Größenverteilung*; Dissertation, Universität Karlsruhe, 1977

12. Reactor safety analysis report of the HTR-MODUL, Siemens/Interatom, Germany, 1988
13. Dingelstadt, R.: *Numerische Simulation des konvektiven Wärmetransports in durchströmten Kugelschüttungen*, Studienarbeit, IKE-8-D-004, 2002

Numerical Investigation of the Shuttle-Like Configuration PHOENIX

Birgit U. Reinartz, Michael Hesse, and Josef Ballmann

Lehr- und Forschungsgebiet für Mechanik, RWTH Aachen,
Templergraben 64, 52062 Aachen, Germany

Abstract. The development of the shuttle-like technology demonstrator PHOENIX is intended to proof the feasibility of a future European reusable space transportation system. The present work focuses on aerodynamic solutions for the clean-configuration of PHOENIX by three-dimensional Euler computations. Details on the structured grid generation process using the interactive MegaCads software are discussed and a converged solution for supersonic inviscid flow is presented. The agreement of the obtained aerodynamic coefficients with wind tunnel and design data is favorable. Preliminary work has started on viscous hypersonic flow computations for different flight conditions.

1   Introduction

Within the frame of the Collaborative Research Center SFB 253 "Fundamentals of the Design of Aerospace Planes", compressible hypersonic flow problems are studied numerically at the "Lehr- und Forschungsgebiet für Mechanik" using the RANS flow solver FLOWer.

Currently, the SFB 253 is involved in the German development program ASTRA (selected systems and technologies for future space transport system applications). This program is the German contribution to the European development of reusable transport systems co-ordinated by DLR. Its current focus is the construction and testing of the technology demonstrator PHOENIX, which will serve as predecessor to the unmanned HOPPER. HOPPER is very similar in appearance to the US Space Shuttle and its reusable main stage is intended to take-off horizontally from a skid and land horizontally on an airfield. The contribution of the SFB concerns the numerical simulation of the flow around the clean configuration of PHOENIX which is a small scale model of HOPPER with a total length of just under seven meters and a wing span of 3.8 meters.

Substantial work was performed to generate a multi-block grid for numerical applications starting from the CAD surface data of PHOENIX . Additionally, inviscid flow computations are performed for different angles of attack to tailor the grid to the flow field and determine the aerodynamic coefficients for specified reference flow conditions. Currently, the grid is being adapted to the requirements of viscous, turbulent flow. Future investigations will include the simulation of full Navier-Stokes flow at high-enthalpy conditions for different flight Mach numbers and different angles of attack. Therefore, a real gas model assuming local thermodynamic equilibrium has been implemented into the code and is currently being validated.

2 Physical Model

2.1 Conservation equations

The governing equations for high-speed turbulent flow are the unsteady, compressible, Favre-averaged Navier-Stokes equations in integral form

$$\frac{\partial}{\partial t} \int_V \mathbf{U} \ dV + \oint_{\partial V} \left(\mathbf{F}^c - \mathbf{F}^d \right) \mathbf{n} \ dS = 0 \tag{1}$$

where

$$\mathbf{U} = [\ \bar{\rho} \ , \ \bar{\rho}\tilde{\mathbf{v}} \ , \ \bar{\rho}\tilde{E} \ ]^{\mathrm{T}} \tag{2}$$

is the array of the mean conserved quantities: density, momentum density, and total energy density. The tilde and the bar over the variables denote the mean value of Favre-averaged and Reynolds-averaged variables, respectively. The quantity V denotes an arbitrary control volume with the boundary ∂V and the outer normal $\mathbf{n}$. The flux array is divided into its inviscid part and its diffusive part

$$\mathbf{F}^c = \begin{pmatrix} \bar{\rho}\tilde{\mathbf{v}} \\ \bar{\rho}\tilde{\mathbf{v}}\circ\tilde{\mathbf{v}} + \bar{p}\mathbf{1} \\ \tilde{\mathbf{v}}(\bar{\rho}\tilde{E} + \bar{p}) \end{pmatrix} \quad \text{and} \quad \mathbf{F}^d = \begin{pmatrix} \mathbf{0} \\ \bar{\boldsymbol{\sigma}} - \overline{\rho\mathbf{v}''\circ\mathbf{v}''} \\ \tilde{\mathbf{v}}\bar{\boldsymbol{\sigma}} + \overline{\mathbf{v}''\boldsymbol{\sigma}} - \bar{\mathbf{q}} - c_p\overline{\rho\mathbf{v}''T''} \\ -\frac{1}{2}\overline{\rho\mathbf{v}''\ \mathbf{v}''\circ\mathbf{v}''} - \tilde{\mathbf{v}}\ \overline{\rho\mathbf{v}''\circ\mathbf{v}''} \end{pmatrix}$$

where $\mathbf{1}$ is the unit tensor and $\circ$ denotes the dyadic product[1]. Currently, the airflow is still considered to be a calorically perfect gas with constant ratio of specific heats, $\gamma = 1.4$, and a specific gas constant of $R = 287 \ \mathrm{J/(kgK)}$, which yields the following expression for the specific total energy:

$$\tilde{E} = c_v\bar{T} + \frac{1}{2}\tilde{\mathbf{v}}\tilde{\mathbf{v}} + k \ . \tag{3}$$

The last term represents the turbulent kinetic energy

$$k := \frac{1}{2}\frac{\overline{\rho\mathbf{v}''\mathbf{v}''}}{\bar{\rho}} \ . \tag{4}$$

For isotropic Newtonian fluids, the mean molecular shear stress tensor

$$\bar{\boldsymbol{\sigma}} = 2\bar{\mu}\bar{\mathbf{S}} - \frac{2}{3}\bar{\mu}\operatorname{tr}(\bar{\mathbf{S}})\ \mathbf{1} \tag{5}$$

---

[1] Scalar Products of dyadics formed by two vectors $\mathbf{a}$ and $\mathbf{b}$ with a vector $\mathbf{c}$ are defined as usual, i.e., $\mathbf{a}\circ\mathbf{b}\,\mathbf{c} = \mathbf{a}(\mathbf{bc})$, $\mathbf{c}\,\mathbf{a}\circ\mathbf{b} = (\mathbf{ca})\mathbf{b}$.

is a function of the mean strain rate tensor

$$\bar{\mathbf{S}} := \frac{1}{2} \left[\mathrm{grad}(\bar{\mathbf{v}}) + (\mathrm{grad}(\bar{\mathbf{v}}))^\mathsf{T} \right] \tag{6}$$

where the molecular viscosity $\bar{\mu} = \bar{\mu}(\bar{T})$ is determined by Sutherland's law. Similarly, the molecular heat flux is considered a linear isotropic function of the temperature gradient

$$\bar{\mathbf{q}} = -\frac{c_p \bar{\mu}}{Pr} \, \mathrm{grad}(\bar{T}) \tag{7}$$

with the Prandtl number $Pr = 0.72$.

2.2   Turbulence closure

To close the above system of partial differential equations, the Boussinesq hypothesis is used where the remaining correlations are modeled as functions of the gradients of the mean conservative quantities and turbulent transport coefficients. The Reynolds stress tensor thus becomes

$$-\overline{\rho \mathbf{v}'' \circ \mathbf{v}''} = 2\mu_t \left(\bar{\mathbf{S}} - \frac{1}{3} \, \mathrm{tr}(\bar{\mathbf{S}}) \right) - \frac{2}{3} \bar{\rho} k \, \mathbf{1} \quad , \tag{8}$$

with the eddy viscosity μ_t, and the turbulent heat flux is

$$c_p \overline{\rho \mathbf{v}'' T''} = -\frac{c_p \mu_t}{Pr_t} \, \mathrm{grad}(\bar{T}) \tag{9}$$

with the turbulent Prandtl number $Pr_t = 0.89$. Finally for hypersonic flows, the molecular diffusion and the turbulent transport are modeled as functions of the gradient of the turbulent kinetic energy

$$\overline{\mathbf{v}'' \sigma} - \frac{1}{2} \overline{\rho \mathbf{v}'' \, \mathbf{v}'' \circ \mathbf{v}''} = \left(\mu + \frac{\mu_t}{Pr_k} \right) \mathrm{grad}(k) \tag{10}$$

with the model constant $Pr_k = 2$.

The turbulent kinetic energy and the eddy viscosity are then obtained from the turbulence model. In case of laminar computations, both variables are set to zero to regain the original transport equations.

2.3   Turbulence model

Due to the oncoming hypersonic flow, a complex flow field develops around PHOENIX with strong shocks interacting with the boundary layer and with large separation zones, e.g., at the blunt trailing edge. Hence, the applied turbulence model needs to resolve separation as well as strong compression. The performance of the Wilcox's low Reynolds number k-ω model has been assessed for such flow fields in the previous project phase and the necessary extensions were implemented [1]. The two additional transport equations for the turbulent kinetic energy k and the specific dissipation rate ω are as follows:

$$(\overline{\rho}k)_{,t} + \mathrm{div}\left[\,\overline{\rho}k\tilde{\mathbf{v}} - (\overline{\mu} + \sigma^*\mu_{\mathrm{t}})\,\mathrm{grad}(k)\,\right] =$$

$$\alpha^*\left[2\mu_{\mathrm{t}}\left(\bar{\mathbf{S}} - \frac{1}{3}\mu_{\mathrm{t}}\,\mathrm{tr}(\bar{\mathbf{S}})\,\mathbf{1}\right) - \frac{2}{3}\overline{\rho}k\,\mathbf{1}\right] : \mathrm{grad}(\mathbf{v}) - \beta^*\overline{\rho}k\omega \qquad (11)$$

$$(\overline{\rho}\omega)_{,t} + \mathrm{div}\left[\,\overline{\rho}\omega\tilde{\mathbf{v}} - (\overline{\mu} + \sigma\mu_{\mathrm{t}})\,\mathrm{grad}(\omega)\,\right] =$$

$$\alpha\frac{\omega}{k}\left[2\mu_{\mathrm{t}}\left(\bar{\mathbf{S}} - \frac{1}{3}\,\mathrm{tr}(\bar{\mathbf{S}})\mathbf{1}\right) - \frac{2}{3}\overline{\rho}k\,\mathbf{1})\right] : \mathrm{grad}(\mathbf{v}) - \beta\overline{\rho}\omega^2 \qquad (12)$$

with the eddy viscosity

$$\mu_{\mathrm{t}} = D_{\mathrm{low}\,Re} \cdot \overline{\rho}\frac{k}{\omega}\ . \qquad (13)$$

$D_{\mathrm{low}\,Re}$ denotes a damping function that applies in the low Reynolds number areas of high-speed flows.

3   Numerical Method

The computations are performed with the DLR FLOWer code, Versions 116.0+, using a finite-volume formulation on block-structured grids [3]. The code was originally developed for the simulation of flows around airfoils in subsonic and transonic regimes. It has been extended recently for the simulation of hypersonic, turbulent flow problems by implementing upwind discretizations and advanced compressible turbulence models [5].

3.1   Spatial discretization

A finite-volume discretization is applied to (1) which results in a consistent approximation to the conservation laws. The computational domain is divided into non-overlapping hexahedra in general curvilinear coordinates ξ, η, ζ, and the integral formulation (1) is then applied to each cell (i, j, k) separately. Semidiscretization of equation (1) results in a set of equations for the time rates of change of the volume-averaged conserved quantities $\mathbf{W}_{i,j,k}$ which are in balance with the sum of the corresponding area-averaged fluxes, $\mathbf{R}^{\mathrm{c}}_{i,j,k}$ and $\mathbf{R}^{\mathrm{d}}_{i,j,k}$, across the cell faces and the artificial dissipation $\mathbf{D}_{i,j,k}$:

$$\frac{\mathrm{d}\mathbf{W}_{i,j,k}}{\mathrm{d}t} = -\frac{1}{V_{i,j,k}}\left(\mathbf{R}^{\mathrm{c}}_{i,j,k} - \mathbf{R}^{\mathrm{d}}_{i,j,k} + \mathbf{D}_{i,j,k}\right) = \mathbf{Res}_{i,j,k}\ . \qquad (14)$$

Standard central discretization schemes are used for the convective and diffusive terms in the presented supersonic computations. However, to account for the directed propagation of information in the inviscid part of the equations, the AUSM (Advection Upstream Splitting Method) scheme will be used for the approximation of the convective flux functions of hypersonic flows [2]. Higher-order accuracy and consistency with the central differences used for the diffusive terms is achieved by MUSCL (Monotonic Upstream Scheme for Conservation Laws) Extrapolation, and TVD (Total Variation Diminishing) property of the scheme is ensured by a modified van Albada limiter function.

3.2   Time-Stepping Scheme

The system of ordinary differential equations (14) is solved by an explicit five-stage Runge-Kutta time-stepping scheme in combination with different convergence acceleration techniques like multigrid and local time stepping for asymptotically steady-state solutions [4]. Additionally, for inviscid flows, the total enthalpy is a constant throughout the flow field and its numerical deviation is applied as forcing function to accelerate convergence. For turbulent flow, the time integration of the turbulence equations is decoupled from the mean equations and the turbulence equations are solved using a Diagonal Dominant Alternating Direction Implicit (DDADI) scheme. The implicit scheme increases the numerical stability of turbulent flow simulations which is especially important since the low Reynolds number damping terms as well as the high grid cell aspect ratios near the wall make the system of turbulent conservation equations stiff. Due to the CFL-condition for explicit schemes, the CFL number of the Runge-Kutta scheme has an upper limit of 4. Implicit residual smoothing allows to increase the explicit stability limit by a factor of 2 to 3 [4] and, thus, a CFL number of 7.5 is realized for PHOENIX. For the implicitly solved turbulence equations, the CFL number can be ten times higher.

3.3   Boundary Conditions

At the inflow, outflow and other farfield boundaries, a locally one-dimensional inviscid flow normal to the boundary is assumed. The governing equations are linearized based on characteristic theory and the incoming and outgoing number of characteristics are determined. For incoming characteristics, the state variables are corrected by freestream values using the linearized equations. Else the variables are extrapolated from the interior [4]. For turbulent flow, the turbulent freestream values are determined by specifying the freestream turbulence intensity Tu_∞: $k_\infty = 0.667 \cdot Tu_\infty v_\infty$ and $\omega_\infty = k_\infty/(0.001 \cdot \mu)$.

For inviscid flow, it is sufficient to set the normal velocity component to zero at slip surfaces. In the viscous case, the no-slip condition is enforced at solid walls by setting all velocity components to zero. Additionally, the turbulent kinetic energy and the normal pressure gradient are set to zero. The specific dissipation rate is set proportional to the wall shear stress and the surface roughness. The energy boundary condition is directly applied through the diffusive wall flux: either by zeroing the contribution of the diffusive flux for adiabatic walls or by prescribing the wall temperature or wall heat flux when calculating the energy contribution of wall faces. At the symmetry plane of the half configuration, the conservation variables are mirrored onto the ghost cells to ensure symmetry.

4   Results

4.1   Grid Generation

The grid generation is an integral part of the numerical solution process and, for complex geometries such as PHOENIX, it cannot be performed independently. Therefore, a preliminary solution of the transport equations on the current grid is computed which is then followed by a redistribution of points to improve the grid where necessary until a satisfactory solution is achieved. For FLOWer, no solution adaptive remeshing procedure is available and the manual process is very time consuming and constitutes a major part of the computational work. The applied grid generation software MegaCads has been developed by the DLR as part of the MEGAFLOW project from which the FLOWer code originated, too [3]. MegaCads is an interactive grid generation system which generates structured, three-dimensional, multi-block grids for flow computations. The flexibility of the grid generation process is enhanced by the possibility to create only partly overlapping blocks and patch more than one block to any given block boundary. The underlying feature is the ability of the grid generator (and of the flow solver) to process block-structured grids. Furthermore, where appropriate, neighbouring blocks can be merged to reduce the overall block number and improve handling. Following this approach, the three-dimensional Euler grid contains 20 blocks for the half configuration of PHOENIX and consists of 2.15 million grid points.

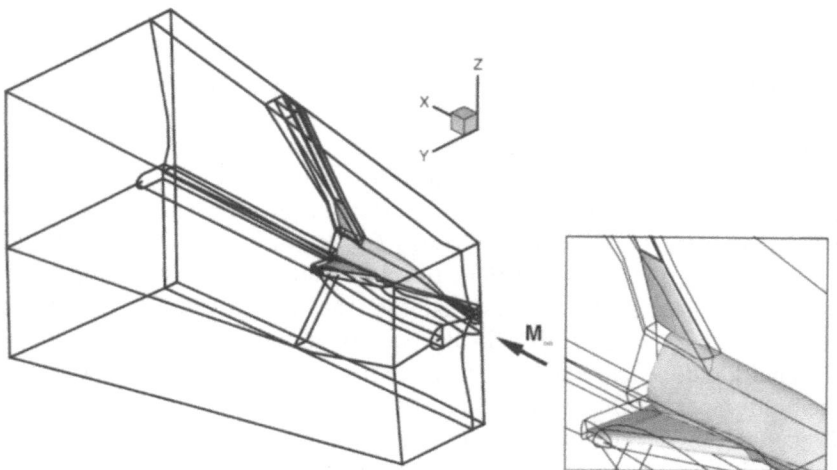

Fig. 1. Multi-block grid with 20 blocks and 2.15 million grid nodes.

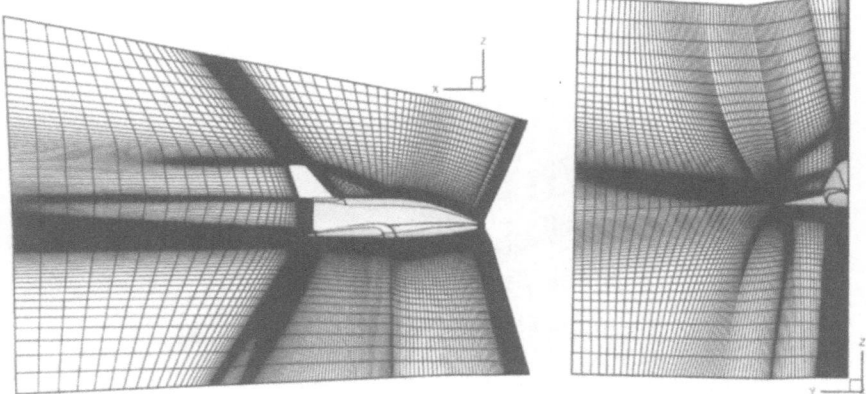

Fig. 2. Grid point distribution along symmetry plane (left) and in one grid plane along the wing center projected on the y-z-plane (right).

Figure 1 presents a schematic overview of the distribution of blocks assembled into a single block-structured grid. In Fig. 2, the node distribution along with the block boundaries in the symmetry plane and in one grid plane (with one constant curvilinear coordinate) projected onto the y-z-plane are shown. On the surface of PHOENIX, there are approximately 96 points distributed in the circumferential direction and 150 in the direction of the roll axis.

Since FLOWer is not able to deal with hanging nodes, the most restrictive requirement of the grid generation process is to match the number of points between adjacent blocks. If a large number of grid points is needed to resolve a certain subdomain like for example the vertical tail, the corresponding grid lines of those points have to be continued throughout the complete domain. In the case of the vertical tail, the grid lines extend to the front where they end at the inflow plane and create a region of very fine grid spacing. However, the maximum allowable time step depends on the grid cell volumes and therefore, if local time stepping is used, the inflow information travels much slower in that block than in any adjacent block with larger volumes causing a non-uniformity that is not physical. Euler calculations were performed to discover similar effects in the grid point distribution and if possible to remedy them.

Currently, sublayers for the resolution of turbulent boundary layers are inserted into the Euler grid. Afterwards, the computations of the viscous high-speed flow will be started.

4.2   Inviscid Flow Computation

Euler calculations for the complete demonstrator PHOENIX at a freestream Mach number of $M_\infty = 2.21$ and different angles of attack have been performed. The calculations were conducted switching between 2-level W-cycle multigrid and single grid computations to damp high wavelength oscillations due to unfavorable grid spacing and to stabilize the inviscid flow solution

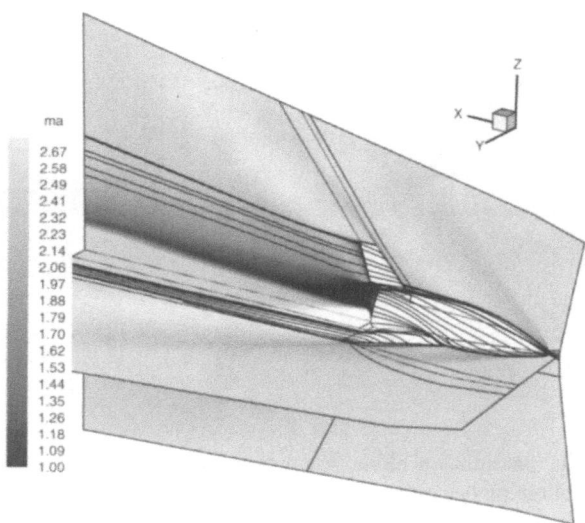

Fig. 3. Streamlines and Mach number distribution in the symmetry plane with additional streamlines on the surface of PHOENIX at $M_\infty = 2.21$ and angle of attack $\alpha = 10°$.

around the vertical tail, respectively. No steady state solution has been found for the flow behind PHOENIX's blunt trailing edge, thus the density residual shows no consistent convergence. Therefore, the convergence behavior of the lift was used instead. As a precaution, 3000 additional iterations were computed after the asymptotic convergence of the lift in order to ensure that the other flow variables are also converged. A perspective view on the flow field in the symmetry plane is shown in Fig. 3. Due to the angle of attack, the bow shock is weaker on the top side of PHOENIX than on its underside. The flow accelerates again over the top until it meets the vertical tail shock and then accelerates again over the back of the tail. The flow along PHOENIX's underside is followed by a strong expansion around the blunt tail. A shock is generated where the expansion meets the subsonic region behind PHOENIX. The additional streamlines simulate an oil flow picture over the surface where only those streamlines are continued beyond the surface which lie in the symmetry plane.

The influence of the grid spacing on the flow solution is emphazised in Figs. 4 and 5. The figures show the grid and the density distribution for the same perspective view of PHOENIX. The inconsistent spacing in the symmetry plane behind PHOENIX is reflected in the frayed density contours. In regions where the variation in the flow is less intense, e.g., behind the wing tip in the wing plane, the flow solver obtains a consistent distribution.

The aerodynamic coefficients at different flight conditions are important design criteria for determining the shape and the trajectory of PHOENIX. For inviscid flow, only the pressure force acting on the surface has to be

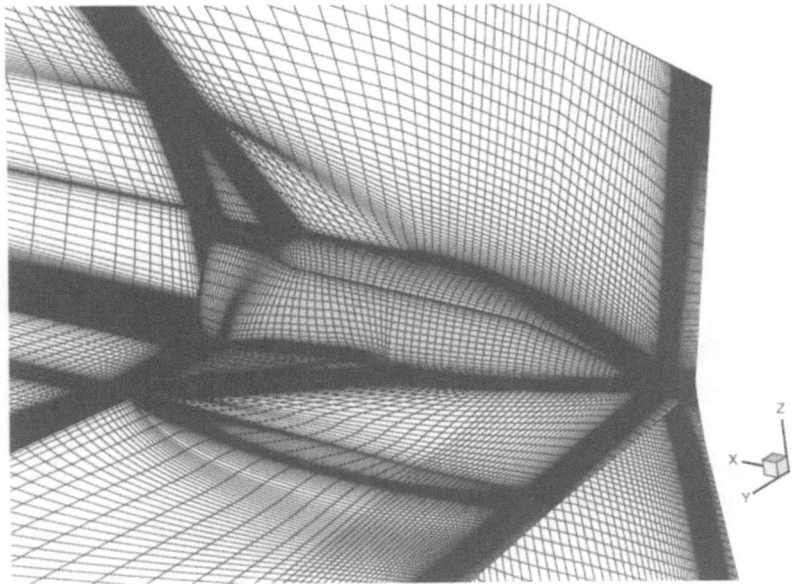

Fig. 4. Spacing of the Euler grid on the surface and in two grid planes (symmetry plane and wing plane) around PHOENIX.

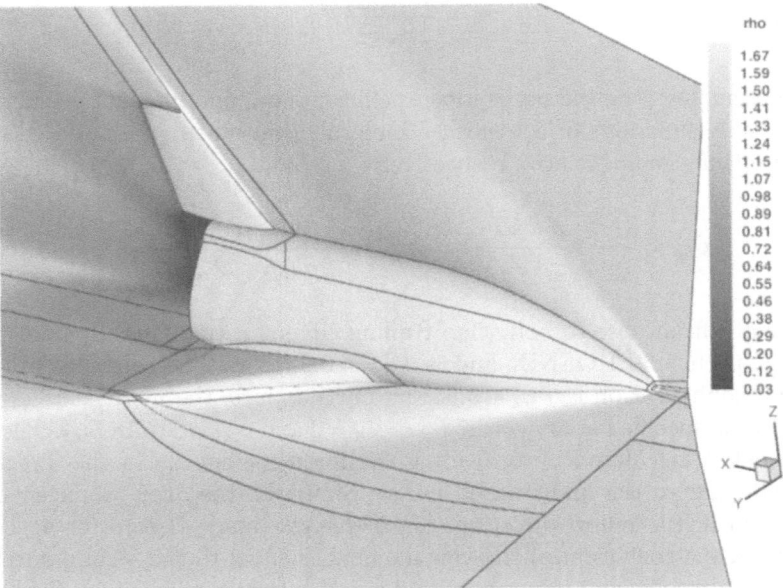

Fig. 5. Density distribution in two grid planes around PHOENIX at $M_\infty = 2.21$ and angle of attack $\alpha = 10°$.

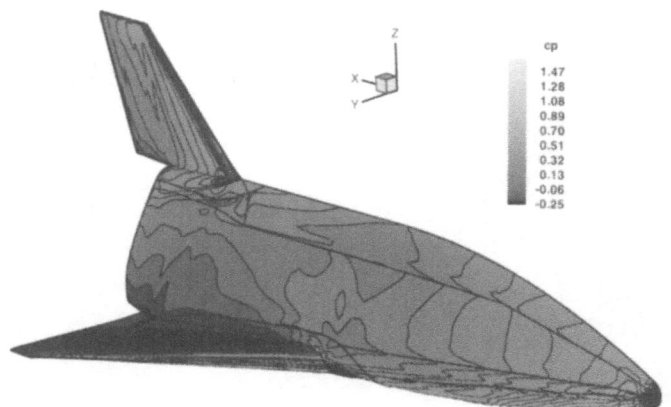

Fig. 6. Contour lines of the pressure coefficient on the surface of PHOENIX at $M_\infty = 2.21$ and angle of attack $\alpha = 10°$

considered. Contour lines of the pressure coefficient distribution are shown in Fig. 6 where the pressure coefficient c_p is defined as deviation from the undisturbed static pressure normalized by the dynamic pressure $\frac{1}{2}\rho_\infty q_\infty^2$:

$$c_p = \frac{p - p_\infty}{\frac{1}{2}\rho_\infty q_\infty^2} \quad . \tag{15}$$

For inviscid flow, the vectors of force coefficients $\mathbf{c}_{fp}$ and moment coefficients $\mathbf{c}_m$ are only functions of the surface integrals over static pressure and over pressure times moment arm, respectively:

$$\mathbf{c}_{fp} = \frac{\oint_A p\, d\mathbf{A}}{\frac{1}{2}\rho_\infty q_\infty^2 \cdot A_{ref}} \qquad \mathbf{c}_m = \frac{\oint_A (\mathbf{r} - \mathbf{r}_{ref}) \times p\, d\mathbf{A}}{\frac{1}{2}\rho_\infty q_\infty^2 \cdot A_{ref} L_{ref}} \quad . \tag{16}$$

The length from nose to body flap trailing edge is used as aerodynamic reference length for PHOENIX and is $L_{ref} = 6.2112\, m$. The reference area of the half configuration is $A_{ref} = 5.66\, m^2$. Furthermore, the moment reference point $\mathbf{r}_{ref}$ is specified as: $x_{ref} = 0.7\, L_{ref}$, $y_{ref} = 0$ and $z_{ref} = 0.0718\, L_{ref}$. The lift coefficient is calculated by projecting the force coefficients in the direction perpendicular to the undisturbed inflow. Similarly, the drag coefficient is a projection in the inflow direction. Since the yaw angle is zero for all computations, the coefficient of the yaw moment is equal to the y-component of the moment coefficients. The computed aerodynamic coefficients of the Euler solutions are given in Table 1. For $\alpha = 10°$, comparison with aerodynamic design data of Astrium consolidated with additional wind tunnel data yields a relative error of 4% for the lift and 7% for the drag.

Table 1. Aerodynamic coefficients for inviscid flow computation.

Angle of attack α	Lift coefficient c_l	Drag coefficient c_d	Moment coefficient c_m
10°	0.1924	0.0973	$-0.240 \cdot 10^{-2}$
15°	0.2876	0.1422	$-0.791 \cdot 10^{-2}$

4.3   Future Investigations

Future investigations will contain a full Navier-Stokes analysis of PHOENIX under high-enthalpy conditions. The schedule of the computational test series is listed in Table 2. The extension of the gas model to incorporate local

Table 2. Computational test series for viscous flow.

Mach number M_∞	Angle of attack α	Environmental conditions	
10.0	20°	47 km	turbulent
12.1	25°	shock tunnel	turbulent
15.0	15°	60 km	turbulent
17.0	40°	68.3 km	laminar

thermodynamic equilibrium effects is a further prerequisite for high-enthalpy analysis. However, first test computations for laminar and turbulent flow will be performed assuming ideal gas flow to determine convergence behavior and favorable numerical parameters like for example number of multigrid-levels and -cycles.

5   Computational Considerations

The used Euler grids generally have a number of cells between two and three millions. The number will increase by approximately one and a half million cells once the grid is refined for Navier-Stokes computations. So far, the memory requirements in the 3D cases are around 1 gigabyte. For the turbulent computations and for the computations in chemical equilibrium or non-equilibrium additional equations have to be solved increasing the memory requirement to 1.5–2 gigabyte. The vectorization level of the FLOWer program package is 99.7%. For an Euler problem, answer time lies between two to three weeks on the NEC SX-5, depending on how busy the machine is. A single batch job performs approximately 3000 time steps and requires 13 hours of CPU-time, after which it is resubmitted into the batch queue. FLOWer is able to use block based parallelization using the CLIC3D library interface to execute the necessary MPI commands. However, during the previous project phases, no experience was gained for the parallel version of FLOWer since mainly test problems with only a few grid blocks were computed. For the multi-block grid used for PHOENIX, application of the parallel version of FLOWer is considered where the exchange of boundary data between neighbouring blocks would be replaced by send and receive procedures.

6   Summary

Inviscid computations for the shuttle-like technology demonstrator PHOENIX are presented. The applied 3D grid consists of 20 blocks with a total of 2.15 million grid points and uses structured, curvilinear coordinates. Using preliminary inviscid solutions, the grid is tailored to the supersonic flow field around PHOENIX with the interactive grid generation system MegaCads. However, due to the restrictions of structured grids not all inconsistencies in the grid spacing could be remedied. The numerical difficulties caused by the perturbed grid are discussed. An interaction between multigrid and single grid computations has been necessary to stabilize the solution process of the flow solver FLOWer. Converged solutions are presented for a flight Mach number of 2.21. The obtained aerodynamic coefficients are in good agreement with design and wind tunnel data provided by Astrium. Final preparations are in process to start with the full Navier-Stokes computations of PHOENIX. Sublayers necessary for the resolution of viscous, turbulent boundary layers are currently added to the Euler grid and the flow solver is extended by a reactive gas model for high-enthalpy flow. The vectorized computations are performed on a single processor of the NEC SX-5 vector machine with a memory requirement of around one gigabyte and an answering time of two to three weeks for a single Euler computation.

References

1. Coratekin T., van Keuk J., Ballmann J: On the Performance of Upwind Schemes and Turbulence Models in Hypersonic Flows. AIAA Paper 2001–1752 (2001)
2. Kroll N., Radespiel R.: An Improved Flux Vector Split Discretization Scheme for Viscous Flows. DLR-Forschungsbericht **93/53** (1993)
3. Kroll N., Rossow C.-C., Becker K., Thiele F.: The MEGAFLOW Project. Aerospace Science and Technology **4** (2000) 223–237
4. Radespiel R., Rossow C., Swanson R. C.: Efficient Cell-Vertex Multigrid Scheme for the 3D Navier-Stokes Equations. AIAA J. **28/8** (1990) 1464–1472
5. Reinartz B. U., van Keuk J., Coratekin T., Ballmann J.: Computation of Wall Heat Fluxes in Hypersonic Inlet Flows. AIAA Paper 2002-0506 (2002)

Aeroelastic Analysis of Helicopter Rotor Blades Using HPC

Hubert Pomin, Andree Altmikus and Siegfried Wagner

Institut für Aerodynamik und Gasdynamik (IAG), Universität Stuttgart,
Pfaffenwaldring 21, 70550 Stuttgart, Germany

Abstract. In this paper, two different comprehensive rotor analyses are described and applied to a high speed forward flight test case using the Stuttgart NEC–SX5 HPC facility. Each analysis consists of an unsteady Euler/RANS CFD module coupled with a CSD module based on Timoshenko and Euler-Bernoulli beam theory, respectively. Deformable Chimera grids are applied for spatial discretization and resolution requirements call for the use of state-of-the-art vector super computers. Qualitatively, stand-alone aerodynamic results using a-priori trim data agree fairly well with wind tunnel measurements. Substantial improvement can be achieved by including the time-accurate structural response of the blades. Recent results indicate that the incorporation of an automated trim procedure is beneficial with respect to the agreement of computed and experimental global rotor parameters.

1 Introduction

Rotorcraft flows rank among the most challenging applications in aerospace CFD. In high speed horizontal flight, unsteady three-dimensional transonic effects are dominant on the advancing side of the rotor and contribute significantly to the overall noise level of the helicopter (high speed impulsive noise – HSI), while regions of highly complex separated and reverse flow may exist on the retreating side. At lower advance ratios and especially in descent, massive blade vortex interaction (BVI) may occur which affects the unsteady lift distributions of the blades and is also known to be a major source of rotor noise and helicopter vibration. Thus, proper capturing and conservation of the rotor wake is crucial for the success of the numerical analysis.

Taking into account the interdependence of blade dynamics and flow field can significantly enhance the predictive capabilities of the overall analysis for all helicopter flight scenarios. The torsional response of the blade has a significant impact on the aerodynamic coefficients. In forward flight, apart from the elastic twist, inadequate representation of especially the flapping motion of the articulated and/or elastic blade will produce wrong local effective angles of attack, resulting in inaccurate and possibly misleading results of the numerical analysis.

The partitioned procedures approach maintains flow and structure solvers as entirely separate programs which exchange information along their common physical boundaries, the rotor blade surfaces, throughout the analysis. If the communication of the modules takes places on a $360°/ < \#blades >$

basis, the procedure is refered to as a "weak coupling" approach where the CFD output is used to iteratively correct and eventually eliminate the influence of a simplified aerodynamic model required in the structure solver. In recent years, this method has been applied by e.g. Servera et al. [14] and Pahlke & van der Wall [11]. In the "strong coupling" category, information between flow and structure solvers is exchanged on a *per time step* basis and the CFD analysis is the sole and transient source of aerodynamic load information for the structure module, which consequently does not require a full a-priori trim of the rotor based on external semi-empirical aerodynamics (e.g. 2D airfoil tables with corrections). Such a modular "strong coupling" procedure was developed by Hierholz [6] and Buchtala [5] using the Euler flow model and has been applied to hover test cases on periodic monoblock grids based on the Navier-Stokes equations by Pomin & Wagner [12].

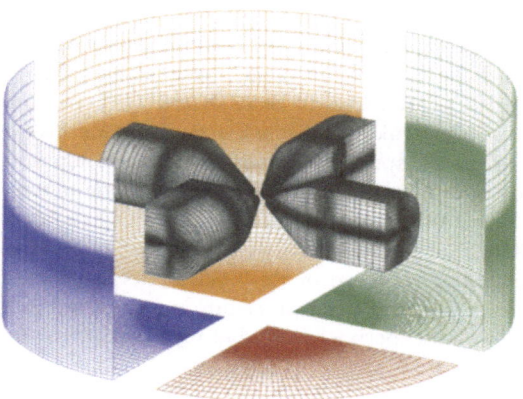

Fig. 1. Chimera overset grids system

In this work, Chimera grids (Fig. 1) are used for "strongly coupled" numerical investigations of a model rotor in high speed forward flight conditions. The objective is to assess the ability of the present aeroelastic analysis tools to capture the essential flow field and blade dynamics characteristics and to quantify rotor performance data. The results are compared with uncoupled computations and available experimental data.

2   Numerical Procedures

The computations presented in this paper are carried out using the IAG codes INROT [15,17,13] and DYNROT [5] as well as the DLR flow solver FLOWer [9] in conjunction with the EUROCOPTER flight mechanics simulation tool HOST [4]. Both coupled procedures are briefly described in the following sections.

2.1   INROT & DYNROT

Aerodynamics INROT solves the Reynolds-averaged Navier-Stokes equations (RANS) in a steadily rotating frame of reference using absolute flow quantities. A time-dependent transformation from physical to computational space is employed to permit arbitrary rigid and elastic motion of the rotor blades.

In order to close the system of equations, perfect gas and Newtonian fluid properties are assumed and for turbulent viscous cases, the Baldwin-Lomax eddy viscosity model is applied. Laminar-turbulent transition points can be prescribed as a function of both radial and azimuthal position. The cell-centered finite volume upwind scheme is based on an implicit time-discrete formulation of the governing equations providing up to third order temporal accuracy which is linearized via a Newton method. For the computation of the inviscid fluxes, an approximate Riemann solver is applied in conjunction with a Low Dispersion scheme which ensures up to third order spatial accuracy in computational space. The viscous fluxes are computed using second order central differences and a LU-SGS operator is applied to integrate the system in time.

For realistic rotor simulations, pitch, flap and lag hinges can be included at arbitrary positions and the Chimera overset grids technique [15] as well as a periodic grid approach [12] are available to ensure proper conservation of the rotor wake. An algebraic algorithm is used to update the near-body grids according to the elastic blade deformation provided either as user input or, in aeroelastic simulations, by the structure solver.

Blade Dynamics In DYNROT, the rotor blade is modeled as a quasi one-dimensional and – at present – geometrically linear Timoshenko beam. In contrast to an Euler-Bernoulli beam, the Timoshenko model takes into account the rotatory inertia of the blade sections and possible shear deformation of the rotor blade. The equations of motion are obtained analytically via an extended form of Hamilton's principle for non-conservative systems. A broader perspective on the structure model is available in Ref. [5].

Fluid-Structure Coupling Fluid and structure solvers are maintained as entirely individual programs which communicate via TCP/IP socket connections and exchange information throughout the computation along their common physical boundary, the rotor blade surface. The partitioned procedures approach greatly simplifies code development and management. Higher order accuracy in time can be preserved, provided a suitable coupling scheme is employed. In the present work, the following implicit-implicit scheme developed by Hierholz [6] and Buchtala [5] is used (Fig. 2):

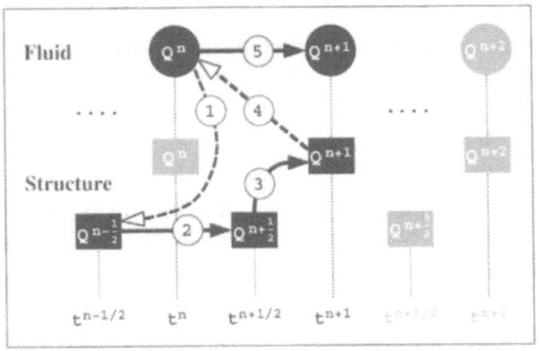

Fig. 2. Implicit-implicit staggered scheme

– Aerodynamic loads of time level t^n are sent to the structure module

- Integration of structure state from $t^{n-1/2}$ to $t^{n+1/2}$ using the midpoint rule
- Predictor step for determination of surface coordinates at time level t^{n+1}
- Surface coordinates of time level t^{n+1} are sent to the fluid module
- Grid deformation, calculation of surface velocities at t^{n+1} acc. GCL, time integration of RANS/Euler equations to t^{n+1}

This staggered coupling algorithm features an offset of half a time step between fluid and structure integration and can provide second order accuracy in time for the overall method [5].

2.2   FLOWer & HOST

Aerodynamics FLOWer also solves the three-dimensional, unsteady Euler or Reynolds-averaged Navier-Stokes equations which are formulated in a hub-attached, non-inertial rotating frame of reference. The discretization of space and time is separated by the method of lines using a cell-vertex finite volume formulation. Spurious oscillations of the central difference scheme are suppressed by first and second order artificial dissipation. The time integration makes use of the dual time stepping technique with a second order implicit time integration operator [8]. For turbulent flow simulations, a variety of turbulence models is available. Details of the overall algorithm can be found in [7]. FLOWer features the Chimera technique [10,3] allowing for arbitrary relative motion of aerodynamic bodies. Body fitted grids around each blade are embedded in a background grid, in which the blade vortex sheets are convected from one blade grid to the next. The body fixed monoblock grids are algebraically deformed to take into account elastic blade deformation.

Structure Modeling HOST represents a computational environment for simulation and stability analysis of complete helicopters. It enables the study of individual components like isolated rotors as well as complete configurations with related substructures. An articulated flexible blade is assumed for the isolated rotor calculations. The interal aerodynamic analysis is based on the blade element theory where the user can choose from various downwash models.

   In this study, the blade is modeled as a quasi one-dimensional beam. The model allows deflections in flap and lag direction and elastic blade torsion along the blade axis. In addition to the assumption that only small deformations occur, the restrictions of no tension elongation and no shear deformation are valid. Possible offsets between local cross-section center of gravity, tension center and shear center are accounted for. The blade model is based on a geometrically non-linear multi-body formulation. The representation consists of rigid segments, which are connected through virtual joints [16]. At each joint elastic rotations about the lag, flap and torsion axes are allowed. Since the immediate use of these rotations as DOF's would yield a rather large system of equations, the number of equations is reduced by a modal Rayleigh-Ritz approach. The differential equations of the structure are not given in matrix form, but are derived from the kinetic and potential energy terms, directly applying the Lagrange equation.

In order to avoid linearization of the system, explicit time integration methods are applied. Several explicit schemes with differing orders of accuracy are available in HOST, but special focus is put upon the second order Adams-Bashforth-Moulton method (ABM2).

Staggered Coupling Scheme The two codes are strongly coupled by an implicit-explicit staggered fluid-structure coupling (FSC) scheme. The fluid and structure partitions interact due to time accurate external input of boundary conditions, provided at synchronization points. The temporal accuracy has been ensured in this case by the following scheme: the explicit dynamic solver updates the structure consistently with a half time-step negative offset with respect to the flow solution from $t^{n-1/2}$ to $t^{n+1/2}$. This is done by applying the second order Adams-Bashforth-Moulton formula, a classic predictor-corrector method. The resulting staggered scheme is depicted in Fig. 3. Details concerning the implementation and a mathematical proof of 2nd order temporal accuracy in terms of exchanged energy can be found in [1].

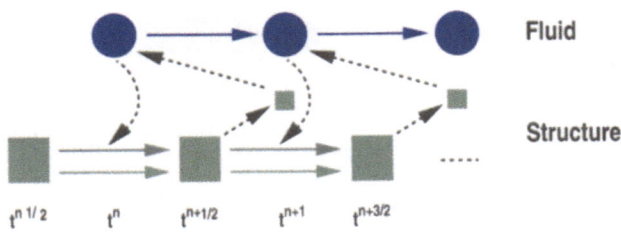

Fig. 3. Implicit-explicit staggered scheme

Automated Trim Procedure In the HOST-FLOWer environment, a strongly coupled simulation is started from an initial trim calculation with HOST which delivers the control angles θ_0, θ_{1c}, θ_{1s} and α_q. However, the resulting rotor state cannot be expected to exactly match the trim objective in terms of performance coefficients at this stage, because the aerodynamic loads are highly dependent on the structural movement. Therefore a re-trimming capability is incorporated into the strong coupling approach. The applied trim method is in principle a combination of weak coupling which naturally delivers a trimmed rotor with appropriate control angles, and strong coupling. The procedure is schematically illustrated in Fig. 4. The strongly coupled simulation is started from an initial HOST standalone trim using the built-in 2D lifting-line method with 2D airfoil tables. During the coupled CFD simulation, the sectional aerodynamic loads are stored at each time step. When convergence with a (nearly) periodic state is obtained after several rotor revolutions for a given set of control input, the stored loads are harmonically decomposed and serve as input for a new trim computation in analogy to the weak coupling approach where the internal 2D loads are corrected by the 3D CFD airloads. Hence, a new set of control input data is obtained followed by a restart of the strongly coupled simulation. This weak/strong coupling cycle is repeated until global trim is achieved. However, all non-linearities or impulsive reactions show up in the unsteady solution due to the strong coupling. Details concerning this trim procedure can be found in Ref. [2].

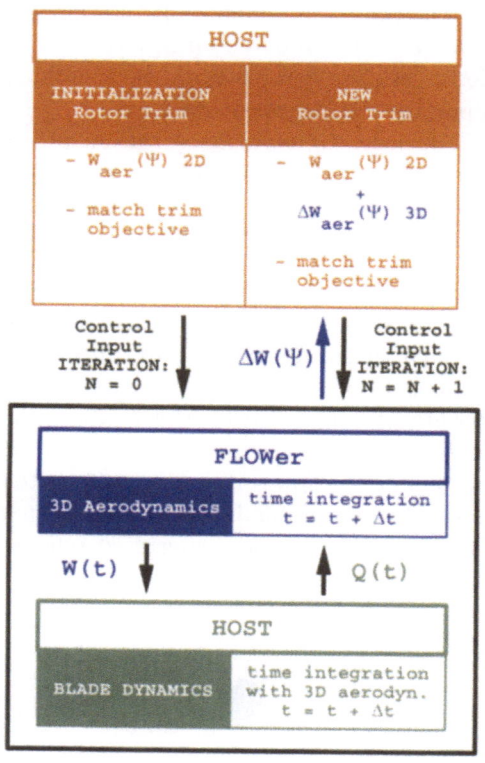

Fig. 4. Automated trim procedure

3   Results

Results are presented for a popular high speed forward flight test case of the four-bladed ONERA/EUROCOPTER 7A rotor which has a diameter of $2R^{\star}_{\mathrm{Tip}} = 4.2$m and a geometric solidity of $\sigma \approx 0.0849$. The fully articulated 7A model blades are characterized by $\Lambda = 15$, a rectangular planform, a square tip and feature a linear aerodynamic twist of $-3.95°$/m and a prelag of $-4°$. The lag, flap and pitch hinges are located at $(0.0357, 0.0362, 0.0743)R_{\mathrm{Tip}}$.

In the course of the ONERA S1MA wind tunnel tests, the rotor was trimmed to satisfy the so-called "Modane law" ($\beta_{1s} = 0$, $\beta_{1c} + \theta_{1s} = 0$). The specifications of the high speed forward flight test case are $Ma_{\mathrm{Tip}} = 0.656$, $\mu = 0.4$ and a tip rotation Reynolds number of approx. $2.14 \cdot 10^6$. Given in components of the wind tunnel system, the measured global rotor force coefficients are $Zb = 200C_Z/\sigma = 12.56$ and $C_X/\sigma = 0.1$ [14]. The horizontal force corresponds to a propulsive force coefficient of $Xb = 1.58$.

3.1   7A Model Rotor in High Speed Forward Flight (INROT/DYNROT)

The primary rotor control parameters (shaft tilt angle α_q and blade pitch $\theta(\psi)$) must be known to fully define the aeroelastic simulation. While these quantities could be retrieved directly from the experimental logs, it must be considered that the isolated rotor analysis does not take into account wall and support pertubations of the wind tunnel experiment that are known to have a significant influence on rotor trim and performance. Therefore, new sets of control parameters are computed using HOST based on the above trim objective. The trim results obtained for stand-alone rigid blade CFD analyses and the aeroelastic computations are given in Table 1 along with the experimental values. Apart from the above parameters, a stand-alone CFD analysis also requires the prescription of the blade flap and lag angles, $\beta(\psi)$ and $\zeta(\psi)$, which are available as part of the HOST solution.

Table 1. Primary rotor control parameters

	α_q	$\theta_{0.7}$	θ_{1C}	θ_{1S}
HOST Rigid	$-12.72°$	$8.83°$	$1.67°$	$-3.46°$
HOST Elastic	$-12.07°$	$10.19°$	$0.86°$	$-3.31°$
Experiment	$-13.75°$	$10.41°$	$3.43°$	$-3.70°$

While the blade grids provide adequate resolution to compute viscous flow, this feature must be compromised for the background grid due to memory and CPU restrictions. Therefore, a RANS solution is computed on the blade grids only while the Euler model, fully capable of convecting the rotor wake, is used on the background mesh (*hybrid analysis*). In order to minimize wall clock computing time, the background grid is internally decomposed and all internal grid blocks are processed in parallel by microtasking.

In an effort to evaluate rotor wake conservation qualities, computations are carried out using both a cylindrical and a cartesian Chimera background (BG) grid with similar grid point distributions. The total number of grid cells in the hybrid analysis amounts to 15,600,000 (cylindrical BG) and 42,600,000 (cartesian BG), respectively. Key performance data of typical hybrid computations on the NEC SX–5 is included in Table 2, where the GFLOPS values

Table 2. Performance data of hybrid analysis on NEC–SX5

	CPUs	CPU-h	GFLOPS	Memory
Cylindrical BG	8	80/rev	11.5	8.1 GB
Cartesian BG	16	171/rev	27.0	23.5 GB

indicate the output of the *qscan* command. The computations with the 12 fairly small internal cartesian background grid blocks outperform those on the cylindrical grid primarily because the partly unvectorizable Chimera transfer operations are distributed on a larger number of processors for parallel execution. Furthermore, the hole completion procedure is found to require fewer iterations on the cartesian grid. However, the cylindrical grid is used for the majority of the forward flight computations due to limited computing resources. Fully turbulent flow is assumed in the hybrid forward flight computations and the azimuthal increment per time step is set to $\Delta\psi = 0.5°$ for Euler and Navier-Stokes.

The computed and measured time-averaged global rotor coefficients are plotted in Fig. 5, where C_X and C_Z represent the forces in the upstream and vertical tunnel directions, respectively, and C_Q is the rotor torque. The stand-alone CFD computations with prescribed pitch, flap, lag and rotor shaft tilt angles underestimate the experimental rotor thrust by approx. 25%. Much better agreement is achieved using the coupled Navier-Stokes analysis which predicts a vertical force merely 1.8% above the measurements. Note that the thrust values computed by the uncoupled inviscid and viscous analyses are nearly identical, while the results of the coupled computations differ by approx. 9%, indicating significant differences between Euler and Navier-Stokes regarding the predicted blade motion and/or elastic deformation. The rotor torque coefficient computed by the coupled hybrid analysis is approx. 6% below the prediction of the corresponding

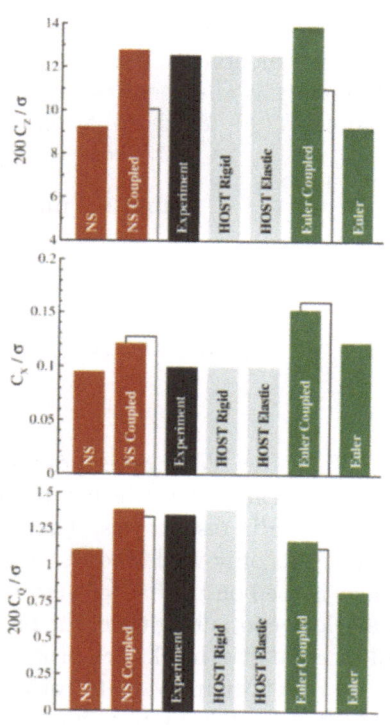

Fig. 5. Global rotor coefficients

HOST soft blade trim computation and within 3% of the experimental value. Additional coupled computations are carried out using the experimental values for α_q and $\theta(\psi)$ and the resulting global rotor coefficients are represented by the white bars in the background of Fig. 5. Since the coupled results obtained with the HOST control parameters are found to correlate much better with the experimental data, only the output of these calculations will be discussed beyond this point.

In Fig. 6, computed and measured normal force coefficients are plotted over ψ for a number of radial stations. The rigid blade CFD analysis with fully prescribed hinge angles taken from the corresponding HOST trim calcu-

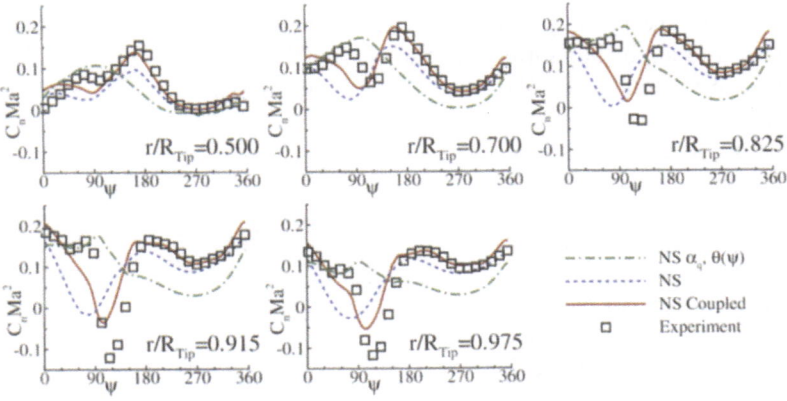

Fig. 6. Sectional normal force

lation is found to reproduce the principal characteristics of the experimental data, although with significant deviations in phase and amplitude. For comparison, results of a rigid blade computation in which the control input is limited to α_q and $\theta(\psi)$ are included in the figure in order to emphasize the impact of especially the flap degree of freedom on the aerodynamic coefficients. Qualitatively and quantitatively, the normal forces computed by the coupled analysis correlate quite well with the experimental data. A moderate phase shift can be observed between the peaks of the computed and measured curves. The characteristic dip in the tip region at around 120° blade azimuth is captured (yet underestimated) by the hybrid aeroelastic analysis, while the measured local normal force maximum preceding it is only reproduced as a trend in the present numerical results.

A comparison of the computed local pitching moment coefficients with experimental data of ONERA is given in Fig. 7. At $0.7R_{\text{Tip}}$, the negative pitching moment measured on the advancing side of the rotor is reproduced, but underpredicted by the viscous analysis, while the Euler simulation returns positive values over the first quarter revolution. Towards the blade tip,

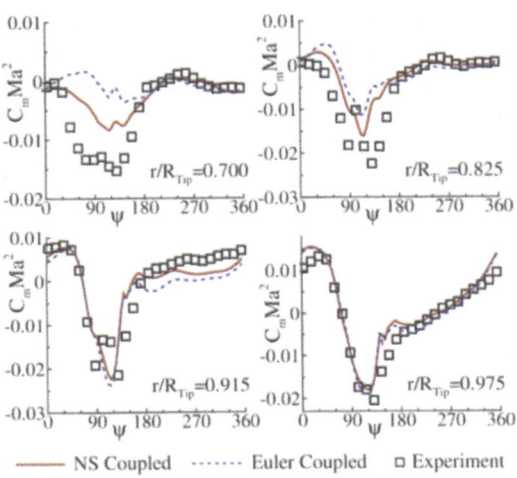

Fig. 7. Sectional pitching moment

the computed sectional pitching moments correlate well with the experimental data and slight advantages of the Navier-Stokes model over the inviscid analysis are visible at 82.5% and 91.5% blade radius.

Chordwise pressure distributions at 97.5% tip radius are presented in Fig. 8. The results of the hybrid aeroelastic analysis correlate well with experimental data and the ability of the procedure to adequately capture unsteady transonic effects is clearly demonstrated. Due to the high advance ratio of the selected test case, a strong shock builds up on the upper surface between 0° and 30° blade azimuth. The predicted shock positions vary between 0.25c and 0.60c and the most significant discrepancies between simulation and experiment can be noticed in the course of the shock breakdown, while very good agreement is observed on the entire retreating side of the rotor in the proximity of the tip.

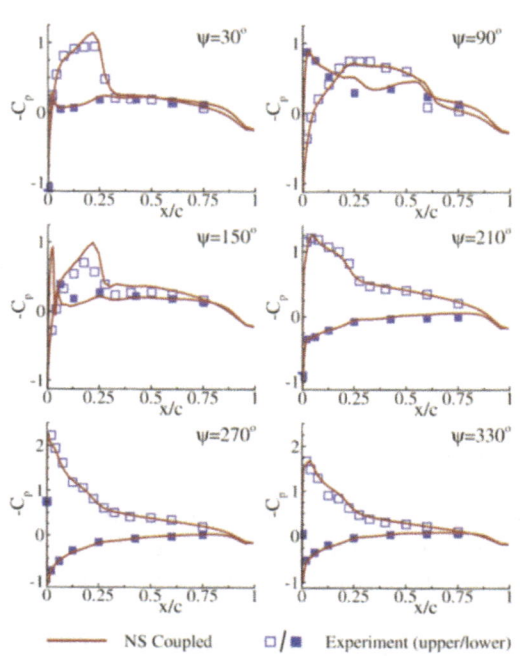

Fig. 8. Pressure distributions at $0.975R_{Tip}$

In Fig. 9, the elastic flap bending contributions computed at the blade tip are plotted versus blade azimuth and a comparison with ONERA data measured in the course of the Modane wind tunnel tests shows fairly good agreement of simulation and experiment with respect to the characteristic 2/rev. behavior and the tip deflection amplitude. Also in Fig. 9, the com-

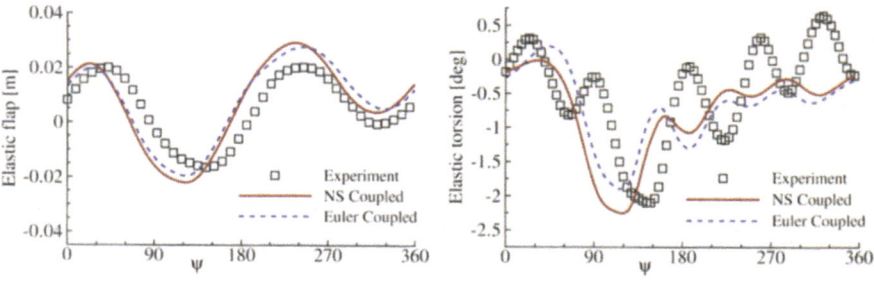

Fig. 9. Elastic tip flap and torsion versus blade azimuth

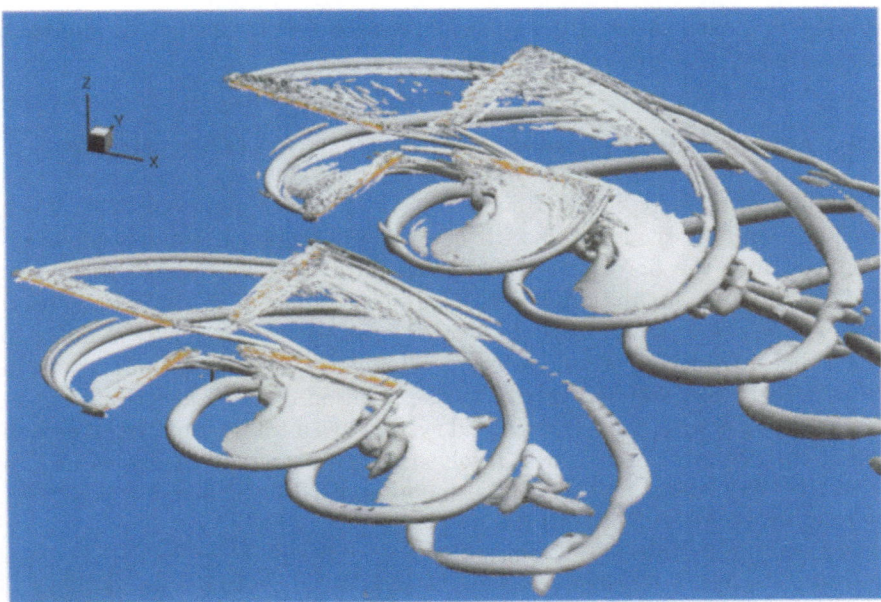

Fig. 10. Rotor wake structures computed on cylindrical (lower left) and cartesian (upper right) background grids

puted torsional deformation of the blade tip is compared with experimental data taken from Ref. [14]. While the low frequency component of the twist distribution is captured quite well by the aeroelastic simulations, the amplitudes of the higher harmonic contributions are underpredicted with respect to the experimental data. Furthermore, the current numerical solutions completely lack the measured local nose-up twist on the advancing side of the rotor at $\psi \approx 90°$. This phenomenon will be investigated more closely in upcoming studies.

The wake structures of the simulations with cylindrical and cartesian background grid are visualized in Fig. 10 using λ_2-iso-surfaces. Both computations are found to provide very similar results with respect to vortex locations and core diameters. Apparently, the sole advantage of the expensive cartesian background grid is improved vortex conservation downstream of the rotor disk which, however, hardly has any effect on the rotor loads.

Finally, it should be stated that the results of the first principles-based forward flight computations presented above are a fairly good approximation, but not an exact representation of the fully trimmed rotor state. Further improvement of the agreement between numerical and experimental data may be achieved – at considerable computational cost – by incorporating a trim capability within the aeroelastic analysis where the rotor trim provided by HOST is used as an initial solution. Recent results of such an approach are presented in the following section.

3.2   Automated CFD-based Trim Analysis (FLOWer/HOST)

Trimming a strongly coupled simulation is a rather elaborate task. Each time the analysis is restarted with new control angles, a disturbance is introduced into the aeroelastic system, as illustrated in Fig. 11, where the FLOWer/HOST simulation is restarted three times with updated control angles. Each time, a minimum of 4 rotor revolutions is required in order to reach a new (nearly) periodic solution. Therefore, the aerodynamic problem is reduced in size and complexity to the solution of the Euler-equations in the following study. The Chimera grid system has a total of approx. 2 million cells. Still, one rotor revolution – resolved by 360 timesteps – takes 14 CPU-h with FLOWer running in sequential mode on a NEC–SX5 vector computer at roughly 1 GFLOPS. Figure 11 shows that the automatic trim procedure is quite efficient since the lift coefficient Zb is significantly improved after the three re-trim operations with respect to the experimental value represented by the bold black line. Table 3 shows the final trim state with respect to experiment and the untrimmed case. The trimmed conditions in terms of lift coefficient Zb, propulsive force coefficient Xb meet the trim objective

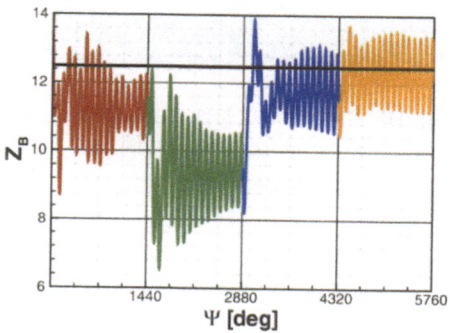

Fig. 11. Influence of re-trim on rotor lift

(experiment) and the $[\beta_{1c} + \theta_{1s} = 0]$-part of the Modane-law is improved.

Despite the beneficial influence of the trim procedure on the global parameters, only very small differences are observed between trimmed and untrimmed solutions in terms of sectional airloads (Figs. 12 & 13). However, this is consistent, since the trim depends on global values, integrated over span. Consequently, the trim does not necessarily improve the sectional results with respect to experimental data. The sign of the sectional moment

Table 3. Influence of trim on control parameters

	Zb	Xb	β_{1s}	$\beta_{1c} + \theta_{1s}$
Experiment	12.56	1.58	0	0
No Trim	11.27	1.81	0.23	0.87
Trim	12.26	1.57	0.25	-0.32

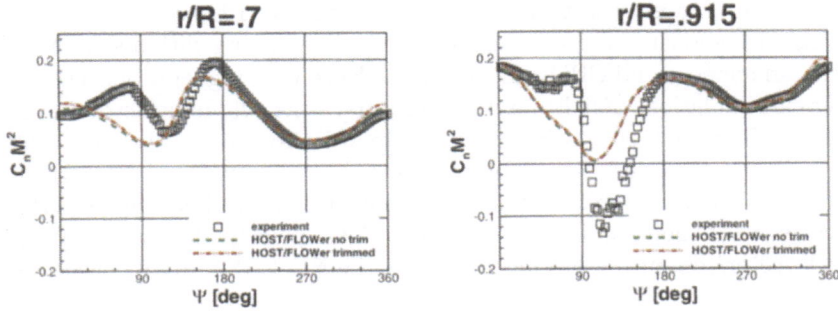

Fig. 12. Influence of trim on sectional lift coefficient

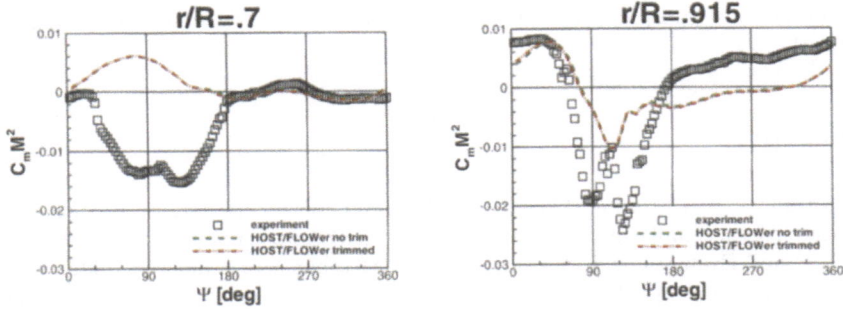

Fig. 13. Influence of trim on sectional moment coefficient

coefficient computed by the Euler analysis is incorrect in the azimuthal range between $\Psi = 0°$ and $\Psi = 180°$ and it is anticipated that including viscous effects will help improve the correlation of measured and computed data in this region.

4 Conclusions

In this paper, we presented two modular CFD/CSD analyses for the aeroelastic simulation of helicopter main rotors. By comparison with experimental data, it could be demonstrated that the principal characteristics of the coupled problem are adequately captured by the numerical procedures. At present, up to 171 CPU hours are required per rotor revolution on fine Chimera grids using 16 vector processors of the NEC–SX5 and a maximum overall performance of 27 GFLOPS is achieved.

Future activities on the INROT/DYNROT side include the application of more sophisticated eddy viscosity models and the upgrade of the beam model to non-linear Timoshenko theory. The comprehensive FLOWer/HOST trim analysis will be rerun in viscous mode which will go along with a substantial increase in memory and CPU-time demands. Even with higher order schemes, numerical simulations of helicopter main rotors, especially in hover or descent flight, require very high spatial resolution in order to capture the rotor wake with sufficient accuracy. These complex vortical structures interact with the individual blades and have an immediate impact on overall helicopter performance as well as noise and vibration levels. Consequently, the use of state-of-the-art vector computing platforms such as the NEC–SX5 and its successors at HLRS is imperative to obtain high quality answers within an acceptable time frame. These authors consider HPC platforms with a reasonable number of high performance vector CPUs (presently 16–32) and adequate memory per node to be the optimum computing facility for the presented applications. For future development of the simulation tools and their application to increasingly complex configurations, an even higher number of nodes is desirable.

References

1. Altmikus, A.R.M. and Wagner, S.: On the Timewise Accuracy of Staggered Aeroelastic Simulations of Rotary Wings. AHS Aerodynamics, Acoustics, and Test and Evaluations Technical Specialist Meeting, San Francisco, 2002.
2. Altmikus, A.R.M., Wagner, S., Beaumier, Ph. and Servera, G.: A Comparison: Weak Versus Strong Modular Coupling for Trimmed Aeroelastic Rotor Simulations. American Helicopter Society, 58th Annual Forum, Montréal, 2002.
3. Benek, J.A., Buning, P.G. and Steger, J.L.: A 3-D Chimera Grid Embedding Technique. AIAA-Paper 85-1523, 1985.
4. Benoit, B., Dequin, A.-M., Kampa, K., Grünhagen, W. v., Basset, P.-M. and Gimonet, B.: HOST, a General Helicopter Simulation Tool for Germany and France. American Helicopter Society, 56th Annual Forum, Virginia Beach, 2000.
5. Buchtala, B.: Gekoppelte Berechnung der Dynamik und Aerodynamik von Drehflüglern. Ph. D. Dissertation, Institut für Aerodynamik und Gasdynamik, Universität Stuttgart, Stuttgart, 2002.
6. Hierholz, K.: Ein numerisches Verfahren zur Simulation der Strömungs-Struktur-Kopplung am Hubschrauberrotor. Ph. D. Dissertation, Institut für Aerodynamik und Gasdynamik, Universität Stuttgart, Stuttgart, 1999.
7. Jameson, A., Schmidt, W. and Turkel, E.: Numerical Solutions of the Euler Equations by Finite Volume Methods Using Runge-Kutta Time-Stepping Schemes. AIAA-Paper 81-1259, 1981.
8. Jameson, A.: Time Dependent Calculation Using Multigrid, With Applications to Unsteady Flows Past Airfoils and Wings. AIAA-Paper 91-1596, 1991.
9. Kroll, N. and Rossow, C.C.: MEGAFLOW – A Numerical Flow Simulation System. ICAS-Congress, Melbourne, 1998.
10. Pahlke, K. G.: Berechnung von Strömungsfeldern um Hubschrauberrotoren im Vorwärtsflug durch die Lösung der Euler-Gleichungen. DLR-Forschungsbericht 1999-22, 1999.

11. Pahlke, K. and van der Wall, B.: Calculation of Multibladed Rotors in High-Speed Forward Flight with Weak Fluid-Structure-Coupling. Proceedings: 27th European Rotorcraft Forum, Moscow, 2001.
12. Pomin, H. and Wagner, S.: Navier-Stokes Analysis of Helicopter Rotor Aerodynamics in Hover and Forward Flight. AIAA-Paper 2001-0998, 2001 (accepted for publication in the AIAA Journal of Aircraft).
13. Pomin, H. and Wagner, S.: Aeroelastic Analysis of Helicopter Rotor Blades on Deformable Chimera Grids. AIAA-Paper 2002-0951, 2002.
14. Servera, G., Beaumier, P. and Costes, M.: A Weak Coupling Method between the Dynamics Code HOST and the 3D Unsteady Euler Code WAVES. Proceedings: 26th European Rotorcraft Forum, Den Haag, 2000.
15. Stangl, R.: Ein Euler-Verfahren zur Berechnung der Strömung um einen Hubschrauber im Vorwärtsflug. Ph. D. Dissertation, Institut für Aerodynamik und Gasdynamik, Universität Stuttgart, Stuttgart, 1996.
16. Toulmay, F., Arnaud, G., Falchero, D. and Villat, V.: Analytical Prediction of the Rotor Dynamics for Advanced Geometry Blades. American Helicopter Society, 52nd Annual Forum, Washington/D.C., June 1996.
17. Wehr, D.: Untersuchungen zum Wirbeltransport bei der Simulation der instationären Umströmung von Mehrblattrotoren mittels der Euler-Gleichungen. Ph. D. Dissertation, Institut für Aerodynamik und Gasdynamik, Universität Stuttgart, Stuttgart, 1999.

Numerical High Lift Research – NHLRes Annual Review 2001

S. Melber, J. Wild, and R. Rudnik

DLR Braunschweig, Institute of Aerodynamics and Flow Technology,
Section Transport Aircraft, Lilienthalplatz 7, 38108 Braunschweig, Germany

Abstract. The project NHLRes is concerned with the simulation of aircraft aerodynamics and thus belongs to the research field of computational fluid dynamics (CFD) for aerospace applications. NHLRes comprises the numerical simulation of the viscous compressible flow around transport aircraft high lift configurations. The investigations are based on the solution of the Reynolds-averaged Navier-Stokes equations using a finite volume parallel solution algorithm with an unstructured data concept. The project consists of two parts representing a typical analysis as well as an optimization task for selected three-dimensional high lift flow problems.

1 Introduction

1.1 Overview of High Lift Research

The efficient design of a transport aircraft high lift configuration with respect to low speed take-off/landing performance and handling qualities represents a complex aerodynamic problem. The flow field around the wing with deployed high lift devices at high incidences is characterized by the co-existence of flow phenomena such as large pressure induced separation, compressibility effects and strong wake/boundary layer interaction. In addition there are some critical areas with respect to the flow topology like the wing/fuselage junction or the engine/pylon/wing intersection, which have a strong influence on the overall aerodynamic performance.

Aside from established experimental investigations of such aerodynamic effects in wind-tunnels or using flight testing, the application of computational fluid dynamics gains increasing importance for the analysis and improvement of high lift aircraft configurations. Due to the high computational resources required, a comparatively limited number of national and international studies [1]–[6] have been carried out to assess the accuracy, performance, and reliability of Reynolds-averaged Navier-Stokes flow solvers applied to realistic three-dimensional geometries. Compared to wind-tunnel experiments the use of advanced CFD methods holds promise to substantially accelerate the aerodynamic design process, save costs, while providing a detailed insight in the complete flow field around the considered configuration.

1.2   The Project NHLRes

One major objective of the project NHLRes is the detailed investigation of the aerodynamics at the wing-root of a transport aircraft wing/fuselage configuration and the impact of certain geometry features on the maximum lift coefficient. Due to the high grid resolution required for studying the interaction of local vortex dominated phenomena with the flow separation on a three-dimensional configuration, grid point numbers of about 10 million or beyond corresponding to about 35 million volume elements characterize this type of application.

The second part of the project covers the optimization of high lift device settings, deflection angles and cove shape design using the same numerical analysis code in conjunction with an optimization module developed at DLR. Based on experiences with the optimization of three-element airfoils, the extension towards the optimization of multi element wings is envisaged within NHLRes as a first step in the direction of three-dimensional optimization.

Both parts of the NHLRes project aim at demonstrating the capabilities and improvements of advanced numerical methods for high lift aerodynamics at the edge of currently available computer resources. The project has been carried out from 01.04.2001 to 01.04.2002 with allocated resources of 4000 CPU-hours on the NEC-SX5 and 1000 CPU-hours for the Hitachi SR 8000. In the present paper the numerical methods, configurations, grid-generation and numerical results of the project NHLRes are presented.

2   Wing-Root Aerodynamics at High Lift

2.1   Overview

A systematic extension of the pilot activities [1]–[4] with the DLR ALVAST wing/fuselage model (Section 2.2) is envisaged for NHLRes. The preceding studies with the high lift configuration have established a basis with a sophisticated modeling of geometric features of the wind-tunnel model and an appropriate grid resolution level. This allows, as a second step, a systematic study and variation of geometric features at the wing/fuselage intersection, e.g. slat end plate devices. As an example for related flow visualization, surface isobars, stream lines on the ALVAST configuration, as well as the slat end vortex are shown in Fig. 1.

The objective of these investigations is to point out the impact of such type of devices on the maximum attainable lift and validate the numerical results against wind-tunnel data in more detail. By computing lift and drag polars, analyzing the three-dimensional flow physics and the stall behavior, the physical understanding of the related flow phenomena should be improved.

After a detailed analysis of three ALVAST configurations with (a) fairing and slat-horn, (b) slat-horn and (c) fairing between wing and fuselage simulated on the DLR NEC-SX5 in Braunschweig differences appeared between

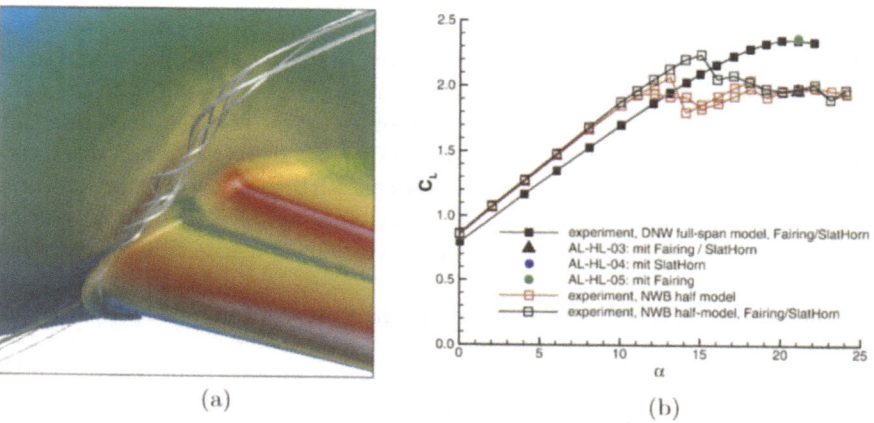

(a)   (b)

Fig. 1. (a) wing/fuselage junction, pressure coefficient on the surface, slat end plate with stream-lines, (b) lift-polar from half/full-model measurements and numerical simulations.

measurements and simulations: From measurements below the maximum lift angle of attack the order of configurations c), b), a) regarding maximum achievable lift is found. In contrast to this the numerical simulations reveal the configuration with the fairing as the one with highest lift (Fig. 1).

The following reasons for this difference between simulations and measurements are conceivable:

– uncertainties of the transferability of numerical and experimental onflow conditions (the chosen angle of attack $\alpha = 21°$ for the simulations is beyond the maximum lift angle of attack in measurements)
– differences in the geometry of wind-tunnel model and CAD-model
– influence of the peniche used in half-model testing
– fuselage vortex influence on maximum lift
– hysteresis effects
– additional unknown physical effects which are not resolved at the moment
– not enough flow-resolution (use of preconditioning)
– insufficient turbulence-modeling
– adaptation influence

Measurements are questionable to answer this question, because semi-span and full-span model tests at high angles of attack show considerable differences (Fig. 1) with regard to the maximum lift and the corresponding angle of attack. On this account first the semi-span test technique is investigated in more detail. Likewise related flow effects and open items of the flow-simulations have to be treated, before the scheduled main geometry variations on the ALVAST configuration are conducted.

2.2 Aircraft Configuration and Computational Grids

The DLR ALVAST wing/fuselage wind-tunnel model [7] with deployed slat and flaps in take-off configuration is selected as the baseline high lift configuration, Fig. 2(a). The geometry specifications are similar to an AIRBUS A320. For the present investigation the take-off configuration is considered, characterized by a continuous slat with a deflection angle of $\delta_s = 20.0°$ and a single slotted flap with a deflection angle of $\delta_s = 19.5°$. The flap is departed in span-wise direction by a thrust gate.

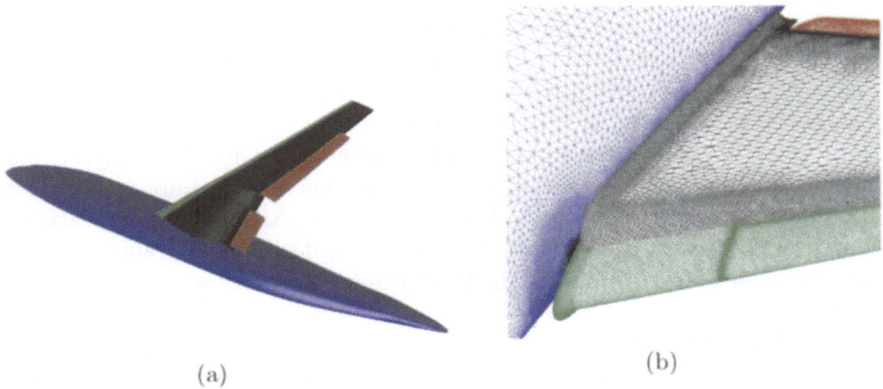

(a)   (b)

Fig. 2. (a) ALVAST high lift configuration, (b) unstructured surface grid in the area of wing/fuselage junction.

The presently used CAD-description captures many details of the wind-tunnel model, like slat-end-plate and slat-stump at the leading edge, and the wing-root fairing on the upper-side of the wing/fuselage junction. The hybrid unstructured grids is generated with the code system CENTAUR [8]. The grid, depicted in Fig. 2(b), consists of two parts: a quasi-structured prismatic cell layer is utilized in order to achieve an appropriate resolution of the viscous effects inside the boundary layer. In contrast to this, tetrahedral cells are used to fill the outer domain of the flow-field. To resolve all features of the wind-tunnel geometry, a typical grid consists of about $10 \cdot 10^6$ points.

The flow computations were performed for a free stream Mach-number of $M_\infty = 0.22$ and a Reynolds-number of $Re_\infty = 2 \cdot 10^6$. Corresponding experimental data from measurements in the German-Dutch low speed wind-tunnel DNW are used for the assessment of the numerical results [9], [10].

2.3 Numerical Methods

The Reynolds-averaged Navier-Stokes equations (RANS) are solved by the hybrid unstructured flow solver DLR TAU, which is based on a three-dimensional finite volume scheme. The governing equations are solved on a dual background grid, which, together with the edge-based data structure, allows to run the code on any type of cells. The solver is part of the MEGAFLOW-project [11].

The flow variables are stored in the centers of the dual grid, i.e. the vertices of the primary grid. The temporal gradients are discretized using a multi-step Runge-Kutta scheme. The inviscid fluxes are calculated either by a Roe- or AUSM-type 2nd-order upwind scheme, or by employing a central method with scalar dissipation, which is used for the present study. The viscous fluxes are discretized using central differences. In order to accelerate the convergence to steady state, local-time stepping and a multigrid technique based on agglomeration of the dual-grid volumes are employed. Optimization for different architectures is achieved by vector- or cache-type coloring of the edges. For parallel computations a domain-decomposition is used providing a subset of dual grids.

Because of the low free-stream Mach-number in the high lift computations preconditioning was used for some cases (Section 2.5) to improve accuracy an enhance the convergence to steady state. The ALVAST configuration is computed by using the Spalart-Allmaras turbulence model and assuming fully turbulent flow, because in the wind-tunnel tests no transition fixing was made.

An adaptation module for unstructured grids is also available in the TAU-code. It detects regions with insufficient grid resolution by gradient sensors of flow variables (e.g. velocity, pressure loss) and performs local grid refinement by bisection of cells. In addition, the adaptation also allows the redistribution of the prismatic layers to capture the viscous boundary-layer in Navier-Stokes computations adequately.

2.4 Computer Resources and Performance

For the simulation of one angle of attack with a typical high lift configuration on hybrid unstructured grid of $10 \cdot 10^6$ points 7.5 iterations per hour are completed on the NEC-SX5 on a single processor. To reach a fully converged solution at high angles of attack with partial flow separation this case needs 5000 iterations, which corresponds to about 650 hours CPU-time and a main-memory requirement of 12 GByte. Such a grid requires a memory of 3.8 GB, one solution needs 0.7 GB disc-space. To investigate the lift-breakdown a minimal amount of five angles of attack is required with an overall of 3000 hours CPU-time.

On a NEC-SX5 the TAU-code (section 2.3, Numerical Methods) reaches a vector-operation-ratio of nearly 99%, 816 MFLOPS floating point performance and 37 MIPS on one processor. In the time-frame from 01.04.2001 to 01.04.2002 the part "Wing-Root Aerodynamics at High Lift" has consumed 3500 CPU-hours on NEC-SX5. These resources are sheared among four configurations. The configurations and related results are discussed in detail in the following sections, some parameters can be found in Table 1.

Table 1. Typical computer-resources on NEC-SX5 and parameters of the configurations.

	CPUs	Mem.	CPU-Time	Variations	α	max. grid-pnts
Peniche	4	4.5 GB	370 h	4	20.99°	5395492
Vortex-motion	4	2.5 GB	270 h	4	8°–28°	5998086
Nacelle-vortex	4	8.0 GB	1640 h	1	12.0°	9614649
Preconditioning	4	0.7 GB	32 h	4	20.99°	802421

2.5   Numerical Results

Peniche: The objective of this subtask is to investigate with numerical methods the differences between semi-span and full-span measurements especially at high angles of attack and the effects referring to the wing-root aerodynamics (Fig. 1). In corresponding literature measurements of the peniche influence, especially [12] to high lift aerodynamics or the search for a optimal geometry of the peniche itself [13] can be found.

The half-model technique is a consequence of the wish to reach high Reynolds-numbers in wind-tunnel testing. In order to account for the given wind-tunnel size, a semi-span model offers the opportunity to increase the model scale of the experiment. In principle, the semi-span model should reach the same results as a full-model for symmetric incoming flow. The symmetry condition of the half model is incorporated by mounting the model on the wind tunnel-floor or sealing. As a consequence, the model fuselage would be positioned partially in the boundary-layer of the wind-tunnel walls. As this undesired effect influences the results, the half-model is mounted on a so-called peniche. The peniche is an extension of the fuselage in the symmetry-plane in order to lift the model out of the wind-tunnel boundary-layer, Fig. 3(a).

In order to address the problem, the isolated (half-)fuselage of the ALVAST configuration is considered, because the fuselage is most directly influenced by the peniche, and a high grid resolution is achievable with a limited overall effort. As variations, different half peniches-heights of 100 mm and a 200 mm are investigated. To improve the resolution of the flow-field the grids has been adapted to the non-dimensional first wall spacing, and three times to pressure loss.

In Fig. 3 a cut of the flow field normal to the fuselage-axis at nearly 90% fuselage length is displayed. The flow streamlines in this cut-plane are shown colored by the vorticity. In the figures the vortices above the fuselage can be seen. This vortice is a result of the flow separation behind the equator of the fuselage at higher angles of attack.

As Figs. 3(b), (c) and (d) reveal, an increasing peniche height causes increasing vortex-size and strength, and thus a clear difference from the ideal

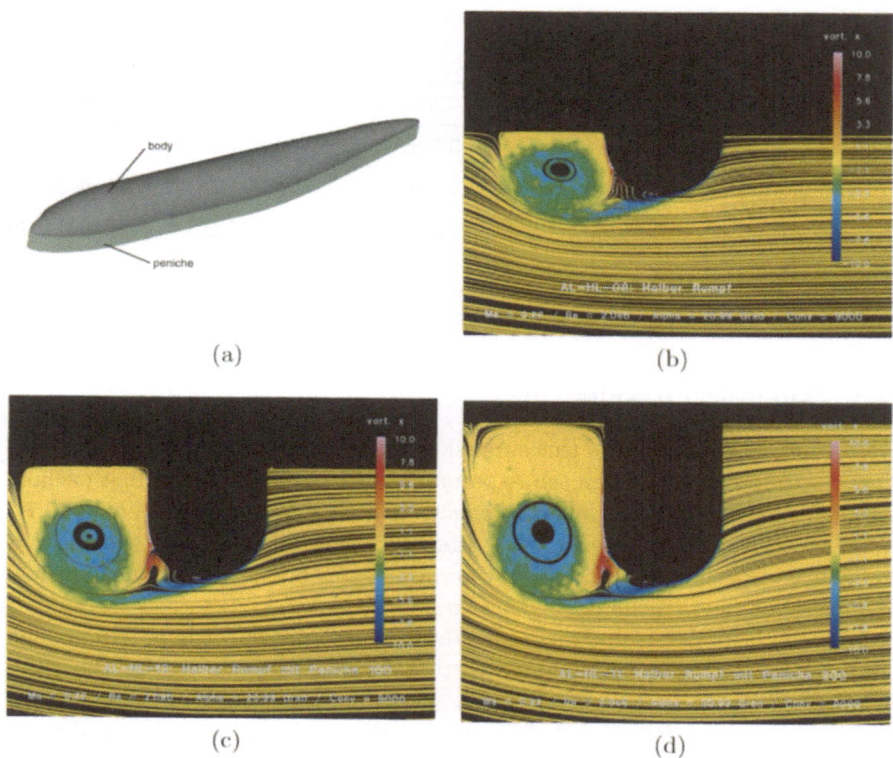

Fig. 3. (a) fuselage configuration with peniche, Streamlines in a plane at 0.9 of fuselage length, $\alpha = 20.99°$, colors: vorticity, (b) half-fuselage without, (c) peniche 100 mm, (d) with peniche 200 mm.

fuselage models. A higher peniche increases the displacement in the flow above the rear fuselage, and hence the flow-speed over the fuselage is also increased. If in addition to the isolated fuselage a wing is although considered, the local angle of attack at the area of the wing-root is increased. This results in a earlier lift breakdown but a higher lift at same angle of attack for a semi-span model in comparison to a full-span model. These findings correlate to the results obtained in the experiments (Fig. 1).

Vortex-Motion: An important aspect of the wing-root aerodynamics is the development of the slat-horn vortex and the level of interaction with the fuselage vortex. Moreover, it is of interest, if the fuselage vortex is able to destroy the slat-horn vortex above the wing.

To reduce the required computer resources for this investigation the wing of the ALVAST high lift configuration is cut-off at the middle of the inboard flap (Fig. 4). To guarantee an appropriate resolution of the vortices, two field adaption steps have been carried out at each angle of attack. In Fig. 4 the

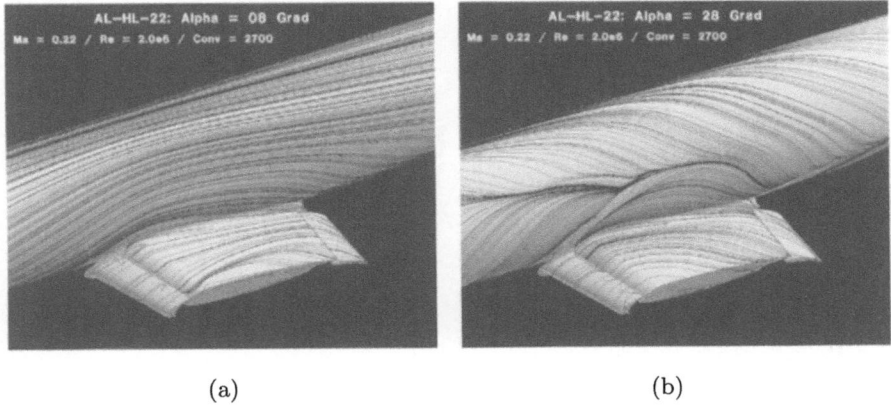

(a)   (b)

Fig. 4. Streamlines on the surface, attachment-lines: green, separation-lines: red, (a) $\alpha = 8°$, (b) $\alpha = 28°$

configuration is shown as an example of the study at two angles of attack, $\alpha = 8°$ and $\alpha = 28°$, with streamlines on the surface, which are comparable to oil-flow pictures in the wind-tunnel.

One problem caused by the shortened wing is, that the influence of the tip-vortex being relatively near to the wing-root cannot be disregarded. In Fig. 4 the streamlines on the wing drift towards the wing tip with increasing angle of attack. This is an indication that the tip vortex rolls over the wing, rotating from outboard to inboard. It induces a downwind over the whole upper-side of the wing and the fuselage in this area and reduces the flow separation trend.

The impact of the slat-horn vortex can be detected in Fig. 4(b) by the surface streamlines. Some lines seems to "cross" the rest of the streamlines, over the wing/fuselage junction beginning at the slat-horn. With increasing angle of attack the vortices strengthen up and are running at an higher angle over the wing-root. The fuselage vortex (see section 2.5, Peniche) is visible as converging stream lines over the wing/fuselage junction. It appears first above the fuselage near the wing-junction, because of a higher angle of attack induced in this area. Likewise the slat-horn vortex it strengthen up with increasing angle of attack. The interaction between the fuselage- and slat-horn vortex can been seen for $\alpha = 28°$ at 30% of wing chord length. The slat-horn vortex is weakened by this interaction, but the path on the surface is not changed. In contrast to this the separation line of the fuselage vortex is deformed.

Summarizing the vortex topology is changed corresponding to angle of attack. At higher angles of attack there is a interaction of the fuselage- and slat-horn-vortices. For the later simulations a prober resolution of the fuselage vortex and associated with it the peniche is necessary.

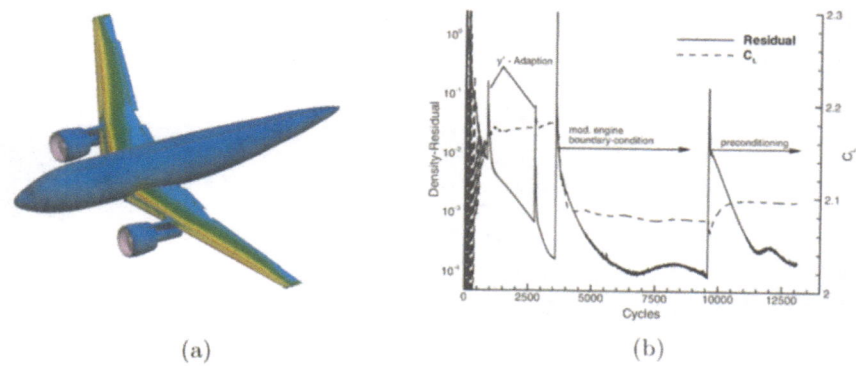

(a)   (b)

Fig. 5. (a) ALVAST high lift configuration with CRUF-engine, pressure coefficient c_p on the surface , (b) convergence-history of density-residual and lift-coefficient for ALVAST high lift configuration with CRUF.

Nacelle-Vortex: The simulation of the flow around realistic high lift aircraft configurations with a high level of detail is one of the near future objectives of CFD in aeronautical applications. In this context, the capabilities of the TAU-code with respect to the topic of engine/airframe integration are investigated on the ALVAST high lift configuration with engines (Fig. 5(a)). Special attention is paid to possible reductions of the maximum lift angle by means of dominating three-dimensional effects due to the engine installation, [14].

In principle the further increase in complexity would require enormous number of grid points leading to a currently unacceptable requirement of computational resources. Therefore the number of grid points is limited to $10 \cdot 10^6$ points in order to stay within reasonable limits. Consequently, the resolution is poorer than that of the preceding studies.

The considered CRUF-engine simulator is a "Counter Rotating Ultra-High-Bypass Fan", representing an advanced engine concept, with a very high mass flow portion passing the outer, cold fan circuit in order to improve the propulsion efficiency of the engine. As an adverse effect, the increased engine diameter causes aerodynamic losses due to the negative influence on the wing aerodynamics. To enable a realistic experimental simulation, the engine is powered by compressed drive air in the wind tunnel. This jet effect is modeled numerically by special inflow and outflow conditions, neglecting the inner parts of the engine. The jets are characterized by total pressure and temperature boundary conditions.

The ALVAST with CRUF engine is computed at an angle of attack of $\alpha = 12°$ in take-off configuration. The setting of the engine is about 90% maximum thrust. To increase the simulation quality preconditioning is used. In Fig. 5(b) the convergence-history is shown. Despite the high complexity a satisfying convergence of the main flow residual of about four magnitudes is achieved with the TAU-code.

In the isobar distribution a spot at the inboard side of the pylon is detected on the slat surface. This spot originates from a vortex crossing the upper-side of the wing. In Fig. 7(b) vorticity-cuts and stream lines are depicted, which show the vortex shedding from the nacelle trailing edge, rolling up and crossing the slat and wing upper-side. This vortex is called "nacelle-vortex".

A more detailed analyses reveals, that the vortex originates form the shear-layer of the outer boundary-layer and the jet behind the trailing-edge. The shear-layer is rolling up from bottom to top forming the nacelle-vortex at the inboard side of the engine, Fig. 7(b). Outboards a similar process occurs, but the vortex is much weaker and its path runs between pylon and wing lower side. The strengthening of the inboard nacelle-vortex is caused by the fuselage influence (channel-effect between engine and fuselage) and the closer coupling between engine and wing (wing-sweep). A weak flow separation can be found between pylon and nacelle, but it plays only a minor role at this angle of attack.

At present wind-tunnel tests with the ALVAST model and the CRUF simulator are under progress to verify the numerical simulations, especially the nacelle-vortex. First results of the PIV-measurements (PIV: particle image velocimetry, a method to measure the speed of the fluid in one plane of the flow field by laser light sheet) prove the existence of the nacelle-vortex at the same position as in the simulations, verifying the numerical result.

Preconditioning: A problem linked with high lift flows is the mix of compressible and incompressible flow regions. The difficulty in solving the compressible equations for low Mach-numbers are associated with the large disparity of the acoustic wave speed and the waves convected with the fluid speed. The application of preconditioning changes the eigenvalues of the system of compressible flow equations to remove this disparity in areas of low Mach-numbers [15]. Preconditioning is implemented in the TAU-code, but has yet not been tested at complex high lift configurations. The objective of this subtask is to test the applicability and efficiency of preconditioning for such cases. Furthermore an improved resolution of flow features like vortices is expected.

The first case is the fuselage configuration from Section 2.5 (Peniche) featuring the strong vortex above the fuselage and a comparatively simple grid. In the stream-lines (Fig. 6), which are shown again in a cut-plane upright to the fuselage-axis at nearly 90% fuselage length, an increased flow-resolution and vortex strength with increasing preconditioning occurs. The second case is the ALVAST high lift configuration with the CRUF-engine. In Fig. 7 the effect of precontioning on this case is depicted: an outboard movement of the vortex-core and a strengthening of the vortex with preconditioning is visible.

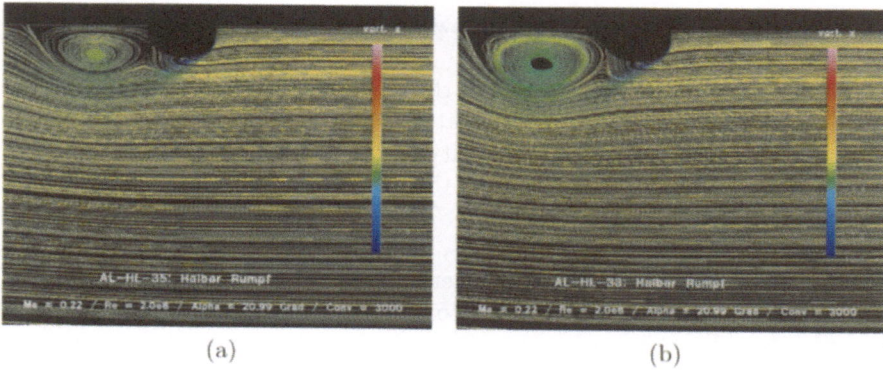

(a)   (b)

Fig. 6. Streamlines in a plane, vorticity (a) preconditioning off, (b) preconditioning on.

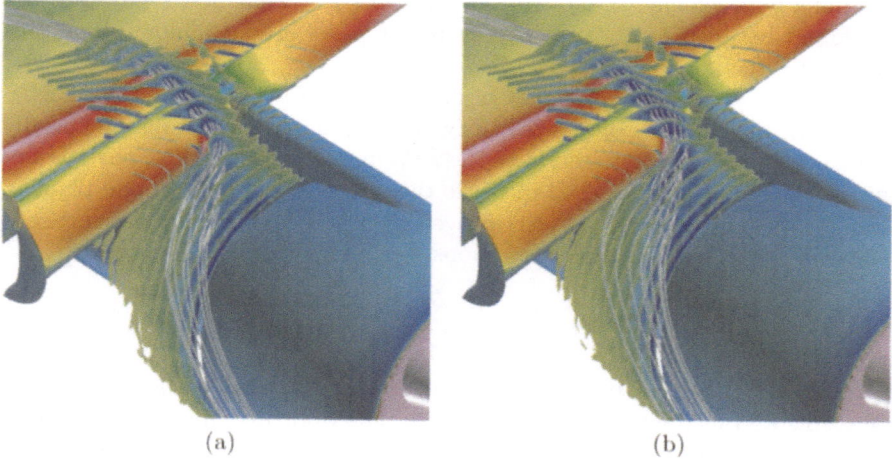

(a)   (b)

Fig. 7. ALVAST high lift with CRUF-engine, pressure coefficient c_P on the surface, vorticity in cut-planes and stream-lines,(a) preconditioning off, (b) preconditioning on.

Both examples demonstrate the applicability of preconditioning for high lift flows with the TAU-code. With the same grid resolution a improved representation of the flow field is achieved.

3   Optimization of High Lift Devices

3.1   Overview

The present reliability of the numerical analysis methods for high lift flows forms the basis to develop and apply automated high lift system design methods. This type of numerical optimization offers the capability to improve the aerodynamic performance, while maintaining or further reducing the system complexity in a shorter turn-around time than that of conventional design approaches. Nevertheless, the development and application of numerical optimization in the framework of high lift systems is in an initial stage.

For two-dimensional flows such a design method has been developed, coupling numerical analyses with optimization strategies for high lift design of multi-element airfoils [16]. Using this method promising results have been obtained for a wide range of configurations. A major limit of further research activities is the high requirement of computational time. The runtime of a two-dimensional optimization amounts to about 150 CPU-hour on an NEC-SX4 vector computer.

Aside from attempts to increase the efficiency of the numerical methods, the next development step is the extension of the design method to 3D high lift flows. The increase in computational time may be alleviated to a certain degree by performing parallel computations of the flow field. The objective of this second part of the research project is therefore to adapt the design method to 3D configurations and to demonstrate the capabilities at the aerodynamic optimization of a generic 3D high lift wing. The design parameters are optimal device positioning and sizing, accompanied by wing planform variations. The analysis module is the above described flow solver TAU.

3.2   Sensitivity analysis

Preparing optimization it is essential to be sure that the calculated flow quantities react probably on geometric changes correctly. This can only be verified in comparison with experimental data. A test case for a variation of high lift devices is the NASA Trapwing [17].

The model used is a simplified high lift wing-body configuration, that has been measured by NASA in two wind tunnels for a range of different device settings. Due to the large scale of the model, the interference with the wind-tunnel walls was too large to be simply corrected. For this reason the first step for the approvement of CFD is a comparison of in-tunnel and free-stream calculations. The grids therefore consists of 6 million points each. Figure 8 shows a comparison of the flow field for these two cases. It can be observed, that there is a significant pressure signature on the wind-tunnel walls.

The corresponding convergence of the TAU-code can be seen in Fig. 9. A converged solution is obtained after 2.000-2.500 multigrid cycles using a 4-level W-cycle and a 3-stage Runge-Kutta time integration. The CPU-time needed is 53.2 h for the in-tunnel case and 36.68 h for the free air case, computed on the NEC-SX5 using 4 processors.

To evaluate the accuracy of flow simulation an assessment of the dependency of the lift coefficient on the angle of attack is shown in (Fig. 9(c)). The experimental lift-slopes for the wind-tunnel/free air comparison and for a variation of the flap-gap are depicted. It can be seen, that with the present grids the differences of free air and in-tunnel flow is not sufficiently cov-

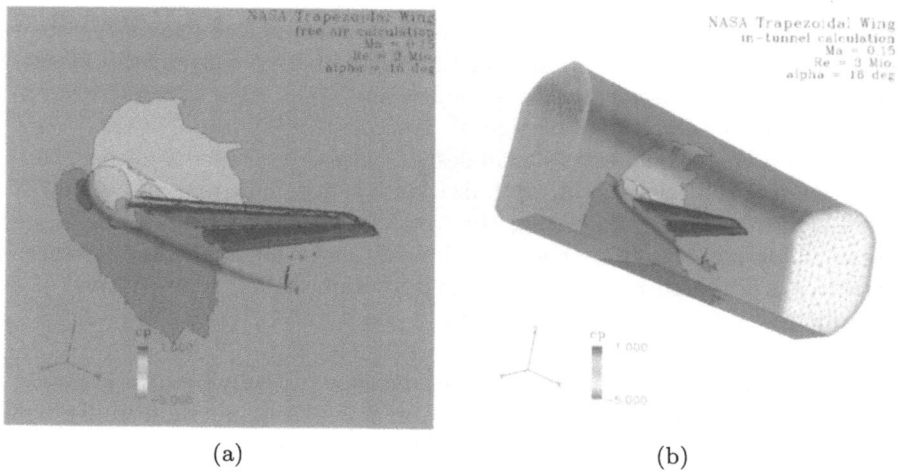

(a)   (b)

Fig. 8. Flow field around the NASA Trapwing configuration for the calculation: (a) in free air, (b) in the wind-tunnel

ered. This might be due to the resolution of the surface grids. Although the grid contains 6 million nodes, the surface is only resolved by approximately 400.000 triangles. The comparison of the different flap-settings shows, that the gap sensitivity is not reproduced properly, although the calculated values are more close to the experiments than in the prior mentioned in-tunnel calculation.

3.3   Extension of present tools – grid generation

At an early point it could be foreseen, that hybrid grids are not applicable for a robust run in an optimization environment. Either they are too time consuming due to high point numbers, or the results achieve not the necessary accuracy. The main reason is the use of surface triangles and extruding prisms in hybrid unstructured grids. As seen in the prior described first part of the NHLRes-project a grid for an adequate resolution of the flow physics for a high lift wing consists of about 10 million grid points. In contrast to this figure, the targeted resolution for the optimization is between 2–4 million grid points.

The envisaged method to overcome this problem is the use of semistructured grids. In contrast to typical hybrid grids the viscous walls are gridded by quadrilaterals, and the extension into the flow field for the resolution of the boundary layer is done by hexaedral elements in spite of prisms, generated by the structured DLR grid generator MEGACADS. The grid generator has accordingly been extended to generate surface triangulations, to convert structured grid blocks into an unstructured database, and to generate the

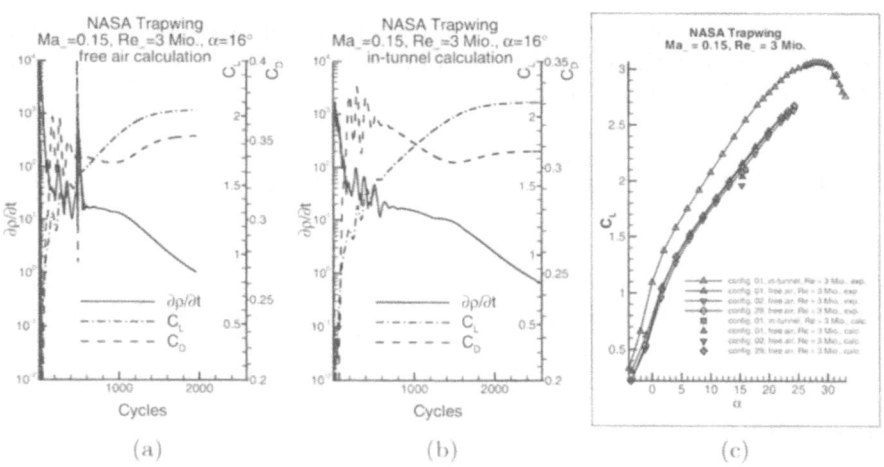

Fig. 9. Convergence of the calculation of the NASA Trapwing configuration: (a) in free air, (b) in the wind-tunnel, (c) Polars of the NASA Trapwing configuration.

interface zone between hexaedral and tetrahedral elements. Present work is concerned with the implementation of 3D volume triangulation algorithms.

4 Summary

The first part the project NHLRes has revealed reasons of the differences between the semi-span and full-span model tests at high angles of attack. Further on, the vortex motion of the slat-horn- and body-vortex and the interaction between these vortices is discussed at different angles of attack. The ability to verify sophisticated 3D-flow features with CFD-tools for complex configurations using CFD-tools in the forefront of wind-tunnel testing was likewise demonstrated for the example "nacelle-vortex". Moreover the improvement of the solution resolution with preconditioning is shown.

Having there the above topics an improved understanding of wing-root aerodynamics at high-lift condition is achieved. Based on this thorough CFD verification activity, the study of configurational variations, according to existing detailed wind tunnel data appears now feasible. Nevertheless, effects of hysteresis in 3D simulations and the influence of grid-adaptation will also have to be addressed.

Concerning the optimization activities preparing studies on the baseline configuration with and without consideration of wind tunnel walls have been carried out. These investigations underline the need for increased accuracy of the flow computations within the optimization loop. Especially a proper resolution of the near wall area would require an unacceptable computational effort with the present approach. To circumvent this problem other types of surface grids based on hexaedral volume cells, that allow for higher accuracy with less points have to be considered. In addition, time savings are expected by using restart options and preconditioning have to be taken into account for future optimization activities.

Acknowledgment

The numerical simulations reported in this paper were performed at the High-Performance Computing-Center Stuttgart. The authors would thank the HLRS for the generous granting of their computational facilities for the project NHLRes.

References

1. Rudnik, R.; Ronzheimer, A.; Raddatz, J.: "Numerical Flow Simulation for a Wing/Fuselage Transport Configuration with Deployed High Lift system" in Notes on Numerical Fluid Mechanics, Vol. 72, pp. 363–370, Vieweg-Verlag, Braunschweig/Wiesbaden, 1999.
2. Rudnik, R.; Melber, S.; Ronzheimer, A.; Brodersen, O.: "Three-Dimensional Navier-Stokes Simulations for Transport Aircraft High Lift Configurations." Journal of Aircraft, Vol. 38, pp. 895–903, 2001.
3. Melber, S.; Rudnik, R.; Ronzheimer, A.: "Structured and Unstructured Numerical Simulation in High Lift Aerodynamics." Workshop on EU-Research on Aerodynamic Engine / Aircraft Integration for Transport Aircraft, 26–27 September 2000, DLR Braunschweig, 2000, pp. 13-1 – 13-10.
4. Melber, S.: "3D RANS Simulations for High Lift Analysis of Transport Aircraft Configurations." Notes on numerical fluid mechanics, Volume 77, Springer Verlag, Berlin, Heidelberg, New York, 2002, pp. 27–34.
5. Rogers, S.E.; Roth, K.; Cao, V.Hoa; Slotnick, J.P.; Whilock, M.; Nash, S.M.; Baker, M.D.: "Computation of viscous Flow for a Boeing 777 Aircraft in Landing Conf." AIAA paper 2000-4221, 2000.
6. Mavriplis, D.J.: "Parallel Performance Investigations of an Unstructured Mesh Navier-Stokes Solver" ICASE Report No. 2000-13, March 2000.
7. Kiock, R.: "The ALVAST Model of DLR." DLR IB 129-96/22, 1996.
8. Kallinderis, Y.: "Hybrid Grids and Their Applications." Handbook of Grid Generation, CRC Press, Boca Raton / London / New York / Washington, D.C., pp. 25-1 – 25-18, 1999.
9. Puffert-Meissner, W.: "ALVAST Half-Model Investigations in the Low-Speed wind-tunnel Braunschweig." DLR IB 129-95/11, 1995.
10. Puffert-Meissner, W.: "ALVAST Half-Model wind-tunnel Investigations and Comparison with Full-Span Model Results." DLR IB 129-96/20, 1996.
11. Kroll, N.; Rossow, C.-C.; Becker, K.; Thiele, F.: "MEGAFLOW – A Numerical Flow Simulation System." 21st ICAS congress, 1998, Melbourne, 13.9.– 18.9.1998, ICAS-98-2.7.4, 1998.
12. Earnshaw, P.B.; Green, A.R.; Hardy, B.C.; Jelly, A.H.: "A Study of the use of Half-Models in High-Lift Wind-Tunnel Testing." AGARD CP 515, Paper 20, 1993.

13. Milholen II, W.E.: "A Design Methodology for Semi-Span Model Mounting Geometries." AIAA-paper 98-0758, 1998.
14. Haines, A.B.: "Scale Effects on Aircraft and Weapon Aerodynamics." AGAR-Dograph 323, 1994, pp. 63.-65.
15. Radespiel, R.; Turkel, E.; Kroll, N.: "Assessment of Preconditioning Methods." DLR IB 95/29, 1995.
16. Wild, J.: "Numerische Optimierung von zweidimensionalen Hochauftriebssystemen durch Lösung der Navier-Stokes-Gleichungen" PhD-Thesis, TU Braunschweig, published as DLR-FB 2001-11, 2001.
17. Johnson, P.L.; Jones, K.M.; Madson M.D.: "Experimental Investigation of a Simplified 3D High Lift Configuration in Support of CFD Validation" AIAA Paper 2000-4217, 2000.

Reactive Flows

Prof. Dr. Dietmar Kröner

Institut für Angewandte Mathematik, Universität Freiburg
Hermann-Herder-Str. 10, 79104 Freiburg

By making use of the computational power provided by massively parallel systems, it is now possible to perform numerical simulations and to study special effects for reactive flows, based on detailed chemical reaction mechanisms, at least in two spatial dimensions. The main progress obtained by the users of the HLRS in the group "Reactive Flows" during the last period concerns the extension and improvement of the modelling, like thermodiffusion, lattice Boltzmann approach and two-phase flows.

Furthermore the users focus on the comparison of the efficiency of different hardware platforms, including clustered systems with high-performance network technologies.

Lange presents some new experiments concerning the direct numerical simulation (DNS) of reactive flows with detailed models for chemical reactions and molecular transport. The simulation has been done with his massively parallel DNS code. This code has already been discussed in the previous issues of the proceedings of the HLRS. Now Lange uses this code as a benchmark for two large hardware clusters equipped with most modern processors and network technology. The processors are arranged in dual-nodes. The influence of the dual nodes on the parallel efficiency is discussed and the performance of the clusters, with single and dual nodes, is compared with the CRAY-T3E system. In the second part of the paper he uses the DNS code for a detailed study of two different premixed flames. It turns out that the influence of combined effects of curvature and preferential diffusion to the chemical composition in curved flame fronts, can be studied on the basis of the data, obtained by the high-resolution experiments and a sufficiently accurate modelling.

In competition to classical CFD (computational fluid dynamics) Zeiser et al. use the lattice Boltzmann method for the detailed simulation of transport in reacting turbulent flows for applications in complex geometries in 3D, like catalytic reactors or adsorption columns. They apply a Monte Carlo method to synthetically generate realistic packings of spherical particles in a tube as models for such reactors. For the flow simulation they numerically solve a transport equation for the density distribution function N_i, which allows the computations of the macroscopic flow quantities, i.e. the density, velocity, and pessure. In addition to the results obtained by the author during the previous period, they are now able to analyse local inhomogenities in the flow properties. In particular, they proved by numerical experiments that the dissipation, caused by the deformation terms, cannot be neglected, as it is done by several other contributions in this field. The major part of the lattice Boltzman code is vectorized. In typical production runs, about 2–2.55 GFlops per CPU are obtained on a NEC SX-5e, that corresponds to 50 % up to 65 % of the peak performance. For the parallelization the shared memory tools have been used. In a series of numerical experiments, for which the number of processors was gradually increased keeping the total work constant, almost linear speedups and efficiencies higher than 90 % are achieved. This was already shown during the last period. Now they reported on some comparisons concerning the performance of the code between different hardware platforms, including the NEC SX-5e and the HITACHI SR8000-F1.

Stroehle et al. consider the 3D, turbulent reacting flow in a coal-fired utility boiler consisting of the continuous gas phase and, in addition to the project of the previous period, a dispersed phase of the coal. For the simulation of the flow they couple a Lagrangian particle model for the discrete phase of the coal with an Eulerian approach for the gas phase with the radiation transport equation and further transport equations for the gas species. The interaction of the particles with the gas phase are modelled by additional source terms in the Eulerian transport equations. The authors prefer the Lagrangian approach since they have seen in other numerical experiments, that the accuracy of the Eulerian code lacks in the near burner regions, where the particle load and the velocity difference between the two phases are very large. For the turbulence, they use the standard k, ϵ-model, neglecting the influence of the particles on to the turbulence. The simulation of the flow of a real boiler of a height of approximately 40 m was performed on the NEC SX-5 with four processors. The underlying body-fitted grid consists of approximately 450 000 cells. The code has been justified by a comparison of the computed numerical data with some measurements of velocity, temperature and gas concentration. The influence of the air flows on the combustion and the location of the flame front are predicted with high accuracy.

Massively Parallel DNS of Flame Kernel Evolution in Spark-Ignited Turbulent Mixtures

Marc Lange

High-Performance Computing Center Stuttgart (HLRS), Stuttgart University
Allmandring 30, D-70550 Stuttgart, Germany
E-mail: lange@hlrs.de

Abstract. A parallel code for the direct numerical simulation (DNS) of reactive flows with detailed models for chemical reactions and molecular transport has been used as a benchmark on two large commodity hardware clusters with current state-of-the-art processor and network technologies. Up to 180 Alpha EV68 833 MHz have been used on the first system and up to 400 AthlonMP 1.4 GHz CPUs have been used on the other one. Both clusters comprise dual-nodes and Myrinet-2000 interconnect. The influence of the dual-nodes on the parallel efficiency is discussed and the performance of these clusters is compared with Cray T3E systems. The other part of the paper deals with the application of DNS to study premixed flames emerging after induced ignition of turbulent mixtures. The discussion focusses on the temporal evolution of the heat-release rate. In addition to the analysis of flames expanding in homogeneous turbulent mixtures, first DNS results of a flame kernel evolving in a turbulent mixture with spatial inhomogeneities of the equivalence ratio are presented.

1 DNS of Turbulent Reactive Flows

Energy conversion in numerous industrial power devices like automotive engines or gas turbines is still based on the combustion of fossil fuels. In most applications, the reactive system is turbulent and the reaction progress is influenced by turbulent fluctuations and mixing in the flow. The understanding and modeling of turbulent combustion is thus vital in the conception and optimization of these systems in order to achieve higher performance levels while decreasing the amount of pollutant emission.

In the last several years, direct numerical simulations (DNS), i.e. the computation of time-dependent solutions of the compressible Navier-Stokes equations for reacting ideal gas mixtures (see e.g. [1–3]), have been one of the most important tools to study fundamental issues in turbulent combustion. Due to the broad spectrum of length and time scales apparent in turbulent reactive flows, a very high resolution in space and time is needed to solve this system of equations. To be able to perform DNS of reactive flows including detailed chemical reaction mechanisms and a detailed description of molecular transport, it is necessary to make efficient use of HPC-systems. The computation of the chemical source-terms and the multicomponent diffusion velocities are the most time-consuming parts in such DNS. Therefore, almost all DNS carried out so far have been (at least) restricted to the use of simplified models (e. g. one global reaction and equal diffusivities) or to two-dimensional simulations. Even with these restrictions it is crucial to make efficient use of the computational power provided by parallel supercomputers to be able to carry out DNS of reactive flows in acceptable time.

2   Structure of the Parallel DNS Code

A code has been developed for the DNS of reactive flows on parallel computers with distributed memory using message-passing communication [4–6]. In favor of being able to include detailed models for the chemical reaction kinetics and the molecular transport, only two-dimensional simulations are performed. The spatial discretization uses a finite-difference scheme with sixth-order central-derivatives, avoiding numerical dissipation and leading to very high accuracy. The integration in time is carried out using a fourth-order fully explicit Runge-Kutta method with adaptive time-stepping. The control of the timestep is based on (up to) three independent criteria: A Courant-Friedrichs-Lewy (CFL) criterion and a Fourier criterion for the diffusion terms are checked to ensure the stability of the integration and an additional accuracy-control of the result for one or more selectable variables is obtained through time-step doubling.

Due to the fully explicit formulation there are only local data dependencies, which leads to a straightforward parallelization based on a regular two-dimensional domain-decomposition. After the initial decomposition, each processor controls a rectangular subdomain of the global computational domain. By using halo-boundaries, an integration step on the subdomain can be carried out independently from the other nodes. After each integration step, the values in the three points wide halo-regions are updated by point-to-point communications with the neighboring nodes. MPI is used for the communication.

Besides the computations on clusters discussed in this paper, the main platform for our production runs have been Cray T3E systems. Table 1 lists the speedups and corresponding parallel efficiencies for a benchmark computation on a Cray T3E-900 with a fixed size of 544×544 grid points, which is a size similar to that of some production runs [7]. The average performance per PE within this simulation (including I/O) for the computation using 64

Table 1. Scaling behavior on a Cray T3E-900 for a simulation with 9 species and 37 reactions on a 544^2 points grid

# PEs	1	4	8	16	32	64	128	256	512
speedup	1.0	4.3	8.1	15.9	30.5	57.9	108.7	189.0	293.6
efficiency / %	100.0	106.6	100.7	99.2	95.3	90.4	84.9	73.8	57.4

processors was 86.3 MFLOP/s. However, the parallel efficiencies obtained in such a benchmark of fixed size are often dominated by the dependency of the single-node performance on the problem size. For a scaled problem with a constant load per PE of 32×32 grid points, a speedup of 197 was achieved using 256 processors corresponding to an efficiency of 79 % with an older version of the code using PVM for the communication [6]. As the MPI-version of our code performs better than the one using PVM [6] and the typical load per PE in production runs is higher than in this test, scaled parallel efficiencies for production runs are typically well above 80 %. The main observation regarding the scaling behavior of our application on the T3E is however, that when increasing the number of processors while keeping the load per processor constant, only a very small decrease in the parallel efficiency occurs above 16 processors, which is the first power of two for which there are processors that have to exchange boundary values with four neighboring processors. Above 16 processors there are always such inner nodes and for a scaled benchmark, on such an inner node the same local computations and the same communication operations between this node and the ones controlling the neighboring subdomains of the computational grid are performed independently from the number of processors used. At least on a Cray T3E using up to 512 processors, global communication and I/O do not play a significant role for the run-time of our application.

3   Performance on Clusters

A big trend in parallel computing today is the use of clustered systems. By using commodity hardware for the nodes as well as for the network, clusters of PCs can provide an excellent performance/price ratio – at least for problems which they are well suited for. Therefore, there is a broad interest in the performance behavior of real world applications on such clusters.

We have studied the performance of several clusters with different types of nodes and network hardware for our application. As a starting point of this investigation, benchmarks for different problem sizes using one and 16 processors of the following clusters have been performed:

Alpha-Myrinet (Alpha-M) Nodes: 64 Compaq DS10 workstations, each having one 466 MHz Alpha 21264 (EV6) processor (2 MB cache) and 256 MB RAM. Interconnect: Myrinet.
Dual-Pentium-Myrinet (DualP3-M) Nodes: 8 IBM dual-processor Netfinity servers, each having two 600 MHz Pentium III processors and 896 MB RAM. Interconnect: Myrinet.
Pentium-Ethernet (P3-E) Nodes: 16 PCs with 660 MHz Pentium III processors and 512 MB RAM. Interconnect: 100 MBit Ethernet.

The results of this study are discussed in detail in [8]. Main results of this study were:

- For problem sizes with per processor loads which are typical for production runs of our application, the single-processor performance of the Pentium III systems is between 10 and 25 percent higher than that of a Cray T3E-1200 PE. The Alpha EV6 is faster than the Cray T3E-1200 PE by a factor between 2.2 and 2.4.
- As can be seen from Table 2, all clusters deliver a reasonable parallel efficiency for 16 processors.
- An analysis of the time spent for communication and synchronization shows a high sensitivity of the efficiency of the intra-node parallelization in the case of using dual-nodes. Due to this problem, the Dual-Pentium3-Myrinet cluster performs even worse than the Pentium3-Ethernet cluster for some problems (see Table 3).

Table 2. Scaled parallel efficiency (in percent) using 16 processors for two different loads per processor

Problem Size per Processor	50×50	100×100
T3E-1200	91.0	92.0
Alpha-M	80.5	90.4
DualP3-M	83.0	81.1
P3-E	77.5	85.7

Table 3. Time for communication and synchronization t_c/s for 10 timesteps using 16 processors

Problem Size per Processor	50×50	100×100
T3E-1200	0.80	2.52
Alpha-M	0.81	1.43
DP3-M (-o)	1.36	6.02
P3-E	1.82	4.30

The main questions arising from this investigation concern the efficiency of using dual-nodes and the scaling behavior for large configurations. These are addressed by measurements performed on two large clusters with current state-of-the-art processor and interconnect technology:

Dual-Alpha-Myrinet2000 (API-M2K) Nodes: 96 API NetWorks CS20 1U servers, each having two 833 MHz Alpha 21264B (EV68) processors (4 MB L2 cache) and 2 GB RAM. Interconnect: Myrinet 2000.

Dual-Athlon-Myrinet2000 (HELICS) Nodes: 256 PCs, each with two 1.4 GHz AMD AthlonMP processors and 1 GB RAM. Interconnect: Myrinet 2000.

Access to the first system has been granted by Cray Inc., the latter system has been installed in March 2002 by the Interdisciplinary Center for Scientific Computing (IWR) of Heidelberg University together with the University Computing Centers in Heidelberg (URZ) and Mannheim (RUM) and is currently one of the most powerful PC-clusters worldwide (HEidelberger LInux Cluster System, see http://helics.iwr.uni-heidelberg.de/). Table 4 lists the performance of one processor of each of these two systems normalized with the problem size relative to the performance of a Cray T3E-1200 PE for the 50^2 grid points problem. Both processors perform at a similar level, but the decrease of the performance with increasing problem size is much stronger on the Alpha than on the Athlon system. Compaq Fortran 1.0-920 has been used on the API-M2K. The benchmarks presented in this paper have been carried out during a time when the HELICS system had just been installed and was still in its testing phase. As at that time no suitable compiler was available on the HELICS, we used another IA32-Linux-cluster (Volvox, see http://www.hlrs.de/hw-access/platforms/volvox/) as a development platform. The Intel Fortran Compiler ifc 6.0β has been used on this system to produce the binaries for the HELICS.

Table 4. Monoprocessor performance for different problem sizes related to the performance of the Cray T3E-1200 solving the 50×50 problem

Problem Size	50×50	100×100	200×200	400×400
API-M2K	4.40	4.18	3.67	2.69
HELICS	4.22	3.87	3.84	3.69

To be able to distinguish the influence of the intra-node parallelization from the influence of the network performance, several of the test cases have been run using both processors of $p/2$ nodes as well as using only one processor on each of the p nodes used, where p is the number of processors to be used in the considered test case. In the following, this is denoted by the suffixes -dual and -single respectively.

In Table 5 the parallelization overhead time t_c, i.e. the time spent for communication and synchronization, is listed for cases with 2, 16, and 64 processors. In our benchmark, symmetric boundary conditions are used in the x-direction and periodic boundary conditions in the y-direction. In the cases using two processors with a load of $N \times N$ grid points per processor, the global grid has $2N$ points in the y-direction and N points in the x-direction. Hence the boundary between the two subdomains is parallel to the x-axis and due to the periodicity in the y-direction each of the two processors has to exchange the values at two sides of its subdomain-boundary with the other processor. In the cases with 16 and more processors of course all processors, except for those at the left and right boundaries, have to exchange the values at all four

Table 5. Parallelization overhead t_c/s for 10 timesteps

Number of Processors	2	2	16	16	64
Problem Size per Processor	50×50	400×400	50×50	100×100	50×50
T3E-1200	–	–	0.80	2.52	0.80
API-M2K-single	–	–	–	1.33	0.81
API-M2K-dual	0.35	21.67	0.56	1.68	0.58
HELICS-single	0.19	3.91	0.33	0.85	0.36
HELICS-dual	0.57	36.55	0.75	4.36	0.80

sides of their subdomain-boundaries with the neighboring processors. Thus, the number of messages per processor to be exchanged for the update of the halo regions in the two-processor case is half of that (for the inner processors) in the runs with 16 and 64 processors. The communication performance in the single-processor per node configurations for the Myrinet 2000 on the HELICS is clearly better than the communication performance of the T3E. The tests with two processors in the dual-case again show that for our test-case, the parallelization overhead for the intra-node parallelization strongly depends on the problem-size. The efficiency of the intra-node parallelization is higher on the API-M2K than on the HELICS. In addition, the strong decrease of this efficiency occurs at a larger problem size on the API-M2K as can be seen from the value of t_c in the case with 100^2 grid points on each of 16 processors which is still good for the API-M2K but quite poor for the HELICS.

The computation times for computing ten timesteps in the case with 400×400 grid points (global) are given in Table 6. From these times, speedups for this problem of fixed size can easily be computed, but the resulting values are dominated by the dependence of the single processor performance from the problem size. The computing times however give an overview of the overall performance of these systems for our application.

From the benchmarks presented so far, it is already expected that these clusters provide an excellent platform for our application also for very large configurations. This is confirmed by the parallel efficiencies for a scaled problem with 50×50 grid points per processor listed in Table 7. As the time

Table 6. Overall time (in seconds) for computing ten timesteps on a 400×400 grid

Number of Processors	1	16	64
T3E-1200	419.64	31.55	8.88
API-M2K-single	191.33	9.02	2.43
API-M2K-dual	191.33	9.38	2.41
HELICS-single	139.54	9.16	2.27
HELICS-dual	139.54	12.67	2.71

Table 7. Scaled parallel efficiency (in %) for a load of 50×50 grid points per processor

Number of Processors	16	64	180	400
T3E-1200	91.0	90.6	–	–
API-M2K-single	–	75.3	–	–
API-M2K-dual	76.7	75.9	74.6	–
HELICS-single	85.2	84.0	83.5	–
HELICS-dual	71.7	70.5	69.9	70.1

spent for global communication in our application during the normal integration in time is negligible, the scaled parallel efficiency should stay almost constant for runs with 16 and more processors as it is the case here. There is some global communication involved when gathering data for I/O, but this is not taken into account here as the accumulated time for gathering data and writing it to disk is only about one percent of the total time of a typical production run.

The results of the benchmarks presented clearly show that the performance of a configuration with p single-processor nodes is superior to a configuration with $p/2$ dual-nodes. However, the latter configuration is typically much less expensive and the price/performance ratio is often better for a system with dual-nodes. The main question in this context is therefore often, if it is worthwhile for a system with a given number of nodes to invest into dual-nodes instead of single-nodes. If we compare the computation time for a fixed problem size of 400×400 grid points when using both processors of the dual nodes with that when only one processor per node is used, we find a performance gain of 55.3 % from the additional CPUs for 64 nodes of HELICS (i. e. the run-time is reduced by a factor of $1/1.553$). When using 256 processors on 128 nodes instead of only one processor on each of the 128 nodes for the same case with 400^2 grid points the performance gain is 69.6 %.

3.1   CPU-Times for Production Runs

For the investigation of turbulent flame kernels presented in the next section, several simulations with 704×704 grid points each have been performed using either 256 PEs of the Cray T3E-900 at the HLRS or using 256 processors of HELICS (128 nodes). The computation time for advancing the solution in one such simulation over a physical time of 0.1 ms from $t = 0.4$ ms to $t = 0.5$ ms on the T3E-900 was 14760 seconds. The computation time on HELICS for the same part of another simulation was 3566 seconds. The initial conditions in these two simulations are identical except for the individual realizations of the turbulent flow field from the same statistical parameters. The differences in the turbulent velocity fluctuations cause slightly different sizes of the timesteps between the two simulations, but the difference in the

number of timesteps needed in the two simulations compared is only about
0.1 %. The run-times given above include writing a full solution at two times
during this interval of the run. The time spent for I/O has however been
less than 3 percent in all production runs including some cases with more
frequent savings. The overall physical time in the simulations ranges between
0.6 ms and $t = 1.2$ ms. A typical run-time for one such DNS is thus of the
order of 40 hours on the T3E and about 10 hours on HELICS. In addition,
some larger DNS with 1200×1200 grid points have been performed.

4   Flame Kernel Evolution in Spark-Ignited Turbulent Mixtures

Induced ignition and the following evolution of premixed turbulent flames is
a phenomenon of large practical importance as it occurs e. g. in Otto engine
combustion. DNS studies of this process have been performed using simple
one-step chemistry in a model configuration of an initially uniform premixed
gas under turbulent conditions which is ignited by an energy source in a small
region at the center of the domain [9]. Our investigation extends these studies by using a detailed reaction mechanism including 37 elementary reactions
among 9 species [10] and a detailed transport model including thermodiffusion. The Soret effect (i. e. diffusion caused by temperature gradients) has
often been neglected in simulations of reacting flows with multicomponent
transport, but as we found it to be of importance in onedimensional simulations of induced ignition of laminar $H_2/O_2/N_2$ mixtures, it has been included
in the DNS presented in this paper.

4.1   Homogeneous Turbulent Mixtures

In the simulations presented in this section, a cold $(T = 298\,\mathrm{K})$ uniform
mixture consisting of 23.3 % H_2, 11.7 % O_2, and 65 % N_2 (mole fractions) has
been superimposed with a turbulent flow field computed by inverse FFT from
a von-Kármán-Pao-spectrum with randomly chosen phases. A root mean
square of the velocity fluctuations of $U = 3\,\mathrm{ms}^{-1}$ and an integral length
scale of $\Lambda = 2$ mm have been chosen for the initialization of the flow field.

The mixture is ignited by an energy source in a small round region
$(r = 0.2\,\mathrm{mm})$ at the center of the domain, which is active during the first
15 µs of the simulation. During this time, the mixture at the center of the
domain heats up and radicals are formed. A shockwave is observed which
propagates outwards towards the boundaries of the domain. Non-reflecting
outflow conditions based on characteristic wave relations [11] are imposed
on all boundaries, which allow the shock wave to leave the domain without
disturbing the solution [8]. Above a minimum ignition energy an expanding
flame kernel is observed.

Three DNS have been performed with this set of parameters, two with 704×704 grid points and a computational domain of $16\,\text{mm} \times 16\,\text{mm}$ (Kernel A, Kernel B) and one with a domain size of $27.3\,\text{mm} \times 27.3\,\text{mm}$ discretized by a 1200×1200 points grid (Kernel C).

Figure 1 shows snapshots of one of these turbulent flame kernels (Kernel C) $0.5\,\text{ms}$ and $1.0\,\text{ms}$ after the begin of the simulation. The first row of images shows the temperature and in the second row the mass fraction of OH is shown. At $t = 1.0\,\text{ms}$, an isolated pocket of fuel is observed, which has formed after mutual annihilation of distinct parts of the flame front. The mechanism of pocket formation and burnout is described in more detail in [5], where it has been observed in initially planar premixed hydrogen-air flames evolving in a turbulent flow. In the third row of images, the temporal evolution of vorticity can be seen. The lack of a vortex-stretching mechanism in two-dimensional simulations of decaying turbulence leads to an inverse cascade with growing structures. There is a very strong damping of the turbulence in the hot region of the burnt gas due to the high viscosity. A comparison shows a good qualitative agreement of our DNS with the results of an experimental investigation of turbulent flames performed under similar conditions [12].

An important quantity in turbulent combustion modeling is the reaction intensity. There are several possibilities to define the reaction intensity, e. g. the local heat-release rate

$$\dot{q} = \sum_{\alpha=1}^{N_S} Y_\alpha \dot{\omega}_\alpha h_\alpha \tag{1}$$

can be used for this purpose. Figure 2 shows the temporal evolution of the spatial maximum of the local heat-release

$$\dot{q}_{\text{max}}(t) = \max_{x,y} \dot{q}(x, y, t) \tag{2}$$

for the three turbulent flame kernels and for the corresponding laminar case. The left part of the figure shows the first $30\,\mu\text{s}$ of this evolution. The existence of an ignition-delay, i. e. the time up to the strong increase of $\dot{q}$, is characteristic for chain-radical explosions. During this first time, important radicals are formed, which cause rapid ignition once their concentration is high enough. After the energy source is switched off at $t = 15\,\mu\text{s}$, the turbulence leads to a faster and more intensive ignition in some regions. The maximum heat-release stays also clearly above the one in the laminar case during the subsequent evolution of all three turbulent flame kernels as can be seen in the right part of Fig. 2.

The temporal evolution of the overall heat-release rate, which is obtained by integrating $\dot{q}$ over the computational domain, is shown in Fig. 3. While there are almost no differences during the first $30\,\mu\text{s}$, there is a much stronger increase in the overall heat-release in the subsequent evolution of the turbulent flame kernels. The decline at the end of the curves for Kernels A and B

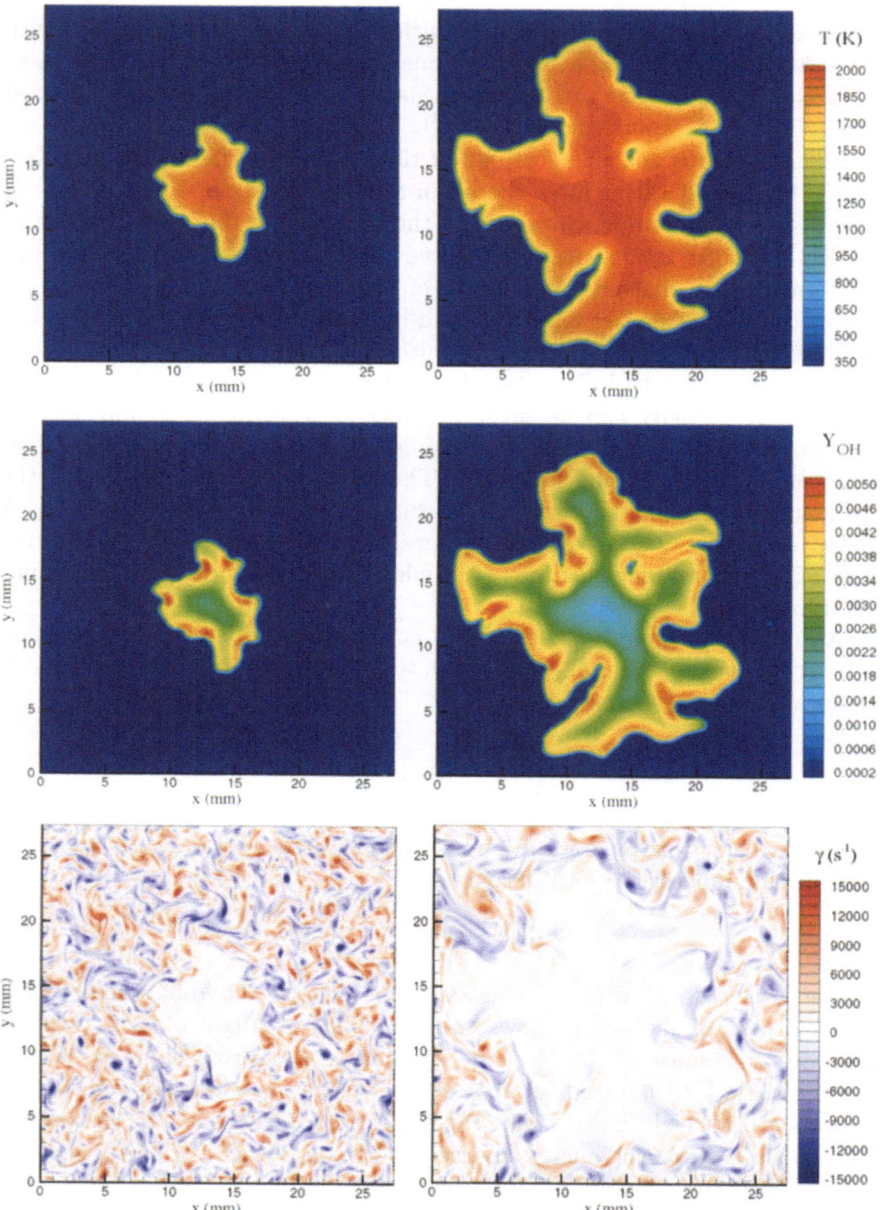

Fig. 1. Temperature, OH mass fraction, and vorticity (top to bottom) at $t = 0.5$ ms (left column) and $t = 1.0$ ms (right column) in a flame (Kernel C) evolving after induced ignition of a homogeneous turbulent mixture

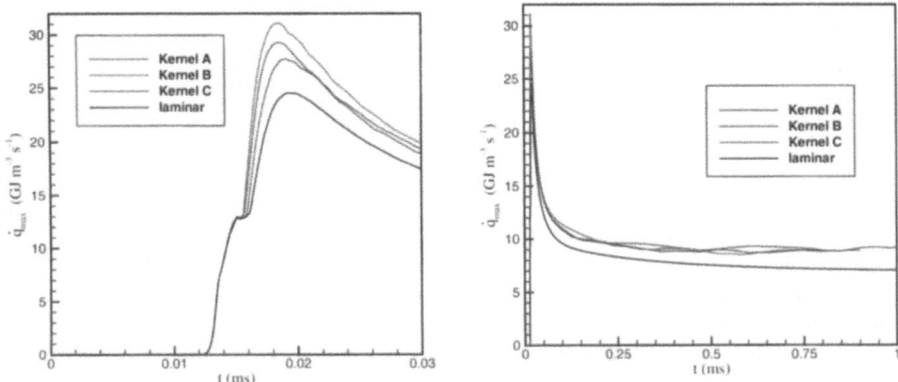

Fig. 2. Temporal evolution of maximum heat-release in three turbulent flame kernels and the corresponding laminar flame

is caused by parts of the flame leaving the computational domain. Figure 4 shows the change of the overall heat-release in time,

$$\frac{\mathrm{d}}{\mathrm{d}t} \int \dot{q}(x, y, t) \mathrm{d}A \quad , \tag{3}$$

as a function of time after the first $200\,\mu\mathrm{s}$. As could already be seen from Fig. 3, the increase of the total heat-release is nearly constant in the laminar case. After the ignition phase, the laminar flame propagates circularly outwards with approximately constant velocity causing an approximately constant increase of flame surface. In the turbulent case, there is an additional flame surface increase due to the wrinkling of the flame caused by the turbulent eddies. On the other hand, mutual annihilation of parts of the flame

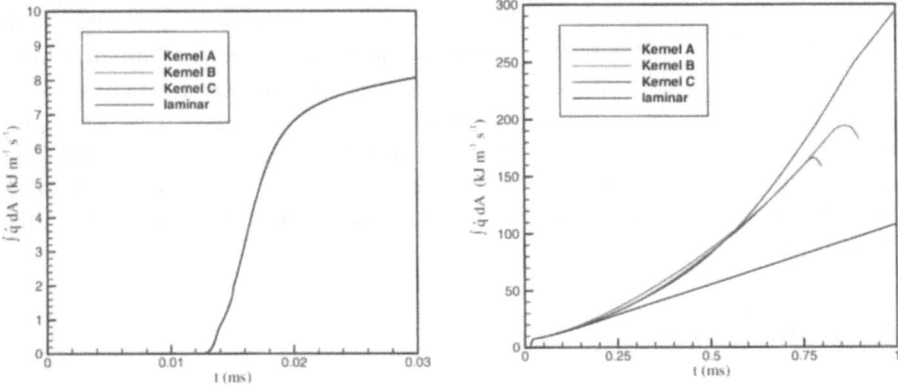

Fig. 3. Temporal evolution of total heat-release in the computational domain for three turbulent flame kernels and the corresponding laminar flame

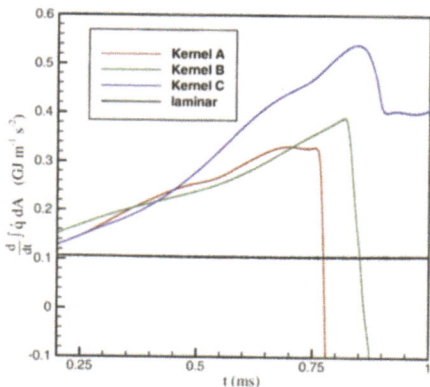

Fig. 4. Change rate of total heat-release in the computational domain over time for three turbulent flame kernels and the corresponding laminar flame

front, like found in regressing cusps or during pocket formation and burnout, leads to a decrease of flame surface. The decline of the increase of overall heat release of Kernel C around $t \approx 0.85$ ms is caused by the mutual annihilation of parts of the flame front behind the pocket which can be seen in Fig. 1.

4.2   Inhomogeneous Turbulent Mixtures

The configuration for the simulations as described in the last section has recently been extended to the more complex situation of flame kernels developing in inhomogeneous turbulent mixtures. A case of practical interest, in which such a situation occurs, are direct injection piston engines. In these systems, fuel and oxidizer are injected separately and partially mix before ignition. The resulting flame propagates into an inhomogeneous mixture with fuel-lean and fuel-rich regions.

As a first test case for flame kernel development in a turbulent mixture with non-uniform stoichiometry, a configuration has been chosen with a mixture with $Y_{H_2} = 0.01$, $Y_{O_2} = 0.19$, and $Y_{N_2} = 0.8$ on the left side and $Y_{H_2} = 0.04$, $Y_{O_2} = 0.16$, and $Y_{N_2} = 0.8$ on the right side of the domain. Compared to the turbulent flame kernels discussed in the last section, a lower root mean square of the velocity fluctuations of $U = 2.2\,\mathrm{ms}^{-1}$ has been chosen while the parameters describing the energy source causing the ignition are the same. Figure 5 shows the fuel mass fraction, the heat release rate and the mass fractions of atomic hydrogen and OH at $t = 0.6$ ms for this simulation.

5   Conclusions

Our previous study of the performance of the DNS code on clusters of commodity hardware [8] has been updated and substantially expanded. A detailed discussion regarding the use of dual-nodes in such systems has been given.

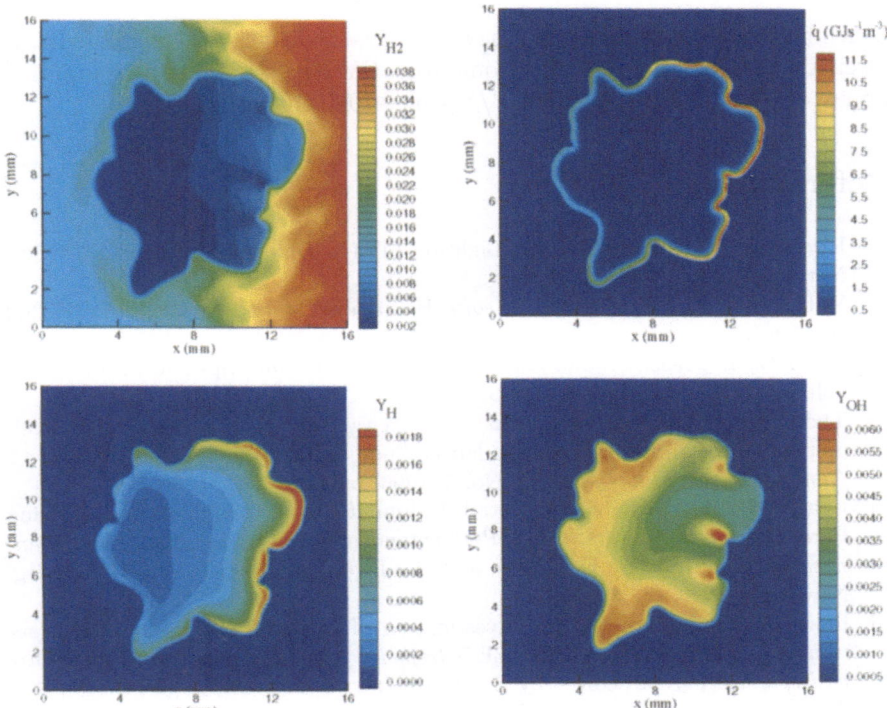

Fig. 5. Fuel mass fraction Y_{H_2}, heat release rate $\dot{q}$, and mass fractions of H and OH at $t = 0.6\,\text{ms}$ after induced ignition of a non-homogeneous mixture under turbulent conditions

The analysis of the scaling behavior has been extended to large configurations with up to 400 processors. Some of the first full large-scale production runs on HELICS, which is currently one of the most powerful clusters worldwide, are included in our performance measurements.

Several DNS of induced ignition and subsequent turbulent flame kernel evolution have been performed. The temporal evolution of the heat-release rate, which is an important parameter in turbulent combustion modeling, has been discussed. In comparison with the corresponding laminar flame, the overall heat-release is clearly enhanced due to the additional flame-surface growth caused by the interaction with the turbulence. The investigation of flame kernel evolution under turbulent conditions is currently extended to cases with spatial inhomogeneities in the chemical composition of the unburnt gas. First results of one such DNS have been presented.

Acknowledgements

The author would like to thank Cray Inc. for providing access to the Dual-Alpha(EV68)/Myrinet-2000 cluster and Dr. M. Wierse for carrying out the benchmark runs on this cluster. The author gratefully acknowledges the Interdisciplinary Center for Scientific Computing (IWR) at Heidelberg for granting him access to the Dual-AthlonMP/Myrinet-2000 cluster (HELICS).

References

1. Bird, R. B., Stewart, W. E., and Lightfoot, E. N., *Transport Phenomena*, Wiley, New York, 1960.
2. Williams, F. A., *Combustion Theory*, Benjamin/Cummings, Menlo Park, 2nd ed., 1985.
3. Warnatz, J., Maas, U., and Dibble, R. W., *Combustion*, Springer, Berlin, Heidelberg, New York, 2nd ed., 1999.
4. Thévenin, D., Behrendt, F., Maas, U., Przywara, B., and Warnatz, J., "Development of a Parallel Direct Simulation Code to Investigate Reactive Flows," *Computers and Fluids*, Vol. 25, No. 5, 1996, pp. 485–496.
5. Lange, M. and Warnatz, J., "Detailed Simulations of Turbulent Flames Using Parallel Supercomputers," *High Performance Computing in Science and Engineering '98*, edited by E. Krause and W. Jäger, Springer, Berlin, Heidelberg, New York, 1999, pp. 343–352.
6. Lange, M. and Warnatz, J., "Investigation of Chemistry-Turbulence Interactions Using DNS on the Cray T3E," *High Performance Computing in Science and Engineering '99*, edited by E. Krause and W. Jäger, Springer, Berlin, Heidelberg, New York, 2000, pp. 333–343.
7. Lange, M. and Warnatz, J., "Direct Simulation of Turbulent Reacting Flows on the Cray T3E," *Proceedings of the 14th Supercomputer Conference '99 in Mannheim*, edited by H.-W. Meuer, 1999.
8. Lange, M., "Direct Numerical Simulation of Turbulent Flame Kernels Using HPC," *High Performance Computing in Science and Engineering '01*, edited by E. Krause and W. Jäger, Springer, Berlin, Heidelberg, New York, 2002, pp. 418–432.
9. Echekki, T., Poinsot, T. J., Baritaud, T. A., and Baum, M., "Modeling and Simulation of Turbulent Flame Kernel Evolution," *Transport Phenomena in Combustion*, edited by S. H. Chan, Vol. 2, Taylor & Francis, 1995, pp. 951–962.
10. Maas, U. and Warnatz, J., "Ignition Processes in Hydrogen-Oxygen Mixtures," *Combustion and Flame*, Vol. 74, 1988, pp. 53–69.
11. Baum, M., Poinsot, T. J., and Thévenin, D., "Accurate Boundary Conditions for Multicomponent Reactive Flows," *Journal of Computational Physics*, Vol. 116, 1995, pp. 247–261.
12. Renou, B., Boukhalfa, A., Puechberty, D., and Trinité, M., "Local Scalar Flame Properties of Freely Propagating Premixed Turbulent Flames at Various Lewis Numbers," *Combustion and Flame*, Vol. 123, 2000, pp. 507–521.

Application of the Lattice Boltzmann CFD Method on HPC Systems to Analyse the Flow in Fixed-Bed Reactors

Thomas Zeiser*, Gunther Brenner, and Franz Durst

Lehrstuhl für Strömungsmechanik, Universität Erlangen-Nürnberg,
Cauerstraße 4, D-91058 Erlangen, Germany

Abstract. The increasing computational power and new numerical methods give the possibility for more detailed investigations of complex physical phenomena which are typical in science and engineering. In the first part of this paper, we show how CFD simulations of the flows through packed beds using the lattice Boltzmann approach can be used to get local and detailed information about the physical transport processes. The second part deals with the performance of the lattice Boltzmann method on different hardware platforms. It shows that this modern numerical method also needs adequate computers, i.e. high performance computers, to give best performance.

1 Introduction

Randomly packed fixed-bed reactors are widely used in the chemical process industries. Despite interesting developments on applications of structured packings in recent years (e.g. [7]), the randomly packed bed is still the state-of-the art reactor type for diverse applications, as e.g. reaction, separation or purification unit. Their design is usually based on pseudo-homogeneous model equations with averaged semi-empirical parameters [2, 3]. However, this design concept fails for low tube-to-particle diameter ratios (= aspect ratios) since local phenomena start to dominate [15, 20]. In order to resolve the local inhomogeneities, the complete 3-D structure of the packing has to be considered.

New numerical methods and the increase of computational power allow us to simulate in detail single phase reacting flows in such reactors, solely based on material properties and a description of the 3-D geometry, thus without the use of semi-empirical data. We apply a Monte Carlo method, based on the algorithm of Soppe [17], to synthetically generate realistic random packings of spherical particles in a confining tube [14, 18]. The subsequent numerical simulation of the flow field is done by means of the lattice Boltzmann method (see Sec. 2). This CFD computation is the part which actually requires large amounts of computational power. The simulation results reveal that not only the local behaviour but also integral values like the pressure drop depend

---

* e-mail: thzeiser@lstm.uni-erlangen.de

remarkably on the local structural properties of the packings, which is neglected when using correlations with averaged values. Furthermore, it can be proven that not only the shear but also the deformation of fluid elements gives a significant contribution to the total dissipation and the pressure drop (Sec. 3). The latter effect (dissipation due to deformation) is neglected in typical capillary theory models [13]. The last part (Sec. 4) deals with performance aspects and justifies the use of HPC systems, especially of vector computers.

2   Brief description of the numerical method

2.1   The lattice Boltzmann approach

The fundamental idea of the lattice Boltzmann methods [9] is to look at the physical background of fluid mechanics from a *statistical point of view*. Within typical Navier-Stokes CFD solvers the conservation equations for mass and momentum (Navier-Stokes equations) are solved. These equations use continuum mechanical quantities like the density, pressure and flow velocity in their differential equations. In contrast, the underlying equation for the lattice Boltzmann technique is a space, momentum and time discrete formulation of the Boltzmann equation, describing the interaction of single fluid particles or an ensemble averaged particle density distribution function $N_i(t_*, r_*)$.

The current version of our lattice Boltzmann code is based on the 3-D 19-speed (D3Q19, see Fig. 1) lattice Boltzmann automata model with the so called single time Bhatnagar-Gross-Krook relaxation collision operator $\triangle_i^{\text{Boltz}}$ proposed by Qian et al. [16]. To further simplify the equations, the time step has been set to unity.

$$N_i(t_* + 1, r_* + c_i) = N_i(t_*, r_*) + \triangle_i^{\text{Boltz}}, \qquad i = 0, \ldots, 18 \qquad (1)$$

$$\triangle_i^{\text{Boltz}} = \omega \left(N_i^{\text{eq}} - N_i \right). \qquad (2)$$

The first equation describes the advection. The particle density distribution function $N_i(t_*, r_*)$ with discrete momentum in direction c_i located at the discrete position r_* at the discrete time t_* is shifted within one iteration $(t_* + 1)$ in direction c_i to its next neighbouring lattice node located at $r_* + c_i$.

The second equation describes the "collision" of the particles, realized by a local redistribution of the particle density distribution function N_i while locally conserving mass and momentum. This is implemented as a single time relaxation approach with a local equilibrium distribution function N_i^{eq} (Eq. 3) and the relaxation fequency $\omega \in \,]0; 2[$ which determines the viscosity of the fluid (see Eq. 7). The local equilibrium distribution function N_i^{eq} is chosen in such a way, as to recover the time-dependent Navier-Stokes equations in the low Mach number limit [16]. The N_i^{eq} has to be calculated at each cell and at each time step from the local fluid density ρ and the local macroscopic flow

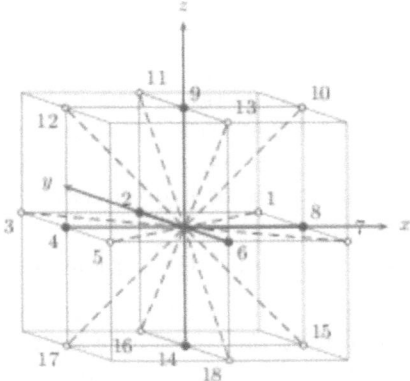

Fig. 1. Discretisation of the velocity space in the D3Q19 model, connections of a cell at r_* to the neighbouring cells and correspondingly chosen directions c_i.

velocity u (see Eq. 3–5). The factor t_p is a well defined constant of the model which only depends on the lattice direction [16].

$$N_i^{\text{eq}} = \rho t_p \left\{ 1 + 3c_i u + \frac{9}{2}(c_i u)^2 - \frac{3}{2}u^2 \right\} \tag{3}$$

All macroscopic flow quantities can simply be calculated from the local particle density distribution function $N_i(t_*, r_*)$ by the following relations:

$$\text{density } \rho = \sum_{i=0}^{18} N_i, \tag{4}$$

$$\text{flow velocity } u = \sum_{i=0}^{18} c_i N_i / \rho, \tag{5}$$

$$\text{pressure } p = c_s^2 \rho, \text{ with: } c_s = 1/\sqrt{3}, \tag{6}$$

$$\text{viscosity } \nu = \frac{1}{6} \left\{ \frac{2}{\omega} - 1 \right\}. \tag{7}$$

2.2   Wall boundary conditions

The lattice Boltzmann equation is solved for all fluid nodes on an equidistant Cartesian mesh. Arbitrary complex geometries can be represented on this grid with the help of the marker-and-cell approach by simply changing the status of single cells (voxels) from fluid (free) to solid (occupied).

Wall boundary conditions can thus be implemented easily within the lattice Boltzmann framework by the so-called *bounce-back* rule which basically means that particle density distributions which hit a solid wall during the advection step are simply put back to the original cell but with opposite momentum (i.e. $N_{i'}(t_* + 1, \boldsymbol{r}_*) = N_i(t_*, \boldsymbol{r}_*) + \triangle_i^{\text{Boltz}}$ with $\boldsymbol{c}_{i'} = -\boldsymbol{c}_i$). This results in a non-slip boundary condition at the wall and allows an easy and efficient handling [5, 12] of arbitrary complex geometries. For low and moderate Reynolds numbers, the staircase approximation of the geometry does not have a significant influence on the hydrodynamical results.

2.3   Lattice Boltzmann model with improved incompressibility

In the standard lattice Boltzmann model as described above, density and pressure are directly coupled by the equation of state (Eq. 6), i.e. a pressure drop automatically results in a decrease of the density. In order to ensure a constant mass flux, the velocity must therefore increase equivalently, leading to an unexpected behaviour and a non-divergence-free velocity field.

To improve the incompressibility, He & Luo [11] suggested to split the pressure p into a constant part p_0 and a slightly changing perturbation δ_p. With this, a distribution function P_i for the pressure can be defined.

$$P_i(t_* + 1, \boldsymbol{r}_* + \boldsymbol{c}_i) = P_i(t_*, \boldsymbol{r}_*) + \omega \left[P_i^{\text{eq}} - P_i(t, \boldsymbol{r}_*) \right], \quad i = 0, \dots, 18 \quad (8)$$

with the local equilibrium distribution function P_i^{eq}

$$P_i^{\text{eq}} = t_p \left\{ p + p_0 \left[3 e_i u + \frac{9}{2} (e_i u)^2 - \frac{3}{2} u^2 \right] \right\}. \quad (9)$$

The resulting quadrature formulae for calculating the macroscopic quantities are

$$\text{pressure} \;\; p = \sum_{i=0}^{18} P_i, \quad (10)$$

$$\text{flow velocity} \;\; \boldsymbol{u} = \sum_{i=0}^{18} e_i P_i / p_0, \quad (11)$$

$$\text{equation of state} \;\; p = c_s^2 \rho, \;\; \text{with:} \; c_s = 1/\sqrt{3}. \quad (12)$$

This set of equations recovers the time-dependent incompressible Navier-Stokes equations in the low Mach number limit in the same way as the original lattice Boltzmann equations but with improved incompressibility. The bounce back wall boundary condition can also be used without changes.

Figure 2 compares the dimensionless pressure drop and the cross-sectional velocity in a fixed bed reactor. It clearly shows (Fig. 2 left) that only the improved model gives a realistic picture of the flow distribution, i.e. the same flow velocity before and behind the packing (axial position <0 and >10). The effect of the two models can even more clearly be seen from the pressure drop where the standard model leads to much lower values. A proof that the incompressible model gives the correct overall pressure drop is given in Sec. 3.2.

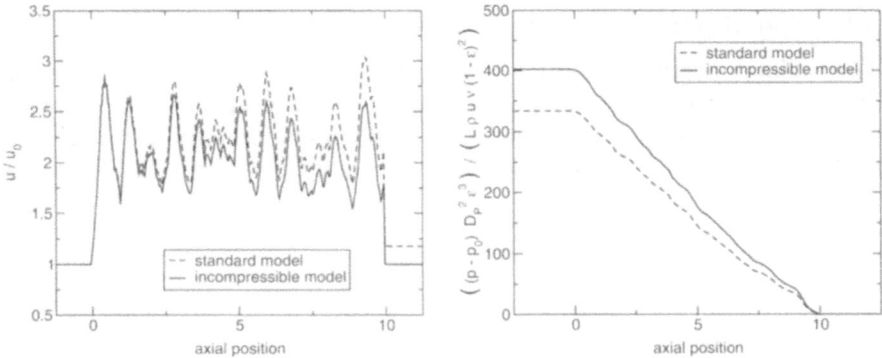

Fig. 2. Cross-section average velocity (left) and pressure (right) in a tubular packed bed computed with the incompressible and the standard model.

3   Analysis of simulation results

In this section, we describe some results obtained by the detailed simulation of the flow through a sample of nine similar random packings of spheres with diameter D_p in a tube of diameter D. More results can be found in [23]. The packings were generated synthetically by a Monte Carlo process based on the algorithm of Soppe [17] which delivers synthetically realistic random packings of spherical particles in a confining tube [14]. The comparison of the relevant features of these packings (e.g. mean porosity and radial porosity profiles) with experimental data from literature shows that this simulation produces realistic packings, matching experimental data on a quantitative level [18]. On the Cartesian marker-and-cell mesh, the diameter of each sphere is discretized by 30 voxels. This resolution ensures that the hydrodynamic properties are correctly predicted [4]. In the present case, an aspect ratio of $D/D_p = 3$ was chosen as this low aspect ratio is of importance in particular applications (see e.g. [1, 8, 19]) and leads to severs inhomogeneities which justify the expensive detailed 3-D simulations.

A typical domain consists of 50 to 800 spheres or approximately $3-40 \times 10^6$ voxels (control volumes). One calculation requires up to 20 GB of memory and takes between a few hours and several days on 1–8 CPUS of the NEC SX-5e.

A special focus of the analysis was set on the contributions of the shear and deformation of the fluid to the total dissipation and their change with variations in the geometry, i.e. the relative position of the randomly placed spheres, and the Reynolds number. These investigations prove that neglecting the deformation of the fluid elements – as done e.g. within the capillary theory models [13] – leads to a significant underestimation of the pressure drop which is typically corrected by unphysical fudge factors like the tortuosity.

3.1   Radial porosity and flow distribution

The radial porosity distribution is a result of the ratio of the tube-to-sphere diameter (aspect ratio). The porosity starts with a value of unity (complete voidage) at the wall and oscillates with decreasing amplitude towards the centre. It varies only very little among the randomly generated packings (see Fig. 3, left).

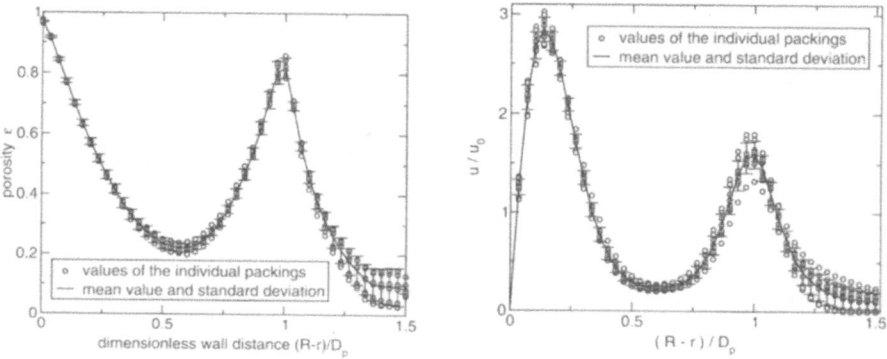

Fig. 3. Radial distribution of the porosity (left) and flow velocity (right). Mean value and standard deviation for nine similar tubular packed beds with $D/D_p = 3$.

These porosity variations lead to a correlated radial variation of the (circumferentially averaged) flow velocity (see Fig. 3, right). For all nine samples a significant peak of the flow velocity near the wall can be observed, while the velocity in the centre of the tube is close to zero, belonging to an area of minimum porosity almost blocked by spheres for this aspect ratio. Also, the position of the second peak is independent of the details of the random packing and due only to the aspect ratio and the ordering effect of the confining wall.

3.2   Global pressure drop

The pressure drop of fixed-bed reactors is a crucial property for their design and operation. Ergun proposed a quadratic correlation function which is valid for a wide range of Reynolds numbers and is widely used for extended packings. If the confining walls have a no longer neglectable influence, they have to be included in the correlation function somehow. One of the first approaches – and according to [10] still one of the best for this type of geometry – is due to Zhavoronkow et al. [24] who used a corrected hydraulic radius. Figure 4 shows the simulated pressure drop in form of the dimensionless friction factor as a function of the Reynolds number together with the correlation of Zhavoronkow et al. These results verify the applicability of the lattice Boltzmann model with improved incompressibility for the present cases.

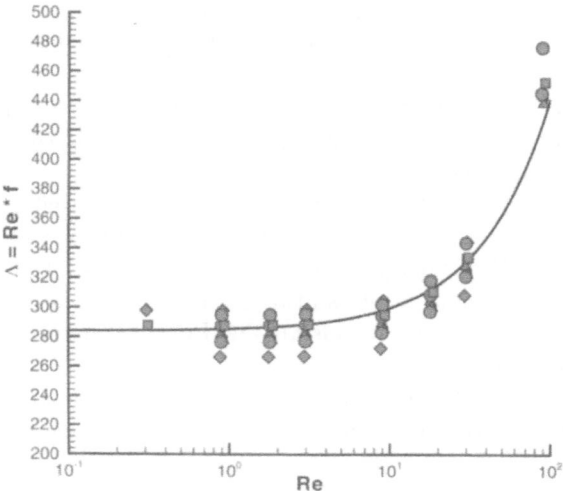

Fig. 4. Comparison of the simulated pressure drop (symbols) for the nine similar random packings with the correlation of Zhavoronkow et al.

3.3   Dissipation due to shear and deformation

The pressure drop is universally related to the dissipated fluid energy per unit time and unit volume. In a general form, the total dissipation Φ of an incompressible flow passing through a porous medium can be expressed by (see e.g. [6])

$$\Phi = -\tau_{ij}\frac{\partial u_j}{\partial x_i} = -\mu \left(\frac{\partial u_j}{\partial x_i} + \frac{\partial u_i}{\partial x_j} + \underbrace{\frac{2}{3}\delta_{ij}\frac{\partial u_j}{\partial x_i}}_{\substack{=0 \text{ for incom-}\\ \text{pressible fluids}}} \right) \frac{\partial u_j}{\partial x_i} \tag{13}$$

$$= -\mu \left\{ \underbrace{2\left[\left(\frac{\partial u_1}{\partial x_1}\right)^2 + \left(\frac{\partial u_2}{\partial x_2}\right)^2 + \left(\frac{\partial u_3}{\partial x_3}\right)^2\right]}_{\Phi_d \text{ (deformation)}} \right.$$

$$\left. \underbrace{+ \left(\frac{\partial u_2}{\partial x_1} + \frac{\partial u_1}{\partial x_2}\right)^2 + \left(\frac{\partial u_3}{\partial x_2} + \frac{\partial u_2}{\partial x_3}\right)^2 + \left(\frac{\partial u_1}{\partial x_3} + \frac{\partial u_3}{\partial x_1}\right)^2}_{\Phi_s \text{ (shear)}} \right\}. \tag{14}$$

The total dissipation Φ can thus be expressed as a sum of two parts: the dissipation caused by shear forces Φ_s and the dissipation caused by deformational strain Φ_d.

By taking the local velocities of the flow field obtained with the lattice Boltzmann simulations and exploiting Eq. 14, the spatial distribution of shear and deformation can be computed.

Comparing the ratio of the integrated dissipation caused by shear and caused by deformation for all nine samples, it is significant that – independent of the details of the underlying geometry – the contribution to the total dissipation caused by deformation is always approximately 50% of the contribution caused by shear (see Fig. 5 left). The Reynolds number dependence is shown in Fig. 5 right.

One can therefore conclude that a significant error is made when the contribution of the deformation is neglected, as is usually the case within the capillary theories [13].

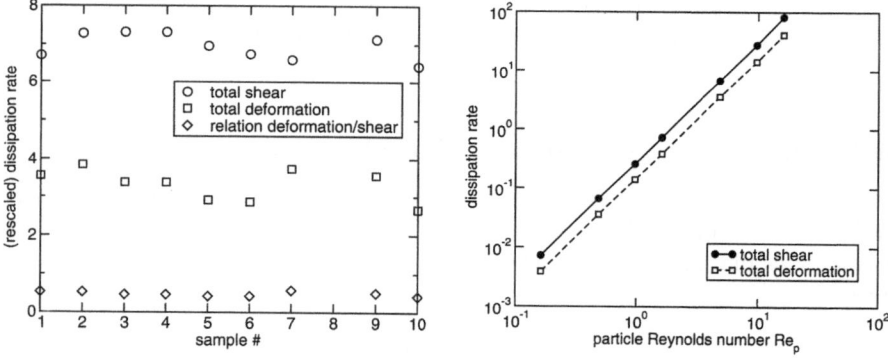

Fig. 5. Total dissipation and its fractions due to deformation and shear for the nine random packings at fixed Reynolds number (left) and for one geometry at different Reynolds numbers (right).

4   Performance on different hardware platforms

In last year's report as well as in [21, 22] we have shown the good vectorisation and shared memory parallelisation using OpenMP or the NEC specific SMP directives on the NEC SX-5/5e.

Here, we therefore present some first and preliminary results which show that it is reasonable to use high performance vector computers for lattice Boltzmann simulations. These investigations have not been done with our normal production code but with a lattice Boltzmann flow solver with strongly

Table 1. Performance results for a domain size of $100 \times 100 \times 100$ cells.

CPU / Machine	Peak Performance	Memory Bandwidth	Lattice Site Updates
INTEL Itanium, 733 MHz, 2 MB Cache, SD-RAM; INTEL-Compiler	2.9 GFlop/s	2.1 GByte/s	0.76×10^6
1 processor SGI Origin 3400; RRZE	1 GFlop/s	1.6 GByte/s	0.77×10^6
INTEL Pentium-4, 1.5 GHz, Rambus RD-Ram; PGI-Compiler 3.2-4	3 GFlop/s	3.2 GByte/s	1.28×10^6
INTEL Pentium-4, 1.5 GHz, Rambus RD-Ram; INTEL-Compiler	3 GFlop/s	3.2 GByte/s	1.66×10^6
1 processor NEC SX-5e	4 GFLop/s	32 GByte/s load 16 GByte/s store	15.3×10^6
1 node (8 CPUS) HITACHI SR8000-F1	12 GFlop/s	32 GByte/s	20.7×10^6

reduced functionality. This flow solver only includes as much as necessary to do simple but meaningful benchmark calculations which are nevertheless representative for our production code. The results will slightly change once special optimisations for cache-based systems will be made – but these changes can never bridge the gap between the vector- and cache- or RISC-based systems.

Table 1 and 2 show the number of lattice site updates obtained on different platforms. This number is a measure for the execution speed and thus the maximal domain size which can be solved in a reasonable time on the respective systems. One can see that the peak performance is not at all a reasonable measure but the combination of CPU, memory and compiler. It shows that the memory bandwidth is more crucial than the CPU performance. This trend becomes even more dominant with increasing complexity of the geometry, i.e. if the memory is not accessed consecutively. With increasing domain size, the lattice site update rate decreases significantly on typical RISC-based machines whereas it keeps constant or increases on vector-based systems (no table given for that effect).

These results justify – despite the large memory requirements – the use of HPC systems with large and fast memories.

5   Conclusions

With the help of simulations of the flow field in randomly packed beds of spheres using the lattice Boltzmann method, we were able to analyse local

Table 2. Performance results for a domain size of $160 \times 34 \times 34$ cells and complex geometry.

CPU / Machine	Peak Performance	Memory Bandwidth	Lattice Site Updates
INTEL Itanium, 733 MHz, 2 MB Cache, SD-RAM; INTEL-Compiler	2.9 GFlop/s	2.1 GByte/s	0.54×10^6
1 processor SGI Origin 3400; RRZE	1 GFlop/s	1.6 GByte/s	0.60×10^6
INTEL Pentium-4, 1.5 GHz, Rambus RD-Ram; PGI-Compiler 3.2-4	3 GFlop/s	3.2 GByte/s	0.98×10^6
INTEL Pentium-4, 1.5 GHz, Rambus RD-Ram; INTEL-Compiler	3 GFLop/s	3.2 GByte/s	1.16×10^6
1 processor NEC SX-5e	4 GFLop/s	32 GByte/s load 16 GByte/s store	11.8×10^6
1 node (8 CPUS) HITACHI SR8000-F1	12 GFlop/s	32 GByte/s	13.2×10^6

inhomogeneities in the flow properties such as velocity and dissipation due to deformation and shear. Further, we were able to prove that the dissipation caused by the deformation of the fluid gives a Reynolds number-dependent contribution to the total dissipation which must not be neglected.

These results underline our statement that detailed simulations are necessary to examine all flow effects caused by the radial and circumferential inhomogeneities of the packings as well as the differences among several samples with almost identical global parameters (porosity, aspect ratio, etc.).

High performance computers are utilised very well with lattice Boltzmann methods. This modern CFD method shows its performance not until high performance computers with a large memory bandwidth and good compilers are used.

Acknowledgements

This work is supported by the German Research Foundation (DFG). The geometrical structures were generated by the Monte Carlo Packing Generation Tool (McPackGen) which is based on the work done by Yong-Wang Li during his stay at our institute in the framework of an Alexander von Humboldt (AvH) foundation fellowship. The benchmark computations were done in cooperation with Gerhard Wellein from the Regionales Rechenzentrum Erlangen in the framework of the Competence Network for Technical, Scientific High Performance Computing in Bavaria (KONWIHR). The authors would also like to acknowledge the support from and helpful discussions with numerous colleagues, especially Jörg Bernsdorf (NEC CCRLE), Hannsjörg Freund (TC1) and Florian Huber (LSTM/TC1).

References

1. ABB Lummus Global Inc. Fixed catalytic bed reactor. International patent WO 99/48604, 1999.
2. R. Adler. Stand der Simulation von heterogen-gaskatalytischen Reaktionsabläufen in Festbettreaktoren – Teil 1. *Chem.-Ing.-Tech.*, 72(6):555–564, 2000.
3. R. Adler. Stand der Simulation von heterogen-gaskatalytischen Reaktionsabläufen in Festbettreaktoren – Teil 2. *Chem.-Ing.-Tech.*, 72(7):688–699, 2000.
4. J. Bernsdorf, G. Brenner, and F. Durst. Numerical analysis of the pressure drop in porous media flow with lattice Boltzmann (BGK) automata. *Comput. Phys. Commun.*, 129(1-3):247–255, 2000.
5. J. Bernsdorf, F. Durst, and M. Schäfer. Comparison of cellular automata and finite volume techniques for simulation of incompressible flows in complex geometries. *Int. J. Numer. Meth. Fluids*, 29:251–264, 1999.
6. R. B. Bird, W. E. Stewart, and E. N. Lightfoot. *Transport Phenomena*. Wiley, New York, 1960.
7. H. Calis, J. Nijenhuis, B. Paikert, F. Dautzenberg, and C. van den Bleek. CFD modelling and experimental validation of pressure drop and flow profile in a novel structured catalytic reactor packing. *Chem. Eng. Sci.*, 56(7):1713–1720, 2001.
8. J. Carberry. Some aspects of catalytic reactor engineering. *Trans. Inst. Chem. Eng.*, 59:75–82, 1981.
9. S. Chen and G. D. Doolen. Lattice Boltzmann method for fluid flows. *Annu. Rev. Fluid Mech.*, 30:329–364, 1998.
10. B. Eisfeld. *Pseudokontinuierliche Modellierung der Strömung in Schüttschichtreaktoren*. PhD thesis, Technische Universität Cottbus, 1999.
11. X. He and L.-S. Luo. Lattice Boltzmann model for the incompressible Navier-Stokes equation. *J. Stat. Phys.*, 88(3/4):927–944, 1997.
12. T. Inamuro, M. Yoshino, and F. Ogino. A non-slip boundary condition for lattice Boltzmann simulations. *Phys. Fluids*, 7(12):2928–2930, 1995.
13. M. Kaviany. *Principles of Heat Transfer in Porous Media*. Mechanical Engineering Series. Springer, Berlin, Heidelberg, New York, 2nd edition, 1999.
14. Y.-W. Li, T. Zeiser, H. Freund, M. Steven, G. Brenner, E. Klemm, G. Emig, and F. Durst. Direct simulation of the structure and consequential flow field in a packed bed. *AIChE Journal*, (submitted; revised version currently prepared).
15. J. Papageorgiou and G. Froment. Simulation models accounting for radial voidage profiles in fixed-bed reactors. *Chem. Eng. Sci.*, 50:3043–3056, 1995.
16. Y. H. Qian, D. d'Humières, and P. Lallemand. Lattice BGK models for Navier-Stokes equation. *Europhys. Lett.*, 17(6):479–484, Jan. 1992.
17. W. Soppe. Computer simulation of random packings of hard spheres. *Powder Technol.*, 62:189–196, 1990.
18. M. Steven. Detaillierte Simulation und Analyse der Struktur von Katalysatorschüttungen und der lokalen Transportprozesse. Diplomarbeit, Universität Erlangen-Nürnberg, 2001.

19. D. Thoenes. Current problems in the modelling of chemical reactors. *Chem. Eng. Sci.*, 35:1840–1853, 1980.
20. D. Vortmeyer. Modellierung chemischer Festbettreaktoren. In D. Behrens, editor, *Reaktionstechnik chemischer und elektrochemischer Prozesse*, volume 94 of *Dechema Monographie*, pages 79–96, Weinheim, 1983. Verl. Chemie.
21. T. Zeiser, G. Brenner, P. Lammers, and J. Bernsdorf. Performance aspects of lattice Boltzmann methods for application in chemical engineering. In C. Jenssen, T. Kvamdal, H. Andersson, B. Pettersen, A. Ecer, J. Periaux, N. Satofuka, and P. Fox, editors, *Parallel Computational Fluid Dynamics 2000, Trends and Applications. Proceedings of the Parallel CFD 2000 Conference, May 22-25, Trondheim, Norway*, pages 407–414, Amsterdam, 2001. Elsevier.
22. T. Zeiser, H. Freund, J. Bernsdorf, G. Brenner, and F. Durst. CFD calculations of flow, dispersion and chemical reactions in fixed-bed tubular reactors using the lattice Boltzmann method. In M. Breuer, F. Durst, and C. Zenger, editors, *High Performance Scientific and Engineering Computing: Proceedings of the 3rd International FORTWIHR Conference on HPSEC*, volume 21 of *Lecture Notes in Computational Science and Engineering*, pages 53–62. Springer, 2002.
23. T. Zeiser, M. Steven, H. Freund, P. Lammers, G. Brenner, F. Durst, and J. Bernsdorf. Analysis of the flow field and pressure drop in fixed bed reactors with the help of lattice Boltzmann simulations. *Phil. Trans. R. Soc. Lond. A*, 360(1792):507–520, 2002.
24. N. Zhavoronkov, M. Aerov, and N. Umnik. Hydraulischer Widerstand und Packungsdichte einer Kornschicht. *Zhur. Fiz. Khim*, 23:342–361, 1949. Zitiert nach Eisfeld (1999); Originalartikel in russischer Sprache, Übersetzung E. Hofman.

A Vectorised Lagrangian Particle Model
for the Numerical Simulation
of Coal-Fired Furnaces

Jochen Ströhle[1], Frank Rückert[1], Benedetto Risio[2], Uwe Schnell[1], and
Klaus R.G. Hein[1]

[1] Institute of Process Engineering and Power Plant Technology (IVD), University
   Stuttgart, Pfaffenwaldring 23, 70569 Stuttgart (Germany)
[2] Recom Services, REaction and COMbustion Modeling, Nobelstr. 15,
   70569 Stuttgart (Germany)

Abstract. A Lagrangian particle tracking model is applied to the simulation of
the discrete phase in coal-fired furnaces. The interaction of particles with the gas
phase leads to additional source terms in the Eulerian transport equations of mass,
momentum and enthalpy. The radiative heat exchange of the particles is directly
coupled with the discrete ordinates method using an extra source term in the discretised radiative transport equation. The routines for the particle model are vectorised
using loops over the number of all particles. They are implemented in the 3D combustion code AIOLOS, optimised for vector and parallel computers to achieve high
numerical efficiency. The code is applied to the numerical simulation of a utility
boiler, showing a high performance and vectorisation rate, combined with good
agreement with measurements.

1 Introduction

Since coal is still one of the major energy sources worldwide, the improvement
of the efficiency of power stations with coal-fired boilers is an important task.
Large improvements in computer technology and detailed physical models
have enabled computational fluid dynamics (CFD) to be a fast and economic
tool for the optimisation of industrial furnaces [1]. The powerful computer
platforms now available for numerical simulations are generally based on
vector processors requiring a certain structure of the computer code.

The simulation program AIOLOS has been developed for the numerical
calculation of three-dimensional, stationary and dynamic, turbulent reactive
flows in pulverised coal-fired utility boilers. In submodels treating fluid flow,
turbulence, homogeneous and heteroegeneous combustion, and heat transfer,
equations for calculating the conservation of mass, momentum, and energy
are solved. A detailed description of the code can be found in [2,3].

The continuous gas phase and radiative heat transfer are accurately modelled in an Eulerian framework, which has successfully been adapted for vectorisation [4]. The dispersed phase of the coal can be modelled by two different
approaches. In the Eulerian approach, the dispersed phase is treated as a continuum using common transport equations equal to the gas phase. Because

of its high numerical efficiency, this method has been successfully applied to simulations of coal-fired utility boilers [2]. However, its accuracy lacks in the near burner region where the particle load and the velocity difference between the two phases is very high. Therefore, the Lagrangian approach is favoured, following a number of representative particles through the furnace and calculating their effect on the continuous phase and on heat transfer. In order to minimise the statistical error, a high number of particle traces has to be calculated, implying a large computational effort. Therefore, the simulations should be performed on supercomputers, requiring the vectorisation of the complete code, including the particle routines.

In the following, the basic equations for the gas and particle phase are briefly described. After that, the vectorisation method for the particle routines are presented. Finally, the code is applied to the numerical simulation of a coal-fired utility boiler.

2   Eulerian Treatment of Fluid Flow and Radiative Transfer

In the Eulerian approach, the computational domain is discretised into a large number of control volumes, over which balances of mass, momentum, energy and other values are performed. The conservation of mass of a stationary gas flow is described by the following continuity equation:

$$\frac{\partial(\rho_G u_{G,j})}{\partial x_j} = S_{m,P} \tag{1}$$

In this equation, ρ_G and $u_{G,i}$ are the density and velocity of the gas. The mass source term $S_{m,P}$ is due to mass transferred from particles to the gas phase. The momentum transport of a stationary gas flow is solved by the following Navier Stokes equation:

$$\frac{\partial(\rho_G u_{G,j} u_{G,i})}{\partial x_j} = -\frac{\partial p}{\partial x_i} + \frac{\partial \tau_{ij}}{\partial x_j} + \rho g_i + S_{u,P,i} \tag{2}$$

In this equation, p is the pressure, τ_{ij} is the stress tensor derived from the turbulence model, and g_i is the gravity. The momentum source term $S_{u,P,i}$ is caused by the forces of the particles on the gas flow. The convective and diffusive transport of energy is described by the transport equation of the specific gas enthalpy h_G:

$$\frac{\partial(\rho_G u_{G,j} h_G)}{\partial x_j} = \frac{\partial}{\partial x_j}\left(\frac{\mu}{\sigma_h}\frac{\partial h_G}{\partial x_j}\right) + S_{h,C} + S_{h,R} + S_{h,P} \tag{3}$$

The heat source terms $S_{h,C}$, $S_{h,R}$ and $S_{h,P}$ are due the energy release during homogeneous reactions, the radiation source and the heat transferred by convection and mass transfer between the two phases, respectively. The mass fractions of the gas species are balanced with similar transport equations and source terms due to chemical reactions of both, the dispersed phase and the gas phase. For the combustion of hydrocarbons and carbon oxide the Eddy Dissipation Concept [5] using a two-step reaction scheme is applied. The turbulence closure is done by a standard k, ϵ-model neglecting the influence of the particles on turbulence.

Radiation is the predominant mode of heat transfer in industrial furnaces. The change of the radiative intensity I along a distance ds due to absorption, out-scattering, emission and in-scattering is described by the radiative transfer equation:

$$\frac{dI}{ds} = -(\kappa + \sigma_s)I + \kappa\,I_b + \frac{\sigma_{s,}}{4\pi}\int_{4\pi} I'\,\Phi(s' \to s)\,d\Omega + S_{I,P} \qquad (4)$$

In this equation, I_b is the blackbody radiative intensity at the temperature of the medium. The absorption coefficient of the gas phase κ is calculated by the weighted-sum-of-grey-gases model of Smith et al. [6] that accounts for the variation of the radiative properties with the electromagnetic spectrum. Emission and absorption of particles is considered by the source term $S_{I,P}$, whereas the scattering coefficient σ_s is calculated by geometric optics as a function of particle load, projected surface area and an efficiency factor. The scattering phase function $\Phi(s' \to s)$ from the incoming direction s' to the outgoing direction s is calculated by the Delta-Eddington approximation of Joseph and Wiscombe [7] that accounts for strongly anisotropic scattering of large particles with a large portion in the forward direction. The radiative transfer equation (4) is solved by the discrete ordinates method [8] that has been applied to general coordinates according to Liu et al. [9]. A detailed description of the radiation model can be found in [10].

The concentration of particles can also be treated by the Eulerian approach, neglecting the slip between the particle and gas phase. However, this assumption is not valid in the near burner region of coal-fired furnaces, where large differences between particle and gas velocities can be observed. Therefore, the particles are more accurately modelled by the Lagrangian approach.

3   Lagrangian Treatment of Particles

In the Lagrangian approach, a number of representative particles is traced from the inlet through the furnace until they leave the domain or are burned out. The evolution of the particle properties are specified by balances of momentum, mass and energy around a single particle. The change of velocity $u_{P,i}$ of a particle with mass m_P is described by the following momentum balance:

$$\frac{d}{dt}(m_P u_{P,i}) = F_{D,i} + F_{G,i} \qquad (5)$$

Several external forces act on the particle, of which only the drag force $F_{D,i}$ and gravity force $F_{G,i}$ are significant:

$$F_{G,i} = m_P g_i \left(1 - \frac{\rho_G}{\rho_P}\right) \tag{6}$$

$$F_{D,i} = \frac{C_D}{2} \rho_G A_{p,P} (u_{P,i} - u_{G,i})|u_{P,i} - u_{G,i}| \tag{7}$$

In these equations, ρ_G and ρ_P are the gas and particle density, respectively, C_D is the drag coefficient, and $A_{p,P}$ is the projected surface area of the particle. The instanteneous gas velocity $u_{G,i}$ is defined as:

$$u_{G,i} = \overline{u}_{G,i} + u'_{G,i} \tag{8}$$

While the time averaged velocity $\overline{u}_{G,i}$ can be taken directly from the Eulerian treatment of the gas phase, the fluctuation velocity $u'_{G,i}$ is derived from the turbulence model, using a random value of a Gaussian distribution. The instanteneous gas velocity is kept constant, until the eddy life time or the transition time in the turbulent eddy is exceeded. Equation (5) is integrated over time to yield the new particle velocity $u_{P,i,n+1}$ after a certain time step Δt. The length of the time step depends both on the residence time of the particle in a turbulent eddy and on the transition time in the actual control volume. When a particle reaches a wall, it is assumed to be reflected elastically. A detailed derivation of the solution of the momentum balance is given by Knaus [11].

The combustion process of the coal particle is modelled by two global heterogenous reactions. During pyrolysis the raw coal (subscript R) releases volatiles leaving ash and char. The burnout of char (subscript C) is described as a function of the specific surface area a_P and a rate coefficient k_C that considers both chemical reaction and diffusion. Both reactions lead to a mass loss of the particle:

$$\frac{dm_P}{dt} = -m_{P,R} \, k_R - m_{P,C} \, a_P \, k_C \tag{9}$$

The mass is transferred to the gas phase, resulting in source terms for the Eulerian transport equations of mass and the gas species. A detailed description of the reaction model can be found in [12].

The change of particle temperature T_P with time due to convection, radiation, and the heat of reaction Δh_C^0 released during char combustion, is derived from the following energy balance:

$$m_P \, c_{p,P} \frac{dT_P}{dt} = m_{P,C} \, a_P \, k_C \Delta h_C^0 + h \, A_P (T_G - T_P)$$

$$+ \epsilon_P \, A_P \left(\int_{4\pi} I \, d\Omega - \sigma T_P^4\right) \tag{10}$$

In this equation $c_{p,P}$ is the specific heat capacity of the particle, h is the heat transfer coefficient, T_G is the gas temperature, A_P is the surface area, and ϵ_P is the emissivity of the particle. Assuming only small changes, the particle temperature $T_{P,n+1}$ after a time step Δt is calculated analytically from the initial temperature $T_{P,n}$ by integrating Eq. 10 over time.

On their way through the computational domain, the particles cross several computational cells where they produce source terms in the transport equations of the gas phase according to the particle-source-in-cell method [13], and also in the radiative transport equation.

4   Vectorisation of the Code

The Eulerian treatment of the gas phase is very suitable for vectorisation, because most of the loops cover all computational cells, leading to long loops without any internal dependencies. For the solution of the discretised transport equations, special solvers like the Red-Black-Gauss solver are implemented to avoid internal dependencies on neighboring cells [4]. In the discrete ordinates method, the radiative intensity field is solved best along the considered direction. Therefore, the loop over all cells is split into several loops over the cells on each diagonal of the computational domain orthogonal to the direction of the ray.

The vectorisation of the particle tracking routines is realised by internal loops over the number of all existing particles, enclosed by an external loop over the number of time steps. This means that all particles are treated simultaneously for each time step that can vary in length for each particle. Hence, all properties of the particles have to be stored in arrays with the dimension of the total number of computed particles. However, the number of existing particles can decrease with time, since some of them leave the furnace or are burned out. After each time step, the remaining particles are determined and the number of existing particles is recalculated. In order to enable the correct memory access, the particle values are addressed indirectly by a pointer that is reordered for all existing particles, as demonstrated in the following example:

```
do i=1,np                  loop over all computed particles
   if(lexist(i))then       inquiry, if particle exists
      nexist=nexist+1       set counter for existing particles
      iadr(nexist)=i        set pointer for existing particles
   endif
enddo


do n=1,nexist              loop over all existing particles
   i=iadr(n)               set address for existing particle
   phi(i)= ...             perform operation for existing particle
enddo
```

Similar procedures are necessary when some operations do not concern all existing particles. This is the case for the wall treatment since only a small fraction of the particles hits a wall within a time step. Therefore, a pointer is set for all particles leaving the computational domain, and a loop over these particles is performed to model an elastical reflexion. Another example is the calculation of the instanteneous gas velocity in Eq. 8, which is only updated for those particles that have exceeded the residence time in a turbulent eddy, as desrcribed in Section 3.

5   Results

The models described above have been implemented in the combustion simulation code AIOLOS. In order to validate the models and to check their numerical efficiency, AIOLOS has been applied to the numerical simulation of a coal-fired utility boiler with an electric output of 397 MW, which is outlined in Fig. 1.

The bituminous coal and the combustion air is introduced through 12 internally air-staged swirl burners, distributed on three levels. On each level one burner is arranged on each wall, slightly shifted from the central axis of the wall inducing a rotating main flow. Beneath each burner, two air nozzles supply the side walls with oxygen to prevent corrosion. Burnout air is added above the highest burner level. The furnace is provided with four measuring

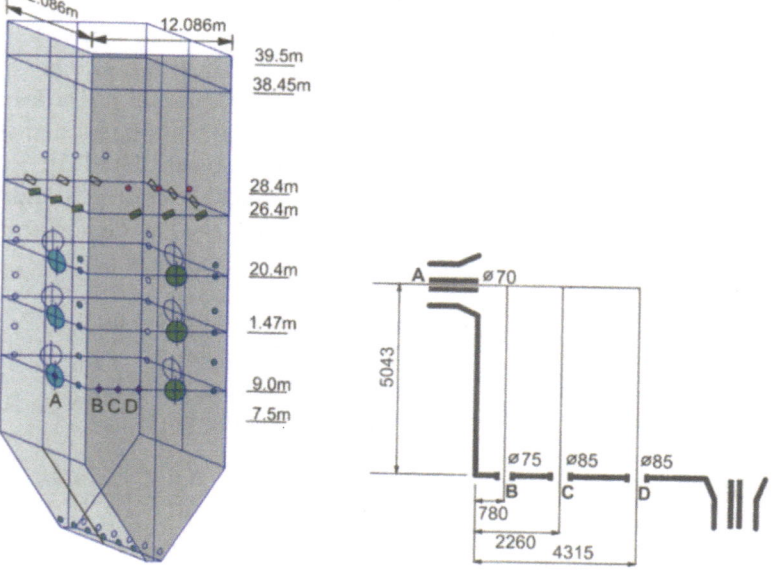

Fig. 1. Furnace scheme (left) and measuring ports on the burner level (right); data are in mm in figure on the right

ports in the burner near field on boiler level 7.5 m and six ports above the over fire air nozzles on level 28.4 m. Measurements of particle velocities, temperature of particle clouds and gas concentrations have been performed by Sabel et al. [14].

A body-fitted grid with approximately 450,000 computed cells was applied to discretise the furnace. The simulations were performed on a NEC SX-5 parallel vector computer using four processors. The efficiency of the vectorisation procedure is highlighted by the degree of vectorisation of 99.0 %. However, this value can reach 99.7 % for a simulation without particle tracking, revealing that there is still some potential for improvement. One complete run, requiring 10,000 iterations for convergence, amounted to approximately 5 hours. The share of the particle tracking routines was about 20 % since they were executed only every 500th iteration.

In the following, the simulation results are compared to the measurements in the burner near field along three paths from the ports B, C and D (see Fig. 1, right) on the front wall up to the axis of the burner on the left side wall.

As shown in Fig. 2, the incoming jets of the side wall air, the staged air, and the primary air-coal mixture cause three peaks at about 0.7 m, 4.2 m and 4.7 m, respectively, in the measured axial velocity of port B that are well reproduced by the simulations. At port C, the jets from the burner have been

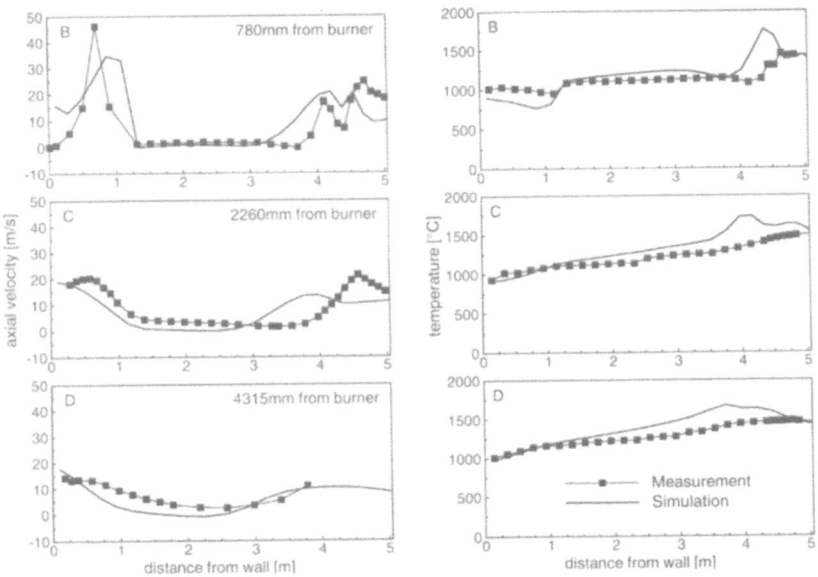

Fig. 2. Computed (lines) and measured (symbols) axial velocity and temperature profiles in the burner near field

combined to one large peak which is slightly underpredicted with a small shift away from the burner axis, while the agreement at port D is quite good.

The computed temperature profiles show the same tendencies as the measurements, with an increasing temperature away from the wall and a distinct minimum close to the side wall nozzle. Very good agreement is achieved close to the wall and on the burner axis, but significant higher values are predicted by the simulations near the jet of the swirled air where the combustion process takes place. These discrepancies might be evoked either by uncertainties in the models, or by radiation losses within the measuring system.

In Fig. 3, the computed O_2 and CO_2 concentrations in the burner near field show good agreement with the measurements. Again, the air jets affect the profiles by a maximum in the O_2, and a corresponding minimum in the CO_2 concentration at port B. The discrepancies near the side wall air jet might be invoked by uncertainties in the boundary conditions of the inlet, since fouling leads to a reduction of the inlet area. Up to 3.7 m from port C and with minor extent at port D, the O_2 concentration is slightly overpredicted, resulting in a corresponding underprediction of CO_2, which might be caused by the simplified modelling of the heterogeneous reactions. But the drop to zero oxygen with the corresponding rise in CO_2 concentration, indicating the location of the flame front, is almost exactly captured by the simulation. In front of the burners, all available oxygen has been consumed,

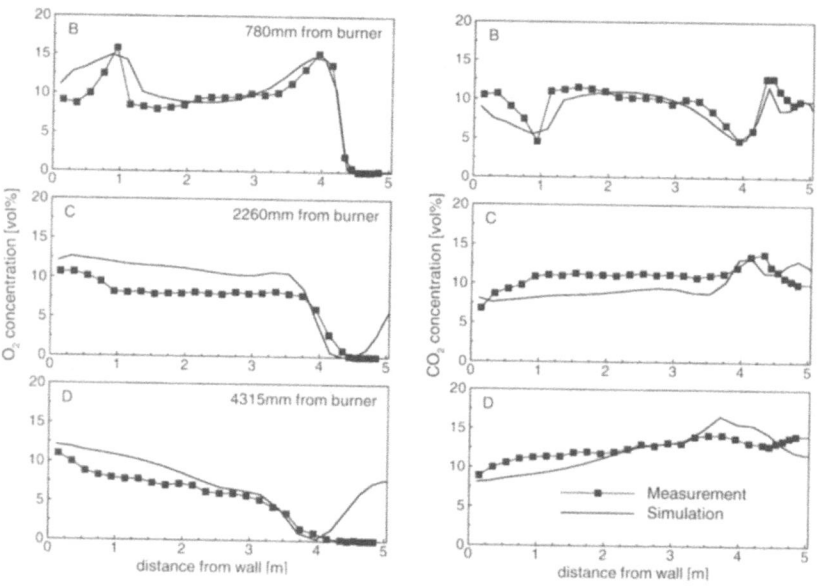

Fig. 3. Computed (lines) and measured (symbols) O_2 and CO_2 concentrations profiles in the burner near field

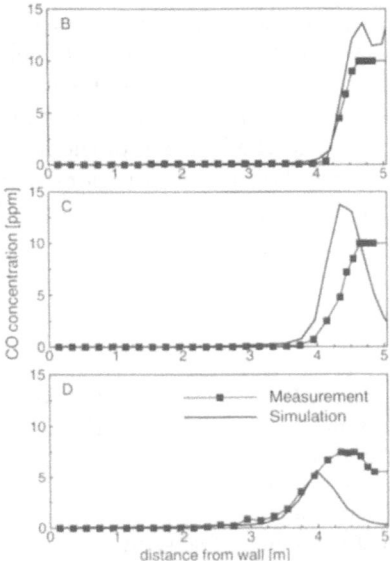

Fig. 4. Computed (lines) and measured (symbols) CO concentration profiles in the burner near field

resulting in a high peak of CO that is also accurately predicted by the simulations, as shown in Fig. 4.

Simulations of the same configuration have been performed before using the same code, but with an Eulerian treatment of the particles and with a much finer Cartesian grid [14]. Good agreement of computed velocities with the measurements at the side wall air and staged air flows was reached, but with a significant underprediction of the axial velocity at the primary air flow, which is not the case in the present study. In this region, where the particle load is very high, the advantages of the Lagrangian treatment of particles are evident. Furthermore, the Eulerian approach can only poorly reproduce the CO concentrations in front of the burners, since the coal particles are assumed to follow the gas flow and release the volatiles within the jet of the primary and secondary air where much oxygen is present, so that most of the CO fast reacts further to CO_2. In contrast, the Lagrangian approach effects the coal particles to be transported almost straight on along the burner axis where the released volatiles produce a high CO peak.

6   Conclusions

Numerical simulations of a coal-fired utility boiler have been performed using a Lagrangian particle model for the discrete phase, coupled with an Eulerian approach for the gas phase and radiation. The particle routines are vetorised by loops over all existing particles. A high degree of vectorisation of 99 % and a low computing time of 5 hours have proofed the efficiency of the method. The computed results are compared to measurements of velocity, temperature and gas concentrations in the burner near field. The influence of the air flows on the combustion process and the location of the flame front are accurately predicted by the simulations. Especially in regions of high particle load, the Lagrangian particle model enables more accurate results than the Eulerian approach.

References

1. B. Risio, U. Schnell, and K.R.G. Hein. Towards a reliable and efficient furnace simulation tool for coal fired utility boilers. In E. Krause and W. Jäger, editors, *High-Performance-Computing in Science and Engineering '98 (Transactions of the High Performance Computing Center Stuttgart)*, pages 353–374. Springer-Verlag Berlin Heidelberg, 1999.
2. U. Schnell. Numerical modelling of solid fuel combustion processes using advanced cfd-based simulation tools. *Progress in Computational Fluid Dynamics*, 1(4):208–218, 2002.
3. R. Schneider. *Beitrag zur numerischen Berechnung dreidimensional reagierender Strömungen in industriellen Brennkammern*. PhD thesis, University of Stuttgart, 1998.
4. B. Risio, R. Schneider, U. Schnell, and K.R.G. Hein. Parallel computation of tubulent combustion processes on individually discretized domains. In *Proc. ECCOMAS 96*, pages 374–380, Paris, 1996.
5. B.F. Magnussen and B.H. Hjertager. On Mathematical Modeling of Turbulent Combustion With Emphasis on Soot Formation and Combustion. *Proceedings of the Combustion Institute*, 16:719–729, 1976.
6. T. F. Smith, Z. F. Shen, and J. N. Friedman. Evaluation of coefficients for the weighted sum of gray gases model. *Journal of Heat Transfer*, 104:602–608, 1982.
7. J. H. Joseph and W. J. Wiscombe. The delta-eddington approximation for radiative flux transfer. *Journal of Atmmospheric Science*, 33:2452–2459, 1976.
8. W. A. Fiveland. Discrete-ordinates solutions of the radiative transport equation for rectangular enclosures. *Journal of Heat Transfer*, 106:699–706, 1984.
9. J. Liu, H. M. Shang, and Y. S. Chen. Predictions of radiative transfer in general body-fitted coordinates. *Numerical Heat Tranfer, Part B*, 31:423–439, 1997.
10. J. Ströhle, H. Knaus, U. Schnell, and K.R.G. Hein. A Radiation Model for the Numerical Simulation of Coal-Fired Furnaces Using Body-Fitted Grids. *Combustion Science and Technology*, 153:127–139, 2000.
11. H. Knaus. *Simulation turbulenter reagierender Zweiphasenströmungen in industriellen Feuerungen mit komplexen Geometrien*. VDI Verlag, Düsseldorf, 2001.
12. U. Schnell, R. Schneider, H.C. Magel, B. Risio, and J. Lepper. Numerical Simulation of Advanced Coal-Fired Combustion Systems with In-Furnace NOx Control Technologies. In *Third International Conference on Technologies and Combustion for a Clean Environment*, July 3-6, Lisbon (Portugal), 1995.

13. C. T. Crowe, M. P. Sharma, and D. E. Stock. The particle-source-in-cell (psi-cell) model for gas-droplet flows. *Journal of Fluids Engineering*, 99:325–332, 1977.

14. T. Sabel, B. Risio, Berger R., U. Schnell, K.R.G. Hein, M. Käss, U. Priesmeier, and H. Thierbach. Full scale measurements and mathematical modelling studies for the investigation of the combustion behaviour of a modern bituminous coal-fired boiler. In *Fifth Europ. Conf. on Industrial Furnaces and Boilers*, Porto, Portugal, 2000.

FE-Simulation of Subsequently Piled Foundations

Hans Mathäus Hügel

TU Hamburg-Harburg, Arbeitsbereich Geotechnik und Baubetrieb,
Harburger Schloßstraße 20, D-21079 Hamburg

Abstract. Existing foundations of buildings can be piled subsequently to improve their bearing behaviour. The effort of subsequent piling depends on the initial state of soil as well as the load history. Model tests were carried out to study the influencing quantities, especially the loading history due to piling. The models tests were numerically simulated based on the FEM and on hypoplastic constitutive models for the soil.

1   Boundary Value Problem

In the course of a research project at the University of Kaiserslautern [9] the effect of subsequent piling of existing foundations of buildings on the bearing behaviour were examined with model tests and numerical simulations based on the Finite Element Method. Piling of an existing foundation means the installation of e.g. concrete piles under a foundation block in order to improve its bearing capacity. All model tests and simulations were carried out for foundations on a sandy soil. The bearing behaviour of the pure foundation depends essentially on the mechanical behaviour of the soil. In the present problem sand is used which stiffness and strength parameters depends on current void ratio, mean stress level and stress history. For piled foundations the change of stress state due to the piling process have to be considered additionally.

2   Model tests

A reinforced concrete foundation with dimensions $0.8 \times 0.8 \times 0.3$ m (pure or subsequent piled) is loaded vertically. The diameter of the piles is 6 cm, the length between 0.8 and 1.2 m, see Fig. 1. The models were carried out in a sand box with dimensions $4.5 \times 4.5 \times 5$ m. The piles were installed carefully in order to minimize a disturbance of the existing foundation. The initial void ratio of the sand and the pile length were varied in test series [9].

The following quantities were measured during the tests:

— vertical load F_z at load piston,
— normal force N in three levels of each pile with special force sensors,
— normal force N at top of each pile,

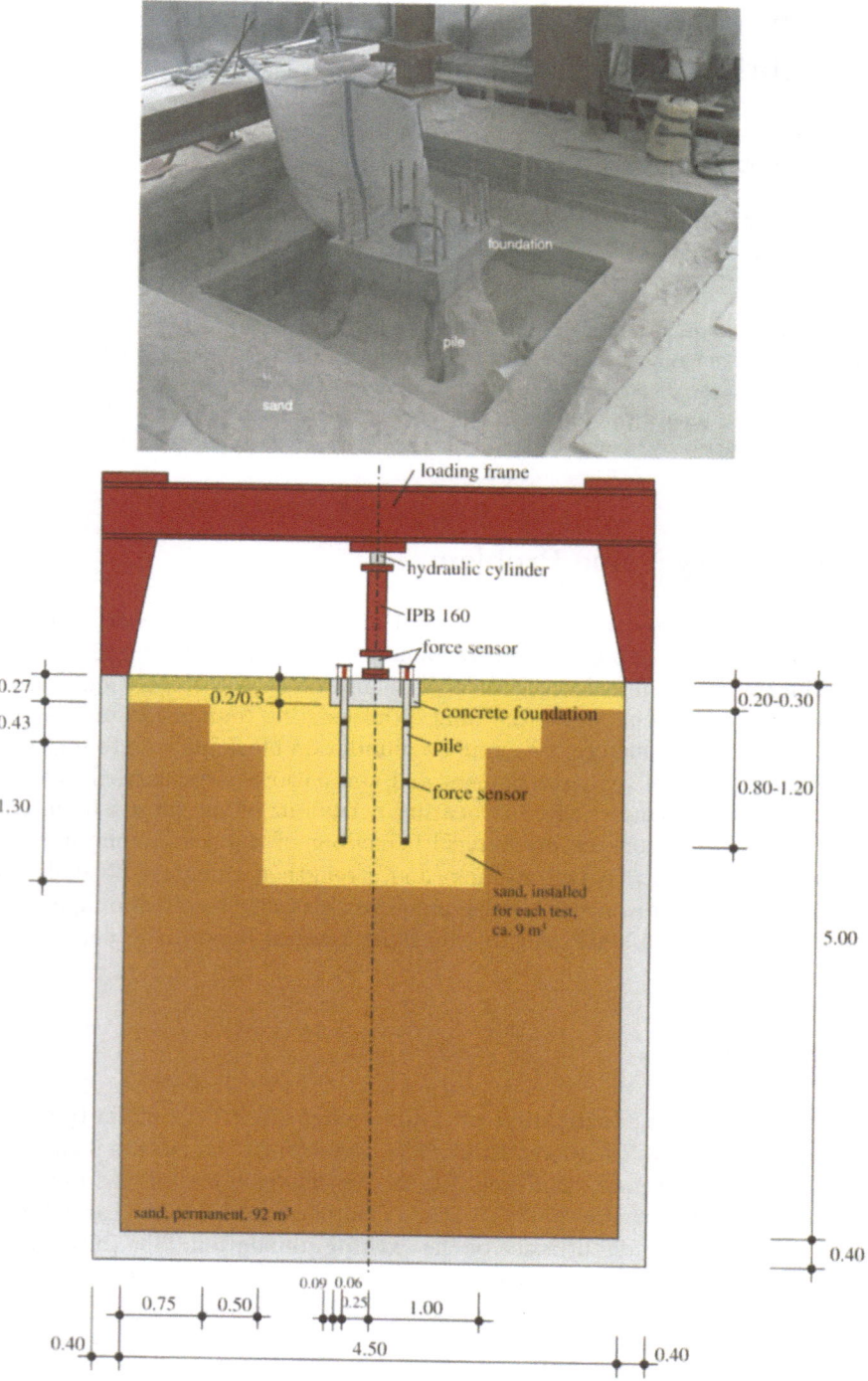

Fig. 1. Photo and sketch of model test, all length in [m]

– vertical displacement u_z at load piston,
– vertical displacements u_{zi} at pile tops.

Between the model tests the biggest part of the sand remained in the sand box in order to avoid a time-consuming complete filling, see orange coloured area in Fig. 1.

3   Numerical Simulations

Based on the Finite Element Method numerical simulations of the model tests and for large scale problems were carried out using the commercial FE-code *Abaqus/Standard 5.8*[1] in a single processor version and *Abaqus/Standard 6.2* in a multi processor version. The simulation run under IBM SP-256 and IBM SP-SMP at the Super Computing Center (SSC) of University of Karlsruhe.

The numerical model is based on classical continuum mechanics. Phenomenological constitutive models for hardened concrete (foundation) and hardening concrete (piles) as well as sand were used accordingly. Two material models were implemented in *Abaqus/Standard* [5] using the user subroutine *umat*.

3.1   Geometry

The numerical model to simulate the model tests were generated for two situations:

– Double symmetric model for vertically loaded foundation and regular distributed initial void ratio $e(\mathbf{x}, t = 0)$, see Fig. 2a. The double symmetric foundation block isn't able to rotate.
– Single symmetric model for vertically loaded foundation with initial imperfections: eccentric load and un-regular distributed initial void $e(\mathbf{x}, t = 0)$, see Fig. 2b. The single symmetric foundation block is able to rotate about the 2-axis.

During the model tests a slight rotation as well as an un-symmetric loading of the piles were registered.

3.2   Material Model for Sand

The mechanical behaviour of sand is described with a rate-type hypoplastic constitutive model, see [8] for an introduction to the theory of hypoplasticity. The model for sand [2,3] is able to describe the stiffness and strength of sand depending on the mean stress p (barotropy) and the void ratio e (pyknotropy).

---

[1] Abaqus is a trademark of Hibbitt, Karlsson and Sorensen Inc., Rhode Island, USA

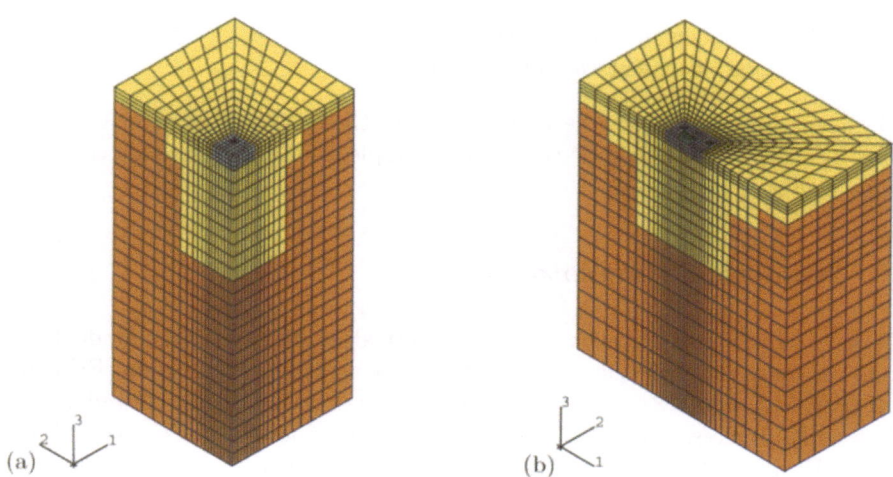

Fig. 2. Double symmetric model (a) and single-symmetric model (b) for simulation of subsequent piling of an existing foundation (colours: orange = permanent sand, yellow = filled sand, grey = foundation, green = loading plate, blue = piles)

Examples of numerical applications of this model can be found in [6,7,11,4]. The constitutive equation reads:

$$\overset{\circ}{\mathbf{T}} = \frac{f_s}{\mathrm{tr}\,(\hat{\mathbf{T}}\hat{\mathbf{T}})}\left[F^2\mathbf{D} + a^2\hat{\mathbf{T}}\,\mathrm{tr}\,(\hat{\mathbf{T}}\mathbf{D}) + f_d a F(\hat{\mathbf{T}} + \hat{\mathbf{T}}^*)||\mathbf{D}||\right] \qquad (1)$$

with Cauchy stress tensor $\mathbf{T}$, deviatoric stress tensor $\mathbf{T}^*$, normalized Cauchy stress tensor $\hat{\mathbf{T}} = \mathbf{T}/\mathrm{tr}\,\mathbf{T}$, Jaumann stress rate tensor $\overset{\circ}{\mathbf{T}} = \dot{\mathbf{T}} - \mathbf{W}\mathbf{T} + \mathbf{T}\mathbf{W}$, rate of deformation tensor $\mathbf{D}$ and spin tensor $\mathbf{W}$. a and F are scalar functions depending on stresses and one material parameter, the critical internal friction angle φ_c, only:

$$a = \frac{\sqrt{3}(3 - \sin\varphi_c)}{2\sqrt{2}\sin\varphi_c} \qquad (2)$$

$$F = \sqrt{\frac{1}{8}\tan^2\psi + \frac{2 - \tan^2\psi}{2 + \sqrt{2}\tan\psi\cos(3\theta)}} - \frac{\tan\psi}{2\sqrt{2}} \qquad (3)$$

with

$$\tan\psi = \sqrt{3}||\hat{\mathbf{T}}^*||$$

$$\cos(3\theta) = -\sqrt{6}\frac{\mathrm{tr}\,(\hat{\mathbf{T}}^3)}{[\mathrm{tr}\,(\hat{\mathbf{T}}^2)]^{3/2}}.$$

f_d and f_s are scalar factors introducing barotropy and pyknotropy:

$$f_d = \left(\frac{e - e_d}{e_c - e_d}\right)^\alpha \qquad (4)$$

$$f_s = \frac{\dfrac{h_s}{n}\left(\dfrac{e_i}{e}\right)^{\beta}\dfrac{1+e_i}{e_i}\left(-\dfrac{\operatorname{tr}\mathbf{T}}{h_s}\right)^{1-n}}{3+a^2-a\sqrt{3}\left(\dfrac{e_{i0}-e_{d0}}{e_{c0}-e_{d0}}\right)^{\alpha}} \tag{5}$$

e_c, e_d and e_i are void ratios for critical, most dense and most loose state depending on mean stress level. An explicit formulated compression law is introduced:

$$\frac{e_c}{e_{c0}}=\frac{e_d}{e_{d0}}=\frac{e_i}{e_{i0}}=\exp\left[-\left(-\frac{\operatorname{tr}\mathbf{T}}{h_s}\right)^{n}\right] \tag{6}$$

Finally the hypoplastic model consists eight material parameters (see [3] for a detailed calibration routine):

φ_c: Internal friction angle for critical states
h_s: Granulate hardness
n: Exponent in the compression law
e_{d0}:   Void ratio for most dense state and $\operatorname{tr}\mathbf{T}=0$
e_{c0}:   Void ratio for critical states and $\operatorname{tr}\mathbf{T}=0$
e_{i0}:   Void ratio for most loose state and $\operatorname{tr}\mathbf{T}=0$
α: Exponent controlling pyknotropy effect
β: Exponent controlling pyknotropy effect

The material parameters for the used medium quartz sand were calibrated using results from oedometer and triaxial tests and read:

$$\begin{aligned}
&\varphi_c = 32° & e_{i0} &= 1.20 & \alpha &= 0.11\\
&h_s = 3036~\mathrm{MN/m^2} & e_{c0} &= 1.09 & \beta &= 1.05\\
&n = 0.28 & e_{d0} &= 0.56 &&
\end{aligned}$$

Further five material parameters have to be included if intergranular strains are considered, see [10,11]. In Fig. 3 results of triaxial tests and corresponding numerical simulations for sand based on the hypoplastic model are presented. The cell pressure σ_3 is varied between 100 and 408 kN/m². The initial void ratio for the calculations were defined corresponding to the laboratory tests.

Measured and calculated curves in Fig. 3 coincide essentially so that the stiffness and strength of the sand can be described very realistic with the hypoplastic model. Both laboratory tests and calculations lead to an internal peak friction angle of $\varphi_p = 43°$ and to dilatant behaviour of an initial dense sand sample.

The constitutive model is implemented in *Abaqus/Standard* with the user routine *umat*. An explicit integration scheme with a sub-stepping algorithm was chosen for the integration of stresses on the local level [6] for the rate independent model. The tangent jacobian is calculated for the mid-point. Compared to elasto-plastic constitutive models hypoplastic models don't use any potential function to describe the hardening/softening or limit states of the material. Therefore the well-known integration schemes including return

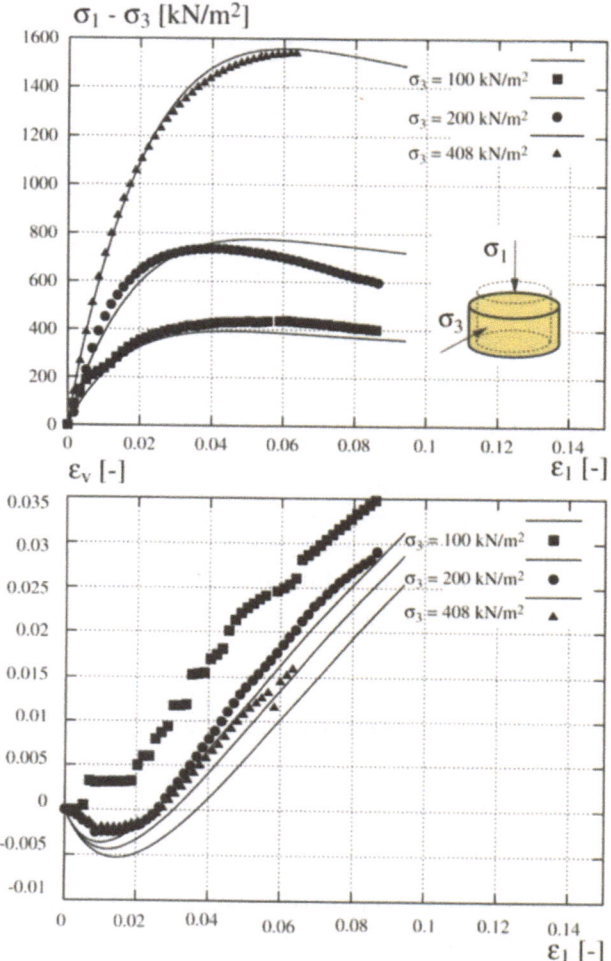

Fig. 3. Results of triaxial tests (dotted) and numerical simulations (line) for initially dense sand: (a) deviatoric stress $\sigma_1 - \sigma_3$ vs. axial strain ε_1, (b) volumetric strain $\varepsilon_v = \varepsilon_1 + 2\varepsilon_3$ vs. axial strain ε_1

mapping algorithms for such models cannot be used here. A consequence of this fact are smaller time steps (i.e. load increments) during the calculation to reach an accurate integration of stresses on the local level. Up to a certain degree this shortcoming can be handled by local sub-stepping algorithms. In any case the application of hypoplasticity in FEM leads to a remarkable increase of CPU-time compared to usual elasto-plastic models. Moreover an un-symmetric FE-solver have to be applied to get convergent global equilibrium iterations, even if nonlinear boundary conditions caused by soil/structure contact do not exist.

The hardened concrete was modeled elastic, the harding concrete with time depending elastic parameters $E(t)$ and $\nu(t)$.

3.3   Initial Conditions

In general the fields of initial stresses $\mathbf{T}(\mathbf{x}, t = 0)$ and initial void ratio $e(\mathbf{x}, t = 0)$ have to be defined for sand (the definition of initial values for $\mathbf{T}$ and e is needed since $\mathring{\mathbf{T}}$ depends on $\mathbf{T}$ and e – the hypoplastic model is not defined defined for vanishing smean stress $p \rightarrow 0$). Furthermore the history of filling the sandbox (partially permanent and filled sand) shown in Fig. 1 have to be taken into account by considering so-called inter-granular strains, see [10] for details.

A critical moment is the proof of the equilibrium of the initial state especially in the presence of contact surfaces sand/foundation and/or sand/pile.

3.4   Contact Conditions

The contact between sand and foundation and pile was modeled rigid or applying contact surfaces with a Coulomb friction model. The double symmetric FE-model has 5 contact surfaces with corners causing numerical instability during contact iteration. In certain cases this can be handled by coupling the corresponding degrees of freedom at corners. Considering contact surfaces made the problem very sensitive and reduces the time increments in general.

The modeling of contact surfaces around the piles which are installed under the existing and loaded foundation is more difficult. Where the installation of piles makes no problems for rigid contact, the allowance of relative motions along these contact surfaces makes problems. Different strategies to model this problem were studied. The best option seems to be the definition of contact surfaces already in the initial state where there is no pile. This means that the sand body has contact surfaces where the piles are installed later. But also with this technique convergent equilibrium iterations are not guaranteed.

3.5   Loading History

Beginning with the initial state the vertical load F_z on the concrete foundation is increased incrementally up to a given vertical displacement $u_{z,v}$. Then the piles are installed by removing the corresponding elements and adding new elements with material behaviour of a hardening concrete. After 28 days of hardening the load F_z on the piled foundation is increased up to a maximal value $F_{z,\max}$. For the purpose of comparison the foundation and subsequently piled foundation is simulated.

4   Results

4.1   Model tests

In Fig. 4 test results of model tests are presented. Tests for a pure foundation and tests with the subsequent piled foundation are distinguished. The
installation of the piles can be recognized at a vertical displacement of the
foundation of $u_{z,v} = 10$ up to 20 mm. Piled foundations show an improved
bearing behaviour. The different load-displacement curves result of varied
pile lengths and installation processes. Furthermore the test results show an
un-symmetric loading of the four piles due to rotation of the foundation block
and/or the one by one installation of the piles [9].

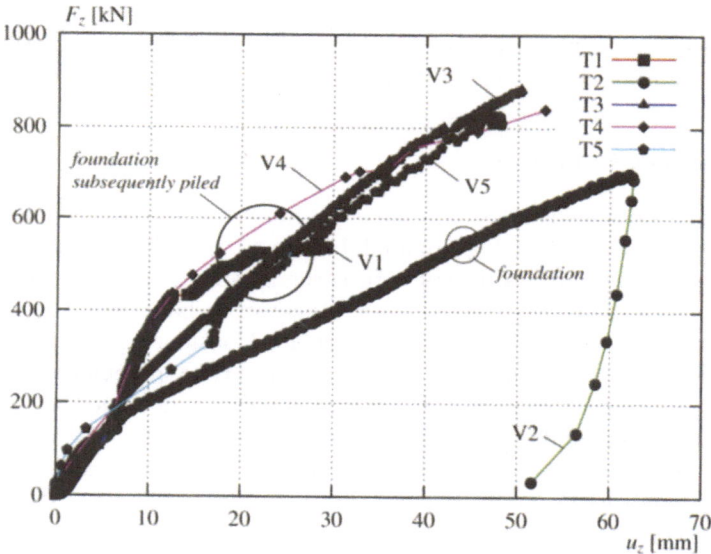

Fig. 4. Measured load-displacement-curves for foundation and subsequently piled
foundations [9]

4.2   Simulation of Model Tests

Double Symmetric Model: Figure 5 shows measured and calculated load-
displacement curves for the pure foundation.

The bearing behaviour of the pure foundation can be simulated realistically including the loading/unloading behaviour. The influence of the initial
void ratio on the bearing behaviour of the foundation can be clearly seen in
Fig. 5.

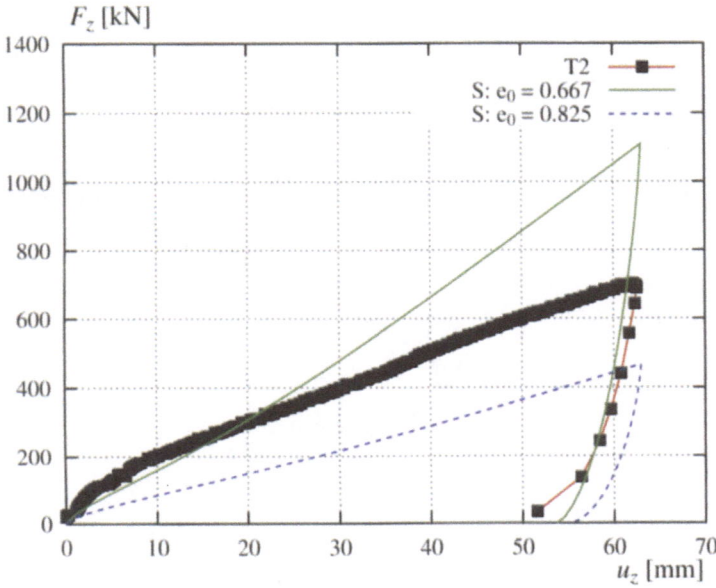

Fig. 5. Measured (dotted) and calculated (line) load-displacement curves for the pure foundation (initial void ratio e_0 is varied for numerical simulation)

A comparison of measured and calculated load-displacement curves of a subsequently piled foundation presented in Fig. 6 shows a good agreement. Differences exist where the piling process took place.

Figure 7 shows the vertical stress $\sigma_{33}(\mathbf{x}, t = 0)$ and $\sigma_{33}(\mathbf{x}, t)$ after reaching the maximal vertical load $F_{z,\max}$.

Single Symmetric Model: Simulations with the single symmetric model were carried out to enable rotations of the foundation block. An eccentricity of the load F_z of 5 mm and an un-regular distributed initial void ratio, see Fig. 8a, were defined.

Both imperfections doesn't lead to a remarkable rotation of the foundation during vertical loading, see distribution of vertical displacement u_z for maximal vertical load $F_{z,\max}$ in Fig. 8b. Reasons for this behaviour can be an inaccurate discretization of the piles and the surrounding soil, especially the pile tips, and an inaccurate simulation of the one by one installation of the four piles. Consequently the complete sandbox have to be discretized with a refined mesh in the region of the piles. Doubling the existing single symmetric model would lead to 60.000 d.o.f., mesh refinement to even more.

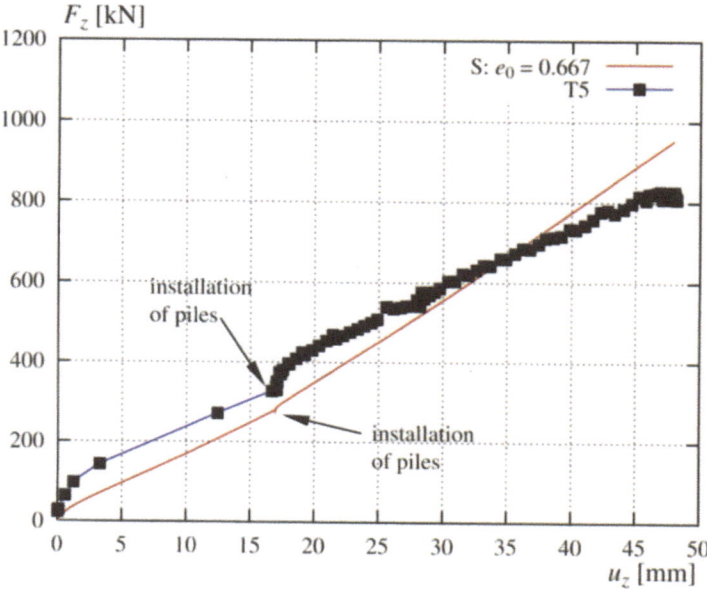

Fig. 6. Measured (dotted) and calculated (line) load-displacement curves for the subsequently piled foundation

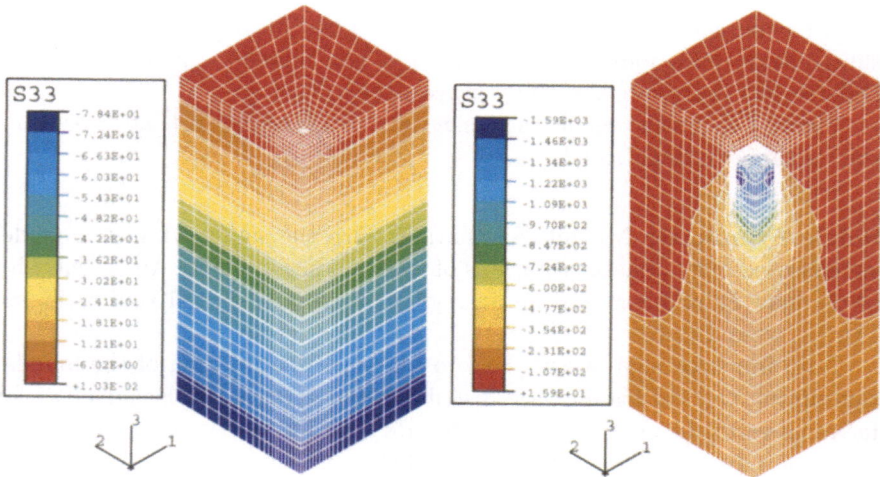

Fig. 7. FE-Simulation: initial stresses $\sigma_{33}(\mathbf{x}, t = 0)$ and stresses $\sigma_{33}(\mathbf{x}, t)$ after reaching the maximal load $F_{z,\max}$

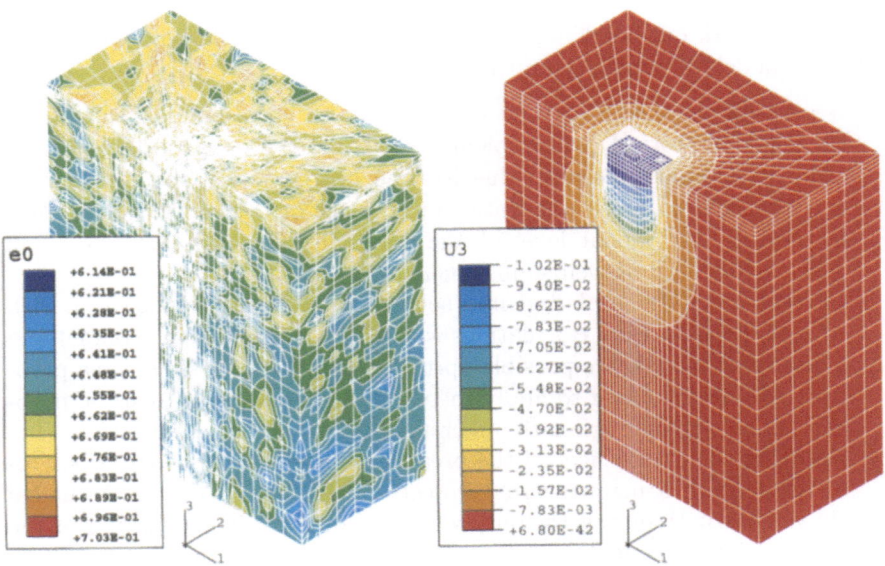

Fig. 8. FE-Simulation: initial void ratio $e(\mathbf{x}, t = 0)$ and vertical displacement $u_3(\mathbf{x}, t)$ for maximal vertical load $F_{z,\max}$

5   Computational Aspects

The largest presented FE-model (single-symmetric) consist of about 30.000 degrees of freedom. The corresponding FE-job had 2 GByte program and data size and lead to up to 92 hours of CPU-time (due to extreme usage os SP-SMP the real time was much higher). The double symmetric models run under a single-processor version of *Abaqus/Standard 5.8* on SP-256. The single symmetric model run under SP-SMP under a parallel version of *Abaqus/Standard 6.2*. Only up to four CPU's were used to avoid very jobs. A parallelization of *Abaqus/Standard* is efficient up eight cpu's. A more efficient alternative is the explicit FE-code *Abaqus/Explicit*, see next section for details.

6   Summary and Outlook

The simulations presented in this paper documents the problems of modelling geotechnical boundary value problems as mentioned in the last report [7]. Although the model tests could be simulated widely realistic, local phenomena cannot be predicted by FEM correctly. Therefore a complete discretization of the sandbox and a local refinement of the FE-mesh in the region of the piles would be necessary. To solve the corresponding FE-model the need of supercomputers is undisputable.

Further work on modelling geotechnical problems with hypoplastic constitutive models is necessary on the following fields:

- Recent developments in time integration and corresponding consistent tangent operators [1] will implemented in the the user routine *umat* in order to estimate the error due to time integration of the constitutive equation and to reach quadratic convergence of the Newton-Raphson's method on the global level of the FE-code. This will be realised in *Abaqus/Standard* as well as in *Abaqus/Explicit*.
- Current simulations of dynamic soil compaction are carried out with *Abaqus/Explicit*. First calculations with 3D-models including contact and hypoplastic models are possible up to 500.000 d.o.f. for short time problems. We are planning to apply the explicit FE-code also to static problems using dynamic relaxation technique.
- Applying the experiences of the presented project done at University of Kaiserslautern a new research project will be carried out at Technical University of Hamburg-Harburg where a complex 3D-simulation of the deformation behaviour of a container terminal in the harbour of Hamburg is investigated. This problem is characterized by different inclined concrete and steel piles, an irregular sheet-pile wall, in-stationary water conditions due to sea conditions and a inhomogeneous subsoil. The essential part of modelling such a problem is a sufficient discretization of the interaction between soil, water and the structure causes. In combination with constitutive models of the rate type this is a challenge in computer capacity and speed. Unfortunately this problem cannot be modeled with the explicit FE-code because a simulation of activating and reactivating elements is not implemented yet. The problem will be solved with *Abaqus/Standard*. For this task a parallel FE-code in combination with a supercomputer is inalienable.

Acknowledgment

The presented research project *Nachträgliche Fundamentverdübelung*, carried out at Universität of Kaiserslautern, was financed by Deutsche Forschungsgemeinschaft (DFG). The simulation were carried out at Super Computing Center (SSC), University of Karlsruhe. We want to thank *Clemens Howar* and *Dr. Paul Weber* for the co-operation.

References

1. Fellin W., Ostermann A.: Consistent tangent operators for constitutive rate equations. Accepted for publication in Numerical and Analytical Methods in Geomechanics, 2002
2. Gudehus G.: A comprehensive constitutive equation for granular materials. Soils and Foundations **36**, No. 1 (1996), 1–12, Japanese Society of Soil Mechanics

3. Herle I.: Hypoplastizität und Granulometrie einfacher Korngerüste. Veröffentlichungen des Instituts für Bodenmechanik und Felsmechanik der Universität Karlsruhe, Heft **142**, 1997

4. Herle I.: Sensitivitätsanalyse der Konsolidierungsberechnung. Bautechnik **78**, *627–634*, 2001

5. Hibbitt, Karlsson, Sorensen: User manual for FE-code *Abaqus/Standard*. HKS Inc., Rhode Island, USA, 2001

6. Hügel H.M.: Prognose von Bodenverformungen. Veröffentlichungen des Instituts für Bodenmechanik und Felsmechanik der Universität Karlsruhe, Heft **136**, 1996

7. Hügel H.M.: Numerical modelling of geotechnical boundary value problems. In High Performance Computing in Science and Engineering '01, Springer-Verlag, Heidelberg, 2001

8. Kolymbas D.: Introduction to hypoplasticity. Series Advances in Geotechnical Engineering and Tunnelling, No. **1**, Balkema, Rotterdam, 2000

9. Meißner H., Hügel H.M.: Tragverhalten von Fundamenten bei nachträglicher Verdübelung. Arbeitsbericht zum DFG-Projekt ME 501/13, Fachgebiet Bodenmechanik und Grundbau der Universität Kaiserslautern, 2001

10. Niemunis A., Herle I.: Hypoplastic model for soils with elastic strain range. Mechanics of Frictional and Cohesive Materials **2**, *279–299*, 1998

11. Von Wolffersdorff P.A., Schwab R.: Schleuse Uelzen I – hypoplastische Finite-Elemente-Analyse von zyklischen Vorgängen. Bautechnik **78**, *771–782*, 2001

Computer Science

Prof. Dr. Christoph Zenger

Institut für Informatik, Technische Universität München,
Boltzmannstraße 3, D-85748 Garching

In this section, three contributions with a close connection to Computer Science are presented.

The first two articles report on ongoing research published also in previous volumes of this series. "ParWave: Granularity in Parallel Wavelet Packet Video Coding" by Feil et al. investigates new compression schemes for video coding. These schemes need a large amount of computer power, and can be executed efficiently on modern high performance computers. Compression of video data is a topic of widespread active research with important applications in industry.

The second article, "Benchmarking Collective Operations with SKaMPI" by Worsch et al., reports recent results of a benchmark initiative at the University of Karlsruhe comparing the efficiency of MPI implementations on various high performance computer architectures. These results are of interest for the optimal design of parallel computer architectures, but they are also useful in order to decide which architecture is useful or even optimal for certain classes of algorithms.

The third article, "Computer Simulation of Internal Protein Dynamics: Are proteins intrinsically anharmonic?" by Hayward et al., is a typical example of the modern field of biocomputing or bioinformatics with the rapidly growing area of numerical simulation in biological processes. Molecular dynamics simulation is used as a tool to study dynamics of proteins on high performance computers.

It should be emphasised that many contributions in other sections of this volume are also related to Computer Science methods and tools, but are not presented here, because the application aspects of these articles seemed to be predominant.

ParWave: Granularity in Parallel Wavelet Packet Video Coding*

M. Feil and A. Uhl

RIST++ and Dept. of Scientific Computing, Salzburg University, Austria

Abstract. In this work, we discuss different granularities for parallel wavelet packet video coding using block-based motion compensation and the performance of the corresponding MPI implementations on the HLRS Cray T3-E. Two inter-frame based parallelization methods (group-of-picture parallelization and frame-by-frame parallelization) are compared against intra-frame parallelization. We highlight the advantages and drawbacks of all three approaches.

1 Introduction

Image and video coding methods that use wavelet transforms [11] have been successful in providing high rates of compression while maintaining good image quality and have generated much interest in the scientific community as competitors to DCT based compression schemes. With the finalization of the wavelet based JPEG2000 standard and the inclusion of a wavelet algorithm for synthetic/natural hybrid coding in MPEG-4 (VTC – visual texture coding) there is no doubt left that wavelet compression has to be considered state of the art nowadays.

Wavelet packets [12] represent a generalization of the method of multiresolution decomposition and comprise the entire family of subband coded (tree) decompositions. Whereas in the wavelet case the decomposition is applied recursively to the coarse scale approximations only (leading to the well known (pyramidal) wavelet decomposition tree), in the wavelet packet (WP) decomposition the recursive procedure is applied to all the coarse scale approximations and detail signals, which leads to a complete WP tree (i.e. binary tree and quadtree in the 1D and 2D case, respectively) and more flexibility in frequency resolution. In Fig. 1 we display three different decomposition schemes for the 2-D case.

There are several possibilities how to determine the frequency subbands suited well for a coding application. The WP "best basis algorithm" [2], as used in this work, performs an adaptive optimization of the frequency resolution of a complete WP decomposition tree by selecting the most suitable frequency subbands for signal compression. Note that WP decomposition is defined for JPEG2000 part II. In the area of 2-D motion-compensated video

---

* This work has been partially supported by the Austrian Science Fund (project 13903) and by the Austrian Academy of Sciences.

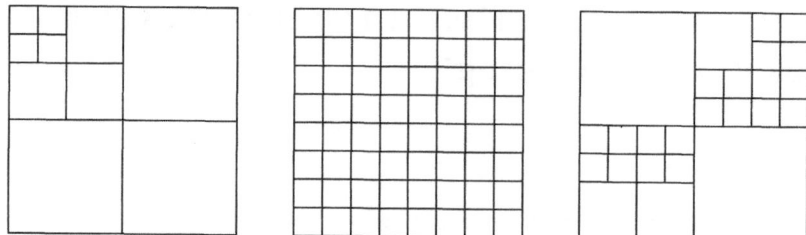

Fig. 1. Pyramidal wavelet decomposition, wavelet packet decomposition with uniform time-frequency resolution, and wavelet packet decomposition with arbitrary subband structure (each rectangle symbolizes a subband i.e. a single or multiple filtered and down-scaled version of the input image)

coding, wavelet packets have been applied successfully [8,1], however, at a significantly higher computational cost as compared to pyramidal coding. Therefore, a thorough examination of parallelization possibilities for such algorithms is desirable to close the gap to real-time processing.

A significant amount of work has been already done concerning parallel algorithms for the fast wavelet transform (see e.g. [13,9]) and for the parallel wavelet packet decomposition and its specific features and demands [5].

However, only little reasearch work has been done with respect to parallel 2-D wavelet-based video coding. Within the project "ParWave", we have focused onto a fully equipped motion-compensated wavelet packet video codec. In contrast to previous work [3] where simple arithmetic difference is used for motion compensation, we now employ a full search block-matching motion-compensation, which covers about 50% of the entire execution time. In Section 2, we describe the building blocks of the video codec in use. Sections 3 and 4 cover corresponding parallel algorithms and implementations. Examing the sequence of frames in the video stream to be encoded, we identify three obvious granularity levels: intra-frame parallelization, frame-to-frame parallelization, and group-of-picture parallelization. Independent of the results corresponding to parallelization efficiency, granularity has major impact with respect to the following important aspects of a video compression system:

- Hardware requirements: the more coarse grained the parallelization is, the higher are memory requirements.
- Coding delay: the more coarse grained the parallelization is, the higher is the coding delay.

We may derive immiediately that coarse grained parallelization will not be suited for real-time and on-line appilcations like video conferencing, since especially coding delay is not acceptable for these types of applications.

2   Wavelet Packet Video Coding with Motion Compensation

The WP video coder we use for practical experiments is composed of the following stages:

1. The video stream is processed with respect to a group of picture (GOP) structure: each GOP consists of 1 I-frame (which is intra-frame compressed, i.e. like a still image) and $gop\_size - 1$ P-frames (see Fig.2). The P-frames are inter-frame compressed, which means that based on a motion estimation process with respect to the reconstructed I-frame a displaced frame difference (DFD) is computed between I and P-frame.

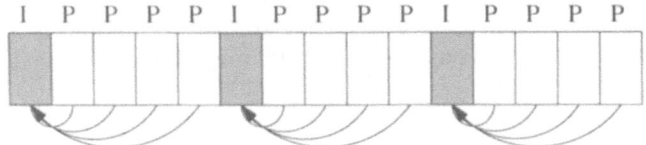

Fig. 2. A sequence of GOPs

2. DFDs are generated by full-search block motion estimation and compensation (BMC) algorithms using the classical non-overlapping BMC [10].
3. The succesive transformation, applied onto I-frame and DFDs, is realized through 2-D adaptive wavelet packet decomposition with Coifman's best basis algorithm (BBA).
4. The transform coefficients are encoded with recently developed extensions to SPIHT [7] which use importance maps instead of lists and are therefore much simpler to parallelize.

Table 1 shows average psnr results for the well known "Coastguard" test video sequence. The used decomposition strategies are the pyramidal wavelet transform, wavelet packets with the best additive cost measure (Norm l^1

Table 1. Avg. psnr performance for the "Coastguard" sequence (frame 100..199, BMC)

GOP_size	decomposition strategy				decomposition strategy		
	dwt	bba_norml	bba_shan.		dwt	bba_norml	bba_shan.
1	22.11	22.19	22.08		24.18	24.56	24.50
10	22.60	**22.74**	22.63		24.94	**25.33**	25.25
50	21.73	21.86	21.78		23.93	24.12	23.98

(a) bitrate=24kBps	(b) bitrate=48kBps

cost) and wavelet packets with the best non-additive cost measure (Shannon entropy cost). Especially for video sequences with distinct intensive motion (e.g. camera pans in "Coastguard") WP techniques may clearly outperform the classical pyramidal approach as seen in this example.

3   Fine Granularity: Intra-frame Parallelization

In order to distribute the computations associated with a single frame, several distinct operations and their corresponding parallelization strategies need to be considered: motion estimation and compensation (BMC), wavelet packet transform and best basis selection, and zerotree encoding. Finally, the concatenation of these techniques needs to be optimized. Note, that parallel BMC only needs to be considered if intra-frame parallelization of the video codec is investigated. The two more coarse grained techniques involve frames as the smallest datatype, consequently BMC may simply be distributed with the corresponding frame.

3.1   Parallelization of Motion Estimation and Compensation

According to the literature, we distinguish two variations of data distribution (see Fig. 3):

Stripe-based data partitioning The blocks of the prediction frame are partitioned into stripes of equal size and distributed among the PEs.

Block-based data partitioning The blocks of the prediction frame are distributed in a round-robbin fashion among the PEs, i.e. if b blocks with size (b_x, b_y) each are distributed among p PEs then block $j \in [0 \ldots b)$ is associated with PE $i = (j|p)i \in [0 \ldots p)$.

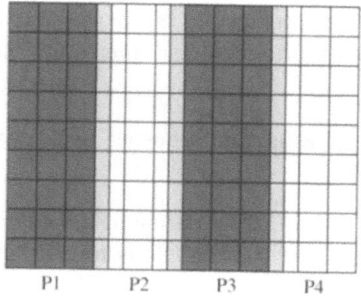
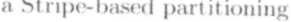
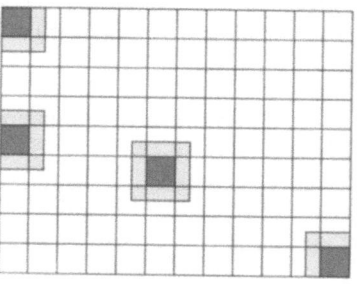

a Stripe-based partitioning   b Block-based partitioning

Fig. 3. Data partitioning for BMC parallelization

In order to avoid interprocessor communication during motion estimation an additional overlapping part of the data is sent to each PE during the initialization phase of both data partitioning schemes (depicted gray in the figure). The size of this overlapping part depends on the number of PEs p and the maximal displacement vector of the BME algorithm (s_x, s_y):

- For stripe-based data partitioning, each PE needs overlapping data of size $s_x * f_y * k$ with frame size (f_x, f_y) and $k \in (1, 2)$ depending whether the stripe is located at the edge or the center of the frame.
- For block-based data partitioning, each PE gets a maximal overlapping part of size $2s_x b_y + 2s_y b_x + 4s_x s_y$ for each block again depending on the location of the block. The overall overlapping amount is then approximately $(2s_x b_y + 2s_y b_x + 4s_x s_y) * \frac{b}{p}$.

Practical experiments (see Fig. 4 for for "Coastgard " and "Mobile" – framesizes 176×144 and 720×576 pixels, respectively) confirm that communication advantage in favour of stripe-based data partitioning results in a better overall performance despite of the distribution imbalance which occurs in some real-world cases (i.e. when $(\frac{f_x}{b_x}|p) \neq 0$).

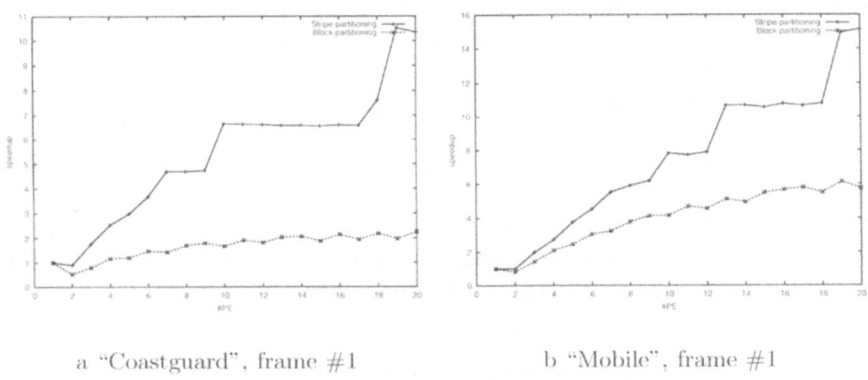

a "Coastguard", frame #1   b "Mobile", frame #1

Fig. 4. Parallel block motion estimation on Cray T3E

3.2 Parallelization of the WP transform

For the parallelization of a 2D WP transform we find two possible partitioning approaches.

Subband based (SB) partitioning On an architecture with a number of processor elements equal to a power of 4 this kind of partitioning is straight forward: each PE p (out of a pool of 4^a PEs) gets input data corresponding to a rectangle area of size 2^{Xmax-a} by 2^{Ymax-a} $(2^{Xmax}$

by 2^{Ymax} is the size of an input frame) and calculates the corresponding subbands situated in this area. During the first a levels each subband is shared by more than one PE. This implies additional communication between neighbouring PE after the calculation step to redistribute the results. During level a and higher, subbands and also the four children of each of them reside on each PE and no additional communication is needed. The generalization for an architecture with a number of PEs equal to a power of 2 does not need major changes, because without loss of generality we can assume that the computational work of two neighbouring PEs in a $2^a(a = 2 * b, b \in \mathbf{N})$ setting is merged into one if 2^{a-1} PEs are used. Finally we developed a WP decomposition for a parallel architecture with an arbitrary number of P PEs. The PE pool is rearranged according to the host -node principle: $2^{\lfloor \log P \rfloor}$ hosts are processing the data analog to the previous algorithm. Each of the remaining $P - 2^{\lfloor \log P \rfloor}$ nodes is assigned to one host to support it for calculation. Both share the computational work during the computations corresponding to one level: the node receives the input subband from its host, computes one half of the required subbands and returns the output. A sophisticated strategy for the assignment of nodes to appropriate hosts has to fulfill several properties (e.g. uniform distribution of the calculation work, avoidance of exaggerating communication etc.) and was presented in all details in [4].

Stripe (ST) partitioning In this partitioning scheme the input data is split into stripes of equal sizes and distributed among the PEs which perform the filtering in parallel on local data. Communication is needed in every step in order to provide the required border data to each PE.

In practical experiments [6] we show that the predominance of SB or ST partitioning depends on several parameters: the size of the frame and the decomposition filters, the level of decomposition, the underlying hardware (communication and computational speed). Therefore, for a parallel WP video coder both parallelization methods have to be taken into account.

3.3   Parallelization of zerotree coding

The parallelization of the original SPIHT algorithm reveals some drawbacks as e.g. complicated bit-stream handling, additional communication needs and non-neglectable sequential code parts. This is a direct consequence of the fact that the SPIHT algorithm is inherently sequential.

Therefore, we use for parallelization a modified zerotree algorithm developed by Kutil [7]. The basic idea is to substitute the lists of coefficient positions involved in the algorithm by bitmaps indicating the membership of each coefficient to a certain list. As a result, list iteration, which is used frequently to process the list entries, is turned into a normal scan of coefficients that follows a certain spatial direction. Thus, the data driven parallelisation can be performed easily by a loop parallelisation of the coefficient scan.

Although the resulting bit-stream will not be compatible with SPIHT, the parallelisation of the altered algorithm will, of course, be compatible with its sequential version.

3.4   Concatenation of Parallel BMC, WP Transform, and Zerotree Coding

Since each of the three coding steps of a WP video coder uses a distinct data partitioning method for parallelization resp. puts several partitioning possibilities at disposal, we now have to deal with an efficient concatenation of all parts.

Regarding parallel motion estimation, stripe-based data partitioning is in all cases the method of choice; zerotree coding can be parallelized in only one reasonable data partitioning fashion. Therefore two scenarios have to be take into account:

- WP Video Coder with stripe-based partitioned wavelet transform. All three coding steps use the same data partitioning and can be executed one after another without the need of redistributing any data.
- WP Video Coder with subband-based partitioned wavelet transform. Parallel WP transform with subband-based data partitioning is in most real-world settings the more efficient parallelization method but it shows a substantially different data partitioning strategy as the preceding parallel ME or the following parallel zerotree coding; a general rearranging of the processed data is needed. Since this change from stripes to subband parts and vice ver sa does not follow any regular pattern (e.g. communication only between a few selected PEs or data portions which remain on the same PE), one all-to-all communication step is required. In MPI V2.0 this kind of communication is implemented efficiently in the routine MPI_ALLTOALL(); the amount of redistributed data is given by the size of the frame, $f_x * f_y$.

3.5   Experimental results

Figure 5 shows the overall experimental speedup results for parallel WP video coding of one frame of the "Coastguard" and "Mobile" sequences. For comparison purposes, we also show results for a wavelet video coder based on the classical pyramidal discrete wavelet transform (DWT) using the stripe based data partitioning method. The overall ranking is clear: although having the lowest computational demand, DWT-based video coding shows the best speedup, followed by the stripe-based WP video codec. The subband-based WP video codec always gives the poorest results.

For small frame sizes, speedups are low and irregular. The larger frame size of the "Mobile" sequence allows for some speedup, however, scalability is low. In the following figures, we compare the three transform variants in

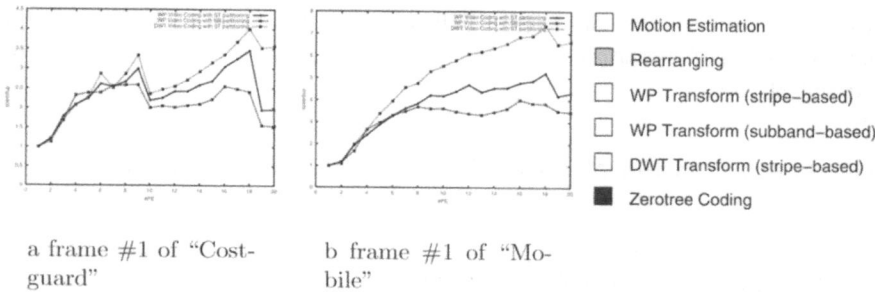

☐ Motion Estimation

▨ Rearranging

☐ WP Transform (stripe–based)

☐ WP Transform (subband–based)

☐ DWT Transform (stripe–based)

■ Zerotree Coding

a frame #1 of "Cost-
guard"

b frame #1 of "Mo-
bile"

Fig. 5. Speedup Results for fine-grained Wavelet Video Coding on Cray T3E are shown in (a) and (b); the legend on the right side explaines the textures used in Figs. 7 and 8.

some detail. We show the consumed processing time for each part of the video coder, i.e.:

- For a WP Video Coder with stripe-based partitioned wavelet transform (Fig. 6a,7a):
  1. motion estimation
  2. wavelet packet transform
  3. zerotree coding
- For a WP Video Coder with subband-based partitioned wavelet transform (Fig. 6b,7b):
  1. motion estimation
  2. rearranging from ME to WP transform
  3. wavelet packet transform
  4. rearranging from WP transform to zerotree coding
  5. zerotree coding

We employed the motion estimation algorithm using full search for a block size of 16*16 pixels and a search window of 32*32 pixels. The resulting DFDs

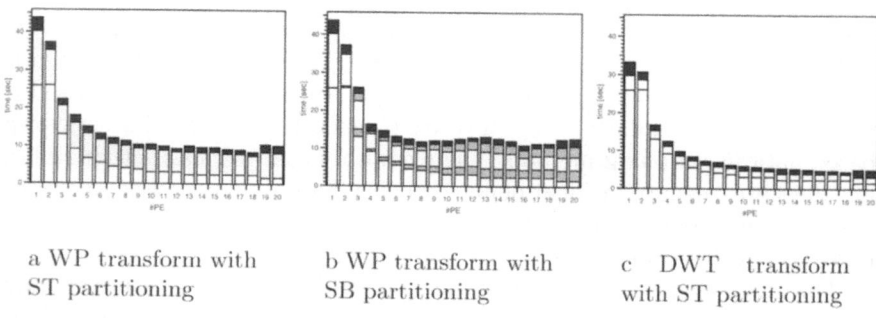

a WP transform with
ST partitioning

b WP transform with
SB partitioning

c DWT transform
with ST partitioning

Fig. 6. Parallel Wavelet Video Coding on Cray T3E: frame #1 of "Mobile"

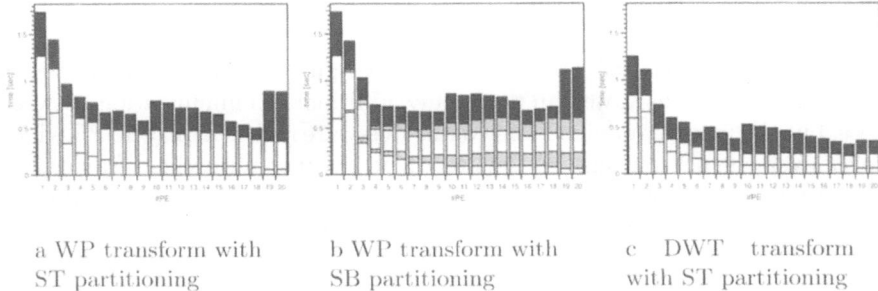

a WP transform with   b WP transform with   c   DWT   transform
ST partitioning   SB partitioning   with ST partitioning

Fig. 7. Parallel Wavelet Video Coding on Cray T3E: frame #1 of "Coastguard"

were decomposed into WP using the Biorthogonal 7,9 filter; the optimal representation was chosen by the best-basis algorithm with entropy information cost. Clearly, BMC shows constant time demand for all three types of codecs. With respect to WP decomposition, SB-based partitioning is, as stated in the last section, the more efficient parallelization method for the stand-alone WP transform. However, the preceding and following all-to-all communications tear down this advantage by far and the ST-based approach performs superior. Interestingly, the zerotree coding part is different whether DWT or WP decomposiiton is applied (which is clear since it involves data dependent operations (see Fig. 7).

4   Coarse Granularity: Inter-frame Parallelization

4.1   Frame-to-frame Parallelization

In this level of granularity we assign a single frame to each PE in a round-robin fashion (see Fig. 8). Since each P-frame is referring to its preceding I-frame, we find dependence between the PEs processing the I-frames (grey in Fig. 8) and all other PEs which process P-frames (depicted white).

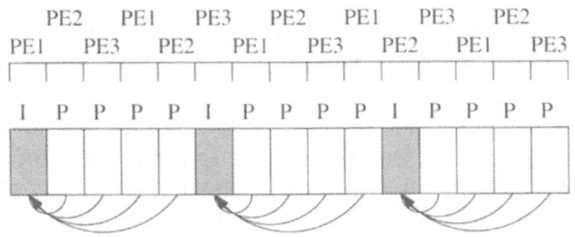

Fig. 8. Frame-to-frame parallelization

4.2   Group-of-picture Parallelization

In group-of-picture based parallelization we assign independent GOPs to each
PE (Fig. 9). Obviously, we don't have any additional communication between
PEs, but the memory demand and coding delay are rather high (one PE which
takes over a synchronising role has to cache the incoming video frames and
rearrange the outgoing bitstream). Therefore, this kind of parallelization is
more suitable for off-line video processing than real-time video processing.

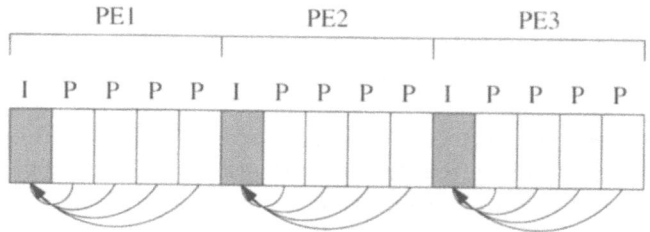

Fig. 9. GOP parallelization

It is clear that the GOP based algorithm has a strong tendency to load
unbalance if there are not enough GOPs (which is equal to not enough input
frames) to engage all PEs. In the worst case we find the same execution time
for $p_{size} * n + 1$ GOPs ($n \in \mathbf{N}$) as for $p_{size} * (n + 1)$ GOPs.

4.3   Experimental Results

In this section we show speedup results for the following experimental setting:
an incoming video stream of 200 frames is to be encoded. Each GOP (of size
10 resp. 30) starts with one I-frame (WP transformed and zerotree encoded),
followed by 9 resp 29 P-frames which are computed by block motion estimation between the reconstructed I-frame and the actual frame and afterwards
decomposed and encoded. As input video stream we used the "Coastguard"
sequence.

Figure 10 displays the corresponding results. For a GOP-size of 10 frames
(Fig. 10a) we find both GOP-based and frame-based algorithms equally performing. The plateau of the GOP-based algorithm after 10 PEs is due to the
non-uniform distribution of 20 GOPS on 11...19 PEs, but interestingly also
the frame-based algorithm exhibits this effect. This results from dependencies
among I- and P-frames which for certain settings are solved very efficiently
in the initialization phase, but remain decisive for other settings. Whereas
a setting of 10 PEs is able to arrange the incoming frames efficiently (after
the first I-frame we find nearly no parallelization overhead), on 11 PEs at
least one (or even more) PEs are idle during the computation for each GOP
because of waiting for the I-frame compression to finish.

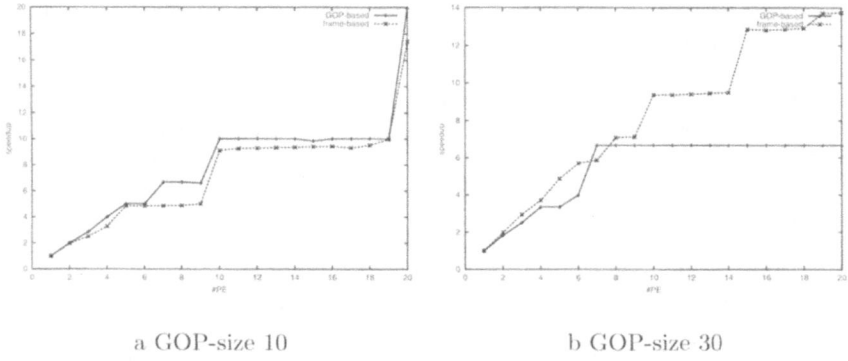

a GOP-size 10   b GOP-size 30

Fig. 10. Speedup of coarse grained parallelizations

When the GOP-size is raised to 30 frames (Fig. 10b), the plateau mentioned above is already detectable after 7 PEs for the GOP-based algorithm whereas the frame-based algorithm still scales quite effectively because the delaying I-frame appears less frequently. Comparing the behaviour of both coarse grained parallelization strategies we see that the frame based algorithm performs the better the *larger* a GOP is, but is quite independent of the overall number of input frames. The GOP based algorithm instead is sensitive to the *number* of GOPs (and therefore tightly coupled to the number of input frames): if there are enough GOPs available, the GOP based parallelization performs optimal in terms of speedup.

5 Conclusion

The granularity of choice is the GOP-based approach if the available memory is large enough to buffer the incoming video stream and if real time processing is not required. Frame based parallelization works very well likewise if the group-of-picture size is large enough to bypass dependencies among I- and P-frames. Fine grained parallelization of a wavelet packet video coder is sensible for moderate parallel systems and reasonably large frame sizes only. In contrast to previous image coding standards which exhibit an inherent parallelism due to their simple structure (like e.g. JPEG), the complicated algorithms involved in current tree-based wavelet video coding schemes seem to inhibit the design of intra-frame parallel implementations which scale to larger parallel systems.

References

1. P.Y. Cheng, J. Li, and C.J. Kuo. Video coding using embedded wavelet packet transform and motion compensation. In R. Juday and S.K. Park, editors, *Visual*

Information Processing V, volume 2753 of *Proceedings of SPIE*, pages 64–75, Orland, Florida, April 1996.

2. R.R. Coifman and M.V. Wickerhauser. Entropy based methods for best basis selection. *IEEE Transactions on Information Theory*, 38(2):719–746, 1992.

3. M. Feil, R. Kutil, R. Norcen, and A. Uhl. ParWave: Parallel wavelet video coding. In E. Krause and W. Jäger, editors, *High Performance Computing in Science and Engineering 2001*, pages 512–526, Stuttgart, Germany, 2002. Springer-Verlag.

4. M. Feil and A. Uhl. Multicomputer algorithms for wavelet packet image decomposition. In *Proceedings of the International Parallel and Distributed Processing Symposium (IPDPS'2000)*, pages 793–798, Cancun, Mexico, 2000. IEEE Computer Society.

5. M. Feil and A. Uhl. Wavelet packet image decomposition on MIMD architectures. *Real-time Imaging*, 2002. To appear.

6. M. Feil and A. Uhl. Wavelet packet zerotree image coding on multicomputers. In *Proceedings of the 10th Euromicro Workshop on Parallel and Distributed Processing*, pages 353–359. IEEE Computer Society, 2002.

7. R. Kutil. Parallelisable zero-tree image coding with significance maps. In R. Sakellariou et al., editors, *Parallel Processing. Proceedings of EuroPar'01*, volume 2150 of *Lecture Notes on Computer Science*, pages 674–677. Springer-Verlag, 2001.

8. D. Marpe, H.L. Cycon, and W. Li. Complexity constrained best-basis wavelet packet algorithm for image compression. *IEE Proceedings Vision, Image, and Signal Processing*, 145(6):391–398, 1998.

9. Ole Moller Nielsen and Markus Hegland. Parallel performance of fast wavelet transforms. *International Journal of High Speed Computing*, 11(1):55–74, 2000.

10. A.M. Tekalp. *Digital Video Processing*. Prentice Hall, One Lake Street, Upper Saddle River, NJ 07458, USA, 1995.

11. P.N. Topiwala, editor. *Wavelet Image and Video Compression*. Kluwer Academic Publishers Group, Boston, 1998.

12. M.V. Wickerhauser. *Adapted wavelet analysis from theory to software*. A.K. Peters, Wellesley, Mass., 1994.

13. L. Yang and M. Misra. Coarse-grained parallel algorithms for multi-dimensional wavelet transforms. *The Journal of Supercomputing*, 12(1-2):99–118, 1998.

Benchmarking Collective Operations with *SKaMPI*

Thomas Worsch[1], Ralf Reussner[2], and Werner Augustin[3]

[1] LIIN, Universität Karlsruhe, Germany, <worsch@ira.uka.de>
[2] DSTC, Monash University, Australia, <rreussner@dstc.monash.edu.au>
[3] LIIN, Universität Karlsruhe, Germany, <augustin@ira.uka.de>

Abstract. This article concentrates on recent work on benchmarking collective operations with *SKaMPI*. The goal of the *SKaMPI* project is the creation of a database containing performance measurements of parallel computers in terms of MPI operations. Its data support software developers in creating portable *and* fast programs. Existing algorithms for measuring the timing of collective operations are discussed and a new algorithm is presented, taking into account the differences of local clocks. Results of measurements on the HLRS Cray T3E/900 are presented and compared with other machines.

1 Introduction

The MPI standard defines a set of powerful collective operations useful for co-ordination and communication between many processes. Knowing the quality of the implementations of collective operations is of great interest for application programmers. In particular, one has to decide, whether to use predefined collective operations, which usually lead to more readable programs, or to implement collective operations by using point-to-point primitives. Similarly, it is often unclear, whether to use complex collective operations, like MPI_Reduce_Scatter, or to use more primitive collective operations (like in this case MPI_Reduce and MPI_Scatterv).

SKaMPI is the Special Karlsruher MPI-Benchmark [10]. It measures the performance of MPI implementations (which principally includes the performance of the underlying hardware). While other benchmarks (e.g., the one used in the famous top-500 list of super-computers) try to express a machine's performance in one figure to ease the ranking of machines, *SKaMPI*'s primary goal is to support software developers. By providing detailed data about the performance of each MPI operation, a software developer can judge the consequences of design decisions regarding the performance of the system to be built.

Until recently, *SKaMPI* measured the timing of collective operations with an approach relying on barrier synchronization. While this approach was state of the art when *SKaMPI* was designed in 1996, more recent work [3,5,4] shows that this approach – by including the time spent for the barrier operation – has a systematic error in case of overlapping collective operations. The recent approach to measure collective operations presented in [3] is restricted to measurements of the MPI_Bcast operation and is relatively slow.

The contribution of this paper is a general approach to measure the performance of arbitrary collective operations in a relatively efficient way. The main results are taken from [1].

This article is organized as follows. In Section 2 we motivate the measurement of collective operations and discuss existing approaches for defining and benchmarking the timing of collective operations. A new benchmark algorithm is described in Section 3. This new algorithm is compared to the algorithm previously used in *SKaMPI* in Section 4. Results from measurements on the Stuttgart Cray T3E/900 and the IBM RS 6000 SP at Karlsruhe are shown. The usage of the Cray T3E/900 is described in section Section 5. Section 6 concludes and discusses future work.

2   Background

MPI and Performance Portability. Software developers are faced with severals problems when designing and implementing code for parallel environments. Best performance is one of the primary development goals. This implies that the performance of a program has to be measured and tuned after its implementation during numerous cost- and time-intensive sessions. And software development is more profitable when the product can be used on several platforms, i.e., is running correctly and efficiently with a minimum of additional work. MPI [11] is a standard for a library to program message passing machines. It is a big step forward towards portability, since programmers now can rely on one interface standard, instead of several vendor-dependent interfaces. The MPI standard comprises several very similar functions, to support a wide range of machines and to exploit their particular advantages. As a consequence, MPI often offers many alternatives when designing and implementing a parallel algorithm. These alternatives offer a great potential for optimization.

The knowledge of the performance of several MPI functions allows the software developer to choose the right way of implementing a program for a given machine, with less tuning sessions. Performance issues can already be considered during the design stage. In fact, developing for performance can proceed even when the considered target machine is not readily accessible. Even a simple workstation can be used for cross-platform development, further lowering development cost.

The *SKaMPI* project supports the achievement of these aims of performance and performance portability in two ways: We offer a user configurable benchmark suite and a report generator, available on the web. Thus users can measure the performance of accessible machines in terms of MPI, generate a report and draw their own conclusions from this. And additionally we provide a public result database, where we store *SKaMPI*'s results from many machines.

Benchmarking Collective Operations. Collective operations are of particular interest for software developers [2,6]. They are coordination and communication operations in which usually more than two processes participate. Examples are broadcasts, where one process sends data to others, or gather, where one process collects data from others. The ability to perform complex coordination and communication patterns makes collective operations to one of the few higher-level constructs in the otherwise relatively non-abstract, close to hardware message-passing programming style. Hence, MPI provides a large variety of collective operations for complex data exchange operations [11]. Furthermore, they provide a large potential for MPI implementors to use optimized algorithms, which can make a complex collective operation faster than using semantically equivalent point-to-point operations in a straightforward manner.

Timing of Collective Operations. For point-to-point communication routines a generally accepted way for defining their timing is to take the time from sending a message from one process A to another process B and back until it is completely received by A (time for a "ping-pong").

However, since several processes are involved in a collective operation different definitions for the timing of collective operations are possible. It may be the time needed

1. for the collective operation at a designated process,
2. for the collective operation at the process which takes the longest time,
3. between the first process starting and the last process finishing it.

The approach previously used in *SKaMPI* is a variation of the first alternative. We measure the time used by a collective operation on the root node plus the time needed by the root node to finish a subsequent barrier synchronization. While the barrier synchronization ensures that the collective operation has finished on every node, it also has the severe drawback that the barrier operation can overlap with the measured collective operation. Therefore one cannot simply subtract the time used by a barrier alone to get the time used by the collective operation, and even worse, the barrier can distort the timing behavior of the collective operation. If one executes several measurements (and usually this is done to get more reliable results), then the barrier might not only overlap with the collective operation before, but also with the one following.

Problems of Benchmarking Collective Operations. However, without a barrier synchronization after the measured collective operation, one cannot ensure that all processes have finished the collective operation. This problem also occurs when the maximum of the execution times on all participating processes is taken (definition 2 in the above list). To see why the algorithm should wait until the collective operation finished on *all* processes, consider the case of a broadcast operation which sends a message to all its recipients as if they were ordered sequentially. The message is sent from process 0 to process 1, from 1 to 2, from 2 to 3, and so forth. This results in a communication time increasing linearly with the number of processes involved. The root process can carry on faster than in an algorithm with tree-like communication, though of course in general the latter is prefered.

As a side remark: Implementing these kind of benchmarking approaches can tempt MPI implementors to use bad algorithms (like the chain-structured one described above), because they lead to better looking benchmarking results than more sophisticated algorithms.

As a result of this discussion, we see that definition three is most suitable to define the timing of a collective operation. It circumvents the problems connected to definitions one and two and also has not the problems related to performing a barrier synchronization after each measured routine.

Existing Algorithms. The algorithm for measuring collective operations presented in [3] is able to measure the time between starting a collective operation on a root node and its completion on a selected other node. By varying the other node during repeated measurements, one gathers the times used by the collective operation for the root node to each other node.

The result of the measurement is the maximum of average times. But what we actually want to know is the average of the maximum of the times. Changing the order of doing a maximum and a mean value operation is only permitted if the communication patterns used by the MPI implementation are always the same. Imagine a MPI library alternating between two different implementations: the first one takes 1 time unit on node A and 5 on node B and the other one takes 5 units on A and 1 on B. Because every operation takes 5 time units the result should be 5, though the average time for every node is 3 and the maximum of these times would be 3 as well.

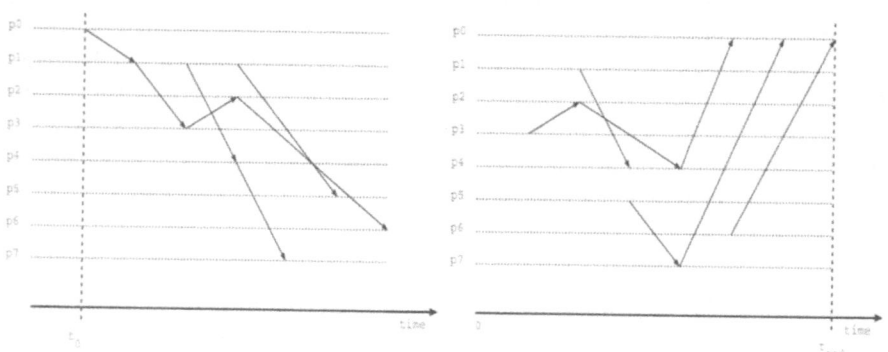

Fig. 1. A collective operation starting from a root process p0 (on the left) and one ending at p0.

Recently we started some investigations in testing the assumption mentioned above and other prerequisites of this algorithm [1]. The approach described above is limited to collective operations, where a designated root node starts the operation (e.g., MPI_Bcast). Such a case is shown on the left of Fig. 1, where a collective operation starts from a root node (p0). The operation ends on the different nodes at different times.

But in many collective MPI operations there is no designated root node starting it. As shown on the right in Fig. 1, a MPI_Gather operation has no common starting time on the different nodes.

An operation like MPI_Alltoall neither has a common starting nor a common end point. In this paper we present a novel approach overcoming this problem.

3   A new Time-Frame based Algorithm for Measuring Collective Operations

To avoid the problems mentioned above, one should try to make sure that all processes involved start at the same time using some sort of global time or clock synchronization.

A very elaborate and complex approach of clock synchronization is presented in [8]. This approach makes use of a log file, storing local times, which allows an off-line approximation of global times. Since our benchmark needs global times at run time and the creation of a log file in advance complicates the benchmark considerably, we preferred a simpler approach.

Augustin [1] therefore suggests a new approach using global time points which are obtained by approximating the offset o_i of the local time with respect to a global time on each process i. By definition let $o_0 = 0$, i.e. the local time of process 0 is the global time.

Computing the difference between two local times. Consider two processes named s and t in Fig. 2. Their local times are represented by the two horizontal arrows. Let's call s-time the local time of s and t-time the local time of t. Let Δ_{st} denote the amount of time by which process s is behind process t. In other words Δ_{st} is the t-time corresponding to s-time 0.

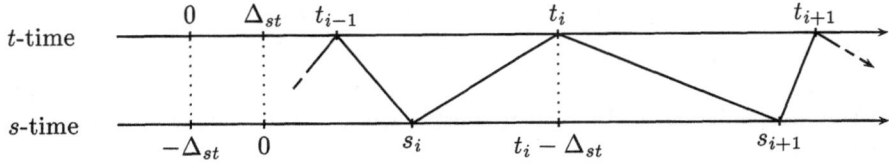

Fig. 2. Communication and timing principle for determining the local time offset.

Let the two processes send messages to each other in a ping-pong manner. A message leaving s at s-time s_i arrives at t at t-time t_i, and a message leaving t at t-time t_i arrives at s at s-time s_{i+1}.

Since t-time t_i corresponds to s-time $t_i - \Delta_{st}$ and a message cannot arrive before it has been sent, it is clear that $s_i < t_i - \Delta_{st} < s_{i+1}$. From this follow $\Delta_{st} < t_i - s_i$ and $\Delta_{st} > t_i - s_{i+1}$. If several messages i are sent, these inequations are true for all of them, and therefore $\max_i(t_i - s_{i+1}) < \Delta_{st} < \min_i(t_i - s_i)$. The times $t_i - s_i$ depend on the communication from s to t and the times $t_i - s_{i+1}$ on the communication from t to s. Minimizing $t_i - s_{i+1}$ and maximizing $t_i - s_i$ means to consider only the fastest communication. Under the assumption that its time does not depend on the direction it is reasonable to approximate Δ_{st} as $\Delta_{st} \approx (\max_i(t_i - s_{i+1}) + \min_i(t_i - s_i))/2$.

A simplified sketch of the algorithm to compute Δ_{st} on process t is given in Algorithm 1. Simple ping-pong messages are used in order to quickly get a sequence of points in time for which their order is known.

Algorithm 1. Sketch of a non-optimized algorithm to compute Δ_{st}:

```
if (pid = s) then                        else /* (pid = t) */
    Δlb ← −∞; Δub ← +∞;
    for i ← 0...N do
        slast ← MPI_Wtime();
        MPI_Send(....);                      for i ← 0...N do

        MPI_Recv(&tlast,....);                   MPI_Recv(....);
        snow ← MPI_Wtime();                      tlast ← MPI_Wtime();
        Δlb ← max(Δlb, tlast − snow);            MPI_Send(&tlast,....);
        Δub ← min(Δub, tlast − slast);
    od                                       od
    Δ ← (Δlb + Δub)/2;
```

Our experience with several computing environments shows that this indeed gives a good approximation of the difference between the local times on different processes. But one should be aware of the fact that the computed value is a good approximation only around the time when the measurements were done. There are machines where the local clocks are running at different speeds, i.e. the local time offsets change over time. They vary so much that after one hour (a time which a full *SKaMPI* run may need) the time offset computed at the beginning is not an adequate approximation any longer. It is therefore necessary to adjust the local time offsets every once in a while.

The use of global time in *SKaMPI*. In order to make use of global time some problems have to be overcome. They and their solutions are the topic of the next subsection. In this subsection we will discuss how global time is used in *SKaMPI*.

By definition the local time of process 0 in MPI_COMM_WORLD is the global time. For each $s \neq 0$ the algorithm sketched in Algorithm 1 is used to compute the offset o_s of s-time with respect to global time; process 0 sends the value o_s to process s. In the following we will denote global times by T (possibly with an index). Using its offset o_s each process s can easily translate global times to local times and vice versa.

In *SKaMPI* a *single measurement* is a single experiment to determine the time needed by an MPI-operation for a specific set of parameters, e.g. an MPI_Bcast for 8 processes. The outcome of a single measurement is a single number, e.g. $18.3\,\mu s$.

A single measurement of a collective MPI-operation is now roughly done as follows (for optimizations see the next subsection).

Algorithm 2. Using global time for benchmarking a collective operation:

1. Process 0 selects a future time slot characterized by T_{start} and T_{end}.
2. T_{start} and T_{end} are sent to all processes taking part in the operation.
3. At T_{start} each process calls the collective operation. If for some reason T_{start} has already passed on a processor this is an error.
4. Once the call returns on a process p, it checks whether the current global time T_p is less than T_{end}. If not, this is an error.
5. If on a process p an error has occured, T_p is set to a value indicating it.
6. Using an MPI_Gather all T_p are collected on process 0 and evaluated:
   - If all T_p are valid, their max is the result of the single measurement.
   - If at least one T_p indicates an error, the results are discarded.

Solutions to problems of global time. As mentioned above, the *SKaMPI* benchmark has to recompute the local time offsets from time to time. But the full computation for approximating the local time offsets (which is done when *SKaMPI* has been started) does take some time and should be done only if it's really necessary. Therefore an adaptive updating of the local time offsets has been implemented. After each measurement (i.e. a sequence of single measurements needed to determine a value with a sufficiently small deviation) Algorithm 1 is started again, but will be stopped prematurely if the time for the used ping-pong messages is within 5 % of the shortest time they needed in the initial (long) synchronization run.

There are several problems concerning the length of time slots:

- At the start a reasonable length for the time slots is completely unknown. *SKaMPI* is supposed to run on a wide variety of platforms; so even the order of magnitude of the time for a collective operation is unknown.
- If the time slots are too short, too many single measurements exceed T_{end} and have to be discarded.
- If the time slots are too long, precious time is wasted.
- Not every exceeding of T_{end} should lead to a prolongation of the time slot. It may simply be due to a rare operating system event.
- If a change of the length of time slots occurs all processes have to be informed about it. This administrative overhead should be kept small.

In order to mitigate these problems *SKaMPI* uses the following approach. So-called *rounds* of single measurements are used. A round is characterized by the number n of single measurements, the length ℓ of the time slots for all of them and the global starting time T_{start} of the first measurement. This implies that for $i = 0, \ldots, n-1$ the time slot of the ith single measurement of the round is $(T_{\text{start}} + i\ell, T_{\text{start}} + (i+1)\ell)$.

Only after a whole round is completed the measured times are collected and evaluated by process 0. If too many single measurements were invalid (because it took too long on a process or because a process could not start in time) the length of time slots is doubled and (if necessary) further increased so that all measurements of the previous round could have fitted 'comfortably' in their slots (i.e. one measurement would have used about $\frac{2}{3}$ of a slot).

If more than 50 % of all invalid measurements occur in a row, this is taken as a hint that possibly only one single event (e.g. an operating system interrupt) has delayed a whole sequence of measurements. Therefore the number of measurements in a round is reduced, but not below 4. To get a good estimate of the magnitude of the results (and therefore of the length of the time slots), the initial length is 0 and the first round is limited to 4 single measurements.

The numbers above have been found empirically and seem to give a good balance between administrative overhead on one hand and overhead due to invalid measurements on the other hand.

4   Results

Comparisons of measurements with the old and the new version of *SKaMPI*. We will now present benchmark results obtained on the Cray T3E of the HLR Stuttgart (T3E for short) and on the IBM SP-256 of the University of Karlsruhe (SP for short). As examples MPI_Barrier and a self-written collective waiting operation are used. Further results for other MPI operations can be found in [1].

The latter has been implemented in two variants, one where process i waits $(i+1) \cdot 100$ ns ("WaitPattern-up") and one where process i waits $(n-i) \cdot 100$ ns where n is the size of the communicator ("WaitPattern-down"). Obviously the time for the whole operation should be $n \cdot 100$ ns.

But a comparison of the results e.g. for WaitPattern-up (Fig. 3) clearly shows the difference between the old and the new measurement algorithm. The influence of the MPI_Barrier on the times measured is obvious.

Even measuring WaitPatter-up and Waitpatter-down with the same old measurement method shows surprising differences (upper graphs in Fig. 3). This clearly demonstrates that it is not possible to correct the result by simply subtracting the time for an MPI_Barrier operation.

Figure 4 shows very interesting measurements of MPI_Barrier. The old way of measuring it results in a sequence of two MPI_Barrier operations. As can be seen in the figure for the SP, this does in general *not* result in twice

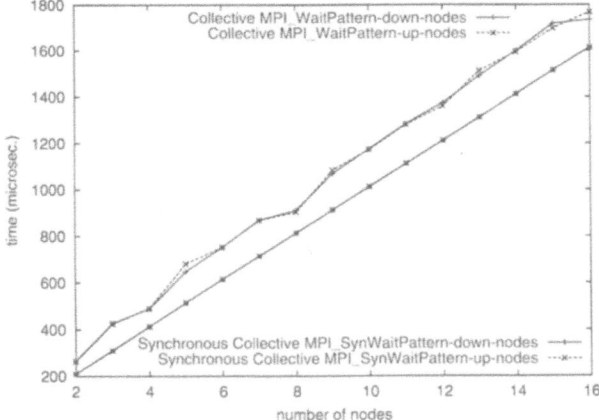

Fig. 3. Collective waiting operation measured with the old (upper graphs) and the new (lower graphs) *SKaMPI* on the SP. Graphs labeled "synchronous" always correspond to the new version.

the time measured by the new algorithm for one `MPI_Barrier`. This is due to the fact that the two operations partially overlap, at least on this machine.

On the other hand the figures for the T3E not only show that it has a very low latency and a much more regular pattern with steps at powers of 2, but that the times differ exactly (within the measurement precision) by a factor of 2, indicating that there is no overlap between the two operations.

Evaluation of the overhead of the new algorithm for benchmarking collective operations. In order to get an estimate of the overhead introduced with the new benchmarking algorithm for collective operations an instrumented version of the new *SKaMPI* was used.

An excerpt of an example evaluation of the overheads of the new method is shown below. It was obtained from a benchmark run with 76744 single measurements on 32 nodes on the SP. The following fractions of the total time (which increased by a factor of 1.96 over the old version) are measured:

	time [s]	percentage
communication time	326.81	24.6
synchronization time	136.01	10.3
wait time	320.15	24.1
late communication	73.38	5.5
too long communication	191.39	14.4
unnecessary communication	103.35	7.8
root comm	29.43	2.2
other overhead	145.42	11.0
total time	1325.94	100.0

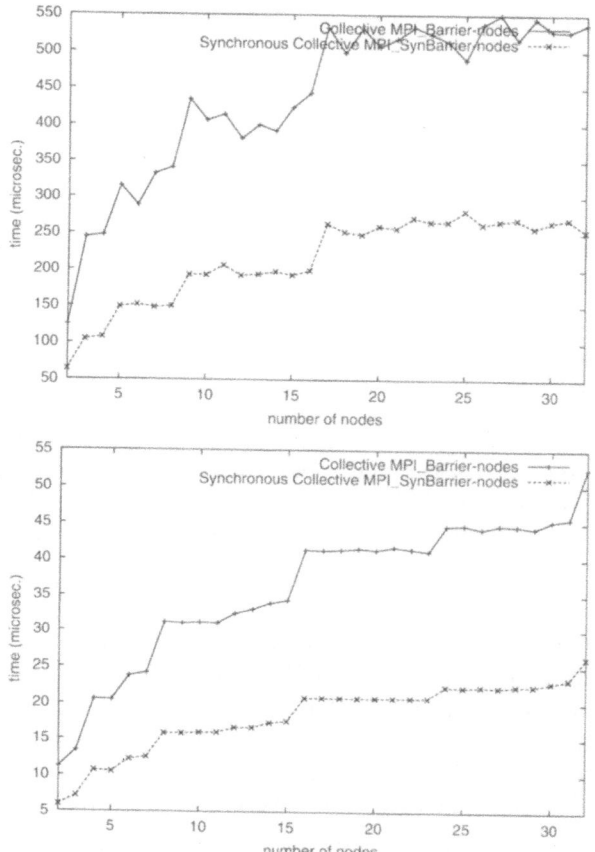

Fig. 4. MPI_Barrier measured with both *SKaMPI* versions on the SP (above) and the T3E (below).

communication time: This is the time spent for communications contributing to the benchmark numbers.

synchronization time: The time spent for computing and adapting the offsets of the local times.

wait time: The time wasted because time slots were too long and *all* processes were only waiting for the next time slot to come.

late communication: The time wasted because a single measurement started too late on a process and had to be discarded.

too long communication: The time wasted because a collective operation took too long on a process and the measurement had to be discarded.

unnecessary communication: The time wasted because all measurements of a round had to be done, though not all of them were needed for the desired precision of the measurement.

root communication: The time needed for communicating the measured
   times to process 0.
other overhead: all the rest.

The particular numbers vary quite a bit. Sometimes the wait time is significantly smaller while the time for unnecessary communications is significantly larger. In the example above the communication time contributes only about $1/4$ to the total running time. Of course, the old version of *SKaMPI* also has some overheads. Therefore altogether the new version of *SKaMPI* usually needs about a factor 2 to 2.4 longer than the old version. This is a significant amount, but it seems to be quite difficult to further improve the algorithm and its parameters in a way which is independent of a particular machine.

5   Typical usage of the Cray T3E

As the general development platform we mainly use the Karlsruhe IBM SP. The Stuttgart Cray T3E is primarily used for two important purposes. On the one hand we perform compatibility and portability checks with *SKaMPI*'s code after each change. As described in previous reports (e.g., [9]) this lead to significant changes of *SKaMPI*, which were not motivated by our tests of the IBM SP or our workstations. These runs typically use 4 to 32 processing elements and take up to 1 hour.

On the other hand we use the Cray T3E for performance measurements and for selected comparisons against other MPI implementations (as, for example, available on the WWW at `http://liinwww.ira.uka.de/~skampi/`). These runs can use up to 256 processing elements. Their running times vary heavily (between 30 min. and more than 2 hours).

6   Conclusions and future work in the *SKaMPI* project

After reviewing different definitions for the performance of a collective operation and discussing existing approaches and their limitations, we presented a novel approach for measuring the performance of collective operations. Our approach relies on approximations of the time differences between local clocks. This information is used to start a collective operation simultaneously on all processes. The measurements of time differences are controlled by an adaptive validation mechanism. The new approach is compared to the previous approach used in *SKaMPI*. Results of measurements on the Stuttgart Cray T3E and the Karlsruhe IBM SP are shown.

The performance of collective operations is interesting not only for application programmers but also for evaluating the performance of a connection network and the quality of an MPI implementation, since many message streams must be coordinated and conflicts and bottlenecks are to be avoided. Similarly, point-to-point communication is affected by traffic of other network nodes. For example, the b_{eff} benchmark is taking this into account when measuring the performance of point-to-point communication operations [7]. As further extensions to *SKaMPI* we plan to include measurements of the performance of more complicated patterns of communication, such as ring communications, simultaneous bidirectional communication, etc.

References

1. W. Augustin.   Messverfahren für kollektive Operationen.   Diplomarbeit, Fakultät für Informatik, Univ. Karlsruhe, December 2001.
2. V. Bala, J. Bruck, R. Cypher, P. Elustondo, A. Ho, C. Ho, S. Kipnis, and M. Snir. CCL: A portable and tunable collective communication library for scalable parallel computers. *IEEE Transactions on Parallel and Distributed Systems*, 6(2):154–164, 1995.
3. B. R. de Supinski and N. T. Karonis. Accurately Measuring MPI Broadcasts in a Computational Grid. In *Proc. 8th IEEE Symp. on High Performance Distributed Computing*, Redondo Beach, CA, 2000. IEEE.
4. W. Gropp and E. Lusk.   Reproducible measurements of MPI performance characteristics. In J. J. Dongarra, E. Luque, and T. Margalef (eds.), *EuroPVM/MPI 1999*, Barcelona, LNCS 1697, pp. 11-18. Springer, 1999.
5. R. Hempel. Basic message passing benchmarks, methodology and pitfalls, Sep. 1999. Presented at the SPEC Workshop (www.hlrs.de/mpi/b_eff/hempel_wuppertal.ppt).
6. P. Mitra, D. G. Payne, L. Shuler, R. van de Geijn, and J. Watts. Fast collective communication libraries, please. TR-95-22, Dept. Comp. Sc., Univ. Texas, 1995.
7. R. Rabenseifner. The parallel communication $b_{\textit{eff}}$ benchmark.
8. R. Rabenseifner. *Die geregelte logische Uhr, eine globale Uhr für die tracebasierte Überwachung paralleler Anwednungen*. Diss., Univ. Stuttgart, 2000.
9. R. H. Reussner. Recent Advances in SKaMPI. In E. Krause and W. Jäger (eds.), *High Performance Computing in Science and Engineering 2000*, Transactions of the HLRS Stuttgart, pp. 520–530, Springer, 2001.
10. R. H. Reussner, P. Sanders, L. Prechelt, and M. Müller. SKaMPI: A detailed, accurate MPI benchmark. In V. Alexandrov and J. J. Dongarra (eds.), *EuroPVM/MPI 1998*, Liverpool, LNCS 1497, pp. 52–59, Springer, 1998.
11. M. Snir, S. Otto, S. Huss-Lederman, D. Walker, and J. Dongarra. *MPI – The Complete Reference*, volume 1. MIT Press, Cambridge, MA, 2nd edition, 1998.

The Glass Transition in Proteins

Jennifer A. Hayward, Torsten Becker and Jeremy C. Smith

Lehrstuhl für Biocomputing, IWR, Im Neuenheimer Feld 368, Universität Heidelberg, D-69120 Heidelberg, Germany.

Abstract. The temperature dependence of the internal dynamics of an isolated protein, bovine pancreatic trypsin inhibitor, is examined using normal mode analysis (NMA) and molecular dynamics (MD) simulation. It is found that this model exhibits marked anharmonic dynamics, at temperatures much lower than previously detected in proteins, as evidenced by departure from the harmonic model mean-square displacement. A new method for determining mean-square displacements from elastic incoherent neutron scattering experiments is demonstrated and it is found to be more accurate than the method currently used by experimentalists.

1   Introduction

The temperature dependence of the internal motion of proteins and its relation to function have been much studied of late, using several experimental and theoretical techniques, including elastic incoherent neutron scattering, [1–10]. These studies have revealed a smooth change in the slope of the mean-square displacement of atoms in proteins in the temperature range ~170–240K. At temperatures below the transition, the dynamics is harmonic and vibrational. Above the transition anharmonic motions are also present. This change in dynamics resembles that seen in the liquid-glass transition [11]. The anharmonic dynamics involve diffusive motion and allow sampling of different conformational substates [12–15]. Some studies have found a correlation between the onset of the anharmonic motion and activity [16,9,17–19].

In the present paper we use computer simulation to examine the dynamical transition. The role of solvent is of particular interest. Neutron scattering experiments and computational work has indicated that the activation of anharmonic dynamics is strongly influenced by solvent [8,17,20–24]. However, the question as to whether solvent is *required* for the anharmonic dynamics to take place is as yet unanswered *i.e.*, it is unclear as to whether an isolated protein in the absence of solvent would have any dynamical transition, or whether it remains harmonic and vibrational up to room temperatures. Here, to answer this question, molecular dynamics (MD) simulations are performed of a model system, an isolated bovine pancreatic trypsin inhibitor molecule (BPTI), as a function of temperature. To set a "harmonic reference" normal mode analysis (NMA) is performed. In NMA a harmonic approximation to the potential energy function is made, and the resulting atomic trajectories are a superposition of vibrational modes. In contrast in MD simulations the full, anharmonic potential energy function is used. The temperature dependence of the mean-square displacements is calculated from both NMA and MD. Deviation of the MD displacements from the NMA values is evidence for anharmonic dynamics.

It has been found that the mean square displacements are not being accurately determined from elastic incoherent neutron scattering experiments [25]. The fact that there are a range of atomic mean-square displacements in a protein can reduce the value of the mean-square displacement obtained from an experiment by up to 30% from the correct value. Here, a we present another method for determining the mean-square displacement from neutron scattering experiments that avoids this problem.

2   Methods

2.1   Molecular dynamics simulations

The model system of BPTI consists of 892 atoms, 58 residues and 4 internal water molecules. To mimic typical experimental conditions, the exchangeable hydrogen atoms were replaced by deuterium, leaving 315 hydrogen atoms. This model system is an approximation to a dry powder sample, commonly used in neutron experiments.

The system was simulated using CHARMM [26] version 27 with parameter set 22 [27]. The four water molecules were represented by the TIP3P potential [28] and all hydrogen and deuterium atoms were included. The simulations were performed in the NVE ensemble. The maximum temperature fluctuations at the highest temperature were 300±20K. A timestep of 0.002ps was used with SHAKE [29] applied to constrain bonds containing hydrogen or deuterium atoms. Non-bonded and electrostatic interactions were truncated using the shifting function [30] at 13.0Å.

Simulations were performed at 19 different temperatures: 80 K, 100 K, 120 K, 140 K, then in steps of 10 K to 280 K, and finally 300 K. The starting structure for the first temperature simulated, 80 K, was the energy-minimized BPTI crystal structure [31]. The starting structures for the rest of the simulations at increasing temperatures were the final structures from the preceding temperature. The systems at each temperature were equilibrated for 50 ps and trajectories collected at every 0.1 ps for 2000 ps i.e., a total simulation time of 2050 ps per temperature i.e., 38.95 ns in total. Each simulation required 4 CPU hours running in parallel on 64 processors on the IBM RS/6000 SP-256 from the High-Performance Computing-Center Stuttgart.

2.2   Normal modes

Normal modes were computed using the VIBRAN module of CHARMM using the same model system and potential function as for the MD simulations. At low temperatures (<150 K) proteins are trapped in multiple minima leading to structural inhomogeneity [6,12]. This leads to dynamical inhomogeneity reflected in broadening of neutron and infrared spectra [32]. To account for this effect on the mean-square displacements 20 structures were taken from the 300K simulation trajectory at regular intervals and were energy minimized to a root-mean-square (RMS) energy gradient of $<10^{-7}$kcalmol^{-1}Å^{-1}. Normal modes were computed in each minimum. The eigenvalues and eigenvectors from the 20 different normal mode analyses were used to calculate the mean-square displacements. Both classical and quantum mechanical mean-square displacements were calculated and compared. The classical mean-square displacement, $\langle u^2 \rangle_{\text{class}}$, is given by,

$$\langle u^2 \rangle_{\text{class}} = \frac{2}{N_{\text{atom}}} \sum_{\alpha=1}^{N_{\text{atom}}} \sum_{k=7}^{3N_{\text{atom}}} (|c_\alpha^k|)^2 \left(\frac{k_{\text{B}}T}{m_\alpha \nu_k^2} \right)^2 \tag{1}$$

where N_{atom} is the number of atoms, c_α^k is the eigenvector for atom α in mode k, k_{B} is the Boltzmann constant, T is the temperature, m_α is the mass for atom α and ν_k is the eigenvalue, or frequency, for mode k. The equivalent quantum mechanical expression is,

$$\langle u^2 \rangle_{qm} = \frac{2}{N_{\text{atom}}} \sum_{\alpha=1}^{N_{\text{atom}}} \sum_{k=7}^{3N_{\text{atom}}} (|c_\alpha^k|)^2 \left[\frac{\hbar}{2m_\alpha \nu_k} \coth\left(\frac{\hbar \nu_k}{2k_{\text{B}}T} \right) \right]^2. \tag{2}$$

Two times the standard deviation in the harmonic mean-square displacements over the 20 calculations are shown as error bars on the figures in Section 3.

2.3   Time dependence

We also examine here the time dependence of the dynamical transition behaviour. This is of particular interest since it was demonstrated, using elastic neutron scattering from instruments of different energy resolutions, that dynamical transition behaviour of two enzymes in cryosolvents is strongly time dependent [33]. The time-dependent mean-square displacements for each atom α, $\langle u_\alpha^2(t) \rangle = \langle (\boldsymbol{R}_\alpha(t) - \boldsymbol{R}_\alpha(0))^2 \rangle$, were calculated from the MD configurations as follows,

$$\langle u_\alpha^2(t) \rangle = \langle (\boldsymbol{R}_\alpha(m) - \boldsymbol{R}_\alpha(0))^2 \rangle \approx \frac{1}{N_t - m} \sum_{k=0}^{N_t - m - 1} (\boldsymbol{R}_\alpha(k+m) - \boldsymbol{R}_\alpha(k))^2, \tag{3}$$

where the steps in the trajectory are denoted by $k = 0, .., N_t - 1$ and N_t is the total number of timesteps.

For comparison with the appropriate neutron-derived quantity the mean-square displacements are weighted by the scattering cross-sections,

$$\langle u^2(t) \rangle = \frac{1}{N_{\text{atom}}} \sum_{\alpha=1}^{N_{\text{atom}}} b_\alpha^2 \langle u_\alpha^2(t) \rangle, \tag{4}$$

where b_α is the incoherent scattering cross section for atom α.

2.4   Experimental mean-square displacement

In this section we will examine two methods for determining mean-square displacements from incoherent elastic neutron scattering experiments. This is done by using the atomic trajectories from the MD simulations as input into the equations for the neutron scattering data analyses. The mean-square displacements calculated directly from the MD simulations are then considered to be the "true" mean-square displacements and they are directly compared with those calculated from the experimental data analyses.

The incoherent elastic scattering of proteins can be calculated from the atomic trajectories from the MD simulations by

$$S_{\text{inc}}(q, \omega = 0) = \frac{1}{N_{\text{atom}}} \sum_{\alpha} b_{\alpha}^2 e^{-\frac{q^2}{6} \langle u_{\alpha}^2 \rangle}, \tag{5}$$

where $\langle u_{\alpha}^2 \rangle$ is the $t \to \infty$ mean-square displacement of atom α. Equation 5 describes the so-called Gaussian approximation for single atoms. The range of applicability of this approximation in a protein has been examined. It was found to be valid at values of $q < 1.2 \text{Å}^{-1}$ ([25]) .

In experimental data treatments Eq. 5 is simplified even further by replacing the sum of Gaussian functions by a single Gaussian:

$$S_{\text{inc}}(q, \omega = 0) = \frac{1}{N_{\text{atom}}} \sum_{\alpha=1}^{N_{\text{atom}}} b_{\alpha}^2 \exp(-\langle u_{\alpha}^2 \rangle \, q^2/6) \approx b^2 \exp(-\langle u^2 \rangle \, q^2/6), \tag{6}$$

where $\langle \Delta u^2 \rangle$ is the average mean-square displacement and b^2 denotes an average scattering cross section. By plotting the logarithm of Eq. 6 against q^2 the $\langle \Delta u^2 \rangle$ can be calculated from slope of the resulting curve.

However, Eq. 5 involves a sum of Gaussians and a sum of Gaussians will not necessarily be Gaussian. A better approximation to Eq. 5 can be obtained from the cumulant expansion:

$$S_{\text{inc}}(q, \omega = 0) = e^{-\frac{1}{6}q^2 \langle \Delta u^2 \rangle} \left(\frac{1}{N} \sum_{l}^{N} e^{-\frac{1}{6}q^2 (\langle \Delta u_l^2 \rangle - \langle \Delta u^2 \rangle)} \right) \tag{7}$$

$$S_{\text{inc}}(q, \omega = 0) = e^{-\frac{1}{6}q^2 \langle \Delta u^2 \rangle} \left(\sum_{m=0}^{\infty} \frac{1}{m!} \left(\frac{-q^2}{6} \right)^m \mu(m) \right) \tag{8}$$

$$S_{\text{inc}}(q, \omega = 0) \approx e^{-\frac{1}{6}q^2 \langle \Delta u^2 \rangle} \left(1 + \frac{q^4}{72}\sigma^2 \right) \tag{9}$$

where $\mu(m)$ are the m'th central moment of the distribution of $\langle \Delta u^2 \rangle$s, i.e. $\mu(m) = \frac{1}{N} \sum_{l}^{N} (\langle \Delta u_l^2 \rangle - \langle \Delta u^2 \rangle)^m$. σ^2 is the variance of the distribution of $\langle \Delta u^2 \rangle$. Equation 9 introduces one more parameter into the analysis of experimental scattering spectra.

3   Results

3.1   Timescale dependence

In Fig. 1 $\langle u^2(t) \rangle$ is shown at timescales, t, ranging from 10 ps to 1000ps for each of the 19 temperatures sampled. The results are from the MD simulations, together with the quantum and classical quantities from the harmonic analysis. The quantum correction adds $\sim$30% to the $\langle u^2 \rangle_{\text{class}}$ at 80K, this fraction dropping to below 10% at 300 K. $\langle u^2 \rangle_{\text{qm}}$ is formally nonlinear in T (see Eq. 2). However, the deviation of $\langle u^2 \rangle_{\text{qm}}$ from linearity in Fig. 1 is clearly very small and demonstrates that $\langle u^2 \rangle \propto$ T is a good approximation in the harmonic regime. At 80 K the classical MD $\langle u^2(t) \rangle$ at 10ps, 100ps, 500ps and 1000ps and the harmonic $\langle u^2 \rangle_{\text{class}}$ coincide. This indicates that at 80 K the dynamics is entirely described by fast vibrational motions in the protein. The lowest frequency vibration in the harmonic analysis is $6.7 \, \text{cm}^{-1}$, corresponding to a period of 5ps. Consequently the modes are adequately sampled in timescales as low as 10ps.

The MD $\langle u^2(t) \rangle$ at temperatures $>$ 100K show a timescale dependence. Already at 150 K $\langle u^2(10ps) \rangle < \langle u^2(100ps) \rangle < \langle u^2(500ps/1000ps) \rangle$, indicat-

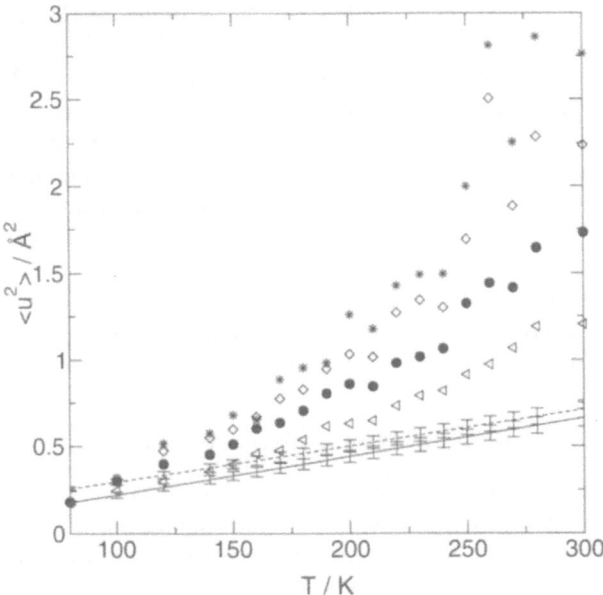

Fig. 1. The mean-square displacement per temperature and for timescales of ($\triangleleft$) 10ps, ($\bullet$) 100ps, ($\diamond$) 500ps, and ($*$) 1000ps. The values are weighted by incoherent scattering lengths. The lines with error bars (—) classical, and (- - -) quantum mechanical, are harmonic mean-square displacements. The error bars are twice the standard deviation in the harmonic mean-square displacements at each temperature point.

ing the presence of slow motions on the 10-1000ps timescale. These differences are considerably larger at higher temperatures, and at 270-300K $\langle u^2(500ps/1000ps)\rangle$ are about twice as large as $\langle u^2(10ps)\rangle$.

The MD profiles for 100ps and longer times deviate significantly from the classical harmonic results for T > 100 K, indicating the presence of anharmonic dynamics. For these timescales, anharmonic motions contribute the major part of $\langle u^2(t)\rangle$ at physiological temperatures. The faster, 10ps-timescale dynamics starts to deviate from harmonic behaviour at a higher temperature, T >∼ 180 K. That the onset of fast anharmonic dynamics is at a higher temperature than for slow dynamics is consistent with experimental results on proteins that demonstrate a timescale dependence, and with the presence of activated dynamical processes in the protein [33]. At timescales greater than 100ps the $\langle u^2(t)\rangle$ profiles are considerably rougher at high temperatures than at shorter timescales. This may reflect the presence of long-timescale structural transitions in the protein that are incompletely sampled in the present simulations.

In powdered dry protein systems deviation from harmonic behaviour has not been reported experimentally. The present results suggest that this may be due to the absence of a harmonic reference profile, such as is provided here by the NMA, in the experimental analyses. Examination of Fig. 1 shows that the MD results alone, considered in the absence of the harmonic baseline, can be well-fitted by straight lines up to 240 K. At higher temperatures a small deviation from linearity is seen. However, at 300 K, this is at most 10%. The 500ps/1000ps increased mobility is due to increased flexibility in the loops at the C-terminal and N-terminal ends of the protein. In the absence of the harmonic baseline this would lead to the conclusion that little anharmonic behaviour is present. However, as the present MD/NMA comparison shows, deviation from harmonic dynamics is present at temperatures as low as ∼100 K.

For completeness, in Fig. 2 the long-time evolution of $\langle u^2(t)\rangle$ at the highest temperature, 300 K, is shown. Diffusion occurs at a steady rate from 200–900ps, but from 900–1000ps $\langle u^2(t)\rangle$ appears to be converging.

3.2   Experimental mean-square displacement

Fig. 3 show a comparison of mean-square displacements obtained by various methods. (◇) is the mean-square displacement taken directly from the simulation and serves therefore as a reference value for $\langle \Delta u^2\rangle$ derived by experimental analysis. (●) was obtained by using the typical neutron scattering data treatment process i.e., from Eqs. 5 and 6 with q-range of $0 < q^2 < 1.4\,\text{Å}^2$. As already demonstrated in [25], use of this method leads to an underestimation of $\langle \Delta u^2\rangle$ of ∼ 20%. Introducing a second parameter in the analysis, however, increases the accuracy of $\langle \Delta u^2\rangle$ substantially. (△) is the resulting $\langle \Delta u^2\rangle$ obtained by fitting Eq. 9 within a q-range of $0 < q^2 < 1.4\,\text{Å}^2$. Now, satisfactory agreement with the reference value is achieved. The error

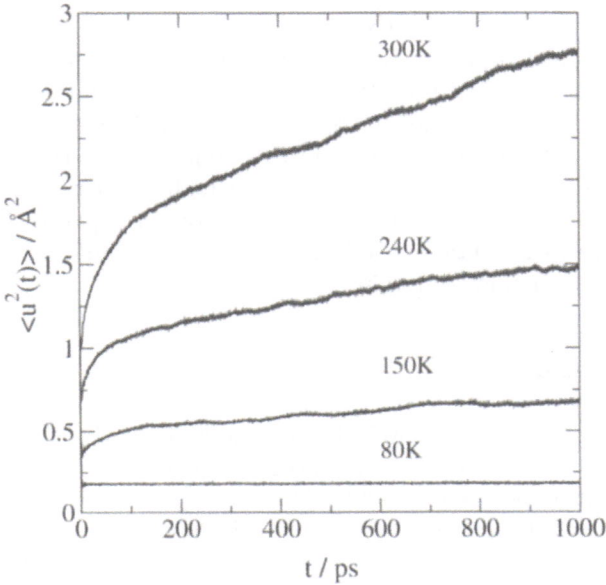

Fig. 2. The long-time evolution of the mean-square displacement, $\langle u^2(t)\rangle$ for the protein at $300\,\mathrm{K}$.

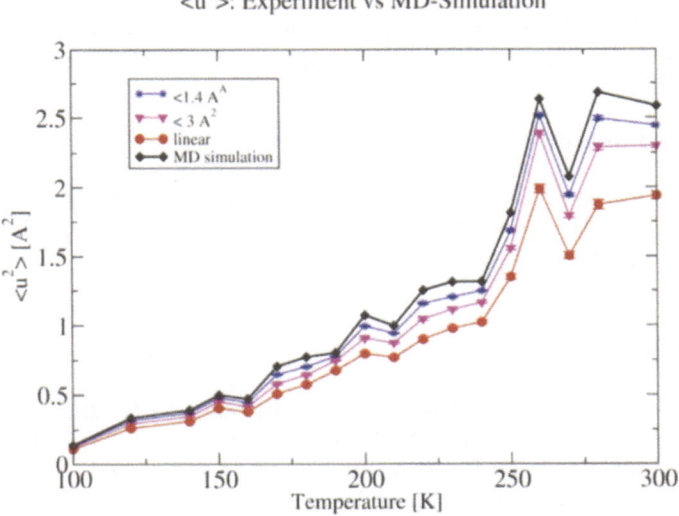

Fig. 3. $\langle\Delta u^2\rangle$ calculated using various methods. Black ($\diamond$): MD-simulation; red ($\bullet$): fit of Eq. 6 (linear) for $0 < q^2 < 1.4\mathrm{\AA}^2$; blue ($\triangle$): fit of Eq. 9 for $0 < q^2 < 1.4\mathrm{\AA}^2$; magenta ($\triangledown$): fit of Eq. 9 for $0 < q^2 < 3\mathrm{\AA}^2$

in $\langle \Delta u^2 \rangle$ decreases to $\sim 6\%$. Increasing the q-range used in this analysis to $0 < q^2 < 3 \text{ Å}^2$ allows the inclusion of more data points, thereby increasing the statistical accuracy of the fitting procedure. The resulting $\langle \Delta u^2 \rangle$ is also shown in Fig. 3. The error now increases to $\sim 10\%$. This is still only half the error obtained with conventional analysis. The increase in the error of $\langle \Delta u^2 \rangle$ with an increasing q-range demonstrates that higher-order corrections in Eq. 8 have to be taken into account. The number of extra parameters that can be used in an experimental analysis, however, will be limited by the amount and quality of available data points.

4   Conclusions

Through comparison between MD-derived and NMA-derived $\langle \Delta u^2 \rangle$ of a small protein over a wide range of temperatures, it has been found that the dynamical transition occurs at $\sim 100K$. This is a much lower temperature than previously detected experimentally. Timescale dependence of the anharmonic dynamics is also observed, as was seen in experiments on enzymes in solution [34,33,24]. Moreover we have demonstrated here that dynamical heterogeneity in a protein molecule leads to errors in the experimentally measured $\langle \Delta u^2 \rangle$. Taking into account one additional parameter related to the variance of $\langle \Delta u^2 \rangle$ substantially improves the experimentally-derived $\langle \Delta u^2 \rangle$.

Because of the relatively large number of simulations performed (18) and the long MD timescales required for each simulation (2050ps each), these results could not have been realised without the use of the IBM RS/6000 SP-256 from the High-Performance Computing-Center Stuttgart.

Acknowledgments

High-Performance Computing-Center in Stuttgart for time on the IBM RS/ 6000 SP-256. Sirus Zarbakhsh for calculating mean-square displacements from the normal modes analysis.

References

1. W. Doster, S. Cusack, and W. Petry, Nature **337**, 754 (1989).
2. J. Fitter, R. E. Lechner, and N. A. Dencher, Biophys. J. **73**, 2126 (1997).
3. J. C. Smith, Quarterly Review of Biophysics **24**, 227 (1991).
4. J. C. Smith, Inelastic and quasielastic neutron scattering: complementarity with biomolecular simulation, in Structure and dynamics of biomolecules: neutron synchroton radiation for condensed matter studies, pages 161–180, Oxford, 2000, Oxford University Press.
5. E. W. Knapp, S. F. Fischer, and F. Parak, J. Phys. Chem. **86**, 5042 (1982).
6. F. Parak and E. W. Knapp, Proc. Natl. Acad. Sci. USA **81**, 7088 (1984).
7. L. Cordone et al., Eur. Biophys. J. **27**, 173 (1998).

8. F. Demmel, W. Doster, W. Petry, and A. Schulte, Eur. Biophys. J. **26**, 327 (1997).

9. B. F. Rasmussen, A. M. Stock, D. Ringe, and G. A. Petsko, Nature **357**, 423 (1992).

10. J. C. Smith, K. Kuczera, and M. Karplus, Proc. Natl. Acad. Sci. USA **87**, 1601 (1990).

11. C. A. Angell, Science **267**, 1924 (1995).

12. R. Elber and M. Karplus, Science **235**, 318 (1987).

13. M. Karplus and G. A. Petsko, Nature **347**, 631 (1990).

14. H. Frauenfelder, S. G. Sligar, and P. G. Wolynes, Science **254**, 1598 (1991).

15. G. R. Kneller and J. C. Smith, J. Mol. Biol. **242**, 181 (1994).

16. F. Parak et al., FEBS Letters **117**, 368 (1980).

17. M. Ferrand, A. J. Dianoux, W. Petry, and G. Zaccai, Proc. Natl. Acad. Sci. USA **90**, 9668 (1993).

18. X. Ding, B. Rasmussen, G. A. Petesko, and D. Ringe, Biochemistry **33**, 9285 (1994).

19. A. Ostermann, R. Waschipky, F. G. Parak, and G. U. Nienhaus, Nature **404**, 205 (2000).

20. J. Fitter, S. A. W. Verclas, R. E. Lechner, H. Seelert, and N. A. Dencher, FEBS Letters **433**, 321 (1998).

21. J. Fitter et al., Eur. Biophys. J. **27**, 638 (1998).

22. J. Fitter, Biophys. J. **76**, 1034 (1999).

23. D. Vitkup, D. Ringe, G. A. Petsko, and M. Karplus, Nature Structural Biology **7**, 34 (2000).

24. V. Réat et al., Proc. Natl. Acad. Sci. USA **97**, 9961 (2000).

25. J. A. Hayward and J. C. Smith, Biophys. J. **82**, 1216 (2002).

26. B. R. Brooks et al., J. Comput. Chem **4**, 187 (1983).

27. A. D. Mackerell et al., J. Phys. Chem. B **102**, 3586 (1998).

28. W. L. Jorgensen, J. Chandrasekhar, J. D. Madura, R. W. Impey, and M. L. Klein, J. Chem. Phys. **79**, 926 (1983).

29. J. P. Ryckaert, G. Ciccotti, and H. J. C. Berendsen, J. Comput. Phys. **23**, 327 (1977).

30. P. J. Steinbach and B. R. Brooks, J. Comp. Chem. **15**, 667 (1994).

31. S. Parkin, B. Rupp, and H. Hope, Crystallogr. D. Biol. Crystallogr. **52**, 18 (1996).

32. A. V. Lamy, M. Souaille, and J. C. Smith, Biopolymers **39**, 471 (1996).

33. R. M. Daniel et al., Biophys. J. **77**, 2184 (1999).

34. R. M. Daniel et al., Biophys. J. **75**, 2504 (1998).